Soil Biology

Volume 25

Series Editor
Ajit Varma, Amity Institute of Microbial Technology, Amity University Uttar Pradesh, Noida, UP, India

For further volumes:
http://www.springer.com/series/5138

Mahendra Rai • Ajit Varma
Editors

Diversity and Biotechnology of Ectomycorrhizae

Springer

Editors
Mahendra Rai, Ph.D.
Professor and Head
Department of Biotechnology
SGB Amravati University
Amravati 444 602, Maharashtra
India
mkrai123@rediffmail.com;
pmkrai@hotmail.com

Prof. Dr. Ajit Varma
Director General
Amity Institute of Microbial Technology
Amity University Uttar Pradesh
& Vice Chairman
Amity Science, Technology & Innovation
Foundation
Block A, Amity Campus, Sector 125
Noida, UP 201303
India
ajitvarma@aihmr.amity.edu

ISSN 1613-3382
ISBN 978-3-642-15195-8 e-ISBN 978-3-642-15196-5
DOI 10.1007/978-3-642-15196-5
Springer Heidelberg Dordrecht London New York

Cover design: SPi Publisher Services

Printed on acid-free paper

Springer is part of Springer Science+Business Media (www.springer.com)

Dedicated to my revered teacher Professor R. C. Rajak, Ex-head and Dean, Department of Bio-Sciences, RD University, Jabalpur, Madhya Pradesh, India

– Mahendra Rai

Preface

Roots – the hidden half of the plants – are important organs for absorption of water and nutrients from the soil. Roots release many carbon compounds as exudate. These compounds provide support to develop a symbiotic relationship with soil-borne fungi, often called as mycorrhiza. The fungal partner (mycobiont) provides the plant with improved access to water and nutrients in the soil due to a profusely branched hyphal network that spreads from the root surface and extends far into the soil. The plant, in return, supplies carbohydrate for fungal growth and survival. The mycorrhizae in general and ectomycorrhizae (ECM) in particular are more beneficial to the plants growing in nutrient-poor soil. Ectomycorrhizal plants are resistant to soil-borne diseases and often tolerate drought stress. In fact, the ECM is responsible for the succession of ecosystems.

There is enormous diversity of ectomycorrhizal fungi in a forest. The ECM can be used as indicators of quality and also for the development of forest ecosystem. ECM could be applied for reforestation as they accelerate the plant growth by supplying water and nutrients. Interestingly, without ECM, healthy woodland community cannot be maintained. Moreover, some ectomycorrhizal fungi produce edible sporocarps (fruiting bodies), which are eaten by the people and thus important for the food industry.

For better performance of the plants, it is necessary to inoculate them at seedling stage by ECM to make their life safe. ECM plays a multifunctional role during symbiosis with higher plants. These fungi have diverse roles as bioremediators, bioprotectors, biofertilizers, and stress indicators. They are the true "mycoindicators" of the forest ecosystem. There are many metal chelating molecules produced by ectomycorrhizal fungi, which have remarkable biotechnological significance. Furthermore, ECM secretes important secondary metabolites.

Molecular approaches are very important for the identification and differentiation of the fungi forming symbiosis with higher plants. Molecular tools are also important to understand how the genes are expressed during symbiosis with higher plants. The ectomycorrhizal fungi can be transformed by using *Agrobacterium tumefaciens*.

The main goal of this book is to provide information to the readers regarding diversity and applications of ectomycorrhizae and the use of modern biotechnological tools in understanding and transforming them.

The volume is divided into three parts, viz. (1) diversity, morphology, and applications, (2) biotechnological aspects of ectomycorrhizal fungi, and (3) functions and interactions. The whole book has been made user-friendly and worth reading.

Part I includes three chapters, out of which the first chapter explains how ectomycorrhizal inoculation benefits the members of family diptercarpaceae. The second chapter discusses the status of ectomycorrhizal fungi in South America, while the third chapter deals with inocula and the techniques of inoculation into the host plants.

Part II incorporates the molecular approaches in the systematics of ECM, gene expression during symbiosis, *Agrobacterium*-mediated gene transfer, biotechnological process, and signaling in ECM symbiosis and RNA-silencing.

In Part III, ectomycoremediation, functions of ECM when challenged with heavy metals, scale issues concerning the role of ECM in functioning of ecosystems as indicators of stress in forests, effect of pesticides on ECM, their secondary metabolites, carbon and nitrogen interactions, interaction of *Cantharellus* with *Dendrocalamus*, and edible ectomycorrhizal fungi have been included.

This volume would be of utmost importance to students, researchers, and teachers of botany, mycology, microbiology, forestry, and biotechnology. The readers should find the book full of information and reader friendly.

In planning this volume, invitations for contributions were extended to leading international authorities working with ectomycorrhizae. The editors would like to express sincere appreciation to each contributor for his/her work and for their patience and attention to detail during the entire production process. We sincerely hope that these eminent contributors will encourage us in the future as well, in the greatest interest of the academia.

We are extremely grateful to the staff members of Springer Heidelberg, especially Hanna G. Hensler-Fritton, Editorial Director Life Sciences/Biomedicine Europe II, Dieter Czeschlik (now retired), and Jutta Lindenborn for their continued interest, critical evaluation, constructive criticism, and support. We wish to acknowledge the help and support given to us by our students, faculty colleagues, and family members for their constant encouragement.

Amravati, Maharashtra, India — Mahendra Rai
Noida, Uttar Pradesh, India — Ajit Varma

Contents

Contributors

Acioli-Santos, Bartolomeu Departamento de Micologia, Universidade Federal de Pernambuco, Av. Prof. Nelson Chaves, s/n, Cidade Universitária, Recife PE, 50670-420, Brasil, bartacioli@cpqam.fiocruz.br, bartacioli@yahoo.com

Al Sayegh Petkovšek, Samar ERICo Velenje, Ecological Research & Industrial Cooperation, Koroška 58, 3320, Velenje, Slovenia, samar.petkovsek@erico.si

Alvarez Crespo, Maria C. Laboratorio de Micología Molecular, Departamento de Ciencia y Tecnología, Universidad Nacional de Quilmes, y Consejo Nacional de Investigaciones Científicas y Tecnicas (CONICET). Roque Sáenz Peña 352, (B1876BXD) Bernal, Provincia de Buenos Aires, Argentina, apardo@unq.edu.ar

Amadou, Bâ IRD. UMR 113 CIRAD/INRA/IRD/SUP-AGRO/UM2, Laboratoire des Symbioses Tropicales et Méditerranéennes (LSTM), TA A-82/J, Campus International de Baillarguet, Montpellier, Cedex 5, France

Antoine, Galiana IRD. UMR 113 CIRAD/INRA/IRD/SUP-AGRO/UM2, Laboratoire des Symbioses Tropicales et Méditerranéennes (LSTM), TA A-82/J, Campus International de Baillarguet, Montpellier, Cedex 5, France

Baptista, Paula CIMO / Escola Superior Agrária, Instituto Politécnico de Bragança, Campus de Santa Apolónia, Apartado 1172, 5301-855 Bragança, Portugal, pbaptista@ipb.pt

Becerra, Alejandra G. Instituto Multidisciplinario de Biología Vegetal (CONICET), Cátedra de Diversidad Vegetal I, Facultad de Ciencias Exactas, Físicas y Naturales, Universidad Nacional de Córdoba, C.C. 495, 5000 Córdoba, República Argentina, abecerra@efn.uncor.edu

Bernard, Dreyfus IRD. UMR 113 CIRAD/INRA/IRD/SUP-AGRO/UM2, Laboratoire des Symbioses Tropicales et Méditerranéennes (LSTM), TA A-82/J, Campus International de Baillarguet, Montpellier, Cedex 5, France

Brearley, Francis Q. Department of Environmental and Geographical Sciences, Manchester Metropolitan University, Chester Street, Manchester M1 5GD, UK, f.q. brearley@mmu.ac.uk

Bücking, Heike Biology and Microbiology Department, South Dakota State University, Brookings 57007, SD, USA, heike.bucking@sdstate.edu

Corrêa, Ana Departamento de Microbiología del Suelo y Sistemas Simbióticos, Estación Experimental del Zaidín, CSIC, Profesor Albareda 1, 18008 Granada, Spain, ambcorrea@gmail.com

Dahm, Hanna Department of Microbiology, Institute of General and Molecular Biology, Faculty of Biology and Earth Sciences, University of Nicolaus Copernicus, 87-100 Toruń, Gagarina 9, Poland, dahm@biol.uni.torun.pl

Daniel, Mousain INRA. UMR 113 CIRAD/INRA/IRD/SUP-AGRO/UM2, Laboratoire des Symbioses Tropicales et Méditerranéennes (LSTM). TA A-82/J, Campus International de Baillarguet, Montpellier, Cedex 5, France

Ezékiel, Baudoin IRD. UMR 113 CIRAD/INRA/IRD/SUP-AGRO/UM2, Laboratoire des Symbioses Tropicales et Méditerranéennes (LSTM), TA A-82/J, Campus International de Baillarguet, Montpellier, Cedex 5, France

Gheghel, Felicia UMR 1136 Interactions Arbres/Microorganismes, INRA 54280, Champenoux, France

Golińska, Patrycja Department of Microbiology, Institute of General and Molecular Biology, Faculty of Biology and Earth Sciences, University of Nicolaus Copernicus, 87-100 Toruń, Gagarina 9, Poland, plgolinska@umk.pl

Iordache, Virgil Department of Systems Ecology, University of Bucharest, Bucharest, Romania, virgil.iordache@g.unibuc.ro

Karwa, Alka Department of Biotechnology, SGB Amravati University, Amravati 444 602, Maharashtra, India, alkajaju@rediffmail.com

Kemppainen, Minna J. Laboratorio de Micología Molecular, Departamento de Ciencia y Tecnología, Universidad Nacional de Quilmes, y Consejo Nacional de Investigaciones Científicas y Tecnicas (CONICET). Roque Sáenz Peña 352, (B1876BXD) Bernal, Provincia de Buenos Aires, Argentina, apardo@unq.edu.ar

Kothe, Erika Microbial Phytopathology, Institute of Microbiology, Friedrich Schiller University, Jena, Germany

Kraigher, Hojka Slovenian Forestry Institute, Večna pot 2, 1000 Ljubljana, Slovenia, hojka.kraigher@gozdis.si

Lima, Cláudia E.P. Departamento de Micologia, Universidade Federal de Pernambuco, Av. Prof. Nelson Chaves, s/n, Cidade Universitária, Recife, PE 50670-420, Brasil

Lino-Neto, Teresa BioFIG, Departamento de Biologia, Universidade do Minho, Campus de Gualtar, 4710-057 Braga, Portugal, tlneto@bio.uminho.pt

Machuca, Ángela Department of Plant Science and Technology, Universidad de Concepción, Campus Los Ángeles, J.A.Coloma, 0201 Los Ángeles, Chile, angmachu@udec.cl

Maia, Leonor C. Departamento de Micologia, Universidade Federal de Pernambuco, Av. Prof. Nelson Chaves, s/n, Cidade Universitária, Recife, PE 50670-420, Brasil

Marin, Miguel Department of Physiology, Genetics and Microbiology, University of Alicante, Carretera San Vicente del Raspeig s/n, 03690 San Vicente del Raspeig (Alicante), Spain, mmarinz@ua.es

Martins-Loução, Maria-Amélia Faculdade de Ciências, Centro de Biologia Ambiental, Universidade de Lisboa, Campo Grande, 1749-016 Lisboa, Portugal

Neagoe, Aurora Research Center for Ecological Services (CESEC), University of Bucharest, Bucharest, Romania

Pardo, Alejandro G. Laboratorio de Micología Molecular, Departamento de Ciencia y Tecnología, Universidad Nacional de Quilmes, y Consejo Nacional de Investigaciones Científicas y Tecnicas (CONICET), Roque Sáenz Peña 352, (B1876BXD) Bernal Provincia de Buenos Aires, Argentina, apardo@unq.edu.ar

Rai, Mahendra Department of Biotechnology, SGB Amravati University, Amravati 444 602, Maharashtra, India, mkrai123@rediffmail.com, pmkrai@hotmail.com

Rajak, Ram C. SGHG Center for Biotechnology, Priydarshini Society, Dumna Road, Jabalpur Madhya Pradesh, India, rcmykes@yahoo.com

Repáč, Ivan Forestry Faculty, Department of Silviculture, Technical University in Zvolen, T.G. Masaryka 24, 96053 Zvolen, Slovak Republic, repac@vsld.tuzvo.sk

Robin, Par Duponnois IRD. UMR 113 CIRAD/INRA/IRD/SUP-AGRO/UM2, Laboratoire des Symbioses Tropicales et Méditerranéennes (LSTM), TA A-82/J, Campus International de Baillarguet, Montpellier, Cedex 5, France, Robin. Duponnois@ird.fr

Sharma, Rohit Microbial Culture Collection, Agarkar Reserch Institute, Pune, 411 007, Maharashtra, India, rsmushrooms@yahoo.co.uk

Taiana, Rivière-Dobigny IRD, Avenue de Maradi, 11416 Niamey Niger, taianariviere@yahoo.co.uk

Tavares, Rui Manuel BioFIG, Departamento de Biologia, Universidade do Minho, Campus de Gualtar, 4710-057 Braga, Portugal, tavares@bio.uminho.pt

Urban, Alexander Department of Systematic and Evolutionary Botany, University of Vienna, Rennweg 14, 1030 Vienna, Austria, alexander.urban@univie.ac.at

Varma, Ajit Amity Institute of Microbial Technology, Amity University, Noida Uttar Pradesh, India, ajitvarma@aihmr.amity.edu

Vieira, Helder Elísio E. Departamento de Micologia, Universidade Federal de Pernambuco, Av. Prof. Nelson Chaves, s/n, Cidade Universitária, Recife PE 50670-420, Brasil

Yves, Prin IRD. UMR 113 CIRAD/INRA/IRD/SUP-AGRO/UM2, Laboratoire des Symbioses Tropicales et Méditerranéennes (LSTM). TA A-82/J, Campus International de Baillarguet, Montpellier, Cedex 5, France

Zak, Marcelo R. Cátedra de Recursos Naturales y Gestión Ambiental, Departamento de Geografía, Facultad de Filosofía y Humanidades, Universidad Nacional de Córdoba, C.C. 495, 5000 Córdoba, República Argentina, marcelzak@gmail.com

Part I
Diversity, Morphology and Applications

Chapter 1
The Importance of Ectomycorrhizas for the Growth of Dipterocarps and the Efficacy of Ectomycorrhizal Inoculation Schemes

Francis Q. Brearley

1.1 Introduction

The Dipterocarpaceae is the most important tree family in the tropical forests of South-east Asia as they are the ecosystem dominants, especially in lowland forests of this region, where they often contribute up to a quarter of all stems and half of the above-ground biomass (Proctor et al. 1983; Davies et al. 2003; Brearley et al. 2004). They also play major role in timber production from this area as they are favoured for their fast growth rates, tall cylindrical boles and high quality wood. In addition, they are a source of non-timber forest products such as oils and resins (Shiva and Jantan 1998). Given the continued exploitation and degradation of lowland tropical forest habitats worldwide, there is interest in reforestation programmes to maintain forest cover and to provide a sustained supply of wood products. Many of these programmes have focussed on fast-growing exotic tree species but, to provide the highest quality timber, reforestation efforts now need to focus on the Dipterocarpaceae.

The majority of trees in tropical forests form arbuscular mycorrhizas but it was first noted by Singh in 1966 that dipterocarps, like many temperate forest trees, formed ectomycorrhizas (EcMs). Since then, many subsequent studies have examined the roots of numerous dipterocarps and found them to be colonised by EcM fungi (Alexander and Högberg 1986; Chalermpongse 1987; Smits 1992; Lee 1998; Hoang and Tuan 2008). There are also minor reports of arbuscular mycorrhizal (Shamsuddin 1979; Ibrahim et al. 1995; Tawaraya et al. 2003) and ectendomycorrhizal (Tupas and Sajise 1976) fungi in dipterocarp roots, but more details are needed on this potential dual colonisation before anything more concrete can be written.

In this chapter, I have (1) given a brief overview of mycorrhizal symbioses in dipterocarps, I now plan to (2) note which are the important fungal species involved

F.Q. Brearley
School of Science and the Environment, Manchester Metropolitan University, Chester Street, Manchester M1 5GD, UK
e-mail: f.q.brearley@mmu.ac.uk

M. Rai and A. Varma (eds.), *Diversity and Biotechnology of Ectomycorrhizae*,
Soil Biology 25, DOI 10.1007/978-3-642-15196-5_1,

in the EcM symbiosis and then (3) focus on the role of this symbiosis in improving plant nutrition, growth and performance, and complete (4) with a special focus on how we might critically use this knowledge to determine if we need to apply EcM inoculation in reforestation programmes.

1.2 Fungal Species Associated with Dipterocarps

We know, from fruit body surveys, that the most speciose groups of fungi found in dipterocarp-dominated forests are the families of Amanitaceae, Boletaceae and Russulaceae (Watling and Lee 1995, 1998, 2007; Watling et al. 1998, 2002; Lee et al. 2002, 2003), and these appear to form a reasonable proportion of the EcM root tips belowground (Lee et al. 1997; Ingleby et al. 1998). However, it is known that above-ground and below-ground views of the EcM community rarely show close concordance (Gardes and Bruns 1996; Yamada and Katsuya 2001), and it has recently been noted that fungi with cryptic fruiting bodies such as members of the Thelephoraceae, which appear to have been overlooked in fruiting body surveys, are also important EcM formers. For example, Ingleby et al. (2000) found a Thelephoraceae species to be common on seedlings planted in soil from Vietnamese forest and I have found two Thelephoraceae species common on nursery-grown seedlings in Malaysia (Brearley et al. 2003, 2007; Brearley 2006). Furthermore, sequences from this group dominated the molecular studies on dipterocarp EcM communities conducted by Sirikantaramas et al. (2003) and Yuwa-Amornpitak et al. (2006). As an example, half of the sequences generated by Sirikantaramas et al. (2003) were from this family. In this regard, the use of molecular identification techniques has shown us that the EcM community in dipterocarp-dominated forests is more similar to many boreo-temperate EcM communities (e.g. Richard et al. 2005; Ishida et al. 2007; Morris et al. 2008) than originally thought. As molecular work continues on the EcMs communities of dipterocarps, it will reveal a much clearer below-ground picture of the dominant species involved in this symbiosis.

1.3 Nutrient Relationships

There are numerous experiments showing that EcMs can improve plant nutrient uptake (Smith and Read 2008) and, in the Dipterocarpaceae, this has also been shown to be the case. For example, Lee and Lim (1989) correlated percentage EcM colonisation (% EcM) with foliar phosphorus (P) in dipterocarp seedlings they studied. Seedlings from a site with lower available soil P showed a positive correlation between foliar P and % EcM suggesting that EcMs enhanced uptake of P at this site. Hadi and Santoso (1988, 1989) presented data suggesting inoculation with EcM fungi increased foliar concentrations of nitrogen (N), P, potassium

(K), calcium (Ca) and magnesium (Mg) of five dipterocarp species (although it is not entirely clear if their data are for nutrient concentrations or total nutrient content in the seedlings). The first experiments to show an unequivocal increase in foliar nutrients in response to EcM colonisation were those of Lee and Alexander (1994) and Yazid et al. (1994) who clearly showed increased P concentrations in response to EcM colonisation in two species of *Hopea* studied. Additionally, in the experiments of Lee and Alexander (1994), whilst there were no suggestions of EcM colonisation increasing shoot N, K or Mg concentrations, there was some indication of improved Ca nutrition. These differential responses may be due to the experimental conditions used but may also represent a certain degree of inter-specific difference in fungal benefits to the host plant or may even be due to an EcM diversity effect on nutrient uptake (Baxter and Dighton 2005).

Improved mineral nutrition has also been shown under field conditions where reduction of EcM colonisation (by Mancozeb fungicide) led to reduced foliar N and P in both *Hopea nervosa* and *Parashorea tomentella* in addition to reduced Ca and Mg in *Hopea nervosa* (Brearley 2003). I have also shown, through stable isotope analysis, that EcM dipterocarps can also obtain more N from organic material (Brearley et al. 2003) with a positive correlation found between % EcM colonisation and uptake of organic N.

An inoculation experiment that purports to show increases in nutrient uptake of *Shorea seminis* when inoculated with two EcM species is that of Turjaman et al. (2006). Total shoot N and P contents were indeed greater in the EcM inoculated seedlings but, surprisingly, on a dry-weight basis, inoculation actually led to a decrease in concentrations of these elements in the shoot (up to 55% in the most extreme case). Smits (1983) also considers that dipterocarp EcMs may be providing thiamin to plants although, in the absence of a strong presentation of data, we should discount this hypothesis for now.

Increased nutrient concentrations within plants are generally likely to lead to improved growth, and the role of foliar nutrients in determining seedling performance is also important during out-planting as it has been shown that higher levels of foliar N allow dipterocarp seedlings to better avoid photodamage when transferred to high irradiance conditions and to allow more rapid acclimation to these conditions (Bungard et al. 2000).

1.4 Inoculation Experiments

Many of the inoculation experiments reported are in the grey literature and have a number of shortcomings. The most serious of these is that they are poorly reported and do not provide sufficient detail for the experiments to be evaluated fully nor repeated by other researchers. Pseudoreplication or the lack of statistical analyses in many cases also makes evaluation of the results problematic. In this chapter, I will focus on papers which, I feel, have mostly overcome these shortcomings or are otherwise noteworthy.

The first reported experiment attempting to inoculate dipterocarp seedlings with EcM fungi and determine seedling responses was that of Hadi and Santoso (1988). They inoculated species of *Boletus*, *Russula* and *Scleroderma* using pieces of chopped fruiting body on the roots of five dipterocarp seedling species. A shortcoming of this experiment was that the roots were not examined to determine the extent of EcM formation by any contaminant fungi. Nevertheless, after 6 months growth, inoculation appeared to at least double seedling height in all fungus/seedling combinations. Furthermore, their approach to inoculation is rarely used as it is difficult to control the inoculum viability and potential supplied to the roots with this method.

Initial experiments, using chopped root inoculum (Lee and Alexander 1994), found that EcM colonisation increased plant dry weights in *Hopea odorata* and *Hopea helferi*. This increase was generally greatest in the absence of additional nutrients supplied to the soil. However, the problem with experiments using root inoculum is that the species of fungi on the inoculant roots cannot be determined and therefore controlled experiments using defined EcM species are needed. In Malaysia, inoculation experiments have focussed on strains of *Pisolithus* species and a member of the Thelephorales (FP160: Lee et al. 2008). In Indonesia, the use of *Scleroderma* for inoculation appears more popular although the range of species being used has increased recently (Turjaman et al. 2007).

1.4.1 Inoculation with Single Species

1.4.1.1 Malaysian Inoculation Experiments

In Malaysia, exotic *Pisolithus* isolates from Brazil (Pt 441 originally from under *Eucalyptus citriodora*) and Thailand (Pt msn) were effective at forming EcMs on four dipterocarp seedling species (although there was a certain degree of host-fungal compatibility with one or other *Pisolithus* isolate having a greater % EcM on each of the seedling species; Lee et al. 1995). Using Chilvers et al.'s (1986) cardboard inoculum technique, Yazid et al. (1994) showed that Pt 441 formed a high percentage of functional EcM colonisation (c. 80%) on *Hopea odorata* and *Hopea helferi* and that this increased the growth (dry weights increased by 7.3 and 3.6 times, respectively) and foliar P concentrations (by at least 40%) after 9 months growth. Similar results in terms of the growth of *Hopea odorata* were seen (although with a lesser growth response) when coconut husk:vermiculite inoculum was added in a ratio of 1:4 to a sterilised soil:sand mix (Yazid et al. 1996). In contrast, problems were found when trying to inoculate a local species: *Pisolithus aurantioscabrosus* (Lee et al. 1995). Why this is, is not clear but may simply reflect the ability of *Pisolithus tinctorius* to form EcMs with a wide host range (Martin et al. 2002), whereas *Pisolithus aurantioscabrosus* has only been reported to be associated with *Shorea parvifolia* and *Shorea macroptera* to date (Watling et al. 1995a, b; Martin et al. 2002).

Initial tests of successful inoculation between *Hopea odorata* and *Hebeloma crustulinifome* (Lapeyrie et al. 1993) were reported, but growth responses were not shown and this exotic European strain does not appear to be used any more. Recent work has focussed on Thelephoraceae FP160 (Lee et al. 2008), which significantly increased stem height, root length and biomass of *Hopea odorata* after 6 months growth in the nursery by 30%, 62% and 40%, respectively (Lee et al. 2008). It currently appears very difficult to bring further tropical dipterocarp-associated EcM species into culture (S.S. Lee, pers. comm.), and the species that are being used form a very limited subset of those available. The best approach here would be a wide-ranging screening using a variety of fungal structures, media and growth conditions although, of course, this will be very labour intensive with potentially little reward. Perhaps, the floating culture technique of Sangtiean and Schmidt (2002) may help South-east Asian researchers to culture some of the later stage EcM species found in these forests. This technique allowed Sangtiean and Schmidt (2002) to carry out culture experiments on *Amanita, Lactarius* and *Russula* species, which are common in South-east Asian forests (see above).

1.4.1.2 Indonesian Inoculation Experiments

In Indonesia, the use of *Scleroderma* species (and especially *Scleroderma columnare*) appears to be favoured, probably from the initial work of Ogawa (1993, 2006) in the early 1990s. Sadly, much of this early work is difficult to evaluate as it is not clearly reported (e.g. Supriyanto et al. 1993; Kikuchi 1997) but, more recently, a number of much better controlled inoculation experiments have been conducted by Turjaman et al. (2005, 2006, 2007). They showed that the growth of *Shorea pinanga* was improved by the addition of spore tablets of *Pisolithus tinctorius* (aka *Pisolithus arhizus*) and *Scleroderma columnare* species. Both fungal species improved the growth of *Shorea pinanga* (150% increase in dry weight after 7 months) although there was some EcM colonisation of the controls. Survival rates (86–87%) were also much higher than the control (16%), which is an equally important factor to take into consideration when planning reforestation schemes. In a follow-up experiment (Turjaman et al. 2006), tabletted spore inoculum was compared with alginated bead mycelial inoculum of *Pisolithus tinctorius* and *Scleroderma columnare*. Percentage EcM colonisation was higher (61–65%) when seedlings were inoculated with spores than with mycelium (35–37%), and there was, again, at least a doubling of dry weight after 7 months growth. In the most recently reported experiment (Turjaman et al. 2007), inoculation of four fungal species on the roots of *Shorea balangeran* increased seedling growth. Whilst we might expect the use of *Boletus* sp., *Scleroderma* sp. and *Strobilomyes* sp. to increase seedling growth as these are known to be EcM forming fungi, the use of *Calvatia* sp. also increased seedling growth, which was unexpected as this is not thought to be an EcM forming genus (Rinaldi et al. 2008).

1.4.2 Other Inoculation Methods

1.4.2.1 Mycorrhizal Tablets

Where sterile facilities are not available to cultivate species aseptically, researchers have used "mycorrhizal tablets". In this case, spores or mycelium are mixed with a carrier (clay or alginate beads) and applied to seedlings' rooting zones to allow hyphal contact and subsequent EcM formation. The first record of this in the South-east Asia region appears to be that of Ogawa (1993). Species used for this method are often those such as *Scleroderma* species, which have the advantage that their spores can be collected much more easily from their enclosed fruit bodies than many other gilled or pored fruit bodied species. Clay tablets at 1:100 (crushed fruit bodies:clay) were used in the experiments by Turjaman et al. (2005, 2006), and these showed increased growth of *Shorea pinanga* and *Shorea selanica* when inoculated as compared with uninoculated controls. However, in such experiments, there is a need to confirm that the tablets do not contain additional nutrients that might improve seedling growth in the absence of a mycorrhizal effect.

1.4.2.2 Mother Tree Inoculation

Other inoculation methods include inoculation from a colonised mother tree in the nursery – in other words, simply letting newly germinated seedlings' roots contact hyphae already radiating out in the soil around established, colonised trees. This method was first used by Roeleffs (1930, in Nara et al. 1999) to inoculate seedlings of *Pinus* species. The technique, also known as inoculation beds, is a low-tech method that allows EcM inoculation before planting-out but it can be rather haphazard in terms of the speed and reliability of inoculation (see Kikuchi 1997). For example, Ogawa (2006) shows a diagram of the spread of *Scleroderma columnare* fruiting bodies through a nursery containing seedlings of *Shorea leprosula* and *Shorea academica* to be between 1 and 2 m per year.

1.4.3 Production of Inoculum

In order to produce inoculum rapidly, conditions for the optimum growth of fungi in culture should be evaluated. For example, Patahayah et al. (2003a) showed that the most rapid growth of *Pisolithus albus* (aka Pt Gemas) was obtained at 25°C when grown on Oddoux medium but at 30°C when grown on MMN or Pachlewski's medium (Patahayah et al. 2003b). We have also shown that this species grows best when N is supplied in an organic form (BSA in the experiments conducted; Brearley et al. 2005). Thelephoreaceae FP160 shows best growth at 25°C (Patahayah et al. 2003b) but has minimal preferences for N source (Brearley et al. 2005). In terms of efficient spread of EcM inoculum in the nursery, Nara et al. (1999) considered that seedlings are often maintained under sub-optimal conditions in potted soil with a high clay content and hence poor aeration, thereby slowing growth of fungal hyphae. They found that, by using a growth medium with particles of 2–4 mm diameter, the optimum growth of EcM mycelium (Th1 on *Shorea roxburghii*) was obtained. It is also important to consider the longevity of the different forms of inoculum. For example, Fakuara and Wilarso (1993) showed that mycorrhizal tablets remained effective up to 4 months in storage (this was the longest period tested). More experiments are clearly needed in this area, with longer test periods, to gain a better idea of spore longevity.

1.4.4 Field Experiments

There is now a need to determine how well inoculated seedlings and their symbiotic EcM species survive in the wild when seedlings are out-planted. This is important as, if considerable effort is being put into inoculation programmes, this is simply being wasted if seedlings or their inoculant fungal species are dying unnecessarily. Furthermore, in terms of reforestation schemes, growth is not necessarily the most important parameter, seedling survival is arguably equally as important.

Chang et al. (1994, 1995) showed that the species of *Pisolithus* in the Malaysian inoculation experiments noted above did not remain competitive when colonised seedlings of *Shorea glauca* were planted into the field; indeed *Pisolithus* had mostly disappeared from the roots after 6 months suggesting that they are either early stage fungi, or are poorly adapted to the biotic or abiotic environments of the Malaysian forest soils. Using Thelephoraceae FP160, Lee et al. (2008) found it to remain competitive on the roots of seedlings (*Hopea odorata* and *Shorea leprosula*) for up to 23 months after out-planting in a sandy tin mine tailings site (after this time contaminant EcM fungi had only colonised up to 15% of the root tips of less than half the seedlings). However, the improved growth of *Hopea odorata* seen in the nursery due to inoculation with this fungus (see above) was not seen in the field (by measurement of root collar diameter) and improved growth of *Shorea leprosula* was only seen for up to 3 months following out-planting. Under field conditions, I found that the reduction in EcM colonisation by fungicide addition to the roots of two species (*Hopea nervosa* and *Parashorea tomentella*) did not lead to changes in seedling growth but that foliar nutrient concentrations were reduced (Brearley 2003). In field experiments in a degraded peat swamp forest in Kalimantan, Turjaman et al. (2007) showed that a spore suspension of *Boletus* sp. and *Scleroderma* sp. applied to the seedling rooting zone led to increased growth of *Shorea balangeran* but application of *Calvatia* sp. and *Strobilomyces* sp. did not. For the two fungal species that were beneficial, it took around 8 months for growth improvements to be seen, perhaps due to the very wet conditions at the start of the experiment (Turjaman et al. 2007). However, it is difficult to determine if the species applied were those that maintained improved seedling growth as the roots at the end of this 40-month experiment were not examined to determine which EcM fungi were present – it would have been a notable improvement to the study design to do this. The study of Tata et al. (2009) did report this examination at the end of their experiments with *Shorea selanica* and *Shorea lamellata*, which were inoculated with spore tablets of *Scleroderma columnare* and planted in natural forests or rubber agroforests in Sumatra. Their results were complex but did not show consistent increases in growth, performance or survival of the two dipterocarp species over a 2-year period. It is notable that, among the 19 EcM types they identified using PCR-RFLP at the end of the experiment, none of them were *Scleroderma* species indicating that the inoculated fungus did not remain competitive on the roots for more than this length of time.

There is clearly a need to further evaluate the growth and survival of inoculated seedlings in the field as positive responses to EcM inoculation in the ecologically simplified, and somewhat benign, nursery environment are unlikely to be representative of that found at out-planting sites. There is an argument to be made to use indigenous species for inoculation schemes as they are anticipated to be the most effective, but we may also need to consider the potential impact of biological invasions if using exotic fungal species (Vellinga et al., 2009).

1.5 Under What Conditions Will EcM Inoculation Be Beneficial?

Whilst the body of this chapter thus far has outlined how inoculation with EcM fungi may improve dipterocarp seedling growth, and the various methods to do so, it is certainly worth considering whether inoculation should indeed be conducted at all. In the first paper on dipterocarp EcMs, Singh (1966) noted that “mycorrhizal infection should not be taken as the ‘cure of all ills’ in the establishment of trees in all sorts of habitats”, and this warning still stands, more than 40 years later. I now pose three key questions for consideration before starting to plan inoculation schemes.

It is often considered that there is a need to inoculate seedlings prior to out-planting but, in fact, in most cases inoculation will occur naturally, and inoculation schemes may not yield any major benefits for seedling growth or survival (although we cannot be confident that the same species, or most beneficial species, of EcM fungi will be formed on each seedlings’ roots every time). The first key question is, therefore, will inoculation be of benefit to the seedlings? The major benefits of inoculation are knowing that a seedling, at out-planting, is mycorrhizal with a known species of fungus which is functionally beneficial, and thus it will not need to wait to form EcMs with an unknown group of fungi present in the soil which may or may not promote seedling growth; this gives it something of an initial advantage over any out-planted non-mycorrhizal seedlings. However, the main benefits of inoculation are more likely to be shown under poor soil conditions, as I outline below.

1.5.1 Successful Inoculation Schemes

For inoculation schemes to be successful, a series of well-defined and consistently repeatable techniques is needed. In other words, a pure culture of inoculum is needed to allow a regular supply, and currently there are very few fungal species being maintained in pure culture in the South-east Asian region. Access to a laboratory with sterile facilities is needed which may be problematic for a number of sites. In the absence of this, the production of mycorrhizal tablets may be useful although vagaries of fungal fruit body production and genetic variation between individual genets will remain unaccounted for. Infrequent production of dipterocarp seeds can make regular production of planting stock difficult although production of cuttings from a selected number of dipterocarp species now appears fairly routine (Moura-Costa and Lundoh 1994; Itoh et al. 2003; Haji Ahmad 2006). It must also be shown that the inoculant fungus has the ability to improve seedling growth or survival over that of non-inoculated seedlings under field conditions. It appears much easier to culture species such as *Pisolithus* or *Scleroderma* but it must be remembered that these species are not necessarily those which are most beneficial to seedling growth or, indeed, are found commonly on dipterocarp seedling roots.

1.5.2 When and Where to Inoculate?

It is often suggested that inoculation may be beneficial for seedlings planted following logging operations. However, in most cases after logging, there are still a number of smaller dipterocarp trees which will have EcM fungi associated with them and, as long as the light conditions are not detrimental to seedling growth, this should allow the rapid formation of EcMs on seedlings by hyphae, sclerotia and spores already present in the soil (Lee et al. 1996; Ingleby et al. 1998). There is little evidence so far that selective logging seriously impoverishes the fungal flora (Watling et al. 1998) although there is an indication of a loss of the some of the rarer EcM species in logged forest (Lee et al. 1996). Out-planted dipterocarp seedlings are almost certain to become colonised within a short period of time as long as they have below-ground access to roots and mycelium radiating from adult trees (Lee 1991; Alexander et al. 1992; Lee and Alexander 1996; Lee et al. 1996). The second key question is, therefore, is inoculation beneficial under all situations? If not, which situations or conditions are most likely to be improved by inoculating seedlings prior to out-planting?

Inoculation is considered more likely to be of benefit when seedlings are planted in areas where suitable EcM inoculum is not available. This may include severely degraded areas such as mine tailings (Lee et al. 2008), burnt areas (Akema et al. 2009), degraded peatlands (Turjaman et al. 2007) and areas previously used for agriculture (Ingleby et al. 2000). For example, Ingleby et al. (2000) found that the inoculation potential of soils which had been under agriculture for over 20 years was essentially absent when compared with an undisturbed forest or plantation in Vietnam. The work of Turjaman et al. (2007) in degraded peat swamp forest is also of relevance here as they showed improved growth of inoculated dipterocarp seedlings when out-planted in a degraded area.

The final key question is, is simply adding colonised soil appropriate as inoculum? In many cases, local soil from the vicinity of dipterocarp trees may be equally as effective as any inoculation schemes although these EcMs are not necessarily the best species to promote seedling growth and there is no way to control which fungal species successfully colonise the seedlings' roots. Smits (1992) outlines a simple method by which large numbers of seedlings can be inoculated by soil colonised by EcM hyphae and spores. He advocates the use of soil collected from beneath an adult tree of the same species, but this is based on his weak assertions (Smits 1983, 1985) of a high degree of host specificity. I suggest it is equally likely that host-specific pathogens will be present (Packer and Clay 2000) and therefore suggest a general soil inoculum but ensuring that it is collected in the vicinity of dipterocarps. In the absence of any other schemes, the inclusion of forest soil should be seen as the minimum to ensure early EcM colonisation of dipterocarp seedlings.

1.6 Conclusions

Whilst EcMs are often thought to be essential for the successful growth of dipterocarp seedlings, there is surprisingly little evidence confirming this assertion under natural conditions. In nursery experiments, mycorrhizal inoculation has regularly been shown to increase seedling growth and nutrient concentrations, but when similar experiments have been conducted in the field, the results are much more equivocal with inoculation often showing minimal improvements in growth if seedlings are planted in natural forests (e.g. Tata et al. 2009). If inoculated seedlings are planted in degraded soils, the improvement in growth is often more marked although these improvements may not be maintained if the inoculated fungus does not remain competively dominant on the seedlings' roots. I therefore suggest that researchers and forest restorationists carefully consider whether EcM inoculation is of benefit in the areas they plan to re-plant.

Acknowledgements I thank Dr. Lee Su See and Dr. Robin Sen for helpful thoughts and comments on the manuscript. My work on dipterocarp ectomycorrhizas was supported by the British Ecological Society Overseas Research Programme.

References

Akema T, Nurhiftisni I, Suciatmih, Simbolon H (2009) The impact of the 1998 forest fire on ectomycorrhizae of dipterocarp trees and their recovery in tropical rain forests of East Kalimantan, Indonesia. JARQ 42:137–142

Alexander IJ, Högberg P (1986) Ectomycorrhizas of tropical angiospermous trees. New Phytol 102:541–549
Alexander IJ, Ahmad N, Lee SS (1992) The role of mycorrhizas in the regeneration of some Malaysian forest trees. In: Marshall AG, Swaine MD (eds) Tropical rain forest: disturbance and recovery. The Royal Society, London, UK, pp 357–367
Baxter JW, Dighton J (2005) Diversity–functioning relationships in ectomycorrhizal fungal communities. In: Dighton J, White JF Jr, Oudemans P (eds) The fungal community: its organization and role in the ecosystem, 3rd edn. CRC Press, Boca Raton, Florida, USA, pp 383–398
Brearley FQ (2003) The role of ectomycorrhizas in the regeneration of dipterocarp seedlings. PhD Thesis, University of Sheffield, UK
Brearley FQ (2006) Differences in the growth and ectomycorrhizal community of *Dryobalanops lanceolata* (Dipterocarpaceae) seedlings grown in ultramafic and non-ultramafic soils. Soil Biol Biochem 38:3407–3410
Brearley FQ, Press MC, Scholes JD (2003) Nutrients obtained from leaf litter can improve the growth of dipterocarp seedlings. New Phytol 160:101–110
Brearley FQ, Prajadinata S, Kidd PS, Proctor J, Suriantata (2004) Structure and floristics of an old secondary rain forest in Central Kalimantan, Indonesia, and a comparison with adjacent primary forest. For Ecol Manage 195:385–397
Brearley FQ, Scholes JD, Lee SS (2005) Nitrogen nutrition and isotopic discrimination in tropical ectomycorrhizal fungi. Res Microbiol 156:184–190
Brearley FQ, Scholes JD, Press MC, Palfner G (2007) How does light and phosphorus fertilisation affect the growth and ectomycorrhizal community of two contrasting dipterocarp species? Plant Ecol 192:237–249
Bungard RA, Press MC, Scholes JD (2000) The influence of nitrogen on rain forest dipterocarp seedlings exposed to a large increase in irradiance. Plant Cell Environ 23:1183–1194
Chalermpongse A (1987) Mycorrhizal survey of dry-deciduo us and semi-evergreen dipterocarp forest ecosystems in Thailand. In: Kostermans AJCH (ed) Proceedings of the third round table conference on dipterocarps. UNESCO Regional Office for Science and Technology, Jakarta, Indonesia, pp 81–103
Chang YS, Lapeyrie FF, Lee SS (1994) The survival and competitiveness of *Pisolithus tinctorius* on outplanted seedlings of *Shorea glauca* in Malaysia. In: Khoo KC, Appanah S (eds) Proceedings of the fifth round table conference on dipterocarps. Forest Research Institute of Malaysia, Kepong, Malaysia, pp 165–169
Chang YS, Lee SS, Lapeyrie FF, Yazid SM (1995) The competitiveness of two strains of *Pisolithus tinctorius* on seedlings of three species of dipterocarps under nursery and field conditions: preliminary results. In: Wickneswari R, Yahya AZ, Shariff AHM, Haji Ahmad D, Khoo KC, Suzuki K, Sakurai S, Ishii K (eds) Proceedings of the international workshop of BIO-REFOR, Kangar, 1994. BIO-REFOR, IUFRO-SPDC, Tokyo, Japan & FRIM, Kepong, Malaysia, pp 208–212
Chilvers GA, Douglas PA, Lapeyrie FF (1986) A paper-sandwich technique for rapid synthesis of ectomycorrhizas. New Phytol 103:397–402
Davies SJ, Nur Supardi MN, LaFrankie JV Jr, Ashton PS (2003) The trees of Pasoh Forest: stand structure and floristic composition of the 50-ha forest research plot. In: Okuda T, Manokaran N, Matsumoto Y, Niiyama K, Thomas SC, Ashton PS (eds) Pasoh: ecology of a lowland rain forest in Southeast Asia. Springer-Verlag, Tokyo, Japan, pp 35–50
Fakuara Y, Wilarso S (1993) Effect of mycorrhizal tablet storage on seedling growth of *Shorea pinanga* Scheff. In: Anon. (ed) BIO-REFOR: proceedings of Tsukuba-workshop. BIO-REFOR, IUFRO-SPDC, Tsukuba Science City, Japan, pp 174–179
Gardes M, Bruns TD (1996) Community structure of ectomycorrhizal fungi in a *Pinus muricata* forest: above and below-ground views. Can J Bot 74:1572–1583
Hadi S, Santoso E (1988) Effect of *Russula* spp., *Scleroderma* sp. and *Boletus* sp. on the mycorrhizal development and growth of five dipterocarp species. In: Mohinder Singh M (ed) Agricultural and biological research priorities in Asia, Proceedings of the IFS symposium of

science Asia 87. International Foundation for Science & Malaysian Scientific Association, Kuala Lumpur, Malaysia, pp 183–185

Hadi S, Santoso E (1989) Accumulation of macronutrients by five dipterocarp species inoculated with different species of mycorrhizal fungi. In: Mahadevan A, Raman N, Natarajan K (eds) Mycorrhizae for Green Asia: proceedings of the first Asian conference on mycorrhizae. Centre for Advanced Studies on Botany, University of Madras, India, pp 139–141

Haji Ahmad D (2006) Vegetative propagation of dipterocarp species by stem cuttings using a very simple technique. In: Suzuki K, Ishii K, Sakurai S, Sasaki S (eds) Plantation technology in tropical forest science. Springer-Verlag, Tokyo, Japan, pp 69–77

Hoang PND, Tuan DLA (2008) Investigating the ectomycorrhizal appearance of seedlings in the Tan Phu forest enterprise's nursery, Dong Nai Province. J Sci Technol Dev 11(1):96–100

Ibrahim Z, Mahat MN, Lee SS (1995) Response of *Hopea odorata* seedlings to biological soil conditioners. In: Wickneswari R, Yahya AZ, Shariff AHM, Haji Ahmad D, Khoo KC, Suzuki K, Sakurai S, Ishii K (eds) Proceedings of the international workshop of BIO-REFOR, Kangar, 1994. BIO-REFOR, IUFRO-SPDC, Tokyo, Japan & FRIM, Kepong, Malaysia, pp 179–182

Ingleby K, Munro RC, Noor M, Mason PA, Clearwater MJ (1998) Ectomycorrhizal populations and growth of *Shorea parvifolia* (Dipterocarpaceae) seedlings regenerating under three different forest canopies following logging. For Ecol Manage 111:171–179

Ingleby K, Thuy LTT, Phong NT, Mason PA (2000) Ectomycorrhizal inoculum potential of soils from forest restoration sites in South Vietnam. J Trop For Sci 12:418–422

Ishida TA, Nara K, Hogetsu T (2007) Host effects on ectomycorrhizal fungal communities: insight from eight host species in mixed conifer-broadleaf forests. New Phytol 174:430–440

Itoh A, Yamakura T, Tan S, Kendawang JJ, Lee HS (2003) Effects of resource plant size on rooting of *Dryobalanops lanceolata* cuttings. J For Res 8:117–121

Kikuchi J (1997) Ectomycorrhiza formation of dipterocarp seedlings. In: Sangwanit U, Thaiutsa B, Puangchit L, Thammincha S, Ishii K, Sakurai S, Suzuki K (eds) Proceedings of the fifth international workshop of BIO-REFOR, Bangkok, 1996. BIO-REFOR, IUFRO-SPDC, Tokyo, Japan, pp 49–52

Lapeyrie FF, Lee SS, Yazid SM (1993) Controlled ectomycorrhizal inoculation of *Hopea odorata* (Dipterocarpaceae) cuttings with *Hebeloma crustuliniforme*. In: Anon. (ed) BIO-REFOR: proceedings of Tsukuba-workshop. BIO-REFOR, IUFRO-SPDC Tsukuba Science City, Japan, pp 189–190

Lee SS (1991) Some views on dipterocarp mycorrhiza research in Malaysia. In: Anon. (ed) BIO-REFOR: proceedings of pre-workshop. BIO-REFOR, IUFRO-SPDC, Bogor, Indonesia, pp 66–70

Lee SS (1998) Root symbiosis and nutrition. In: Appanah S, Turnbull JM (eds) A review of dipterocarps: taxonomy, ecology and silviculture. Center for International Forestry Research, Bogor, Indonesia, pp 99–114

Lee SS, Alexander IJ (1994) The response of seedlings of two dipterocarp species to nutrient additions and ectomycorrhizal infection. Plant Soil 163:299–306

Lee SS, Alexander IJ (1996) The dynamics of ectomycorrhizal infection of Shorea leprosula seedlings in Malaysian rain forests. New Phytol 132:297–305

Lee SS, Lim KL (1989) Mycorrhizal infection and foliar phosphorus content of seedlings of three dipterocarp species grown in selectively logged forest and a forest plantation. Plant Soil 117:237–241

Lee SS, Lapeyrie FF, Yazid SM (1995) Techniques for controlled ectomycorrhizal inoculation of dipterocarp seedlings and cuttings. In: Supriyanto, Kartana JT (eds) Proceedings of the second symposium on biology and biotechnology of mycorrhizae and third Asian conference on mycorrhizae (ACOM III). BIOTROP Special Publication 56, SEAMEO BIOTROP, Bogor, Indonesia, pp 217–221

Lee SS, Alexander IJ, Moura-Costa PH, Yap SW (1996) Mycorrhizal infection of dipterocarp seedlings in logged and undisturbed forests. In: Appanah S, Khoo KC (eds) Proceedings of the fifth round table conference on dipterocarps. Forest Research Institute of Malaysia, Kepong, Malaysia, pp 157–164

Lee SS, Alexander IJ, Watling R (1997) Ectomycorrhizas and putative ectomycorrhizal fungi of *Shorea leprosula* Miq. (Dipterocarpaceae). Mycorrhiza 7:63–81

Lee SS, Watling R, Noraini Sikin Y (2002) Ectomycorrhizal basidiomata fruiting in lowland rain forests of peninsular Malaysia. Bois Fôr Trop 274(4):33–43

Lee SS, Watling R, Turnbull E (2003) Diversity of putative ectomycorrhizal fungi in Pasoh Forest Reserve. In: Okuda T, Manokaran N, Matsumoto Y, Niiyama K, Thomas SC, Ashton PS (eds) Pasoh: ecology of a lowland rain forest in Southeast Asia. Springer-Verlag, Tokyo, Japan, pp 149–159

Lee SS, Patahayah M, Chong WS, Lapeyrie FF (2008) Successful ectomycorrhizal inoculation of two dipterocarp species with a locally isolated fungus in Peninsular Malaysia. J Trop For Sci 20:237–247

Martin F, Díez J, Dell B, Delaruelle C (2002) Phylogeography of the ectomycorrhizal Pisolithus species as inferred from nuclear ribosomal DNA ITS sequences. New Phytol 153:345–357

Morris MH, Smith ME, Rizzo DM, Rejmánek M, Bledsoe CS (2008) Contrasting ectomycorrhizal fungal communities on the roots of co-occurring oaks (*Quercus* spp.) in a California woodland. New Phytol 178:167–176

Moura-Costa PH, Lundoh L (1994) A method for vegetative propagation of *Dryobalanops lanceolata* (Dipterocarpaceae) by cuttings. J Trop For Sci 6:533–541

Nara K, Kawahara M, Okamura K, Sakurai K, Hogetsu T (1999) Prospects and problems pertaining to the application of ectomycorrhizal fungi to dipterocarp seedlings in tropical nurseries. In: Anon. (ed) Proceedings of the international symposium "Can Biological Production Harmonize with Environment?". University of Tokyo, Japan, pp 151–154

Ogawa M (1993) Inoculation method of *Scleroderma columnare* to dipterocarp seedlings. In: Anon. (ed) BIO-REFOR: proceedings of Tsukuba-workshop. BIO-REFOR, IUFRO-SPDC, Tsukuba Science City, Japan, pp 185–188

Ogawa M (2006) Inoculation methods of *Scleroderma columnare* onto dipterocarps. In: Suzuki K, Ishii K, Sakurai S, Sasaki S (eds) Plantation technology in tropical forest science. Springer-Verlag, Tokyo, Japan, pp 185–197

Packer A, Clay K (2000) Soil pathogens and spatial patterns of seedling mortality in a temperate tree. Nature 404:478–481

Patahayah M, Cynthia PC, Lee SS (2003a) Optimizing growth conditions for ectomycorhizal inoculum production of the Malaysian strain of *Pisolithus tinctorius*. Tropical forestry research in the new millenium: international conference on forestry and forest products research 2001, pp 551–552

Patahayah M, Brearley FQ, Lee SS (2003b) Responses of three ectomycorrhizal fungi to different temperatures and media in vitro. Poster presentation at conference on forestry and forest products research 2003, 6–8 October 2003, Kuala Lumpur, Malaysia

Proctor J, Anderson JM, Chai P, Vallack HW (1983) Ecological studies in four contrasting lowland rain forests in Gunung Mulu National Park, Sarawak. I. Forest environment, structure and floristics. J Ecol 71:237–260

Richard F, Millot S, Gardes M, Selosse M-A (2005) Diversity and specificity of ectomycorrhizal fungi retrieved from an old-growth Mediterranean forest dominated by *Quercus ilex*. New Phytol 166:1011–1023

Rinaldi AC, Comandini O, Kuyper TW (2008) Ectomycorrhizal fungal diversity: separating the wheat from the chaff. Fungal Div 33:1–45

Roeleffs JW (1930) Over kunstmatige verjonging van *Pinus merkusii* Jungh. et de Vr. en *Pinus khasya* Royle. Tectona 23:874–907

Sangtiean T, Schmidt S (2002) Growth of subtropical ECM fungi with different nitrogen sources using a new floating culture technique. Mycol Res 106:74–85

Shamsuddin MN (1979) Mycorrhizas of tropical forest trees. In: Furtado JI (ed) Abstracts: fifth international symposium of tropical ecology. University of Malaya, Kuala Lumpur, Malaysia, p 173

Shiva MP, Jantan I (1998) Non-timber forest products from dipterocarps. In: Appanah S, Turnbull JM (eds) A review of dipterocarps: taxonomy, ecology and silviculture. Center for International Forestry Research, Bogor, Indonesia, pp 187–197

Singh KG (1966) Ectotrophic mycorrhiza in equatorial rain forests. Malay For 29:13–18

Sirikantaramas S, Sugioka N, Lee SS, Mohamed LA, Lee HS, Szmidt AE, Yamazaki T (2003) Molecular identification of ectomycorrhizal fungi associated with Dipterocarpaceae. Tropics 13:69–77

Smith SE, Read DJ (2008) Mycorrhizal symbiosis, 3rd edn. Academic Press, London, UK

Smits WTM (1983) Dipterocarps and mycorrhiza – an ecological adaptation and a factor in forest regeneration. Flora Males Bull 36:3926–3937

Smits WTM (1985) Specificity of dipterocarp mycorrhiza. In: Molina R (ed) Proceedings of the 6th North American conference on mycorrhizae. Forest Research Laboratory, Corvallis, Oregon, USA, p 364

Smits WTM (1992) Mycorrhizal studies in dipterocarp forests in Indonesia. In: Read DJ, Lewis DH, Fitter AH, Alexander IJ (eds) Mycorrhizas in ecosystems. CAB International, Wallingford, UK, pp 283–292

Supriyanto, Setiawan I, Omon RM (1994) Effect of *Scleroderma* sp. on the growth of *Shorea mecistopteryx* Ridl. seedlings. In: Suzuki K, Sakurai S, Ishii K (eds) Proceedings of the International Workshop of BIO-REFOR, Yogjakarta, 1993. BIO-REFOR, IUFRO-SPDC, Tokyo, Japan, pp 186–188

Tata HL, van Noordwijk M, Summerbell R, Werger MJA (2009) Limited response to nursery-stage mycorrhiza inoculation of *Shorea* seedlings planted in rubber agroforest in Jambi, Indonesia. New For 39:51–74

Tawaraya K, Takaya Z, Turjaman M, Tuah SJ, Limin SH, Tamai Y, Cha JY, Wagatsuma T, Osaki M (2003) Arbuscular mycorrhizal colonization of tree species grown in peat swamp forests of Central Kalimantan, Indonesia. For Ecol Manage 182:381–386

Tupas GL, Sajise PE (1976) Mycorrhizal associations in some savanna and reforestation trees. Kalikasan 5:235–240

Turjaman M, Tamai Y, Segah H, Limin SH, Cha JY, Osaki M, Tawaraya K (2005) Inoculation with the ectomycorrhizal fungi *Pisolithus arhizus* and *Scleroderma* sp. improves early growth of *Shorea pinanga* nursery seedlings. New For 30:67–73

Turjaman M, Tamai Y, Segah H, Limin SH, Osaki M, Tawaraya K (2006) Increase in early growth and nutrient uptake of *Shorea seminis* inoculated with two ectomycorrhizal fungi. J Trop For Sci 18:243–249

Turjaman M, Saito H, Santoso E, Susanto A, Gaman S, Limin SH, Shibuya M, Takahashi K, Tamai Y, Osaki M, Tawaraya K (2007) Effect of ectomycorrhizal fungi inoculated on *Shorea balangeran* under field conditions in peat-swamp forests. In: Rieley JO, Banks CJ, Radjagukguk B (eds) Proceedings of the international symposium and workshop on tropical Peatland, Yogyakarta, 27–29 Aug 2007. CARBOPEAT, University of Leicester, UK, pp 143–148

Vellinga EC, Wolfe BE, Pringle A (2009) Global patterns of ectomycorrhizal introductions. New Phytol 181:960–973

Watling R, Lee SS (1995) Ectomycorrhizal fungi associated with members of the Dipterocarpaceae in Peninsular Malaysia – I. J Trop For Sci 7:657–669

Watling R, Lee SS (1998) Ectomycorrhizal fungi associated with members of the Dipterocarpaceae in Peninsular Malaysia – II. J Trop For Sci 10:421–430

Watling R, Lee SS (2007) Mycorrhizal mycodiversity in Malaysia. In: Jones EBG, Hyde KD, Vikineswary S (eds) Malaysian fungal diversity. Mushroom Research Centre, University of Malaya & Ministry of Natural Resources and Environment, Kuala Lumpur, Malaysia, pp 201–219

Watling R, Taylor AFS, Lee SS, Sims K, Alexander IJ (1995a) A rainforest *Pisolithus*; its taxonomy and ecology. Nova Hedwig 61:417–429

Watling R, Taylor AFS, Lee SS, Sims K, Alexander IJ (1995b) Pisolithus aurantioscabrosus. In: Agerer R (ed) Colour atlas of ectomycorrhizae. Einhorn-Verlag, Schwäbisch Gmünd, Germany, plate 85

Watling R, Lee SS, Turnbull E (1998) Putative ectomycorrhizal fungi of Pasoh Forest Reserve, Negri Sembilan, Malaysia. In: Lee SS, Dan YM, Gauld ID, Bishop J (eds) Conservation, management and development of forest resources. Forest Research Institute of Malaysia, Kepong, Malaysia, pp 96–104

Watling R, Lee SS, Turnbull E (2002) The occurrence and distribution of putative ectomycorrhizal basidiomycetes in a regenerating south-east Asian rainforest. In: Watling R, Frankland JC, Ainsworth AM, Isaac S, Robinson CH (eds) Tropical mycology, vol 1, Macromycetes. CAB International, Wallingford, UK, pp 25–43

Yamada A, Katsuya K (2001) The disparity between the number of ectomycorrhizal fungi and those producing fruit bodies in a *Pinus densiflora* stand. Mycol Res 105:957–965

Yazid SM, Lee SS, Lapeyrie FF (1994) Growth stimulation of *Hopea* spp. (Dipterocarpaceae) seedlings following mycorrhizal inoculation with an exotic strain of *Pisolithus tinctorius*. For Ecol Manage 67:339–343

Yazid SM, Lee SS, Lapeyrie FF (1996) Mycorrhizal inoculation of *Hopea odorata* (Dipterocarpaceae) in the nursery. J Trop For Sci 9:276–278

Yuwa-Amornpitak T, Vichitsoonthonkul T, Tanticharoen M, Cheevadhanarak S, Ratchadawong S (2006) Diversity of ectomycorrhizal fungi on Dipterocarpaceae in Thailand. J Biol Sci 6:1059–1064

Chapter 2
The Ectomycorrhizal Symbiosis in South America: Morphology, Colonization, and Diversity

Alejandra G. Becerra and Marcelo R. Zak

2.1 Introduction

About 95% of the world's living species of vascular plants belong to families that are characteristically mycorrhizal (Trappe 1977). The symbiotic root–fungus associations result from the coevolution between plants and fungi, which determined mycorrhizae to be the norm in terrestrial plant nutrition, not the exception (Trappe 1977, 1987, Brundrett and Cairney 2002).

Among the seven types of mycorrhizae widely described (arbuscular, arbutoid, ectendo, ecto, ericoid, monotropoid, and orchidaceous), both arbuscular mycorrhizae (AM) and ectomycorrhizae (ECM) are the most abundant and widespread in forest communities (Allen et al. 2003; Smith and Read 2008).

Forest communities cover approximately 33% of the world's land surface (Rumney 1968) being ECM the most frequent and widespread mycorrhizal type in forests and woodlands of cool temperate and boreal latitudes. Forests characterized by the dominance of ECM woody species would have extended both throughout the hemispheres and upwards in mountain areas at the expense of AM woodlands (Malloch et al. 1980). On the other hand, even though various tropical and subtropical trees throughout the world also form ECM (Moyersoen et al. 1998a, b,

A.G. Becerra (✉)
Instituto Multidisciplinario de Biología Vegetal (CONICET), Universidad Nacional de Córdoba, C.C. 495, 5000 Córdoba, República Argentina
and
Cátedra de Diversidad Vegetal I, Facultad de Ciescias Exactas, Físicas y Naturales, Universidad Nacional de Córdoba, C.C. 495, 5000 Córdoba, República Argentina
e-mail: abecerra@efn.uncor.edu

M.R. Zak
Instituto Multidisciplinario de Biología Vegetal (CONICET), Universidad Nacional de Córdoba, C.C. 495, 5000 Córdoba, República Argentina
and
Cátedra de Recursos Naturales y Gestión Ambiental, Departamento de Geografía, Universidad Nacional de Córdoba, C.C. 495, 5000 Córdoba, República Argentina
e-mail: marcelzak@gmail.com

M. Rai and A. Varma (eds.), *Diversity and Biotechnology of Ectomycorrhizae*, Soil Biology 25, DOI 10.1007/978-3-642-15196-5_2,

2001; Pérez-Moreno 1998; Founoune et al. 2002; Onguene and Kuyper 2002), most of them form AM (Alexander 1989). This AM are obligate symbionts belonging to the Glomeromycota phylum (Schüßler et al. 2001), which have been found in trees, shrubs, and herbs in all tropical regions (Janos 1980, 1985; Béreau et al. 1997; Smith and Read 2008).

According to Singer and Morello (1960), ECM forests in South America continue the important ectomycorrhizal complex of Central America into Colombia, where *Quercus* sp. (Fagaceae) appears. There is also a strip formed by *Alnus acuminata* Kunth (Betulaceae) extended from Venezuela to Argentina, while the largest and most important ectomycorrhizal area, the *Nothofagus* (Fagaceae) region extends from 33° to 50°S. Beyond these naturally occurring forests, human activity made possible the existence of many ECM forests of plant species naturally found in the Northern Hemisphere, such as *Pinus* sp., *Eucalyptus* sp., *Populus* sp., *Salix* sp., *Larix* sp., *Cedrus* sp., *Betula* sp., and *Quercus* sp. (Singer and Morello 1960) with their ECM fungi of European and North American origin.

On this basis, the general aim of the present chapter is to review the ECM symbiosis in South America with particular emphasis on its anatomical characteristics. This will provide an organized structure essential for the better understanding of the plant–fungus mutualism in ectomycorrhizal associations in South America, so adding new tools for the management and conservations of threatened ecosystems.

2.2 ECM Studies in South America

The tradition on ECM research in South America is important. The great diversity of available studies covers descriptions of mycorrhizal status, ECM fungi, inoculations, colonization, and morphological characterization. Unfortunately, and due in part to their abundance but often also to their inaccessibility, it is not possible to aim at describing the whole array of studies carried throughout the continent. Therefore, this chapter pretends to summarize the wide range of families and genera of Angiospermae and Gymnospermae forests growing in South America in which ectomycorrhizal studies have been carried.

Most ECM fungi can associate with various host plants, while the opposite is also true, being it possible for a single host to form ECM with a number of different fungi (Moyersoen 1993). Most studies analyzed (Table 2.1) present lists of ectomycorrhizal fungi (most of which belong to the Basidiomycota phylum) and taxonomic descriptions in exotic plants such as *Pinus* (*P. radiata*, *P. elliottii*), *Eucalyptus* (*E. citriodora*, *E. dunii*, *E. robusta*), and *Quercus*. Among endemic plants, most studies have been carried out in *Nothofagus* (*N. obliqua*, *N. dombeyi*, *N. alpina*) forests. Besides, some of the associations found in these forests are established with fungi species from exotic plantations (*P. radiata* en Chile) (Garrido 1986).

Table 2.1 Description of 103 studies on ectomycorrhizal status in South America

Family	Species	Publication with country[a] and ECM characteristics[b] indication
Native plants		
Betulaceae	*Alnus acuminata*	1(A;2) 2(A;2) 3(A;1) 4(A;1) 5(A;1) 6(A;1) 7(A;1,4) 8(A;1,3,4) 9(A;1,3,4) 10(A;2)
Caesalpiniaceae	*Dicymbe corymbosa*	11(FG;2,4) 12(FG;2) 13(FG;4) 14(FG;2)
	D. altsonii	11(FG;2,4) 12(FG;2) 13 (FG; 4) 15(FG;1,2)
	D. jenmanii	11(FG;4)
	Dicymbe sp.	14(FG;2)
Dipterocarpaceae	*Pakaraimaea dipterocarpacea*	17(V;1) 18(V;2)
Fabaceae	*Acacia bonaerensis*	19(U;4)
	Aldinia heterophylla	20(B;2)
	A. insignis	11(G;4) 13(G;4)
	A. kunhardtiana	21(V;1,4)
	A. latifolia	21(V;1,4)
	Calliandra selloi	19(U;4)
	C. tweedii	19(U;4)
	Gleditsia amorphoides	19(U;4)
	Lonchocarpus nitidus	19(U;4)
	Prosopis spp.	19(U;4)
Fagaceae	*Nothofagus alpina*	22(C;1) 23(C;1) 24(C;3) 25(C;2) 26(C;2) 27(C;1,4) 28(C;4)
	N. alessandrii	29(C;3)
	N. antarctica	25(C;2) 26(C;2) 30(C;4) 31(C;4) 32(A;4)
	N. betuloides	23(C;1) 25(C;2)
	N. dombeyi	23(C;1) 24(C;3) 25(C;2) 26(C;2) 28(C;4) 30(C;4) 31(C;4) 32(A;4) 34(A;1) 35(C;1) 36(C;2)
	N. glauca	25(C;2) 38(C;3)
	N. leonii	25(C;2)
	N. nervosa	32(A;4)
	N. nitida	25(C;2) 39(C;1)
	N. obliqua	23(C;1) 25(C;2) 26(C;2) 30(C;4) 32(A;4) 37(C;4) 38(C;3)
	N. obliqua var. *macrocarpa*	25(C;2)
	N. pumilio	

(*continued*)

Table 2.1 (continued)

Family	Species	Publication with country[a] and ECM characteristics[b] indication
		23(C;1) 25(C;2) 26(C;2) 32(A;4) 33(C;1) 40(C;1) 41(C;1) 42(C;1) 43(C;1) 44(C;2,4)
	Nothofagus spp.	1(C;2) 25(C;2) 26(C;2) 45(A-C;2)
Gnetaceae	*Gnetum* sp.	46(B;4) 47(B;4)
Melastomataceae	*Graffenrieda emarginata*	48(E;1) 49(E;4)
Nyctaginaceae	*Neea altisima*	48(B;4)
	N. divaricata	50(E;2,4)
	N. laetivirens	52(B;4)
	N. obovata, N. robusta	21(V;1,4)
	N. tristis	53(FG;4)
	Neea sp.1, *Neea* sp. 2	49(E;4) 51(E;1)
	Neea sp. 3, *Neea* sp. 5	50(E;2,4)
	Neea sp.	46(B;4) 47(B;4) 54(P;4)
	Guapira sancarlosiana	21(V;1,4) 49(E;4)
	Guapira sp.	51(E;1)
Polygonaceae	*Coccoloba excelsa*	21(V;1,4)
	C. latifolia	53(FG;4)
	C. moliis	53(FG;4)
	C. rosea	55(B;1)
	Coccoloba sp.	53(FG;4)
Salicaceae	*Salix humboldtiana*	1(A;2) 56(A;1,2) 57(A;4)
Sapotaceae	*Glycoxylon inphyllum*	47(B;4)
Introduced plants		
Betulaceae	*Betula pendula*	25(C;2) 58(A;1)
	Betula sp.	59(A;2)
Myrtaceae	*Eucalyptus camaldulensis*	60(B;4) 61(B;3)
	E. citriodora	61(B;3) 62(B;2) 63(B;2) 64(B;2,3)
	E. cloeziana	61(B;3)
	E. dunnii	62(B;2) 65(B;2) 66(B;3) 67(B;3) 68(B;3) 69(B;3) 71(B;3) 72(B;3)
	E. globulus	25(C;2)
	E. grandiis	60(B;4) 61(B;3) 63(B;2) 64(B;2,3) 73(B;2) 74(B;2) 75(B;3) 76(B;3)

	E. microcorys	63(B;2) 73(B;2)
	E. robusta	62(B;2) 64(B;2,3)73(B;2)
	E. rostrata	77(A;3)
	E. tereticomis	63(B;2)
	E. saligna	63(B;2)
	E. urophylla	61(B;3) 75(B;3) 76(B;3)
	E. viminalis	63(B;2) 77(A;3) 78(A;1,2)
	Eucaliptus spp.	25(C;2) 79(Co;2) 80(B;2)
Pinaceae	*Cedrus atlántica*	58(A;1)
	C. deodara	81(A;1,2)
	Cedrus sp.	59(A;2)
	Larix decidua	25(C;2) 59(A;2)
	Picea abies	25(C;2)
	Pinus caribaea var. *hondurensis*	82(B;3)
	P. contorta	25(C;2)
	P. elliotti	58(A;1) 59(A;2) 62(B;2) 83(A;1,3)
	P. halepensis	59(A;2) 84(A;3)
	P. patula	25(C;2) 59(A;2) 79(Co;2)
	P. pinaster	25(C;2) 85(A;3) 84(A;3)
	P. pinea	25(A;3) 84(A;3)
	P. ponderosa	85(A;2,3,4) 86(A;2) 87(A;2) 88(A;1,2) 89(A;3)
	P. radiata	25(C;2) 59(A;2) 79(Co;2) 90(C;2) 91(C;2)
	P. sylvestris	25(C;2)
	P. taeda	59(A;2) 62(B;2) 65(B;2) 68(B;3) 84(A;3) 92(B;3) 93(B;3) 94(A;3)
	P. thunbergii	84(A;3)
	Pinus sp.	25(C;2)
	Pinus spp.	1(A;2) 95(U-A;2) 96(B;2)
	Pseudotsuga menziesii	25(C;2) 86(A;2) 88(A;1,2) 97(A;2,3,4) 98(A;2)
	Tsuga heterophylla	25(C;2)
Fagaceae	*Quercus humboldtii*	99(Co;2) 100(Co;2) 101(Co;2) 102(Co;2) 103(Co;2)
	Q. robur	25(C;2)
	Quercus sp.	59(A;2)
	Quercus spp.	79(Co;2)

(*continued*)

Table 2.1 (continued)

Family	Species	Publication with country[a] and ECM characteristics[b] indication
Salicaceae	*Populus nigra*	25(C;2)
	Populus spp.	25(C;2)
	Salix viminalis	25(C;2)
	Salix spp.	25(C;2)

Each line within the table represents a studied species, with the indication of the Family to which it belongs and all publications (numbered – all respective citations are found below under References) in which it has been described. Following the paper number is an indication of the characteristics of the study carried (country and ECM characteristics)

[a]Country: A Argentina; B Brazil; C Chile; Co Colombia; E Ecuador; FG French Guyana; P Perú; U Uruguay; V Venezuela

[b]ECM characteristics studied: 1: anatomical characteristics, 2: list of ECM fungi, 3: ECM inoculations, 4: mycorrhizal status or colonization

References: (1) Singer (1953), (2) Singer and Morello (1960), (3) Becerra (2002), (4) Becerra et al. (2002), (5) Becerra et al. (2005a), (6) Becerra et al. (2005b), (7) Becerra et al. (2005c), (8) Becerra et al. (2009a), (9) Nouhra et al. (2003), (10) Nouhra et al. (2005), (11) Henkel et al. (2002), (12) Henkel (1999), (13) McGuire et al. (2008), (14) Fulgenzi et al. (2007), (15) Henkel et al. (2000), (16) Matheny et al. (2003), (17) Moyersoen (2006), (18) Moyersoen (2008), (19) Frioni et al. (1999), (20) Singer et al. (1983), (21) Moyersoen (1993), (22) Palfner (1997), (23) Palfner (2001), (24) Godoy and Palfner (1997), (25) Garrido (1986), (26) Valenzuela et al. (1999), (27) Palfner et al. (2008), (28) Castillo et al. (2006), (29) Flores et al. (1997), (30) Carrillo et al. (1992), (31) Godoy et al. (1994), (32) Diehl et al. (2008), (33) Palfner (2002a), (34) Beenken (2001), (35) Palfner (2002c), (36) Singer (1971), (37) Castillo et al. (2006), (38) Pérez et al. (2007), (39) Palfner (2002b), (40) Palfner and Godoy (1996a), (41) Palfner and Godoy (1996b), (42) Palfner (1996), (43) Palfner (2002d); (44) Villegas et al. (2007); (45) Wrigth (1988); (46) St John (1980), (47) Singer and Araujo (1979), (48) Haug et al. (2004), (49) Kottke and Haug (2004), (50) Lunt and Hedger (1996), (51) Haug et al. (2005), (52) Janos (1980), (53) Béreau et al. (1997), (54) Alexander and Högberg (1986), (55) Landim de Souza (2003), (56) Becerra et al. (2009b), (57) Fracchia et al. (2009), (58) Nouhra (1997), (59) Nouhra (1999), (60) Pagano and Scotti (2008), (61) Dos Santos et al. (2001), (62) Giachini et al. (2000), (63) Yokomizo (1981), (64) Schwan (1984), (65) Giachini et al. (2004), (66) Oliveira et al. (1997), (67) Lupatini et al. (2008), (68) Rocci (2006), (69) Voigt et al. (2000), (70) Borges de Souza et al. (2004), (71) Borges de Souza et al. (2008), (72) Pinto Bonnassis (2007), (73) Marx (1977), (74) Hentz de Mello et al. (2006), (75) Arruda Bacchi and Krugner (1988), (76) Liparini Pereira et al. (2005), (77) Halbinger et al. (1986), (78) Brandán de Weth (2006), (79) Guzmán and Varela (1978), (80) Barros et al. (1978), (81) Daniele et al. (2005), (82) Gross et al. (2004), (83) Nouhra and Becerra (2001), (84) Tackacs (1961), (85) Barroetaveña and Rajchenberg (2003a), (86) Barroetaveña et al. (2007), (87) Barroetaveña et al. (2005), (88) Barroetaveña (2004), (89) Martínez et al. (2007), (90) Garrido (1983), (91) Garrido (1986), (92) Paloschi de Oliveira et al. (2006), (93) Rocci (2006), (94) Tackacs (1964), (95) Singer (1968), (96) Putzke and Pereira (1994), (97) Barroetaveña and Rajchenberg (2003b), (98) Barroetaveña et al. (2006), (99) Singer (1963), (100) Halling (1989), (101) Franco Molano and Uribe Calle (2000), (102) López-Quintero et al. (2007), (103) Vasco-Palacio et al. (2005)

In South America, most studies on mycorrhizal status and ECM colonization deal with the Fagaceae, Fabaceae, Nyctaginaceae, and Polygonaceae families (Table 2.1). For the *Nothofagus* spp. forests of Chile, a colonization of 1.8–4.8% and 19.6% has been reported by Carrillo et al. (1992) and Palfner et al. (2008), respectively, while for the *Nothofagus* spp. forests of Argentina, it is of 79% (Diehl et al. 2008). Meanwhile, for some genus of the Fabaceae family, Frioni et al. (1999) reported a colonization that ranges between 17 and 36%, while Moyersoen (1993) reported a 65% colonization for *Aldinia kunhardtiana.*

Most studies carried out for the Nyctaginaceae and Polygonaceae families refer to the status of their species in terms of presence/absence of ECM (Singer and Araujo 1979, St John 1980, Janos 1980, Béreau et al. 1997). In the particular case of *Neea obovata*, *N. robusta*, and *Guapira sancarlosiana* (Nyctaginaceae), colonization was 100%, 7%, and 96%, respectively, while it was of 56% in *Coccoloba excelsa* (Polygonaceae) (Moyersoen 1993).

Inoculation technologies appear as an alternative to the use of fertilizers since they reduce both the costs of production and environmental contamination (Garbaye 1990). ECM inoculations in South America have been carried almost exclusively in introduced plants (*Eucalyptus* spp., *Pinus* spp., and *Pseudostuga menziesii*), with the exception of *Nothofagus* spp. and *Alnus acuminata.* For this, various techniques have been applied: spores from sporocarps mainly belonging to the Basidiomycota phylum (*Alpova diplophloeus*, *Laccaria laccata*, *Rhizopogon roseolus*, *Suillus granulatus*, among others); ECM culture micelia (micelia, airlift bioreactors) in *Eucalyptus* spp. and *Pinus* spp.; and natural soil (potential inoculum) for *Pinus* spp., *Pseudotsuga menziesii*, *Eucalyptus dunii*, *E. citriodora*, *Nothofagus alpina*, *N. dombeyi*, and *Alnus acuminata.* All three techniques have been used for introduced trees, while only spores and natural soil have been used for native species.

2.3 Morphological and Anatomical Features of the ECM in South America

The morphological and anatomical description of mycorrhizae and the identification of their fungal partners are prerequisites for recognizing mycorrhizal biodiversity in ecosystems (Agerer 1991; Miller et al. 1991). Meanwhile, the occurrence of natural ECM associations in the indigenous vegetation types from South America has been virtually unexplored.

Table 2.2 summarizes the volume of ECM anatomotypes described in South America for both native and introduced plants. Four anatomical complexes for recognition of fungal relationships were used: (a) structure of outer mantle layers in plan view, (b) structure of rhizomorphs, (c) shape of cystidia, and (d) features of emanating hyphae (Agerer 1999, 2006). Besides, the cross section of mantle area (e) was also considered, which showed useful for comparing the relative

Table 2.2 Anatomical characterization of ECM in native and introduced plants in South American forests

Ectomicorrhiza/host tree[a]/country[b]/ publication[c]	Mantle layers (Plect.: plectenchymatous; Pseud: pseudoparenchymatous)	Rhizomorphs	Cystidia	Emanating hyphae (color, branching, clamps, septa, anastomoses)	Cross section (μm)
Naucoria escharoides/ Aa/A/1	Plect. with parallel arranged hyphae (Mantle type B)	Uniform-loose hyphal bundles	Lacking[d]	Hyaline abundant, branched, with clamps; simple anastomoses	39–56
Cortinarius helodes/ Aa/A/2	Plect. loosely interwoven (type B)	Uniform-loose	Lacking	Hyaline abundant, branched, clampled, anastomoses open	130–220
Gyrodon monticola/ Aa/A/2	Plect. with globular cells (type F)	Boletoid	Spherical to clavate	Hyaline and brown abundant, branched, with clamps	24–57
Russula alnijorullensis/Aa/ A/3	Pseud. with angular cells (type L)	Lacking	Lacking	Colorless, few, clamps lacking	40–60
Cortinarius tucumanensis/Aa/ A/3	Plect. with irregularly arranged hyphae (type B)	Uniform-loose	Lacking	Colorless abundant, clamps present, anastomoses open	62–120
Lactarius aff. omphaliformis/Aa/ A/3	Pseud. with angular cells (type P-Q)	Lacking	Lacking	Colorless to membranaceously yellowish, clamps lacking	11–30
Tometella cf. *sublilacina*/Aa/A/4	Plect. with hyphal net arrangement (type A)	Undifferentiated	Lacking	Yellowish to brownish, branched, with clamps	23–36
Tometella cf. *stuposa*/ Aa/A/4	Pseud., with angular and roundish cells (type L)	Not found[d]	Bottle-shaped with a straight to bent neck	Membranaceously brownish, branched with clamps	45–65
Tometella cf. *ellisii*/ Aa/A/4	Plect. with irregularly arranged hyphae (type B)	Not found	Lacking	Colorless, branched with clamps	70–105
		Not found	Lacking		44–58

Lactarius omphaliiformis/Aa/A/4	Pseud. with epidermoid cells bearing a hyphal net (type Q)			Colorless to membranaceously yellowish, without clamps	
Russula sp. /Aa/A/4	Plect. irregularly arranged hyphae (type B)	Not found	Lacking	Membranaceously yellowish, branched without clamps	25–75
Unidentified A/No/V/5	Plect.	NR[d]	NR	Present	75–86
Unidentified B/No/V/5	Compactly mantle	NR	NR	NR	31–49
Unidentified/Nr/V/5	Compactly mantle	NR	NR	NR	33–46
Russula puiggarii/Nsp.1/E/6	Plect., loosely with gelatinous matrix	NR	Not observed	Whitish to light yellowish	30 thick
Lactarius sp./Nsp.1/E/6	Plect. hyphae of larger and irregular diameter	Undifferentiated	Not observed	Abundant white, septa without clamps	30 thick
Tomentella/*Thelephora* sp.1/Nsp.1/E/6	Plect. loosely, hyphae net-like arranged	NR	NR	Brown, septa without clamps	40 thick
Tomentella/*Thelephora* sp.2/Nsp.1/E/6	Plect., hyphae with clamp	NR	NR	Reddish	40 thick
Ascomycete/Nsp.1/E/6	Plect., hyphae arranged star-like	NR	NR	Black, simple pores with Woronin bodies	30 thick
Thelephora/*Tomentella*/Nsp.2/E/6	Plect. loosely	NR	NR	Brown with clamps	NR
Unidentified/Gs/V/5	Compactly mantle	NR	NR	NR	21–36
Thelephora/*Tomentella*/Gsp./E/6	Plect. loosely arrangement	NR	NR	Colorless with clamps	NR
Unidentified/Ce/V/5	Compactly mantle	NR	NR	NR, Brown mycorrhizae	40–48
Unidentified/Cr/B/7	Plect.	NR	NR	Dark Brown, abundant	40–65
Unidentified/Ah/V/5	Plect.	NR	NR	Abundant	33–60
Clavulinaceae 1/Pd/V/8	NR	NR	NR	NR	NR

(*continued*)

Table 2.2 (continued)

Ectomicorrhiza/host tree[a]/country[b]/publication[c]	Mantle layers (Plect.: plectenchymatous; Pseud: pseudoparenchymatous)	Rhizomorphs	Cystidia	Emanating hyphae (color, branching, clamps, septa, anastomoses)	Cross section (μm)
Clavulinaceae 2/Pd/V/8	Plect. on top of a Pseud.	Not found	Lacking	White, non ramified, thick and rough wall	NR
Sebacina sp./Pd/V/8	Plect. (net on top of a plectenchyma)	Present (not described)	NR	Yellowish	NR
Cortinarius sp./Pd/V/8	Plect.	Frequent, hairy, ramified	Lacking	Whitish, with clamps	NR
Inocybe 1/Pd/V/8	NR	NR	NR	NR	NR
Inocybe 2/Pd/V/8	Plect. covered by a loose net	With smooth margin; clamp connections	Lacking	Whitish	NR
Amanita sp./Pd/V/8	NR	NR	NR	NR	NR
Unidentified/Pd/V/8	Plect. (net on top a plect.)	Lacking	Elongated	White, simple septa	NR
Unidentified 2/Pd/V/8	Plect. or pseud.	Thelephoroid rhizomorphs	Lacking	Brown often ornamented, simple septa	NR
Nothofagirhiza vinicolor/Np/C/9	Plect., loosely, interwoven hyphae (type B, E)	Lacking	Lacking	Pinkish to dark red, very abundant, with clamps	10–20
Russula fuegiana/Np/C/10	Plect. loose net of thin hyphae (type C, D)	Not frequent, agglutinated hyphae	Cylindrical to gradually tapering with apical knobs	Not found	13–30
Thaxterogaster albocanus/Np/C/11	Plect., loosely, interwoven hyphae (type B)	Numerous undifferentiated	Lacking	Frequent, branched	13–27
Austropaxillus boletinoides/Np/C/12	Plect. (type A, B)	Rather frequent, highly differentiated	Cistidia-like hyphal ends of variable shape	Rather frequent	15–25
Gautieria inapire/Np/C/15	Plect., loosely interwoven (type B)	Unconspicuous, simple or slightly differentiated	Rarely with acanthocystidia	Gelatinous, cell walls with short protuberances	1–25
Nothofagirizha reticulosa/Np/C/16	Plect. (type B), mantle surface often parallely blunded by anastomoses and short ramifications	Rhizomorphs-like strands or simple	Lacking	Light brown, verrucose, septa mostly with clamps, anastomoses also closed by a clamp	20–30

Cenococcum geophilum/Np/C/16	Plect. (type G) formed by very densely arranged, star-like pattern	Lacking	Lacking	Numerous black, bristle-like	NR
Russula nothofaginea/Nd/A/19	Pseud., covered by hyphal net embedded in a distinct gelatinous matrix (type Q)	Not found	Lacking	Colorless infrequent	15–20
Cortinarius magellanicus/Nd/C/14	Plect. (type B)	Abundant, with attached spherical sclerotia	Lacking	Colorless-hyaline hyphae, with clamps	30–60
Stephanopus stropharoides/Nd/C/16	Plect. (type B)	Abundant, blue-bruising when squeezed	Lacking	Abundant, with clamps	30–50
Nothofagiriza vesiculosa/Nd/C/16	Pseud. (type K), without clamps, thick-walled	Without	Lacking	Very few short	20–45
Cortinarius austrosalor/Nn/C/13	Plect. (type B)	Differentiated	Lacking	With clamps, open anastomoses	20–30
Austropaxillus boletinoides f. *olivascens*/Nb/C/16	Plect. (type B)	Dark brown, pigmented hyphae	Scattered Clavate	Clamped	15–25
Boletus loyo/Nob/C/16	Plect. (type B)	Scattered, highly differentiated	Occasionally clavate, slightly thickened hyphal ends	Scarse, without clamps	36–45
Boletus putidus/Nob/C/16	Plect. (type A/B)	Highly differentiated, thicker, margin with inflated hyphal elements	Poorly differentiated or weakly clavate cystidia or cystidia-like hyphal ends	Hyphae without clamps	15–25

(*continued*)

Table 2.2 (continued)

Ectomicorrhiza/host tree[a]/country[b]/publication[c]	Mantle layers (Plect.: plectenchymatous; Pseud: pseudoparenchymatous)	Rhizomorphs	Cystidia	Emanating hyphae (color, branching, clamps, septa, anastomoses)	Cross section (μm)
Nothofagirhiza tricystidiis/Nob/C/16	Plect. (type D)	Not found	Three types: slightly tapering with an enlarged barrel-shaped; awl-shaped or worm-shaped with enlarged basal cell; ampullate with an apical knob	Rare, clamps lacking	15–22
Descolea antarctica/Na/C/17	Plect. with awl-shaped cystidia (type D)	Not found	Lateral, tapering, knob-bearing outgrowth	Infrequent, with clamps	15–20
Hymenoscyphoid (probably *Rhizoscyphus ericae*)/Na/C/18	Compact plect.	Lacking	Lacking	Black hyphae, without clamps	NR
Inocyboid/Na/C/18	Loose plect.	Lacking	Lacking	Abundant, with clamps, inocyboid	NR
Russuloid/Na/C/18	Plect.	Lacking	Russuloid gloeocystidia	Lacking	NR
Thelephoroid/Na/C/18	Compact plect.	Whitish or lacking	Thelephoroid	Abundant, with clamps	NR
Tomentelloid/Na/C/18	Pseud. or plect.	Dark brown or lacking	Present or lacking	Scarce or lacking, present or lacking clamp connections	NR
Rhizoscyphus ericae/Gm/E/20	No hyphal mantle; a superficial Hartig net (fingerlike branched hyphae)	Lacking	Lacking	Lacking	NR
Lactarius panuoides/Da/G/21	Mantle densely tomentose to substrigose mass	NR	Occasionally similar hyphae forming capitate cystidia	Abundant, with septa	NR
Russula campinensis/Da/G/21	Well developed mantle; coarse net of irregularly shaped	NR	NR	No evident	NR

Inocybe sp./Sh/A/22	Plect. irregularly arranged hyphae (type A)	Lacking	Lacking	Hyaline clampled	19–35
Tometella sp./Sh/A/22	Plect. irregularly arranged hyphae (type A)	Lacking	Lacking	Hyaline clampled, not branched	25–31
Anatomotype III/Sh/A/22	Plect., hyphae without clamps	Lacking	Lacking	Lacking	7–16
Anatomotype IV/Sh/A/22	Plect.	Lacking	Lacking	Scarse, with clamps	12–16
Anatomotype V/Sh/A/22	Plect.	Lacking	Lacking	Scarse, without clamps	12–20
Anatomotype VI/Sh/A/22	Plect.	Lacking	Lacking	Few light to brown, with clamps	12–15
Anatomotype VII/Sh/A/22	Plect.	Lacking	Lacking	Few hyaline, without clamps	12–30
Leccinum scabrum/Bp/A/23	Plect. (type A)	Boletoid, abundant	Lacking	Rough abundant, clamps lacking	NR
Suillus granulatus/Pe/A/23	Plect. with brownish drops of pigments	Boletoid, abundant	Lacking	Abundant, clamps lacking	NR
Amphinema byssoides/Pp/A/24	Plect.	Lacking	NR	Abundant yellowish, clampled, with spines	NR
Rhizopogon roseolus/Pp/A/24	Plect. with a ring-like hyphal pattern	Branched white, abundant hyaline crystals	Lacking	Lacking	NR
Rhizopogon aff. *ellenae*/Pp/A/24	Plect. with yellow-brown crystals	Abundant, branched, white to pink, with hyaline or yellow crystals	Lacking	Lacking	NR
Rhizopogon inmature type/Pp/A/24	Plect. with squarrosely hyphae with crystals	Thin branched, with crystals	Lacking	Abundant	NR
Rhizopogon coralloid type/Pp/A/24	Plect. with hyaline crystals	Abundant, branched, with crystals	Lacking	NR	NR
Rhizopogon brownish type 42/Pp/A/24	Plect. with abundant crystals	Lacking	NR	Lacking	NR
Tuber type/Pp/A/24	Pseud. with irregular cells	Lacking	Awl-shaped, bristle like	NR	NR

(*continued*)

Table 2.2 (continued)

Ectomicorrhiza/host tree[a]/country[b]/publication[c]	Mantle layers (Plect.: plectenchymatous; Pseud: pseudoparenchymatous)	Rhizomorphs	Cystidia	Emanating hyphae (color, branching, clamps, septa, anastomoses)	Cross section (μm)
Brownish coralloid anatomotype/Pp/A/24	Plect., hyphae with septa	Lacking	Lacking	Lacking	NR
Whitish anatomotype/Pp/A/24	Plect. with a ring-like arrangement	Lacking	Lacking	With septa	NR
Dichotomous yellowish antatomotype/Pp/A/24	Plect.	Lacking	Lacking	Clamped	NR
Brownish antatomotype/Pp/A/25	Plect., hyphae with septa	Lacking	NR	Lacking	NR
Xerocomus chrysenteron/Ca/A/23	Plect. (type A)	Boletoid, abundant	Lacking	Smooth abundant, clamps lacking	40
Amanita muscaria/Cd/A/25	Plect. (type A)	Undifferentiated, abundant	Lacking	Smooth, with clamps	32–57
Tuber type 1/Pm/A/24	Pseud.	Lacking	Awl-shaped, bristle like	Abundant	NR
Rhizopogon type 1/Pm/A/24	Plect.	White branched	NR	Hyaline, with septa	NR
White coralloid antatomotype/Pm/A/24	NR	White branched	NR	NR	NR
Brownish anatomotype/Pm/A/24	Plect.	Lacking	NR	Abundant	NR

Each line within the table represents a studied ectomicorrhiza, with an indication of its host trees, country for which it has been described and citation (numbered – all respective citations are found below under References), followed by a detailed description of the anatomical characteristics of the referred ECM (structure of outer mantle layers and of rhizomorphs, shape of cystidia, features of emanating hyphae and cross section)

[a]Host Tree: Aa *Alnus acuminata*; No *Neea obovata*; Nr *Neea robusta*; N sp. 1 *Neea* sp. 1; N sp. 2 *Neea* sp. 2; Gs *Guapira sancarlosiana*; G sp. *Guapira* sp.; Ce *Coccoloba excelsa*; Cr *Coccoloba rosea*; Ah *Aldinia kunhartiana*; Pd *Pakaraimaea dipterocarpacea*; Np *Nothofagus pumilio*; Nd *Nothofagus dombeyi*; Nn *Nothofagus nitida*; Nb *Nothofagus betuloides*; Nob *Nothofagus obliqua*; Na *Nothofagus alpina*; Ge *Graffenrieda emarginata*; Da *Dicymbe altsonii*; Sh *Salix humboldtiana*; Bp *Betula pendula*; Pe *Pinus elliottii*; Pp *Pinus ponderosa*; Ca *Cedrus atlantica*; Cd *Cedrus deodara*; Pm *Pseudostuga menziessi*

[b]Country: A: Argentina; B: Brazil; C: Chile; E: Ecuador; G: French Guyana; V: Venezuela

[c]References: (1) Becerra et al. (2002), (2) Becerra et al. (2005a), (3) Becerra et al. (2005b), (4) Pritsch et al. (2010), (5) Moyersoen (1993), (6) Haug et al. (2005), (7) Landim de Souza (2003), (8) Moyersoen (2006), (9) Palfner and Godoy (1996a), (10) Palfner and Godoy (1996b), (11) Palfner (1996), (12) Palfner (2002a), (13) Palfner (2002b), (14) Palfner (2002c), (15) Palfner (2002d), (16) Palfner (2001), (17) Palfner (1997), (18) Palfner et al. (2008), (19) Beenken (2001), (20) Haug et al. (2004), (21) Henkel et al. (2000), (22) Becerra et al. (2009b), (23) Nouhra (1997), (24) Barroetaveña (2004), (25) Daniele et al. (2005)

[d]Lacking: the author makes its inexistence explicit, NR (not recorded): the author makes no mention to either its presence or absence, Not found: the author describes it as not found

importance of the fungal partner in different plants for different ecosystems, climates, and soils (Landim de Souza 2003). Table 2.2 does not consider morphological descriptions of seedlings grown under greenhouse conditions since, as Agerer (2006) states, descriptions of ECM that are exclusively based on synthesized material often do not provide features occurring in nature.

Out of the 85 ECM anatomotypes described in the different studies (every ECM not identified to the species level was considered to be a different ECM), 35 were identified up to their species, 20 to their genus, 5 to supraspecific groups, 1 to its phylum, 4 were unidentified ECM morphotypes published under a binomial name (Agerer 1986, 1987–2002, 1994, 1999), and 20 were unidentified ECM anatomotypes lacking a name.

The associated fungi recorded in the 85 ECM descriptions were mostly Basidiomycota (55) and Ascomycota (6), although in 29 descriptions, the fungal division could not be determined.

Morphotypes belonging to the Helotiaceae family (Ascomycota) showed plectenchymatous mantles or, exceptionally, no mantle, as seen in *Graffenrieda emarginata* from Ecuador (Haug et al. 2004). *Cenococcum*, the most cosmopolitan morphotype, showed a black plectenchymatous mantle forming a characteristically star-like pattern and frequently numerous black hyphae. Morphotypes belonging to the Tuberaceae family showed a pseudoparenchymatous mantle with irregular cells and awl-shaped cystidia.

Fungal families in the Basidiomycota were represented in ECM descriptions as follows: Amanitaceae (2), Atheliaceae (1), Bolbitiaceae (1), Boletaceae (4), Clavulinaceae (2), Cortinariaceae (11), Ramariaceae (1), Rhizopogonaceae (6), Russulaceae (10), Paxillaceae (3), Sebacinaceae (1), Suillaceae (1), Thelephoraceae (8), plus four supraspecific groups: Inocyboid (1), Russuloid (1), Thelephoroid (1), and Tomentelloid (1).

Boletoid rhizomorphs were present in most Boletaceae, Paxillaceae, Rhizopogonaceae, and Suillaceae ECM of Boletales families. In general, these morphotypes showed plectenchymatous mantles frequently with ring-like patterns and the presence of crystals.

Among the Russulaceae, and as Agerer (2006) states, russuloid rhizomorphs are combined with plectenchymatous mantles. Meanwhile, Russulaceae species with pseudoparenchymatous mantles do not form rhizomorphs at all.

ECM belonging to the Cortinariaceae family was characterized by an extramatrical organized mycelia and undifferentiated or slightly differentiated rhizomorphs, while ECM of the Thelephoraceae family was characterized by a heterogeneous mantle type assemblage (Agerer 2006). In this ECM, anatomotypes plectenchymatous and pseudoparenchymatous mantles were found. Different from Agerer (2006), the morphotype associated with *Amanita muscaria* (Amanitaceae) showed a plectenchymatous mantle, while the *Sebacina* sp. morphotype (Sebacinaceae) presented rhizomorphs.

The morphotypes of *Amphinema byssoides* (Atheliaceae), *Descolea anarctica* (Bolbitiaceae), and *Gautieria inapire* (Ramariaceae) show similar characteristics than those described by Agerer 2006.

2.4 Conclusion

Although more than 90% of terrestrial plants are associated with mycorrhizal fungi, and two-thirds of them are AM, ECM tree species are also notorious in South American forests. This review summarizes the ectomycorrhizal studies carried out in the neotropical ecozone and provides information about ECM fungi and their anatomical characteristics.

The analyzed studies reveal that most fungal symbionts are Basidiomycota, and that most studies on mycorrhizal status and ECM colonization deal with the Fagaceae, Fabaceae, Nyctaginaceae, and Polygonaceae families. Meanwhile, the three inoculum techniques (spores from sporocarps, culture mycelia, and natural soil) have been used for introduced trees (*Eucalyptus* spp., *Pinus* spp., and *Pseudostuga menziesii*), whereas only spores and natural soil have been used for native species (*Nothofagus* spp. and *Alnus acuminata*). The associated fungi recorded in the 85 ECM anatomotypes reviewed were mostly Basidiomycota and a few Ascomycota, mostly found on native species (77%).

An important byproduct of this review is the realization of the existence of many gaps in the existing knowledge of ECM in South America. This suggests that mycorrhizologists should focus on little known/studied geographic areas, ecosystems, host trees, and fungal groups to reveal those aspects of the ECM symbiosis in South America, which may have practical applications in, for example, afforestation and environmental management programs. This knowledge may also be an important contribution to the conservation community, in a time when firsthand knowledge for urgent decisions is required.

Acknowledgments This work was supported by Conicet, Agencia de Promoción Científica y Tecnológica (PICT 438-2006), Secyt, and the National University of Córdoba. The authors wish to thank all other researchers whom through their continuous efforts contributed with the studies on which this review was based.

References

Agerer R (1986) Studies on ectomycorrhizae II. Introducing remarks on characterization and identification. Mycotaxon 26:473–492

Agerer R (1987–2002) Colour atlas of ectomycorrhizae. Einhorn-Verlag, Munich

Agerer R (1991) Characterization of ectomycorrhiza. In: Norris JR, Read DJ, Varma AK (eds) Techniques for the study of mycorrhiza, vol 23, Methods in microbiology. Academic, London, pp 25–73

Agerer R (1994) Index of unidentified ectomycorrhizae. III. Names and identifications published in 1992. Mycorrhiza 4:183–184

Agerer R (1999) Anatomical characteristics of identified ectomycorrhizas: an attempt towards a natural classification. In: Varma A, Hock B (eds) Mycorrhiza: structure, function, molecular biology and biotechnology, 2nd edn. Springer-Verlag, Berlin, pp 633–682

Agerer R (2006) Fungal relationships and structural identity of their ectomycorrhizae. Mycol Progress 5:67–107

Alexander IJ (1989) Mycorrhizas in tropical forests. In: Proctor J (ed) Mineral nutrients in tropical forest and savanna ecosystems. Blackwell Scientific Publications, Oxford, pp 169–188

Alexander IJ, Högberg P (1986) Ectomycorrhizas of tropical Angiospermous trees. New Phytol 102:541–549

Allen MF, Swenson W, Querejeta JI, Egerton-Warburton LM, Treseder KK (2003) Ecology of mycorrhizae: a conceptual framework for complex interactions among plants and fungi. Annu Rev Phytopathol 41:217–303

Arruda Bacchi LM, Krugner TL (1988) Desenvolvimento ectomicorrízico em mudas de *Eucalyptus grandis* e *Eucalyptus urophylla* inoculadas com *Pisolithus tinctorius* em um vivero comercial. IPEF 40:21–25

Barroetaveña C (2004) Micorrizas y micorrización controlada en plantaciones y viveros forestales de pino ponderosa (*Pinus ponderosa*) y pino oregón (*Pseudotsuga menziesii*) en la región Andino Patagónica. Doctoral thesis, Comahue National University, 225p

Barroetaveña C, Rajchenberg M (2003a) Las micorrizas la producción de plántulas de *Pinus ponderosa* Dougl. ex Laws. en la patagonia Argentina. Bosque 24:17–23

Barroetaveña C, Rajchenberg M (2003b) Las micorrizas la producción de plántulas de *Pseudostuga menziesii* (Mirb.) Franco en la patagonia Argentina. Bosque 24:3–15

Barroetaveña C, Rajchenberg M, Cazares E (2005) Mycorrhizal fungi in *Pinus ponderosa* introduced in central Patagonia (Argentina). Nova Hedwig 80:453–464

Barroetaveña C, Cázares E, Rajchenberg M (2006) Mycorrhizal fungi of *Pseudotsuga menziesii*, an introduced tree species in Central Patagonia (Argentina). Nova Hedwig 83:53–66

Barroetaveña C, Cázares E, Rajchenberg M (2007) Ectomycorrhizal fungi associated with ponderosa pine and Douglas-fir: a comparison of species richness in native western North American forests and Patagonian plantations from Argentina. Mycorrhiza 17:355–373

Barros HF, Brandi RM, Reis MS (1978) Micorriza em eucalipto. Revista Arvore 2:130–140

Becerra AG (2002) Influencia de los suelos Ustorthentes sobre las ectomicorrizas y endomicorrizas en *Alnus acuminata* H.B.K. Master thesis, Facultad de Agronomía, Buenos Aires University, 190p

Becerra A, Daniele G, Domínguez L, Nouhra E, Horton T (2002) Ectomycorrhizae between *Alnus acuminata* H.B.K. and *Naucoria escharoides* (Fr.:Fr.) Kummer from Argentina. Mycorrhiza 12:61–66

Becerra A, Nouhra E, Daniele G, Domínguez L, McKay D (2005a) Ectomycorrhizas of *Cortinarius helodes* and *Gyrodon monticola* with *Alnus acuminata* from Argentina. Mycorrhiza 15:7–15

Becerra A, Beenken L, Pritsch K, Daniele G, Schloter M, Agerer R (2005b) Anatomical and molecular characterization of *Lactarius aff. omphaliformis*, *Russula alnijorullensis* and *Cortinarius tucumanensis* ectomycorrhizae on *Alnus acuminata*. Mycologia 97:1047–1057

Becerra A, Pritsch K, Arrigo N, Palma M, Bartoloni N (2005c) Ectomycorrhizal colonization of *Alnus acuminata* Kunth in northwestern Argentina in relation to season and soil parameters. Ann For Sci 62:325–332

Becerra A, Menoyo E, Lett I, Li CL (2009a) *Alnus acuminata* in dual symbiosis with *Frankia* and two different ectomycorrhizal fungi (*Alpova austroalnicola* and *Alpova diplophloeus*) growing in soilless growth medium. Symbiosis 47:85–92

Becerra A, Nouhra E, Silva M, McKay D (2009b) Ectomycorrhizae, arbuscular mycorrhizae and dark septate fungi on *Salix humboldtiana* Willd. from Central Argentina: a first assessment in two riparian populations. Mycoscience 50:343–352

Beenken L (2001) *Russula nothofaginea* Sing. + *Nothofagus dombeyi* (Mirbel) Oersted. Descr Ectomycorrhizae 5:175–179

Béreau M, Gazel M, Garbaye J (1997) Les symbioses mycorhiziennes des arbres de la Forêt tropicale humide de Guyane francaise. Can J Bot 75:711–716

Borges de Souza LA, Silva Filho GN, Lopes de Oliveira V (2004) Eficiência de fungos ectomicorrízicos na absorção de fósforo e na promoção do crescimento de eucalipto. Pesq Agropec Bras 39:349–355

Borges de Souza LA, Pinto Bonnassis PA, Silva Filho GN, Lopes de Oliveira V (2008) New isolates of ectomycorrhizal fungi and the growth of eucalypt. Pesq Agropec Bras 43:235–241
Brandán de Weth CI (2006) Estudio de micorrizas en zonas disturbadas, no disturbadas y en recuperación, en el Parque Biológico Sierra de San Javier (PSSJ). National University of Tucumán, Tucumán, Argentina
Brundrett MC, Cairney JWG (2002) Ectomycorrhizas in plant communities. In: Sivasithamparam K, Dixon KW, Barret RL (eds) Microorganisms in plant conservation and biodiversity. Kluwer Academic Publishers, Dordrecht, The Netherlands, pp 105–150
Carrillo R, Godoy R, Peredo H (1992) Simbiosis micorrícica em comunidades boscosas del Valle Central en el Sur de Chile. Bosque 13:57–67
Castillo CG, Borie F, Godoy R, Rubio R, Sieverding E (2006) Diversity of mycorrhizal plant species and arbuscular mycorrhizal fungi in evergreen Forest, deciduous Forest and grassland ecosystems of Southern Chile. J App Bot Food Qual 80:40–47
Daniele G, Becerra A, Crespo E (2005) *Amanita muscaria* (Basidiomycota) y su asociación micorrícica con *Cedrus deodara* (Pinaceae) en las sierras de Córdoba, Argentina. Bol Soc Arg Bot 40:45–49
Diehl P, Mazzarino MJ, Fontenla S (2008) Plant limiting nutrients in Andean-Patagonian woody species: effects of interannual rainfall variation, soil fertility and mycorrhizal infection. For Ecol Manag 255:2973–2980
dos Santos VL, Muchovej RM, Borges AC, Neves JCL, Kasuya MCM (2001) Vesicular-Arbuscular/Ectomycorrhiza succession in seedlings of Eucalyptus spp. Braz J Microbiol 32:81–86
Flores R, Godoy R, Palfner G (1997) Morfo-anatomía de la ectomicorriza *Cenococcum geophilum* Fr. en *Nothofagus alessandrii* Esp. Gayana Bot 54:157–162
Founoune H, Duponnois R, Bâ AM, El Bouami F (2002) Influence of the dual arbuscular endomycorrhizal/ectomycorrhizal symbiosis on the growth of *Acacia holosericea* (A. Cunn. ex G. Don) in glasshouse conditions. Ann For Sci 59:93–98
Fracchia S, Aranda A, Gopar A, Silvani V, Fernandez L, Godeas A (2009) Mycorrhizal status of plant species in the Chaco Serrano Woodland from central Argentina. Mycorrhiza 19:205–214
Franco Molano AE, Uribe Calle E (2000) Hongos Agaricales y Boletales de Colombia. Biota Colombiana 1:25–43
Frioni L, Minasian H, Volfovicz R (1999) Arbuscular mycorrhizae and ectomycorrhizae in native tree legumes in Uruguay. For Ecol Manag 115:41–47
Fulgenzi TD, Henkel TW, Halling RE (2007) *Tylopilus orsonianus* sp. nov. and *Tylopilus eximius* from Guyana. Mycologia 99:622–627
Garbaye J (1990) Utilisation des mycorhizes en sylviculture. In: Strullu DG (ed) Les mycorhizes des arbres et plantes cultivées. Lavoisier, Paris, pp 197–250
Garrido N (1983) Las Boletaceae en plantaciones de *Pinus radiata* D. Don en Chile (Fungi, Basidiomycetes). Bol Soc Biol 54:77–88
Garrido N (1986) Survey of ectomycorrhizal fungi associated with exotic forest trees in Chile. Nova Hedwig 43:423–442
Giachini AJ, Oliveira VL, Castellano MA, Trappe JM (2000) Ectomycorrhizal fungi in *Eucalyptus* and *Pinus* plantations in southern Brazil. Mycologia 92:1166–1177
Giachini AJ, Souza LAB, Oliveira VL (2004) Species richness and seasonal abundance of ectomycorrhizal fungi in plantations of *Eucalyptus dunnii* and *Pinus taeda* in Southern Brazil. Mycorrhiza 14:375–381
Godoy R, Palfner G (1997) Ectomicorrizas en *Nothofagus alpina* (P. et E.) Oerst y *N. dombeyi* (Mirb.) Oerst. del sur de chile. Bol Micol 12:55–61
Godoy R, Romero R, Carrillo R (1994) Estatus micotrófico de la flora vascular en bosques de coníferas nativas del sur de Chile. Rev Chil Hist Nat 67:209–220
Gross E, Thomazini Casagrande LI, Caetano FH (2004) Ultrastructural study of ectomycorrhizas on *Pinus caribaea* Morelet. var. *hondurensis* Barr. & Golf. seedlings. Acta Bot Bras 18:1–7

Guzmán G, Varela L (1978) Los Hongos de Colombia III. Observaciones sobre los hongos, líquenes y mixomicetos de Colombia. Caldasia 12:309–338

Halbinger RE, Frontera GM, Correa OS, Gauna FE (1986) Inoculación artificial de hongos micorríticos de los géneros *Boletus* y *Scleroderma* de especies de plantas del género *Eucaliptus*. Rev Agr Micología 9:3–9

Halling RE (1989) A synopsis of Colombian Boletes. Mycotaxon 34:93–113

Haug I, Lempe J, Homeier J, Weiß M, Setaro S, Oberwinkler F, Kottke I (2004) *Graffenrieda emarginata* (Melastomataceae) forms mycorrhizas with Glomeromycota and with a member of the *Hymenoscyphus ericae* aggregate in the organic soil of a neotropical mountain rain forest. Can J Bot 82:340–356

Haug I, Weis M, Homeier J, Oberwinkler F, Kottke I (2005) Russulaceae and Thelephoraceae form ectomycorrhizas with members of the Nyctaginaceae (Caryophyllales) in the tropical mountain rain forest of southern Ecuador. New Phytol 165:923–936

Henkel TW (1999) New taxa and distribution records for *Tylopilus* from *Dicymbe* forests of Guyana. Mycologia 91:655–665

Henkel TW, Aime MC, Miller SL (2000) Systematics of pleurotoid Russulaceae from Guyana and Japan, with notes on their ectomycorrhizal status. Mycologia 92:1119–1132

Henkel TW, Terborgh J, Vilgalys RJ (2002) Ectomycorrhizal fungi and their leguminous hosts in the Pakaraima Mountains of Guyana. Mycol Res 106:515–531

Hentz de Mello A, Antoniolli ZI, Kaminski J, Lorensi Souza E, Lopes Oliveira V (2006) Fungos arbusculares e ectomicorrízicos em áreas de *Eucalipto* e de campo nativo em solo arenoso. Ciência Florestal 16:293–301

Janos DP (1980) Mycorrhizae influences tropical succession. Tropical succession 12:56–64

Janos DP (1985) Mycorrhizal fungi: agents or symptoms of tropical community composition. In: Molina R (ed) Proceedings of the 6th North American conference on mycorrhizae. Oregon State University, Corvallis, pp 98–103

Kottke I, Haug I (2004) The significance of mycorrhizal diversity of trees in the tropical mountain forest of southern Ecuador. Lyonia 7:49–56

Landim de Souza MF (2003) Brazilian Atlantic rainforest remnants and Mycorrhizal symbiosis – implications for reforestation. A case study in Sergipe, Northeast Brazil. PhD. Universität Bremen, 193pp

Liparini Pereira O, Dutra Costa M, Chaer Borges A, Fernandes Araújo E, Megumi Kasuya MC (2005) Compatibility and ectomycorrhiza formation among *Pisolithus* isolates and *Eucalyptus* spp. R Bras Ci Solo 29:337–344

López-Quintero C, Vasco-Palacios AM, Franco-Molano AE (2007) Macrohongos de un Bosque de Roble, *Quercus humboldtii* Bonpl., en la Vereda Contrafuerte, Municipio de Andes (Colombia). In: Reserva natural Regional Cuchilla Jardín Támesis Antioquia, Una mirada a su biodiversidad. Corporación Ambiental, Medellín, pp 21–34

Lunt PH, Hedger JN (1996) A survey of mycorrhizal infection of trees in the Terra Firme rainforest, Cuyabeno, Ecuador. Mycologist 10:161–165

Lupatini M, Bonnassis PAP, Steffen RB, Oliveira VL, Antoniolli ZI (2008) Mycorrhizal morphotyping and molecular characterization of *Chondrogaster angustisporus* Giachini, Castellano, Trappe & Oliveira, an ectomycorrhizal fungus from *Eucalyptus*. Mycorrhiza 18:437–442

Malloch DW, Pirozynski KA, Raven PH (1980) Ecological and evolutionary significance of mycorrhizal symbioses in vascular plants (A review). PNAS USA 77:2113–2118

Martínez DB, Barroetaveña C, Rajchenberg M (2007) Influencia del régimen de fertilización y del momento de inoculación en la micorrización de *Pinus ponderosa* en la etapa de vivero. Bosque 28:226–233

Marx DH (1977) The host range and world distribution of the ectomycorrhizal fungus *Pisolithus tinctorius*. Can J Microbiol 23:217–223

Matheny PB, Aime MC, Henkel TW (2003) New species of *Inocybe* from *Dicymbe* forests of Guyana. Mycol Res 107:495–505

McGuire KL, Henkel TW, Granzow de la Cerda I, Villa G, Edmund F, Andrew C (2008) Dual mycorrhizal colonization of forest-dominating tropical trees and the mycorrhizal status of non-dominant tree and liana species. Mycorrhiza 18:217–222

Miller SL, Koo CD, Molina R (1991) Characterization of red alder ectomycorrhizae: a preface to monitoring belowground ecological responses. Can J Bot 69:516–531

Moyersoen B (1993) Ectomicorrizas y micorrizas vesiculo-arbusculares en Caatinga Amazonica del Sur de Venezuela. Scientia Guianae 3:1–82

Moyersoen B (2006) *Pakaraimaea dipterocarpacea* is ectomycorrhizal, indicating an ancient Gondwanaland origin for the ectomycorrhizal habit in Dipterocarpaceae. New Phytol 172:753–762

Moyersoen B (2008) Biodiversity and specificity of ectomycorrhizal fungi associated with *Pakaraimaea dipterocarpacea* in Guayana, Venezuela. Mycorrhizas in Tropical Forests, Workshop UTPL, 10p

Moyersoen B, Fitter AH, Alexander IJ (1998a) Spatial distribution of ectomycorrhizas and arbuscular mycorrhizas in Korup National Park rain forest, Cameroon, in relation to edaphic parameters. New Phytol 139:311–320

Moyersoen B, Alexander IJ, Fitter AH (1998b) Phosphorus nutrition of ectomycorrhizal and arbuscular mycorrhizal tree seedlings from a lowland tropical rainforest in Korup National Park, Cameroon. J Trop Ecol 14:47–61

Moyersoen B, Becker P, Alexander IJ (2001) Are ectomycorrhizas more abundant than arbuscular mycorrhizas in tropical heath forests? New Phytol 150:591–599

Nouhra E (1997) Boletáceas (Hymenomycetes) de la provincia de Córdoba y sus relaciones micorrícicas con especies arbóreas de la flora autóctona e introducida. PhD thesis, National University of Córdoba, 115pp

Nouhra E (1999) Novedades em Boletáceas Del centro de la Argentina. Kurtziana 27:403–423

Nouhra E, Becerra A (2001) Síntesis micorrícica de *Suillus granulatus* (Eumycota) y plantines de *Pinus elliotti* (Pinaceae). Bol Soc Arg Bot 36:209–215

Nouhra E, Domínguez L, Becerra A, Mangeaud A (2003) Colonización micorrícica y actinorrícica en plantines de *Alnus acuminata* y *Alnus rubra* (Betulaceae). Bol Soc Arg Bot 38:199–206

Nouhra E, Domínguez L, Becerra A, Trappe J (2005) Morphology, molecular analysis and ecological aspects of the South American hypogeous fungus; *Alpova austroalnicola* sp. nov. Mycologia 97:598–604

Oliveira VL, Schmidt VDB, Bellei MM (1997) Patterns of arbuscular- and ecto- mycorrhizal colonization of *Eucalyptus dunnii* in southern Brazil. Ann Sci For 54:473–481

Onguene NA, Kuyper TW (2002) Importance of the ectomycorrhizal network for seedling survival and ectomycorrhiza formation in rain forests of south Cameroon. Mycorrhiza 12:13–17

Pagano MC, Scotti MR (2008) Arbuscular and ectomycorrhizal colonization of two *Eucalyptus* species in semiarid Brazil. Mycoscience 49:379–384

Palfner G (1996) *Thaxterogaster albocanus* Horak & Moser + *Nothofagus pumilio* Poepp. et Endl.) Krasser. Descr Ectomycorrhizae 1:155–160

Palfner G (1997) *Descolea antarctica* Singer + *Nothofagus alpina* (Poepp. et Endl.) Oerst. Descr Ectomycorrhizae 2:7–12

Palfner G (2001) Taxonomische studien an ektomykorrhizen aus den *Nothofagus*-Wäldern Mittelsüdchiles. Bibliotheca Mycologica. Band 190. J. Cramer. Berlin, Alemania, 243pp

Palfner G (2002a) *Austropaxillus boletinoides*. In: Agerer R (ed) Colour atlas of ectomycorrhizae. Einhorn Verlag, Schwäbisch Gmünd, plate 151

Palfner G (2002b) *Cortinarius austrosalor*. In: Agerer R (ed) Colour atlas of ectomycorrhizae. Einhorn Verlag, Schwäbisch Gmünd, plate 152

Palfner G (2002c) *Cortinarius magellanicus*. In: Agerer R (ed) Colour atlas of ectomycorrhizae. Einhorn Verlag, Schwäbisch Gmünd, plate 153

Palfner G (2002d) *Gautieria inapire*. In Agerer R (ed) Colour atlas of ectomycorrhizae. Einhorn Verlag, Schwäbisch Gmünd, plate 154

Palfner G, Godoy R (1996a) "*Notofagirhiza vinicolor*"+ *Nothofagus pumilio* (Poepp. et Endl.) Krasser. Descr Ectomycorrhizae 1:65–70

Palfner G, Godoy R (1996b) *Russula fuegiana* + *Nothofagus pumilo* (Poepp. Et Endl.) Krasser. Descr Ectomycorrhizae 1:131–136

Palfner G, Canseco MI, Casanova-Katny A (2008) Post-fire seedlings of *Nothofagus alpina* in Southern Chile show strong dominance of a single ectomycorrhizal fungus and a vertical shift in root architecture. Plant Soil 313:237–250

Paloschi de Oliveira L, Rossi MJ, Furigo Júnior A, Silva Filho GN, Lopes de Oliveira V (2006) Viability and infectivity of an ectomycorrhizal inoculums produced in an airlift bioreactor and immobilized in calcium alginate. Braz J Microbiol 37:251–255

Pérez F, Palfner G, Brunel N, Santelices R (2007) Synthesis and establishment of *Tuber melanosporum* Vitt. ectomycorrhizae on two *Nothofagus* species in Chile. Mycorrhiza 17:627–632

Pérez-Moreno J (1998) La ectomicorriza una simbiosis mutualista en el sostenimiento de Gaia, el planeta viviente. In: Ferrera-Cerrato R, Pérez-Moreno J (eds) Manejo de agroecosistemas sostenibles. Veracruzana University, México, pp 93–120

Pinto Bonnassis PA (2007) Caracterização de isolados fúngicos ectomicorrízicos na promoção do crescimento e na colonização radicular de *Eucalyptus dunnii* Maiden. Master thesis, Santa Catarina Federal University, 73p

Pritsch K, Becerre A, Põlme S, Tedersoo L, Schloter M, Agerer R (2010) Descrioption and identification of Alnus awminate ectomycorrhizae from argentinean older stands. Mycologia (in press)

Putzke JML, Pereira AB (1994) O fungos da familia Boletaceae conhecidos no Rio Grande do Sul, Brasil (Fungi, Basidiomycotina). Caderno de Pesquisa Sér Bot 6:75–100

Rocci MJ (2006) Tecnologia para produção de inoculantes de fungos ectomicorrízicos utilizando cultivo submerso em biorreator *airlift*. PhD thesis, Santa Catarina Federal University, 188p

Rumney GR (1968) Climatology and the world's climate. The Macmillan Company, New York, 643pp

Schüßler A, Schwarzott D, Walker C (2001) A new fungal phylum, the Glomeromycota: phylogeny and evolution. Mycol Res 105:1413–1421

Schwan KR (1984) Caracterização, incidência e ecologia de micorrizas em viveiro e florestas de *Eucalyptus* spp., na Região de Viçosa, Minas Gerais. Master thesis, Viçosa, Universidade Federal de Vicosa, 55p

Singer R (1953) Four years of mycological work in southern South America. Mycologia 45:865–891

Singer R (1963) Oak mycorrhiza fungi in Colombia. Mycopathol et Mycol Applicata 20:239–252

Singer R (1968) Sand-dune inhabiting fungi of the South Atlantic coast from Uruguay to Bahía Blanca. Mycopathol Mycol Applicata 34:129–143

Singer R (1971) Forest mycology and forest communities in South America. II. Mycorrhiza sociology and fungus succession in the *Nothofagus dombeyi-Austrocedrus chilensis* woods of Patagonia. In: Hacskaylo E (ed) Mycorrhizae, proceedings of the first North American conference on mycorrhizae. Government Printing Office, Washington, DC, pp 204–215

Singer R, Araujo I (1979) Litter decomposition and ectomycorrhiza in Amazonian forest. I. A comparison of litter decomposing and ectomycorrhizal basidiomycetes in latosol-terra-firme forest and white podzol campinarana. Acta Amazonica 9:25–41

Singer R, Morello JH (1960) Ectotrophic tree mycorrhiza and forest communities. Ecology 41:549–551

Singer R, Araujo I, Ivory MH (1983) The ectotrophically mycorrhizal fungi of the neotropical lowlands, especially Central Amazonia. Nova Hedwig Beihefte 77:1–352

Smith SE, Read DJ (2008) Mycorrhizal symbiosis, 2nd edn. Academic Press, New York, 787p

St John TV (1980) A survey of mycorrhizal infection in an Amazonian rain forest. Acta Amazonica 10:527–533

Tackacs EA (1961) Inoculación de especies de pinos con hongos formadores de micorizas. Silvicultura 15:5–17

Tackacs EA (1964) Inoculación artificial de pinos de regiones subtropicales con hongos formadores de micorrizas. I. Formación de micorrizas ectotróficas en *Pinus taeda* L. con *Scleroderma* vulgare Horn. Idia-Suplemento Forestal 12:41–44

Trappe JM (1977) Selection of fungi for ectomycorrhizal inoculation in nurseries. Ann Rev Phytolaphol 15:203–222

Trappe JM (1987) Phylogenetic and ecologic aspects of mycotrophy in the angiosperms from an evolutionary standpoint. In: Safir GR (ed) Ecophysiology of VA mycorrhizal plants. CRC Press, Boca Raton, FL, pp 5–25

Valenzuela E, Moreno G, Garnica S, Godoy R, Ramirez C (1999) Mycosociology in native forests of *Nothofagus* of the X Region of Chile, diversity and ecological role. Mycotaxon 72:217–226

Vasco Palacio AM, Franco Molano AE, López Quintero CA, Boekhaut T (2005) Macromicetes (Ascomycota, Basidiomycota) de la región del medio Caquetá, departamentos de Caquetá y Amazonas (Colombia). Biota Colombiana 6:127–140

Villegas JN, Carrillo RF, Núñez PM, Rodríguez M (2007) Caracterización micorrizogénica de un bosque de *Nothofagus pumilio* (Poepp. et Endl.) Krasser en el Parque Nacional Villarrica, Cordillera de los Andes, IX Región, Chile. Biodiversidad y conservación de bosques nativos. International congress on development, environment and natural resources: multi-level and multi-scale sustainability, vol 3. Bolivia, pp 1448–1455

Voigt EL, Lopes de Oliveira V, Randi AM (2000) Mycorrhizal colonization and phenolic compounds accumulation on roots of *Eucalyptus dunii* Maiden inoculated with ectomycorrhizal fungi. Pesq Agropec Bras 35:1905–1910

Wrigth JE (1988) Interrelaciones entre Macromycetes (Fungi) y *Nothofagus*. Acad Nac Cs Exactas, Físicas y Naturales. Monografía 4:132–155

Yokomizo NKS (1981) Associacáo ectomicorrícica de *Pisolithus tinctorius* (Pers.) Coker e Couch com espécies de *Eucalyptus* L. Héritier. Piracicaba Escola Superior de Agricultura "Luis de Queiróz". Master thesis, 54p

Chapter 3
Ectomycorrhizal Inoculum and Inoculation Techniques

Ivan Repáč

3.1 Introduction

ECM symbiosis plays an important role in physiology, ecology, resistance, production, and other aspects of life of a single trees, populations, and ecosystems of ectotrophic forest tree species. Among the various ways of improving the early growth and survival of forest plantations, controlled mycorrhization by inoculating seedlings with selected fungal strains is an energy-efficient and environment-friendly alternative to fertilization or soil tilling (Kropp and Langlois 1990; Marx 1991; Garbaye and Churin 1997). Trees planted particularly in soils deficient in ECM fungi (e.g., mine spoils, polluted areas, agricultural, and other treeless lands) and even in routine reforestation sites may benefit greatly from ECM inoculation (Marx 1991; Castellano 1996; Querejeta et al. 1998; Duñabeitia et al. 2004; Núñez et al. 2006), although the positive effect of inoculation is not consistent (Castellano 1996; Corrêa et al. 2006; Menkis et al. 2007; Baum et al. 2009). Several types of natural and laboratory-produced inocula and several application techniques have been used for seedling inoculation. Forest soil, litter, humus, or excised ectomycorrhizae were substituted by fungal spores and mycelial inoculum (Marx and Bryan 1975; Le Tacon et al. 1983; Duponnois and Garbaye 1991; Parladé et al. 1999). Vegetative inoculum appears to be the most appropriate and effective method of inoculation (Marx and Kenney 1982; Mortier et al. 1988; Garbaye and Churin 1997; Rincón et al. 2007). Artificial mycorhization with vegetative inoculum consists of several steps – selection of ECM fungi, isolation and maintenance of fungi in pure cultures, preparation of fungal inoculum, and inoculation of seedlings.

A large body of literature exists on ECM inoculum preparation and application in a research scale; a comprehensive review of the literature and of all aspects of this topic is impossible. Although there have been previous studies on inoculum preparation and inoculation techniques, this chapter attempts to bring together a

I. Repáč
Department of Silviculture, Technical University in Zvolen, Forestry Faculty, T.G. Masaryka 24, SK-96053, Zvolen, Slovak Republic
e-mail: repac@vsld.tuzvo.sk

M. Rai and A. Varma (eds.), *Diversity and Biotechnology of Ectomycorrhizae*, Soil Biology 25, DOI 10.1007/978-3-642-15196-5_3,

brief review of the major inoculum types and methods of their application developed and to offer a list of selected references as a source of other literature and detailed description of a particular material, procedure or product.

3.2 Isolation and Manipulation of Fungal Pure Cultures, Nutrient Media

Isolation, cultivation, and maintenance of ECM fungal cultures are crucial phases in artificial mycorrhization of forest tree seedlings (Molina and Palmer 1982). Many, but far from all, fungi can be cultivated in pure culture in the laboratory under controlled conditions. The fungal tissues are most commonly isolated from young and healthy fruit bodies as soon as possible after collection but can also be isolated from surface sterilized ectomycorrhizae, sclerotia, and rhizomorphs (Molina and Palmer 1982; Obase et al. 2009). The pieces of the tissues are usually transferred to sterile glass tubes on agar nutrient media. Tissue transfers and fungal cultures are incubated in the dark at 21–24°C for 2–6 weeks. Between the cultivation cycles the stock cultures are stored approximately for 2 months at 2–4°C. Higher rates of agar cultures are cultivated in Petri dishes and submerged cultures in Erlenmeyer flasks.

One of the most common nutrient media was the Modess modification of the Hagem semisynthetic formula (Molina and Palmer 1982). Modess medium was further modified (e.g., amended with thiamine by Timonen et al. 1997) and is still used (e.g., Niemi et al. 2006). Modified Melin–Norkrans medium (MMN) (Marx 1969) is probably the most used worldwide for experimental procedures (e.g. Fortin et al. 1980; Wong and Fortin 1989; Repáč 1996a; Gáper et al. 1999; Parladé et al. 1999). The original solution was a synthetic solution (no malt extract) with only 25 μg of thiamine and 2.5 g of glucose. MMN altered by Vrålstad et al. (2002) contained only 1.0 g of glucose to reduce possible adverse effects on ECM formation caused by high levels of exogenous carbon. MMN was also modified by Niemi et al. (2002) and Langer et al. (2008). Pachlewski medium (Pachlewski and Pachlewska 1974) was used, e.g., by Duponnois and Garbaye (1991) and Parladé et al. (1999). Moore et al. (1989) cultivated cultures of fungus *Cantharellus cibarius* on special species-specific C-medium. In the study of Guerin-Laguette et al. (2003), stock cultures were maintained on agar medium proposed by Ohta (1990), and mycelium was grown on Tween- or olive oil-supplemented A medium. Fungal biomass increased up to 15-fold as a result of olive oil incorporation.

Basic components of natural media are leachates of various crops such as potato, cabbage, wheat, and others. Although use of natural formulations is less frequent than semi-synthetic or synthetic ones, potato dextrose agar (PDA) was used in the past years, e.g., by Parladé et al. (1996a) and Giomaro et al. (2002). MMN (Marx 1969), BAF (Moser 1960), KHO (Šašek and Musílek 1967), and malt–peptone (50 ml of brewery wort + 5.0 g of peptone to 1,000 ml of H_2O) media are used for manipulation of fungal cultures in our laboratory of Department of Silviculture,

Forestry Faculty, Technical University in Zvolen, Slovakia (project VEGA 1/0516/09). The media are changed in recultivation cycles to support the vitality and resistance of cultures.

3.3 Synthesis of Ectomycorrhizae in Controlled Conditions

Syntheses of ECM in controlled conditions are important techniques for verification of symbiotic relationships (compatibility) between intended fungi and host trees, structural, physiological and biochemical studies, and ultimately the genetic control of ECM formation. Depending on the research goal and synthesizing apparatus either sterile, semi-sterile, or non-sterile techniques can be used. The purpose of this Section (and also Sect. 3.4) is not a comprehensive categorization, but just a review of some of the most often reported or meaningful inoculum types, methods of inoculum preparation, and inoculation procedures. For better understanding, see illustrations of the apparatus of synthesizing techniques, e.g., in the works of Peterson and Chakravarty (1991), Duponnois and Garbaye (1991), and other related literature cited.

3.3.1 Fungal Cultures Preparation

Fungi are applied into synthesis system most often as agar culture transfers. Usually, agar cultures are grown on nutrient agar in Petri plates for 3–4 weeks, then transferred onto fresh solid medium, and incubated for a few days. Several plugs (5–8 mm diameter) are then aseptically removed and placed around the roots in the synthesis apparatus. Mycelial agar plugs served as inoculum in Petri dish synthesis systems, e.g., in experiments of Wong and Fortin (1989), Brunner (1993), Gáper et al. (1999), Bending et al. (2002), Krüger et al. (2004), and Sarjala and Taulavuori (2004). Fortin et al. (1980) used plugs of mycelium in ECM synthesis in growth pouches, Högberg et al. (1999) and Giomaro et al. (2002) in vessels, and Vrålstad et al. (2002) in conical flasks. In Petri dishes, Herrmann et al. (2004) inoculated roots with a nylon sheet carrying 7–8 fungal plugs and Niemi et al. (2006) using two filter paper strips covered by mycelium.

Apart from agar cultures, fungi are usually introduced into synthesis system from liquid cultures (slurry of mycelium). Because the liquid cultures are shaken just before inoculation, the numerous mycelial fragments will yield rapid and uniform colonization of the synthesis substrate (Molina and Palmer 1982). Molina (1979) cultivated mycelium in glass tubes filled with MMN nutrient solution and small chips of broken glass; cultures were shaken periodically to fragment the mycelium. In experiments of Duponnois and Garbaye (1991), Brunner (1993), Parladé et al. (1996a), and Obase et al. (2009), fungal mycelium was grown for 3–4 weeks on a shaker in Erlenmeyer flasks containing liquid medium. The mycelium was then washed in tap water to remove residual nutrients, homogenized in a blender for a few seconds, and resuspended in distilled water. Guerin-Laguette et al. (2003) collected mycelium of *Tricholoma matsutake* grown in liquid medium over nylon mesh filter, washed it thoroughly with and resuspended in 100 ml of a modified nutrient medium (NM). Flores et al. (2008) reported cultivation of *Laccaria bicolor* in flasks with semi-liquid MMN medium (1.5 g agar.l^{-1}) in the dark at 23°C. The mycelia obtained were filtered through a 65-μm net, washed in sterile water to eliminate sugars and to reduce the growth of contaminants in the seedling root systems, fragmented and homogenized by manual agitation, and then resuspended in distilled water (70–80 g mycelium.l^{-1}).

Vermiculite–peat (Marx and Bryan 1975) and alginate-bead (Le Tacon et al. 1983) inocula can be used in ECM synthesis at controlled conditions, even though they are primarily used in the operational inoculations. Preparation of these inoculum types is described in detail in Sect. 3.4.1.3. Mycelial plugs were laid on top of vermiculite–peat mixture (4:5–1:5, v:v) moistened to field capacity with modified liquid Pachlewski medium to prepare vermiculite–peat inoculum for synthesis in sterile conditions (Duponnois and Garbaye 1991). Corrêa et al. (2006) and Martín-Pinto et al. (2006) inoculated this substrate mixture with liquid fungal cultures. Alginate-bead inoculum of *Laccaria laccata* was used to synthesize ectomycorrhizae in aseptic conditions of glass test tubes (Duponnois and Garbaye 1991) and inoculum of *Amanita rubescens* and *Hebeloma sinapizans* in polyethylene pots in a plant growth cabinet (Kozdrój et al. 2007).

3.3.2 *Inoculation Techniques*

3.3.2.1 Sterile Techniques

Petri Dishes

Petri plates are probably the most extensively used method of ectomycorrhiza synthesis. Wong and Fortin (1989) described a Petri dish technique that avoids limitations of previous ones. In Petri dish filled with sugar-free agar medium, two sheets of nylon membrane sandwiched the root and were overlaid with a sheet of filter paper to keep the exposed surface of the roots moist. Cotton rolls were placed along the opposite edge of the Petri dish to absorb water which condensed during incubation. The filter paper was removed and the fungal plugs were placed on the membrane beside root laterals. Peterson and Chakravarty (1991) differentiated between simple system, divided Petri plates, sandwich technique, and nylon mesh method. Sandwich technique (mycelial plugs placed on cellophane sheets) was a method of synthesis experiments conducted by Langer et al. (2008) on *Populus tremula* plantlets. Mixture of vermiculite and peat seems to be a suitable substrate for Petri dish synthesizing system (Duponnois and Garbaye 1991; Bending et al. 2002; Sarjala and Taulavuori 2004). Duponnois and Garbaye (1991) observed pure culture synthesis of *Pseudotsuga menziesii–Laccaria laccata* in autoclaved soil or silica sand. Gáper et al. (1999) studied synthesis of *Picea abies–Inocybe lacera* and *Suillus bovinus* in a mixture of perlite, peat, florex, and keramzit (7:2:2:1). Of course, substrate is always autoclaved or sterilized regardless of the purpose of ECM synthesis; in most instances substrate is disinfected or fumigated in other inoculation experiments, including operational conditions. To protect fungus and root system from direct illumination, Niemi et al. (2006) placed a semi-circle brown paper on the lower part of the lid of 14-cm Petri dish. To reduce the loss of plants during acclimatization, Krüger et al. (2004) placed a 90-mm Petri dish system in a 140-mm Petri dish in which humidity was regulated with moistened paper. Herrmann et al. (2004) transferred rooted micropropagated oak plantlets into 90-mm Petri dish rhizotrons. In this system, the roots grew two dimensionally inside the dish, whereas the shoot developed outside. Between inoculation and transfer to the growth chamber, the plants were kept in plastic bags moistened with wet cotton sheets to reduce adaptation shocks for the shoots.

Flasks and Jars

Some of the early attempts to synthesize ectomycorrhizae under sterile conditions were made by Melin using Erlenmeyer flasks containing sterile sand moistened with a nutrient solution into which an aseptically germinated seedling and single fungus culture were introduced (Peterson and Chakravarty 1991). Hacskaylo improved the method by using vermiculite instead of sand (Molina and Palmer 1982). Peterson and Chakravarty (1991) described Mason jars and Leonard jars, in which shoot of seedling remains in the environment and roots in aseptic conditions inside the jars. Fungal inoculum can be added into flasks as plugs of actively growing mycelium (Vrålstad et al. 2002) or as slurry of fungal hyphae (Brunner et al. 1992) either before, at time, or after seedling introduction. Vermiculite, peat, and particularly their mixture are mostly used as substrate since their advantages had been detected and confirmed for ECM synthesis. Martín-Pinto et al. (2006) transferred 7-day-old *Pinus nigra* seedlings to 45-ml sterile flasks containing 30 ml of substrate (MMN, peat or vermiculite) colonized by the ECM fungi.

Test Tubes

Pachlewski and Pachlewska (1974) achieved good results to synthesize ectomycorrhizae between *Pinus sylvestris* and more ECM fungi in test tubes containing water agar amended with thiamine. Molina (1979) used 300 × 38-mm glass test tubes filled with 110 ml of vermiculite and 10 ml of peat moss moistened with 70 ml of MMN nutrient solution to test a wide range of ECM fungi on *Alnus* species entirely enclosed in tubes. Technique described by Molina (1979), modified with double the amount of peat, was used by Parladé et al. (1996a). Within series of experiments studying the biology of *Pseudotsuga menziesii–Laccaria laccata* symbiosis, Duponnois and Garbaye (1991) used glass test tubes (3 × 15 cm) filled with autoclaved peat–vermiculite (1:1, v:v) moistened to field capacity with modified Shemakanova mineral nutrient solution. Inoculation was carried out either with (1) peat–vermiculite inoculum mixed with substrate (1:10, v:v) or laid on top of the tube (1–2 cm), (2) alginate-bead inoculum (five beads laid on top of the tube), or (3) mycelial suspension (injected with a syringe or deposited with a pipette). The roots were maintained in axenic conditions, while the top of the plant developed outside the tube.

Polycarbonate Boxes

Guerin-Laguette et al. (2003) introduced mycorrhizal synthesis between *Tricholoma matsutake* and *Pinus densiflora* in the following substrate and apparatus: dry soil, vermiculite, and moist *Pinus densiflora* bark (pieces of about 1 cm^2) were mixed together (1:1:1/v:v:v) before addition of distilled water supplemented with olive oil. Wet substrate was autoclaved twice before distribution into sterile 920-ml, wide-mouth, polycarbonate boxes (9.5 × 9.5 × 11.5 cm) and inoculation with mycelial slurry (1 volume of slurry into 5 volumes of substrates). The substrate (approximately 300 ml per box) was applied as five layers (60 ml each), each layer being inoculated with the slurry before application of the following layer, i.e., 10 ml of slurry for the first two layers, 20 ml for the next two, 0 ml for the last one. Five sterile 12-day-old pine seedlings were then aseptically introduced into each box. The boxes were maintained 4 months at 23°C under a photosynthetic active radiation with a 16-h photoperiod.

3.3.2.2 Semi- and Non-sterile Techniques

Growth Pouches

Growth pouches were first introduced in detail by Fortin et al. (1980). Flat, transparent polyester pouches were used for synthesis of ectomycorrhizae between *Pinus strobus* and several ECM fungi. Fifteen milliliters of MMN solution was added, and a piece of glass tube was placed along the side of each pouch to add nutrient solution and water. The root system was laid directly onto the paper pad. As soon as formation of short roots initiated, plugs of actively growing mycelium were placed on the paper pad 3–5 mm from a short root primordium. Brunner (1993) used autoclaved polyethylene pouches 13 × 16 cm including a glass fiber paper with an activated charcoal paper disk attached. Pouches were filled with 10 ml of MMN liquid medium and one *Picea abies* seedling. After 2 months, mycelial disks were placed in the vicinity of the short roots. Two strips of foam were inserted to provide air space. Advantages of this technique are that numerous seedlings can be grown in a very small space, root system can be viewed without disturbing the roots or the fungus, and roots are clean since substrate is not involved (Peterson and Chakravarty 1991).

Pots, Containers, Vessels

A brief review of several experiments can illustrate a variety of vessels, substrates, inoculum types, and methods of their applications used in non-sterile ECM synthesis techniques. Duponnois and Garbaye (1991) used transparent boxes, rootrainers, Hiko and "M" containers filled with soil or vermiculite–peat mixture (1:1, v:v) for growing Douglas fir seedlings in the glasshouse. Liquid, vermiculite–peat or alginate-bead inoculum were applied either by mixing with the substrate before filling the containers or by spreading the inoculum on the well-developed root system. Brunner (1993) synthesized ectomycorrhizae between *Picea abies* and *Hebeloma crustuliniforme* in autoclaved cuvettes of stainless steel (15 × 12 × 2 cm) filled with a vermiculite–peat moss mixture. Homogenized mycelium grown in liquid solution was introduced into cuvettes using an inverted pipette. Högberg et al. (1999) planted *Pinus sylvestris* seedlings in 0.5-l plastic pots filled with sand inoculated with ten mycelial agar plugs placed below the pine roots. Tissue blocks of *Tuber brumale* mycelium served for inoculation of micropropagated plantlets of *Tilia americana* and *Quercus pubescens* in the vessels (12 cm diameter × 18 cm height) filled with 800 ml of peat–vermiculite mixture (1:30, v:v) in experiment of Giomaro et al. (2002). Timonen et al. (1997) planted sterile *Pinus sylvestris* seedlings in plastic pots enclosed in sterile plastic bags to maintain moisture and prevent aerial contamination. After 25 days, the roots of one seedling were inoculated with three agar plugs (5 mm^3) of mycelium. Bogeat-Triboulot et al. (2004) and Flores et al. (2008) used mycelial suspension to inoculate *Pinus pinaster* with *Hebeloma cylindrosporum* in 0.4-l pots filled with soil and *Abies guatemalensis* with *Laccaria bicolor* in 300-cm^3 containers filled with peat moss/vermiculite (1:1, v:v), respectively. Obase et al. (2009) soaked roots of *Populus maximowiczii* in MMN liquid medium containing cultured ECM fungal mycelia. Each seedling was planted in a plastic pot (60 × 60 × 100 mm) filled with heat-sterilized volcanic debris.

Aspray et al. (2006) and Corrêa et al. (2006) inoculated seedlings of *Pinus sylvestris* and *P. pinaster* with vermiculite–peat inoculum of *Laccaria bicolor*, *Lactarius rufus*, and *Pisolithus tinctorius* in 80-ml and 350-ml pots, respectively. Kozdrój et al. (2007) transferred *Pinus sylvestris* seedlings to polyethylene pots containing 150 g of autoclaved soil thoroughly mixed with alginate-bead inoculum of *Amanita rubescens* or *Hebeloma sinapizans* in ratio 15:1 (v:v). The inoculated seedlings were grown after inoculation in either growth (climatic) chamber (Brunner 1993; Högberg et al. 1999; Corrêa et al. 2006; Flores et al. 2008; Obase et al. 2009), culture room (Giomaro et al. 2002), growth cabinet (Bogeat-Triboulot et al. 2004; Kozdrój et al. 2007), or glasshouse (Timonen et al. 1997; Aspray et al. 2006).

3.4 Greenhouse and Nursery Inoculation

3.4.1 Inoculum Types

3.4.1.1 Natural Inoculum (Soil, Humus, Ectomycorrhizae)

Various organic matters, such as soil, litter, variable forms of humus, rotten wood, and ectomycorrhizae collected from forest plantations or mature stands were exploited especially in the beginning of the effort to inoculate forest tree seedlings. Collection of a large amount of viable inoculum of fungi adapted to the sites from which they were taken is relatively reliable and easy. A major drawback of natural inoculum is that the species of ECM fungi in the inoculum cannot be controlled. This inoculum may also contain harmful microorganisms and weeds in addition to the ECM fungi. Forest soil was used as a source of indigenous ECM fungi in experiments of Borchers and Perry (1990), Querejeta et al. (1998), and Wallander et al. (2005). Litter and diverse forms of humus from the forest floor including a range of materials from recently fallen needles on the surface to well-decomposed humus overlying the mineral soil were added to pot substrate (Parke et al. 1983; Repáč 1996a). Aucina et al. (2007) used pine and oak litter to affect development of bareroot seedlings. Rotten wood was collected from a Douglas-fir stand and fragmented by hand into small pieces about 1 cm in diameter for mixing in a growing medium (Kropp 1982). Natural inoculum is usually collected from stand or under individual tree of species which is inoculated. Ectomycorrhizae excised from root systems of trees were also used as inoculum on a limited basis in research trials (Marx and Kenney 1982). A great deal of time and care is required to obtain a sufficient quantity of viable ectomycorrhizae.

3.4.1.2 Basidiospores

Sporocarps and spores of various fungi were used as inoculum to form specific ectomycorrhizae on various forest tree species. Sporocarps are essentially spore inoculum, since their vegetative matrix is killed by dessication during drying or by decomposition when added to soil (Marx and Kenney 1982). Gastromycete, such as genera *Rhizopogon, Scleroderma,* and *Pisolithus*, produces numerous basidiospores that are easier to collect in large quantities than those of mushroom-produced ECM fungi. Advantages of using spores for inoculation are that spores require no extended growth phase under aseptic conditions like vegetative inoculum, spore inoculum is very light, and spores are able to survive storage from one season to the next. Major disadvantages are the lack of standard laboratory tests to determine spore viability, insufficiency of sporocarps of many fungi in any year, delay in ectomycorrhiza formation as compared with vegetative inoculum, and lack of genetic definition.

Basidiospores can be obtained and stored before application in principle either in dry state or blended with water. Basidiospores were obtained by crushing and sieving basidiocarps through a mesh screen (Rincón et al. 2001; Chen et al. 2006) in a closed plastic bag (Marx and Bryan 1975). Using these techniques, well over 1 kg (fresh weight) of basidiospores of *Pisolithus tinctorius* was collected in less than 3 h (Marx and Bryan 1975). The dry spores were stored in either small plastic bags or amber glass bottles in darkness at 5°C for 10 days before use (Marx and Bryan 1975), mixed with vermiculite for seedling inoculation (Rincón et al. 2001) or stored at 4°C for 1–18 months before blended with distilled deionized water (1:10, v:v) for 5 s on low speed (Chen et al. 2006). In experiments of Parladé et al. (1996b, 1999), Rincón et al. (2001), Hortal et al. (2008), and Becerra et al. (2009), spore suspensions were prepared by blending sporocarps with distilled or tap water using a blender at low or high speed until the spores were released. After collection, Parladé et al. (1996b) and Chen et al. (2006) dried sporocarps at 30–40°C for 48 h before blending and sieving, respectively. Núñez et al. (2006) cleaned and sterilized sporocarps of *Tuber melanosporum* by brief superficial flaming. To prepare the inoculum, Roldan et al. (1996), Querejeta et al. (1998), and Núñez et al. (2006) suspended spores in distilled water and stored the suspensions at 2–4°C until use. Required spore concentrations are prepared by serial dilution of spore suspension with water. Spore concentrations are counted using haemocytometer. The approximate number of spores contained per gram of dried sporocarp tissue was 10^7–10^{10} (Parladé et al. 1996b; Rincón et al. 2001) or 1.1 million basidiospores per milligram (Marx and Bryan 1975). Chen et al. (2006) obtained a concentration of 10^5 spores.ml^{-1} and Rincón et al. (2001) 10^3–10^8 spores.10 ml^{-1}.

3.4.1.3 Vegetative (Mycelial) Inoculum

Mycelial Suspension (Slurry)

This type of vegetative inoculum is more often used in a small-scale synthesis in controlled conditions than in nursery experiments (see Sect. 3.3.1). Mycelial slurry of *Pisolithus tinctorius, Suillus granulates,* and *Rhizopogon luteolus* was prepared by blending mycelial mats from liquid cultures with distilled water at high speed for less than 3 s (Cline and Reid 1982). One liter of sterile water contained 40 g of (fresh weight) mycelium. Gagnon et al. (1991, 1995) produced mycelial suspension of *Laccaria bicolor* in a fermentor. Starter content of living propagules was diluted five times with water and mixed with the substrate in a cement mixer to get a final content approximately 6.1×10^4 living propagules or 5.68 mg of dried mycelium per seedling. Gange et al. (2005) prepared mycelial slurry of *Laccaria laccata* maintained in sterile liquid culture on MMN medium by fragmenting the mycelium in a blender for 30 s.

Vermiculite–Peat (Substrate Carrier) Inoculum

Moser (1958) in Austria was one of the first to make a serious attempt to produce vegetative inoculum of ECM fungi. For production of inoculum, mycelium of *Suillus plorans* was first grown in liquid culture and then in sterile peat moss. Moser and other workers tested various organic materials as the final inoculum substrate, e.g., forest litter, sawdust, grain of cereals, corn, bark, and found that they were not as effective as peat moss (Marx and Kenney 1982). Although mycelium tends to grow around rather than into the particles of perlite substrate, Repáč (1996b) and Hönig et al. (2000) reported that a perlite–peat mixture appears to be a possible form of mycelial carrier. Vozzo and Hacskaylo (1971) grew mycelium of several fungi in polypropylene cups containing a 2:1 ratio of sterile peat moss and vermiculite moistened with nutrient solution. In the fundamental study of Marx and Bryan (1975), the inoculum containers were 2-L jars containing the mixture of 1,400 ml of vermiculite, 50 ml of finely divided peat moss, and 750 ml of liquid MMN medium with glucose. The containers were autoclaved for 30 min and each was inoculated with eight mycelium-agar disks of *Pisolithus tinctorius*. After 15 weeks at room temperature, the vermiculite particles were permeated with mycelium. To prepare mass inoculum for infestation of soil, mycelium was removed from the jars, passed through a 5-mm mesh screen, and held with two layers of cheesecloth while being leached with cool running tap water to remove nonassimilated nutrients.

Novel formulations of vermiculite–peat inoculum did not require leaching and drying before use (Garbaye et al. 1988; Marx and Cordell 1989). An important modification of original inoculum formulation was carried out on vermiculite:peat ratio of inoculum. Marx et al. (1982) reported that vermiculite–peat inoculum mixture produced by solid-substrate fermentation contained 5–10% peat moss by volume. Other vermiculite:peat ratios used were, e.g., 4:5–1:5 (Duponnois and Garbaye 1991), 9:1 (Garbaye et al. 1988; Baum et al. 2009), 10:1 (Machón et al. 2006), and 11:1 (Rincón et al. 2001). Blended mycelial starter cultures mixed with the substrate will reduce the time of incubation by half (Marx and Kenney 1982; Rincón et al. 2001). Depending on fungal properties, inoculum is incubated for 4–8 weeks at 24–25°C in the dark. The attempts to measure the quantity of mycelium in the vermiculite–peat inoculum did not give reliable results because of the peat that interferes with colorimetric measurements (Duponnois and Garbaye 1991).

Alginate-Bead Inoculum

Le Tacon et al. (1983) and Mauperin et al. (1987) have shown that mycelium grown in a liquid medium and entrapped in calcium alginate gel is a very efficient inoculum for ECM development and can be used as an alternative to the classical vermiculite–peat mixture. Mycelium in alginate-bead inoculum is better protected, survives longer, and has a longer lasting effect than when grown on a vermiculite–peat mixture (Mortier et al. 1988). For production of alginate-bead inoculum, fungul cultures are grown in liquid medium. The mycelial pellets are washed in tap water, homogenized in a blender for 5–10 s, and resuspended in distilled water containing 10 $g.l^{-1}$ of sodium alginate and 30 $g.l^{-1}$ of powdered sphagnum peat. This suspension is pumped through a pipe with 5-mm holes above a 100 $g.l^{-1}$ $CaCl_2$ solution, each drop forming a bead of reticulated calcium alginate gel 3–4 mm in diameter. The beads are cured in $CaCl_2$ for 24 h at room temperature (for ensuring complete reticulation of the gel), washed in tap water to remove NaCl, stored in air-tight containers at room temperature to prevent drying, and used in the nursery (Mauperin et al. 1987).

Procedure of the French authors was modified by Kropáček and Cudlín (1989). In their work, sodium alginate and peat were substituted by agents Agricol and perlite, respectively, to obtain alginate paste containing the mycelium. Beads were cured in 5% calcium chloride for 30 min, rinsed with distilled water, dried to surface dryness (about 30 min in air filtered box), and stored in air-tight containers at 4°C until use. Alginate-bead inoculum was further used for inoculation of seedlings, e.g., in experiments of Duponnois and Garbaye (1991), Gagnon et al. (1991), Baum et al. (2000, 2002), and Repáč (2007). Parladé et al. (1999) prepared alginate-bead inoculum containing spores of *Rhizopogon* and mycelium of *Laccaria bicolor*. Mixed inoculum was prepared by adding spores plus fragmented mycelium (blended in autoclaved water) at different proportions to Pyrex flasks containing 20 $g.l^{-1}$ of an autoclaved water solution of sodium alginate. The content of flask was gently mixed, dropped into a 0.3-M water solution of $CaCl_2$ to polymerize, then washed with sterile distilled water, and kept at 4°C in plastic bags for 1 week.

3.4.2 *Methods of Inoculum Application*

Fumigation of nursery soil before inoculation improves ECM development because it lowers populations of soil microorganisms that can colonize introduced inocula, feeder root pathogens that damage roots and thus reduce ECM development, and/or indigenous competing ECM fungi. The seedlings can be inoculated with ECM inoculum during one of the following stages:

- Before seeds are sown or seedlings planted
- When seeds are sown or seedlings planted
- After seedlings emerge or seedlings planting

3.4.2.1 Basidiospores

Several inoculation techniques have been used:

- Mixing either crushed sporocarps or dry spores directly into soil or container medium

- Mixing spores with a moistened carrier, such as vermiculite, kaolin or sand, broadcasting onto soil and then mixing into the nursery soil or the growing medium of containers
- Dusting dry spores onto soil around young seedlings and leaching them into the root zone
- Suspending in water and drenching, irrigating, or injecting into growth substrate
- Dusting or spraying on roots of nonmycorrhizal seedlings
- Mixing with the pelletizing matrix and encapsulating or coating seeds before sowing

Marx et al. (1978) broadcast spores of *Pisolithus tinctorius* added to moist vermiculite onto soil and mixed inoculum with the soil using handtools. The spore quantities were 108, 324, or 648 mg of spores.m^{-2} of soil surface. Rincón et al. (2001) mixed dry spores of *P. tinctorius* and *Scleroderma verrucosum* included in vermiculite with potting substrate and filled containers with 10^3–10^8 spores per 175 ml of substrate.

Perhaps the most used method is application of spore water suspension to growth substrate. Marx and Bryan (1975) poured basidiospores of *P. tinctorius* (10 g suspended in 500 ml of distilled water with 1 drop of Tween 20) into microplots which were heavily watered to wash the basidiospores into the soil. Theodorou (1984) inoculated seeds of *Pinus radiata* with basidiospores (5.15×10^4 spores per seed) of *Rhizopogon luteolus* 1–2 days before sowing. Suspension of spores of *R. luteolus* (4.46×10^7 spores.ml^{-1}) was sprayed onto soil, and the spores were mixed with the soil by raking and rolling and by watering 1 day after sowing of inoculated seeds (i.e., 2 days after soil inoculation).

In experiments of Roldan et al. (1996) and Querejeta et al. (1998), water suspension of spores of *Pisolithus arhizus* was applied three times, 1 month apart, 12 weeks after sowing of *Pinus halepensis* to 300-ml plastic bags filled with 3:1 soil/peat mixture. Each plant received 5×10^5 spores per application. Molina et al. (1997) applied spore slurries of six *Rhizopogon* spp. to container-grown *Pseudotsuga menziesii* and *Pinus ponderosa* seedlings over two inoculations. For each inoculation, 10 ml of diluted spore suspension was pipetted onto the peat–vermiculite substrate. Chen et al. (2006) inoculated *Eucalyptus urophylla* seedlings 2 weeks after transplanting to plastic pots with 10 ml of spore suspension at a rate 10^6 spores per seedling. The spore slurry was added to a 2–3 cm deep hole near the plant using a 5-ml pipette. Parladé et al. (1996b), Rincón et al. (2001), and Hortal et al. (2008) inoculated 1-month-old container-grown seedlings with spore suspension of several ECM fungi at the rate of 10^2–10^8 spores per seedling. Núñez et al. (2006) applied spore suspension of *Tuber melanosporum* by injecting manually into each seedling's pot substrate (400 ml of light and dark peat and vermiculite, 2:1:1) approximately 7.5×10^5 spores at 3–8 cm depth. In the work of Becerra et al. (2009), 1 ml of spore suspension of two *Alpova* species containing 10^6 spores was inoculated at the base of *Alnus acuminata* seedlings.

3.4.2.2 Natural and Vegetative Inoculum

In inoculation programs, since hyphae cannot grow from the inoculum to roots, inoculum must be placed in the rooting zone of seedlings where roots can grow into the inoculum (Marx and Kenney 1982). In principle, natural and vegetative inoculum can be applied by three methods:

- Mixed with the rooting medium
- Banded or layered below seeds or seedlings
- Suspended in water and poured onto seedlings or dipping seedlings into the slurry before planting (except of beads)

Bareroot and container-grown seedlings and cuttings of numerous tree species were inoculated with vegetative inocula of many ECM fungi in greenhouse and nursery experiments. *Pisolithus tinctorius* (Marx and Bryan 1975; Marx et al. 1982; Marx and Cordell 1989; Vijaya and Srivasuki 1999; Rincón et al. 2001) and genus *Laccaria* (Mortier et al. 1988; Kropáček and Cudlín 1989; Gagnon et al. 1995; Garbaye and Churin 1997; Parladé et al. 1999; Baum et al. 2002; Gange et al. 2005; Machón et al. 2006) have been the most often tested fungi in inoculation experiments. ECM seedlings or cuttings are produced through inoculation of different growing substrates. Soil and peat mixed with vermiculite or other substrate components – the standard substrates for production of planting stock of forest tree species, are convenient and most frequently used for ECM inoculation.

Unequivocally, the most preferred technique of application of organic matter and vegetative inoculum is mixing it with a growth substrate. Amount of inoculum mixed with substrate in pot experiments is most commonly expressed as a ratio of inoculum and substrate by volume. Organic matter (humus, litter, rotten wood) is added to substrate in a larger ratio (1:4 to 1:1) (Kropp 1982; Parke et al. 1983; Repáč 1996a) than vegetative inoculum. Vermiculite–peat inoculum was mixed with potting substrate at the proportion 1:4 (Rincón et al. 2001), 1:10 (Hortal et al. 2009), as well as 1:64 (Rincón et al. 2001). Baum et al. (2009) expressed a ratio of soil:vermiculite-peat inoculum mixture (10:1) by weight. Parladé et al. (1999) and Baum et al. (2000, 2002) carried out inoculation by mixing alginate beads with potting substrate in the proportion 1:20 (v:v). Before filling containers, Gagnon et al. (1991) mixed alginate beads with the substrate so that each seedling received a volume of 17 ml of beads. Gagnon et al. (1995) mixed liquid inoculum with substrate in a cement mixer to get a final volume of 30 ml of mycelial slurry per seedling.

In nursery bed experiments, vegetative inoculum was broadcast evenly over nursery bed (0.5–2.8 $l.m^{-2}$) and incorporated into the upper 10–12 cm of substrate with handtools (Marx et al. 1978; Mortier et al. 1988; Duponnois and Garbaye 1991; Repáč 1996b; Garbaye and Churin 1997). Marx and Bryan (1975) reported application of *P. tinctorius* inoculum to fumigated soil at a ratio 1:8 (v:v) in upper 10 cm layer of plot.

Cline and Reid (1982) injected 20 ml of mycelial slurry into a root zone of container-grown seedlings at a depth of 6–8 cm using a glass syringe. Garbaye et al.

(1988), Machón et al. (2006), and Gange et al. (2005) added 50 ml of vermiculite–peat inoculum and 20 ml of mycelial slurry, respectively, into containers in contact with the roots of transplanted seedlings. Aucina et al. (2007) placed a layer of pine or oak litter on the surface of the nursery bed soil with seedlings to influence the growth and ECM communities of seedlings. Kropáček and Cudlín (1989) and Repáč (2007) evenly spread alginate-bead inoculum below surface of substrate immediately before seed sowing.

Perhaps the most capable way of expression of amount of vegetative inoculum is in a dry weight of mycelium, enabling the comparison of inoculum rate between experiments of different patterns. Unfortunately, to ascertain quantity of mycelium in vermiculite–peat carrier is almost impossible. Nevertheless, Mortier et al. (1988) reported 1–2 g of mycelium (dry weight) per m^2 in this type of inoculum. Mortier et al. (1988), Duponnois and Garbaye (1991), Vijaya and Srivasuki (1999), and Repáč (2007) referred 2 g, 2–10 g, 4 g, and 5 g of mycelium in dry weight per m^2, respectively, in alginate-bead inoculum. In the pot experiments, each seedling received either 6.82 mg (Gagnon et al. 1991) or 12 mg (Gagnon et al. 1995) of dried mycelium or 20–80 mg of mycelium in fresh weight (Parladé et al. 1999).

3.5 Field Inoculation

Inoculation of seedlings at the time of field planting is time consuming, requires more inoculum, and the introduced fungus must be compatible with native microorganisms and climatic conditions of the planting site (Riffle and Maronek 1982). Thus, nursery inoculated seedlings are more frequently outplanted (Roldan et al. 1996; Garbaye and Churin 1997; Núñez et al. 2006; Rincón et al. 2007) than inoculated at the time or after field planting. Castellano (1996) reviewed most of available literature (including unpublished data) on outplanting performance of ECM-inoculated seedlings and provided insight to the fungus-inoculum type-host-location combinations in field experiments. Inoculum types and application methods in the field are very similar to nursery inoculation of container-grown seedlings. Various natural organic materials and organic wastes are often used and may be considered as a simple source of ECM fungi propagules. Hallsby (1995) planted Norway spruce seedlings to mounds containing forest floor material from the F- and H-layers. Querejeta et al. (1998) added 150 ml of pine forest soil (top 20 cm of mineral soil) to the planting holes of *Pinus halepensis* seedlings at the time of planting. Organic materials, particularly decayed wood or mixtures containing decayed wood were equal or superior to mineral substrates for supporting ECM activity on planted seedlings (Harvey et al. 1997). Larcheveque et al. (2006) incorporated fresh co-composted sewage sludge and greenwastes (20 or 40 $kg.m^{-2}$ of compost) into the soil at each stem of 1-year-old seedlings of *Quercus ilex*, *Pinus halepensis*, and *P. pinea*.

In field experiments with laboratory produced inoculum, e.g., Baum et al. (2002) applied 50 ml of alginate beads containing mycelium of *Laccaria laccata* around each *Populus trichocarpa* cutting (1.5-m long annual shoots, diameter 13–17 mm) in 50-mm soil depth. Duñabeitia et al. (2004) inoculated seedlings few days after outplanting (and reinoculated 5 months later) by watering at 5 cm around the stem with 200 ml of spore slurry of *Scleroderma citrinum*, *Pisolithus arhizus*, and *Leccinum scabrum*. Menkis et al. (2007) wrapped root systems of *Pinus sylvestris* and *Picea abies* seedlings in a filter paper containing mycelium of *Cenococcum geophilum*, *Piceirhiza bicolorata* or *Hebeloma crustuliniforme*, overlaid with damp peat–sand mixture, wrapped in a paper towel, and planted seedlings on poor sandy soil.

3.6 Commercial Inoculum

It is relatively simple to produce sufficient volumes of inoculum for small research studies, but it is extremely difficult to produce sufficient quantities of inoculum to support commercial nursery inoculation. Since 1976, several formulations of vegetative and spore inocula of "super-strain" of *Pisolithus tinctorius* were produced commercially (e.g., inoculum trademarked MycoRhiz®) and used in large scale for inoculation of millions of bareroot and containerized pine and oak seedlings in USA (Marx 1991). In France, a company called Somycel produced vegetative beads inoculum, which substantially increased reforestation effort of coniferous trees followed by nursery inoculation (Kropp and Langlois 1990).

Numerous entities of international, national, or local importance produce and offer ECM fungal inocula for commercial purposes. For example, several European producers of mycorrhizal fungi inocula constituted organization called "Federation of European Mycorrhizal Fungi Producers" (FEMFiP) in 2003, aimed to achieve and maintain high standards for these products in Europe (Federation of European Mycorrhizal Fungi Producers 2010). The most products of FEMFiP contain arbuscular fungi and are intended to horticulture, agriculture, and landscape sectors. The companies PlantWorks Ltd. (Great Britain, product TerraVital), Biorize (France, product Ectorize), and Symbio-m (Czech Republic, product Ectovit) offer ECM inocula for inoculation of forest tree species.

Spore and mycelial inocula are used in operational scale to match the ECM fungi to the host and environment in which they are required to thrive. Natural biostimulants and ingredients supporting the development of ECM symbiosis (natural humates, sea-grass extracts, ground minerals) are usually added to commercial inoculum. Inoculation techniques are in principle the same as described in Sects. 3.4.2 and 3.5. Compared to research-scale inoculum formulations, gel formulation (slurry), prepared by diluting ECM inoculum containing naturally degradable granules of water-retaining gel in water, is more frequently used in a commercial scale. The slurry can be sprayed onto the substrate before seed sowing, into the rooting zone of transplanted seedlings, or root systems of seedlings can be dipped into the slurry. The production and marketing of inocula on a commercial scale increased recently and further potential appears to be quite large.

3.7 Conclusion

There are a variety of ECM inoculum types, inoculum preparation methods, and inoculation techniques to initiate the development of ectomycorrhizae on forest tree seedlings. The selection of fungi, inoculum type, and inoculation method all depends on the intended purpose of inoculation. Better understanding of the structure and functioning of ECM symbiosis and better performance of inoculated seedlings under nursery and field conditions are the ultimate goals of research on artificial mycorrhization, and therefore further development of inoculation methods in operational conditions is desirable. Although sterile conditions are a faraway from natural ones, techniques of pure culture synthesis are necessary for compatibility, structural, physiological, molecular, and other studies.

Application of ECM inoculum to substrate does not guarantee that ectomycorrhizae will develop on a host plant. Success of inoculation depends on the type and age of inoculum used, inoculum dose, timing of inoculation, inoculum placement in the growing medium, etc. (Riffle and Maronek 1982; Mortier et al. 1988; Rodrigues et al. 1999). Besides inoculum and inoculation pattern, interspecific and intraspecific host–fungus variation, environmental conditions, seedling production practices, and other factors are responsible for seedling response to inoculation (Kropp and Langlois 1990; Castellano 1996; Menkis et al. 2007; Duponnois et al. 2008). Despite of (or just because of) inconsistency of effects of introduced fungi on performance of inoculated seedlings, more research is needed on the screening of potential host–fungus species and genotype combinations and the host–fungus–environment interactions to optimize the effect of fungi on plants. Also, researchers might be able to make much progress in simplifying the application of ECM fungi in seedling inoculation. Because of natural complexity and diversity of ECM relationships, we have to realize that their understanding will neither be completed nor their use completely reliable. Nevertheless, when we consider extensive area of treeless lands and adverse forest sites required artificial regeneration, the importance of artificial mycorrhization of seedlings as reforestation and afforestation management tool is obvious.

References

Aspray TJ, Frey-Klett P, Jones JE, Whipps JM, Garbaye J, Bending GD (2006) Mycorrhization helper bacteria: a case of specifity for altering ectomycorrhiza architecture but not ectomycorrhiza formation. Mycorrhiza 16:533–541

Aucina A, Rudawska M, Leski T, Skridaila A, Riepsas E, Iwanski M (2007) Growth and mycorrhizal community structure of *Pinus sylvestris* seedlings following the addition of forest litter. Appl Environ Microb 73:4867–4873

Baum C, Schmid K, Makeschin F (2000) Interactive effects of substrates and ectomycorrhizal colonization on growth of poplar clone. J Plant Nutr Soil Sci 163:221–226

Baum C, Stetter U, Makeschin F (2002) Growth response of *Populus trichocarpa* to inoculation by the ectomycorrrhizal fungus *Laccaria laccata* in a pot and a field experiment. For Ecol Manage 163:1–8

Baum C, Toljander YK, Eckhardt KU, Weih M (2009) The significance of host-fungus combinations in ectomycorrhizal symbioses for the chemical quality of willow foliage. Plant Soil 323:213–224. doi:10.1007/s111040099928x

Becerra AG, Menoyo E, Lett I, Li CY (2009) *Alnus accuminata* in dual symbiosis with *Frankia* and two different ectomycorrhizal fungi *(Alpova austroalnicola* and *Alpova diplophloeus)* growing in soilless growth medium. Symbiosis 47:85–92

Bending GD, Poole EJ, Whipps JM, Read DJ (2002) Characterisation of bacteria from *Pinus sylvestris – Suillus luteus* mycorrhizas and their effects on root-fungus interactions and plant growth. FEMS Microbiol Ecol 39:219–227

Bogeat-Triboulot MB, Bartoli F, Garbaye J, Marmeisse R, Tagu D (2004) Fungal ectomycorrhizal community and drought affect root hydraulic properties and soil adherence to roots of *Pinus pinaster* seedlings. Plant Soil 267:213–223

Borchers SL, Perry DA (1990) Growth and ectomycorrhiza formation of Douglas-fir seedlings grown in soils collected at different distances from pioneering hard-woods in southwest Oregon clear-cuts. Can J For Res 20:712–721

Brunner I (1993) Production of ectomycorrhizal *Picea abies – Hebeloma crustuliniforme* seedlings for ecological studies: effects of synthesis techniques on the morphology of the symbiosis. Water Air Soil Poll 68:231–240

Brunner I, Amiet R, Zollinger M, Egli S (1992) Ectomycorrhizal syntheses with *Picea abies* and three fungal species: a case study on the use of an in vitro technique to identity naturally occurring ectomycorrhizae. Mycorrhiza 2:89–96

Castellano MA (1996) Outplanting performance of mycorrhizal inoculated seedlings. In: Mukerji KG (ed) Concepts in mycorrhizal research. Kluwer Academic Publishers, Netherlands, pp 223–301

Chen YL, Kang LH, Dell B (2006) Inoculation of *Eucalyptus urophylla* with spores of *Scleroderma* in a nursery in south China: comparison of field soil and potting mix. For Ecol Manage 222:439–449

Cline ML, Reid CPP (1982) Seed source and mycorrhizal fungus effects on growth of containerized *Pinus contorta* and *Pinus ponderosa* seedlings. For Sci 28:237–250

Corrêa A, Strasser RJ, Martins-Loução MA (2006) Are mycorrhiza always beneficial? Plant Soil 279:65–73

Duñabeitia M, Rodríguez N, Salcedo I, Sarrionandia E (2004) Field mycorrhization and its influence on the establishment and development of the seedlings in a broadleaf plantation in the Basque country. For Ecol Manage 195:129–139

Duponnois R, Garbaye J (1991) Techniques for controlled synthesis of Douglas-fir – *Laccaria laccata* ectomycorrhizal symbiosis. Ann Sci For 48:641–650

Duponnois R, Kisa M, Prin Y, Ducousso M, Plenchette Ch, Lepage M, Galiana A (2008) Soil factors influencing the growth response of *Acacia holosericea* A. Cunn. ex G. Don to ectomycorrhizal inoculation. New For 35:105–117

Federation of European Mycorrhizal Fungi Producers (2010) http://www.fenfip.com. Cited 15 Feb 2010

Flores R, Díaz G, Honrubia M (2008) Mycorrhizal synthesis and basidioma primordium formation between *Abies guatemalensis* and *Laccaria bicolor*. Nova Hedwigia 87:153–160

Fortin JA, Piché Y, Lalonde M (1980) Technique for the observation of early morphological changes during ectomycorrhiza formation. Can J Bot 58:361–365

Gagnon J, Langlois CG, Garbaye J (1991) Growth and ectomycorrhiza formation of container-grown red oak seedlings as a function of nitrogen and inoculum type of *Laccaria bicolor*. Can J For Res 21:966–973

Gagnon J, Langlois CG, Bouchard D, Le Tacon F (1995) Growth and ectomycorrhizal formation of container-grown Douglas-fir seedlings inoculated with *Laccaria bicolor* under four levels of nitrogen fertilization. Can J For Res 25:1953–1961

Gange AC, Gane DRJ, Chen Y, Gong M (2005) Dual colonization of *Eucalyptus urophylla* S.T. Blake by arbucular and ectomycorrhizal fungi affects levels of insect herbivore attack. Agric For Entomol 7:253–263

Gáper J, Repáč I, Cudlín P (1999) In vitro mycorrhizae syntheses of *Inocybe lacera* and *Suillus bovinus* with *Picea abies*. In: Moreno G et al (eds) Book of abstracts of the XIII Congress of European Mycologists, Alcalá de Henares (Madrid), p 41

Garbaye J, Churin JL (1997) Growth stimulation of young oak plantations inoculated with the ectomycorrhizal fungus *Paxillus involutus* with special reference to summer drought. For Ecol Manage 98:221–228

Garbaye J, Delwaulle JC, Diangana D (1988) Growth response of Eucalypts in the Congo to ectomycorrhizal inoculation. For Ecol Manage 24:151–157

Giomaro G, Sisti D, Zambonelli A, Amicucci A, Cecchini M, Comandini O, Stocchi V (2002) Comparative study and molecular characterization of ectomycorrhizas in *Tilia americana* and *Quercus pubescens* with *Tuber brumale*. FEMS Microbiol Lett 216:9–14

Guerin-Laguette A, Vaario LM, Matsushita N, Shindo K, Suzuki K, Lapeyrie F (2003) Growth stimulation of Shiro-like mycorrhiza forming mycelium of *Tricholoma matsutake* on solid substrates by non-ionic surfactants or vegetable oil. Mycol Prog 2:37–44

Hallsby G (1995) Field performance of outplanted Norway spruce – effects of organic-matter amendments and site preparation. Can J For Res 25:1356–1367

Harvey AE, Page-Dumroese DS, Jurgensen MF, Graham RT, Tonn JR (1997) Site preparation alters soil distribution of roots and ectomycorrhizae on outplanted western white pine and Douglas-fir. Plant Soil 188:107–117

Herrmann S, Oelmüller R, Buscot F (2004) Manipulation of the onset of ectomycorrhiza formation by indole-3-acetic acid, activated charcoal or relative humidity in the association between oak microcuttings and *Piloderma croceum*: influence on plant development and photosynthesis. J Plant Physiol 161:509–517

Högberg P, Högberg MN, Quist ME, Ekblad A, Näsholm T (1999) Nitrogen isotope fractionation during nitrogen uptake by ectomycorrhizal and non-mycorrhizal *Pinus sylvestris*. New Phytol 142:569–576

Hönig K, Riefler M, Kottke I (2000) Survey of *Paxillus involutus* (Batsch) Fr. inoculum and fruitbodies in a nursery by IGS-RFLPs and IGS sequences. Mycorrhiza 9:315–322

Hortal S, Pera J, Parladé J (2008) Tracking mycorrhizas and extraradical mycelium of the edible fungus *Lactarius deliciosus* under field competition with *Rhizopogon* spp. Mycorrhiza 18:69–77

Hortal S, Pera J, Parladé J (2009) Field persistence of the edible ectomycorrhizal fungus *Lactarius deliciosus*: effects of inoculation strain, initial colonization level, and site characteristics. Mycorrhiza 19:167–177

Kozdrój J, Piotrowska-Seget Z, Krupa P (2007) Mycorrhizal fungi and ectomycorrhiza associated bacteria isolated from an industrial desert soil protect pine seedlings against Cd(II) impact. Ecotoxicology 16:449–456

Kropáček K, Cudlín P (1989) Preparation of granulated mycorrhizal inoculum and its use in forest nurseries. In: Vančura V, Kunc F (eds) Interrelationships between microorganisms and plants in soil. Academia, Praha, pp 177–182

Kropp BR (1982) Rotten wood as mycorrhizal inoculum for containerized western hemlock. Can J For Res 12:428–431

Kropp BR, Langlois CG (1990) Ectomycorrhizae in reforestation. Can J For Res 20:438–451

Krüger A, Peškan-Berghöfer T, Frettinger P, Herrmann S, Buscot F, Oelmüller R (2004) Identification of premycorrhiza-related plant genes in the association between *Quercus ro*bur and *Piloderma croceum*. New Phytol 163:149–157

Langer I, Krpata D, Peintner U, Wenzel WW, Schweiger P (2008) Media formulation influences in vitro ectomycorrhizal synthesis on the European aspen *Populus tremula* L. Mycorrhiza 18:297–307. doi:10.1007/s0057200801825

Larcheveque M, Ballini C, Korboulewsky N, Montes N (2006) The use of compost in afforestation of Mediterranean areas: effects on soil properties and young tree seedlings. Sci Total Environ 369:220–230

Le Tacon F, Jung G, Michelot P, Mugnier M (1983) Efficacité en pépiničre forestiére d'un inoculum de champignon ectomycorhizien produit en fermenteur et inclus dans une matrice de polyméres. Ann Sci For 40:165–176

Machón P, Santamaría O, Pajares JA, Alves-Santos FM, Diez JJ (2006) Influence of the ectomycorrhizal fungus *Laccaria laccata* on pre-emergence, post-emergence and late damping-off by *Fusarium moniliforme* and *F. oxysporum* on *Scots pine* seedlings. Symbiosis 42:153–160

Martín-Pinto P, Pajares J, Díez J (2006) In vitro effects of four ectomycorrhizal fungi, *Boletus edulis*, *Rhizopogon roseolus*, *Laccaria laccata* and *Lactarius deliciosus* on *Fusarium* damping off in *Pinus nigra* seedlings. New For 32:323–334. doi:10.1007/s1105600690067

Marx DH (1969) The influence of ectotrophic mycorrhizal fungi on the resistance of pine roots to pathogenic infections. I. Antagonism of mycorrhizal fungi to root pathogenic fungi and soil bacteria. Phytopathology 59:153–163

Marx DH (1991) The practical significance of ectomycorrhizae in forest establishment. In: Hägglund B (ed) Ecophysiology of ectomycorrhizae of forest trees. The Marcus Wallenberg foundation symposia proceedings: 7, Stralins Tryckeri AB, Grycksbo, Sweden, pp 54–90

Marx DH, Bryan WC (1975) Growth and ectomycorrhizal development of Loblolly pine seedlings in fumigated soil infested with the fungal symbiont *Pisolithus tinctorius*. For Sci 21:245–254

Marx DH, Cordell CE (1989) The use of specific ectomycorrhizas to improve artificial forestation practices. In: Whipps JM, Lumsden RD (eds) Biotechnology of fungi for improving plant growth. Press Syndicate of the University of Cambridge, British Mycological Society, New York, pp 1–26

Marx DH, Kenney DS (1982) Production of ectomycorrhizal fungus inoculum. In: Schenck NC (ed) Methods and principles of mycorrhizal research. American Phytopathology Society, St. Paul, MN, pp 131–146

Marx DH, Morris WG, Mexal JG (1978) Growth and ectomycorrhizal development of Loblolly pine seedlings in fumigated and nonfumigated nursery soil infested with different fungal symbionts. For Sci 24:193–203

Marx DH, Ruehle JL, Kenney DS, Cordell CE, Riffle JW, Molina RJ, Pawuk WH, Navratil S, Tinus RW, Goodwin OC (1982) Commercial vegetative inoculum of *Pisolithus tinctorius* and inoculation techniques for development of ectomycorrhizae on container-grown tree seedlings. For Sci 28:373–400

Mauperin C, Mortier F, Garbaye J, Le Tacon F, Carr G (1987) Viability of an ectomycorrhizal inoculum produced in a liquid medium and entrapped in a calcium alginate gel. Can J Bot 65:2326–2329

Menkis A, Vasiliauskas R, Taylor AFS, Stenlid J, Finlay R (2007) Afforestation of abandoned farmland with conifer seedlings inoculated with three ectomyorrhizal fungi – impact on plant performance and ectomycorrhizal community. Mycorrhiza 17:337–348

Molina R (1979) Pure culture synthesis and host specificity of red alder mycorrhizae. Can J Bot 57:1223–1228

Molina R, Palmer JG (1982) Isolation, maintenance, and pure culture manipulation of ectomycorrhizal fungi. In: Schenck NC (ed) Methods and principles of mycorrhizal research. American Phytopathology Society, St. Paul, MN, pp 115–129

Molina R, Smith JE, McKay D, Melville LH (1997) Biology of the ectomycorrizal genus, *Rhizopogon*. New Phytol 137:519–528

Moore LM, Jansen AE, Van Griensen LJLD (1989) Pure culture synthesis of ectomycorrhizas with *Cantharellus cibarius*. Acta Bot Neerl 38:273–278

Mortier F, Le Tacon F, Garbaye J (1988) Effects of inoculum type and inoculation dose on ectomycorrhizal development, root necrosis and growth of Douglas fir seedlings inoculated with *Laccaria laccata* in a nursery. Ann Sci For 45:301–310

Moser M (1958) Die künstliche Mykkorhizaimpfung an Forstpflanzen. I. Erfahrungen bei der Reinkulture von Mykorrhizapilzen. Forstwiss Centralbl 77:32–40

Moser M (1960) Die Gattung Phlegmacium (Schleimköpfe). Verlag Julius Klinkhardt, Bad Heilbrunn, Germany, 440 pp

Niemi K, Vuorinen T, Ernstsen A, Häggman H (2002) Ectomycorrhizal fungi and exogenous auxins influence root and mycorrhiza formation of Scots pine hypocotyl cuttings in vitro. Tree Physiol 22:1231–1239

Niemi K, Julkunen-Titto R, Häggman H, Sarjala T (2006) *Suillus variegatus* causes significant changes in the content of individual polyamines and flavonoids in Scots pine seedlings during mycorrhiza formation in vitro. J Exp Bot 58:391–401. doi:10.1093/jxb/erl209

Núñez JAD, Serrano JS, Barreal JAR, González JAS (2006) The influence of mycorrhization with *Tuber melanosporum* in the afforestation of a Mediterranean site with *Quercus ilex* and *Quercus faginea*. For Ecol Manage 231:226–233

Obase K, Tamai Y, Yajima T, Miyamoto T (2009) Mycorrhizal synthesis of four ectomycorrhizal fungi in potted *Populus maximowiczii* seedlings. Mycoscience 50:143–145

Ohta A (1990) A new medium for mycelial growth of mycorrhizal fungi. T Mycol Soc JPN 31:323–334

Pachlewski R, Pachlewska J (1974) Studies on symbiotic properties of mycorrhizal fungi of pine (*Pinus sylvestris* L.) with the aid of the methods of mycorrhizal synthesis in pure cultures on agar. For Res Inst Press, Warsaw, Poland, 228 pp

Parke JL, Linderman RG, Trappe JM (1983) Effects of forest litter on mycorrhiza development and growth of Douglas-fir and western red cedar seedlings. Can J For Res 13:666–671

Parladé J, Álvarez IF, Pera J (1996a) Ability of native ectomycorrhizal fungi from northern Spain to colonize Douglas-fir and other introduced conifers. Mycorrhiza 6:51–55

Parladé J, Pera J, Alvarez IF (1996b) Inoculation of containerized *Pseudotsuga menziesii* and *Pinus pinaster* seedlings with spores of five species of ectomycorrhizal fungi. Mycorrhiza 6:237–245

Parladé J, Alvarez IF, Pera J (1999) Coinoculation of containerized Douglas-fir (*Pseudotsuga menziesii*) and maritime pine (*Pinus pinaster*) seedlings with the ectomycorrhizal fungi *Laccaria bicolor* and *Rhizopogon* spp. Mycorrhiza 8:189–195

Peterson RL, Chakravarty P (1991) Techniques in synthesizing ectomycorrhiza. In: Norris JR, Read DJ, Varma AK (eds) Methods in microbiology, vol 23. Academic Press Limited, London, pp 75–106

Querejeta JI, Roldán A, Albaladejo J, Castillo V (1998) The role of mycorrhizae, site preparation, and organic amendment in the afforestation of a semi-arid mediterranean site with *Pinus halepensis*. For Sci 44:203–211

Repáč I (1996a) Inoculation of *Picea abies* (L.) Karst. seedlings with vegetative inocula of ectomycorrhizal fungi *Suillus bovinus* (L.:Fr.) O. Kuntze and *Inocybe lacera* (Fr.) Kumm. New For 12:41–54

Repáč I (1996b) Effects of forest litter on mycorrhiza formation and growth of container-grown Norway spruce (*Picea abies* [L.] Karst.) seedlings. Lesnictví 42:317–324

Repáč I (2007) Ectomycorrhiza formation and growth of *Picea abies* seedlings inoculated with alginate-bead fungal inoculum in peat and bark compost substrates. Forestry 80:517–530

Riffle JW, Maronek DM (1982) Ectomycorrhizal inoculation procedures for greenhouse and nursery studies. In: Schenck NC (ed) Methods and principles of mycorrhizal research. American Phytopathology Society, St. Paul, MN, pp 147–155

Rincón A, Alvarez IF, Pera J (2001) Inoculation of containerized *Pinus pinea* L. seedlings with seven ectomycorrhizal fungi. Mycorrhiza 11:265–271

Rincón A, de Felipe MR, Fernández-Pascual M (2007) Inoculation of *Pinus halepensis* Mill. with selected ectomycorrhizal fungi improves seedling establishment 2 years after planting in a degraded gypsum soil. Mycorrhiza 18:23–32

Rodrigues LS, Kasuya MCM, Borges AC (1999) Viability of ectomycorrhizal fungus mycelium entrapped in calcium alginate gel. Mycorrhiza 8:263–266

Roldan A, Querejeta I, Albaladejo J, Castillo V (1996) Growth response of *Pinus halepensis* to inoculation with *Pisolithus arhizus* in a terraced rangeland amended with urban refuse. Plant Soil 179:35–43

Sarjala T, Taulavuori K (2004) Fluctuation in free and conjugated polyamines in Scots pine seedlings after changes in temperature and daylength. Acta Physiol Plant 26:271–279
Šašek V, Musílek V (1967) Cultivation and antibiotic activity of mycorrhizal basidiomycetes. Folia Microbiol 12:515–523
Theodorou C (1984) Mycorrhizal inoculation of pine nurseries by spraying basidiospores onto soil prior to sowing. Aust For 47:76–78
Timonen S, Tammi H, Sen R (1997) Outcome of interactions between genets of two *Suillus* spp. and different *Pinus sylvestris* genotype combinations: identity and distribution of ectomycorrhizas and effects on early seedlings growth in N-limited nursery soil. New Phytol 137:691–702
Vijaya T, Srivasuki KP (1999) Effect of inoculum type and inoculation dose on ectomycorrhizal development and growth of Acacia nilotica seedlings inoculated with *Pisolithus tinctorius* in a nursery. Biologia 54:439–442
Vozzo JA, Hacskaylo E (1971) Inoculation of *Pinus caribaea* with ectomycorrhizal fungi in Puerto Rico. For Sci 17:239–245
Vrålstad T, Schumacher T, Taylor AFS (2002) Mycorrhizal synthesis between fungal strains of the *Hymenoscyphus ericae* aggregate and potential ectomycorrhizal and ericoid hosts. New Phytol 153:143–152
Wallander H, Fossum A, Rosengren U, Jones H (2005) Ectomycorrhizal fungal biomass in roots and uptake of P from apatite by *Pinus sylvestris* seedlings growing in forest soil with and without wood ash amendment. Mycorrhiza 15:143–148
Wong KKY, Fortin JA (1989) A Petri dish technique for the aseptic synthesis of ectomycorrhizae. Can J Bot 67:1713–1716

Part II
Biotechnological Aspect of ECM

Chapter 4
Systematics and Ecology of Tropical Ectomycorrhizal Fungi Using Molecular Approaches

Rivière-Dobigny Taiana

4.1 Evaluation of Tropical ECM Species Composition

4.1.1 Introduction

Traditionally, surveys on ECM species communities rely on morphological descriptions of sporocarps due to previous species collections. Such macroscopic approaches are inaccurate since sporocarp identifications only provide an incomplete picture of ECM fungal diversity (Egger 1995; Gardes and Bruns 1996; Grogan et al. 2000; Jonsson et al. 2000). Indeed, while sporocarps are necessarily associated with ectomycorrhizas, a fungus forming ectomycorrhizas may not always form sporocarps (Horton and Bruns 2001). Furthermore, investigations on ECM fungi are delicate as morphological characters of root tips are usually not sufficient for species recognition (Bruns et al. 1998). This has long represented a considerable limitation to our understanding of ECM diversity, and thus an adequate understanding of ECM ecology (Debaud et al. 1999; Kretzer et al. 2003). Fortunately, molecular tools have proved to be of great help to solve taxonomic discrepancies and have, therefore, considerably improved our knowledge on their ecology (Horton and Bruns 2001). Indeed, molecular techniques enable identification of ECM genera or species through analyses conducted on DNA extracted from root tips. The latter approach is often achieved through comparisons with DNA sequences database of ribosomal internal transcribed spacer (ITS) or mitochondrial sequences (Gardes et al. 1991; Kendall 2007). Also, cloning and sequencing T-RFLP (Terminal Restriction Fragment Polymorphism) and DGGE (Denaturing Gradient Gel Electrophoresis) strategies have been employed to describe ECM communities in soil (Landeweert et al. 2003; Koide et al. 2005).

R-D. Taiana
IRD, avenue de Maradi, BP 11416 Niamey, Niger
e-mail: taianariviere@yahoo.co.uk

M. Rai and A. Varma (eds.), *Diversity and Biotechnology of Ectomycorrhizae*, Soil Biology 25, DOI 10.1007/978-3-642-15196-5_4,

To our knowledge, the mycorrhizal status of numerous tropical indigenous tree species remains undocumented even if large areas of tropical and subtropical forests are dominated by ECM trees (Redhead 1980; Alexander and Högberg 1986), suggesting a key role of these symbiosis in the functioning of some tropical forest ecosystems (Onguene and Kuyper 2001). For instance, the Dipterocarpaceae family in South East Asia comprises 470 tree species, which largely dominate tropical rainforests and represent a major source of commercial timber (Maury-Lechon and Curtet 1998). These tree species have been shown to be associated with various fungal genera such as *Russula*, *Boletus*, *Cortinarius*, *Lactarius*, *Laccaria*, *Pisolithus*, *Amanita*, *Scleroderma*, *Suillus*, *Strobilomyces*, and *Cantharellus* (Smits 1992; Watling and Lee 1998; Natarajan et al. 2005; Rivière et al. 2007; Peay et al. 2010). In the same manner, surveys in African dry woodlands and tropical rainforests revealed that the dominant group of ECM trees includes caesalpinioid legumes (12 genera in the *Amherstieae* and one genus, *Afzelia*, in the Detarieae) and Phyllanthaceae (one genus, *Uapaca*) (Thoen and Bâ 1989; Newbery et al. 1997; Sanon et al. 1997; Onguene and Kuyper 2001), and associated ECM fungi belong to several genera, mainly *Russula*, *Lactarius*, *Amanita*, *Boletus*, and *Cantharellus* (Buyck et al. 1996; Eberhardt and Verbeken 2004). ECM trees can form local patches that contribute from 45 to 70% of the basal rainforest area, such as in the Korup National Park in Cameroon, where three species *Microberlinia bisulcata*, *Tetraberlinia bifoliolata*, and *Tetraberlinia moreliana* are dominant (Newbery et al. 1997).

South American tropical rainforests are reported to be mainly dominated by vesicular-arbuscular mycorrhizal associations (Picone 2000). For example, in Brazil, the mycorrhizal status of the Araucaria forest and the Atlantic rainforest true 29 native species belonging to 19 families revealed no evidence of ectomycorrhizal colonization (Andrade et al. 2000). Some authors reported ECM occurrence in native Brazilian trees such as Thomazini (1974) and Singer and Araujo (1979) on Caesalpinioidae species. Surveys in the Pakaraima Mountains of western Guyana revealed previously undocumented forests dominated by ECM leguminous trees, with a rich assemblage of ECM mycobionts. These fungi belonged to the Boletaceae, Amanitaceae, Russulaceae, Cortinariaceae, Cantharellaceae, Clavulinaceae, and Entolomataceae Basidiomycete families (Henkel et al. 2002).

There is increasing evidence that some tropical ECM communities are more diversified than previously supposed, and so, many questions are still unstudied. Indeed, ECM diversity is still poorly documented in tropical rainforests and so, with regards to forests destructions going on, sampling efforts are urgently needed. In this study, we examined ECM Basidiomycete species' taxonomic structure in natural tropical rainforests located in West Africa and Western India. We used both morphological and molecular methods to identify ECM fungal species occurring on the root tip, and importantly, in order to link the above- and belowground ECM diversity, sporocarp and ectomycorrhizas sequences were compared. In order to propose species' identification through molecular typing of sporocarps vs. ectomycorrhiza, 40 reference sequences from available databases were included in our phylogenetic analyses.

4.1.2 ECM Systematics and Diversity Assessment

ECM species diversity is poorly documented in natural tropical rainforests. The primary difficulty lies in the identification of the species. Indeed, many species' descriptions are still going on mainly based on morphological descriptions of the sporocarps, and so, tropical investigations, which forget root tip morphotypes, are most of the time incomplete. In this context, molecular methods offer a great opportunity to evaluate ECM taxonomic diversity including ECM root DNA extracts. We present here a survey of ECM Basidiomycete communities conducted in two tropical rainforests, one located in Africa (Western Upper Guinea) and the other in Asia (Western Ghats, India). To solve taxonomic discrepancies, molecular analysis involved sequencing a fragment of the mitochondrial large subunit rRNA gene and comparative phylogenetic analysis using sequences from European specimens as taxonomic benchmarks. Here, we focus on ML5/ML6 sequences for three main reasons. We are aware that the ML5/ML6 region often provides identical sequences for closely related species, thus limiting the interest of this conservative DNA fragment at the species level. The ITS fragment, on the contrary, is more accurate for species identification (Bruns et al. 1998). Here, we had to choose ML5/ML6 fragment for three main reasons (1) with regards to the high morphological diversity of sporocarps and numerous ectomycorrhizas found in the field, it first appeared pivotal to obtain rapid but nonambiguous sequences allowing to identify the samples at the genus level, and yet many sequences are available on Genbank; (2) as recommended by Gardes and Bruns (1993), we previously sequenced the ITS fragment of 37 Guinean *Russula* and 15 Indian *Russula* specimens, but the sequences were too variable to be unambiguously aligned and none of the sequences matched with known species; and (3) we consistently observed that ITS primers often amplify plant DNA from ectomycorrhizas, whereas the ML5/ML6 primers never did.

Inventory of the plant species composition is a prerequisite to the study on ECM fungal community; consequently, it is essential to describe flora diversity before ECM fungal one. Study sites were located in Southern Guinea and in Western India. Southern Guinea forests cover hills and mountains ranging in altitude from 500 m in the Ziama forest (8°51′N, 9°31′W) to 1,752 m on the Mount Nimba forest (7°60′N, 8°49′W). They are typical evergreen or semi-evergreen rainforests with a mean annual rainfall of 2,500–3,000 mm and a dry season from January to March. These Guinean forests shelter ECM trees belonging to the Caesalpiniaceae (*Afzelia bella*, *Paramacrolobium coeruleum*, *Anthonotha fragans*, *A. macrophylla*, *Cryptosepalum tetraphyllum*, *Pelligriniodendron diphyllum*, *Gilbertiodendron limba*) and the Phyllanthaceae (*Uapaca heudelotii*, *U. esculenta*, *U. guineensis*, and *U. chevalieri*). Two additional species, *Uapaca somon* and *Afzelia africana*, were found in the dryer and lower woodlands bordering the Mount Nimba rainforest.

The Kadamakal Reserve Forest is located in the Western Ghats, India, in the district of Kodagu (Karnataka) near the village of Uppangala (12°30′N; 75°39′W). Its altitude ranges from 400 to 600 m. Annual rainfall is about 5,200 mm with a marked dry season from December to March. Vegetation is a dense moist evergreen forest dominated by three species, *Dipterocarpus indicus*, *Kingiodendron pinnatum*, and *Humboltia brunonis* (Pascal 1984). Two ectomycorrhizal Dipterocarpaceae species dominate the high canopy, *Vateria indica* and *Dipterocarpus indicus*, which together represent 41.2% of the basal area (Pélissier et al. 1998). To optimize the assessment of the floristic diversity, the Uppangala forest was sampled following a previously designed transect (3 plots from 180 to 370 m long and 20 m wide; see Pélissier et al. 1998).

Sporocarps belonging to Basidiomycete families that were typically ectomycorrhizal were collected each August during 4 successive years in Guinea. In India, samples were collected during 2 successive years. In each spot, sporophores and ectomycorrhizas were collected in order to subsequent DNA extraction, PCR, and sequencing (see Rivière et al. 2007 for protocols). A total of 198 sequences were edited and assembled. In addition, 40 reference fragments from known boreal and temperate species were downloaded from NCBI databank. These data were used as external taxonomic benchmarks as well as references to investigate any phylogeographic pattern. Most of these sequences were retrieved from the Bruns et al. (1998) database (http://plantbio.berkeley.edu/~bruns/). Five sequences (*Tulasnella irregularis*, *Sebacina* sp., and three *Cantharellus* spp.) were used as outgroups (Bruns et al. 1998).

4.1.2.1 ECM Status of Tropical Trees

In Guinea, the presence of ectomycorrhizas was confirmed on six caesalpinioid species, *A. africana*, *A. bella*, *A. fragans*, *A. macrophylla*, *G. limba*, and *P. coeruleum*, and five *Uapaca* species (*U. guineensis*, *U. esculenta*, *U. heudelottii*, *U. somon*, and *U. chevalieri*), as previously reported (Thoen and Bâ 1989). Furthermore, we report for the first time the ECM status of two Caesalpinioideae species of the Amherstieae tribe, i.e., *Cryptosepalum tetraphyllum* and *Pelligriniodendron diphyllum*. In India, sporocarps were collected under the two species *V. indica* and *D. indicus*, which are known as ECM tree species (Rivière 2004; Natarajan et al. 2005; Rivière et al. 2007).

4.1.2.2 Ectomycorrhizas Diversity

Based on morphological identification, considering the diversity of both the sites, sporocarps were assigned to five families (Amanitaceae, Russulaceae, Sclerodermataceae, Tomentellaceae, and Tricholomataceae) and one superfamilial group (boletoids). Based on molecular typing, the most represented family was the Russulaceae (75 samples of both sporocarps and ectomycorrhizas), followed by the Amanitaceae (26), the boletoids and Chalciporus group (32), including the genera *Boletus*, *Xerocomus*, *Leccinum*, *Tubosaeta*, *Strobilomyces*, and *Chalciporus*, the Sclerodermataceae (32), the tricholomatoids (10), and the telephoroids (21), of which only one sporocarp was identified as a *Tomentella* species. Four subsequent phylogenies were reconstructed in order to independently focus on the four previously identified clades in Fig. 4.1, i.e., (1) Russulaceae

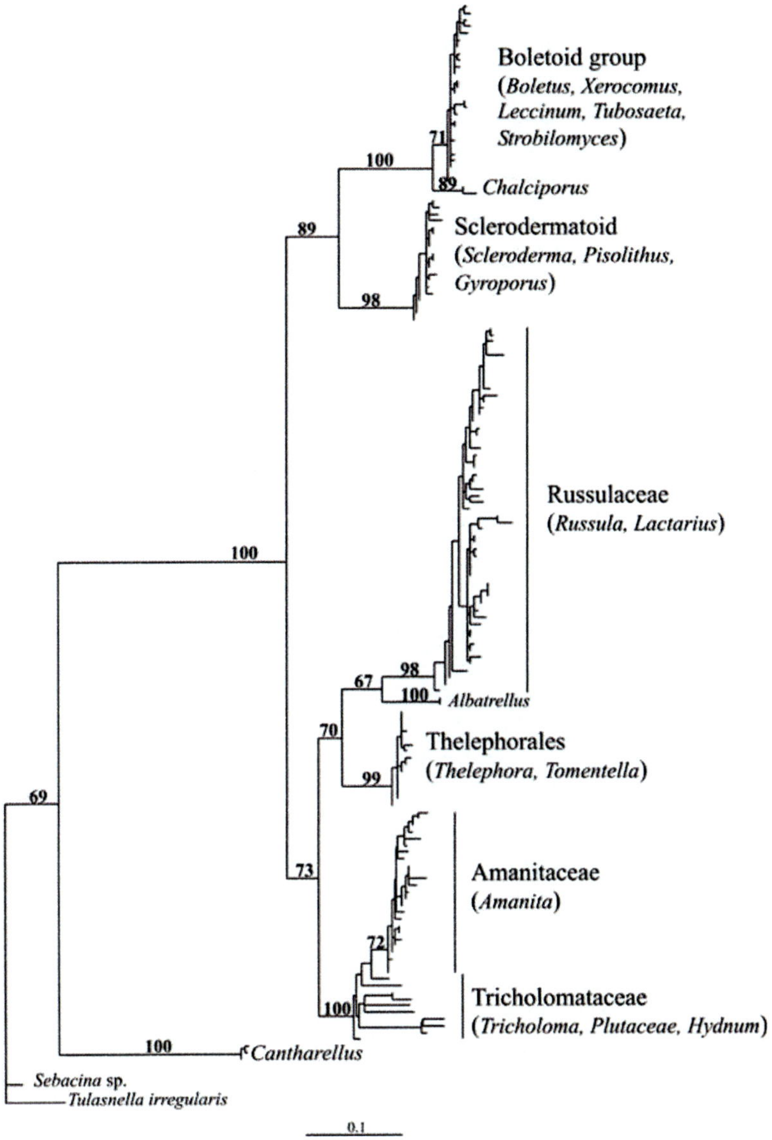

Fig. 4.1 Maximum likelihood ML5-ML6 tree using a HKY85 + γ + Inv model ($\alpha = 0.5684$, proportion of invariable sites = 0.3421, rate categories = 4) for 160 sequences and 389 sites. Bootstrap support values >50% are indicated at the relevant nodes. Main taxonomical groups included in the phylogeny are indicated. *Sebacina* sp. and *Tulanesnella irregularis* were chosen as outgroups according to Bruns et al. (1998)

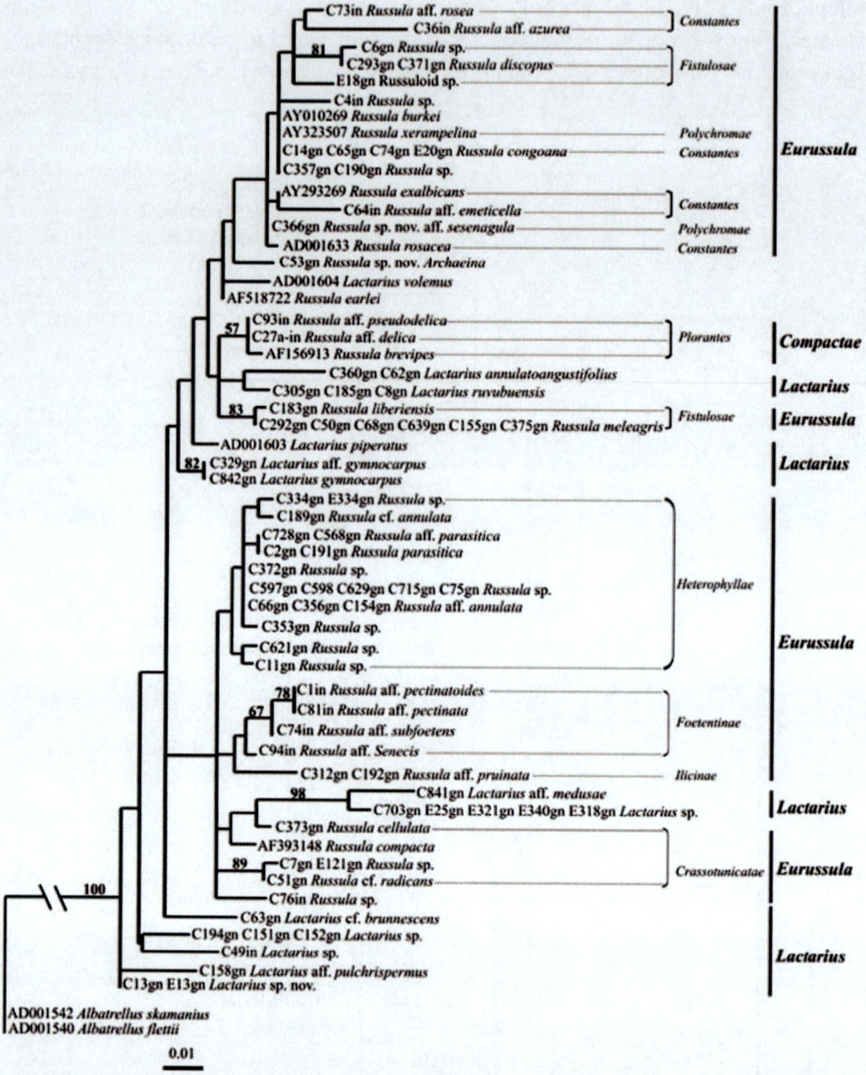

Fig. 4.2 Russulaceae maximum likelihood ML5-ML6, using a HKY85 + γ + Inv model ($\alpha = 0.5897$, proportion of invariable sites = 0.6539, rate categories = 4) for 57 different sequences and 391 sites. Bootstrap support values >50% are indicated at the relevant nodes. Identical sequences are included in the same terminal node. The *brackets* to the *right* of the tree indicate the clades including species of the same section and *vertical lines* indicate sections of the same subgenera (in *bold*). Names and grouping follow Singer (1986), Romagnesi (1985), and Miller and Buyck (2002) classifications. *gn* sample from Guinea, *in* Indian sample. Equality between numbers means perfect homology between their sequences

(Fig. 4.2), (2) boletoids, (3) *Scleroderma*, and (4) Amanitaceae plus Tricholomataceae groups (Rivière et al. 2007).

Among the 55 ectomycorrhizal sequences, a strict correspondence between sporocarps and ectomycorrhizas was obtained for only 27 of them. This is most probably due to the fact that many but not all ECM fungi produce fruitbodies, thus being observable macroscopically. By way of illustration, remarkable percentage of nonsporulating ECM species have been described by Jonsson et al. (1999a), where up to 70% of the ECM fungal species were only sampled from root tips without associated sporocarps in a *Pinus sylvestris* plantation. Moreover, sporocarp production also varied over the season, thus escaping sampling (Erland et al. 1999; Kõljalg et al. 2000). For instance, 20 ectomycorrhizas, but no sporocarp, were found to be closely related to thelephoroid taxa. It is well-known that sporocarps are seldom produced by Telephorales, but they commonly form mycorrhizas on roots of young trees (Kõljalg et al. 2000). Our study has once again confirmed that belowground fungal diversity is dissimilar from that of aboveground sporocarps (Gardes and Bruns 1996; Dahlberg et al. 1997; Jonsson et al. 1999a, b), thus underlying the importance of molecular analyses for the assessment of ECM fungal diversity.

The Russulaceae family, with 45 sequences, was the largest group sampled in our study. Russulaceae includes two genera *Lactarius* and *Russula*. *Lactarius* samples did not group together and appeared polyphyletic. At a lower taxonomical level, the *Eurussula* subgenus appeared to be polyphyletic, too, whereas the subgenus *Compactae* and each section *Heterophylla* or *Foetentinae* were found monophyletic. In the *Scleroderma* phylogeny, 32 sequences have been obtained including 16 ectomycorrhizas, which represent the highest mycorrhizas vs. sporocarps ratio in our sample. The boletoid group includes the genera *Boletus*, *Boletellus*, *Xerocomus*, *Leccinum*, *Tubosaeta*, and *Strobilomyces*. All *Leccinum* sequences as well as the two *Strobilomyces* sequences were grouped within a single clade. All other sequences were intermingled in the phylogeny, including mainly *Boletus* and *Xerocomus* species. The single *Chalciporus* sequence was located close to *Chalciporus piperatoides*, and roots the rest of the boletoid group *sensu stricto*. Finally, in the fourth analysis, all Amanitaceae cluster together in a highly supported monophyletic clade (Fig. 4.1), whereas Tricholomataceae (including Plutaceae, *Hydnum*, and *Tricholoma* genera) were paraphyletic.

Some differences were noted in the fungal diversity between the two sites, as well as between the tropical and more northern areas, as described in the literature. Russulaceae genotypes sequenced from the Guinean site, which include vouchered new species, are in accordance with the observations of Buyck (1994a, b, 1997), suggesting a high species diversity of this family in Africa. However, no species from India have been found to be common with those of Guinea. This lack of shared species (or sequences) between Africa and India suggests that there has been no recent gene flow between the two continents. Berbee and Taylor (1993, 2001) suggested that the divergence within ECM fungi occurred around 180 Mya ago, whereas separation of the Indian–Malagasy block from Africa is usually dated around 120 Mya. Based on morphological identifications, Russulaceae described in the Uppangala forest site may be related to known species from Europe

(Natarajan et al. 2005). However, none of the *Russula* or *Lactarius* species collected in India share the same sequence with already known African or European taxon. Enhanced molecular monitoring of tropical *Russula* diversity based on nuclear sequences is urgently needed to resolve such an interesting ectomycorrhizal group.

A single Boletaceae sample was recovered from India (identified as *Leccinum* sp.), whereas 17 genotypes were found in Guinea. Accordingly, Natarajan et al. (2005) described only two *Suillus* and one *Strobilomyces* species in India. In addition, very few Cortinariaceae specimens or species belonging to the Suilloid group have been found so far in tropical forests (e.g., *Cortinarius* in Cameroon and in India; Onguene and Kuyper 2001; Peintner et al. 2003; Natarajan et al. 2005). Our study supports this trend, as none of the sequences fell within Cortinariaceae or Suilloids available in Genbank database and used here as genetic benchmarks. Nevertheless, tropical forests are still undersampled relative to northern boreal forests, and further surveys are clearly needed to confirm this result.

In Guinea, a great diversity of sequences belonging to the Russulacae family was obtained. Morphological taxonomies of this family (Romagnesi 1985; Singer 1986; Buyck 1994a, b, 1997) are rarely congruent, reflecting the ambiguities of characters used in the classification of this group. In a similar way, some of the sporocarps collected were morphologically identified as already described species, but many others could not be linked to already known morphotypes. The latter may represent putative new species and further studies are required [species identification in progress, all specimens housed at the Natural History Museum, Paris and CASB (Centre of Advanced Study in Botany, Madras)]. So far, three specimens have been formally identified as new species, *Russula* sect. *Archaeinae* sp. nov. (C53), *Russula* sp. nov. aff. *sesenagula* (C366), and *Lactarius* nov. (C13). However, recognition of the remaining ones as new taxa would be premature at this stage.

Here, *Lactarius* appears as a paraphyletic group, whereas Shimono et al. (2004) support the monophyly of all *Lactarius* species based on LSU rDNA. Once again, this discrepancy is probably due to the low variability of the ML5/ML6 fragment at this level. Within the *Russula* genus, the subgenus *Eurussula* appears as a polyphyletic group, being split into four separated clades. This is not congruent with phylogeny based on nuclear regions (Eberhardt 2002; Miller and Buyck 2002; Eberhardt and Verbeken 2004). These differences can have several nonexclusive origins (1) various resolution levels of the markers used among studies, due to different rates and modes of molecular evolution, (2) complex relationships between nuclear and mitochondrial genomes, or (3) the reduced level of resolution at the species level of the locus used in our study (Doyle 1992; Bull et al. 1993; Bruns and Szaro 1992). It is known that mitochondrial genomes evolved at least partially independently from the nuclear genome, thus sometimes leading to incongruent phylogenetic inferences (Moncalvo et al. 2000). Other potential sources of incongruence between these genomes may be due to ancestral polymorphisms or horizontal transfers (Wall 2003). Such phenomena are not rare in plants and may obscure ECM Basidiomycete relationships as well. Unfortunately, too few molecular investigations have been performed so far to conclude (Hibbett et al. 2000;

Moncalvo et al. 2000; Binder and Hibbett 2002; Miller and Buyck 2002; den Bakker et al. 2004).

Tropical ECM fungi diversity found is inseparable from tree species diversity (Kabir et al. 2009), which is today endangered by strong human pressure that threatens tropical rainforests. The composition of ECM associations and the changes they undergo are still very poorly known in tropical regions and thus suggesting that further investigations are required of both above- and belowground species' composition over extended periods of time.

4.2 Genet Distribution of *Russula* sp. *foetentinae* in a Tropical Rainforest

4.2.1 Introduction

ECM fungi are symbiotic partners of most woody plants in forest ecosystems and have a crucial role in terms of nutritional transfer between soil and host plant roots (Smith and Read 1997). ECM symbiosis are characterized by bidirectional movement of nutrients where carbon flows to the fungal partner while inorganic nutrients move to the plant. The fungal vegetative network forms extensive mycelia that radiate from ECM fungal root tips to explore the soil for resources and permit the transfer of nutrients to associated plants (Finlay et al. 1998; Perez-Moreno and Read 2000). Together with sexual sporocarps, ECM fungi use mycelial spread to propagate and colonize new habitats. Therefore, the spatial distribution of individual mycelial systems is of considerable importance to understand ECM population dynamics in forest ecosystems (Dahlberg and Stenlid 1990).

Molecular techniques, such as those described above, make it possible to determine structure of fungal populations on a fine scale. Indeed, genotypes of fungal "individuals" using DNA-fingerprinting techniques are commonly used in population surveys (Burnett 2003). The concept of genet, a group of individuals of a given genotype, is a useful tool to understand not only the spatial distribution of the populations but also the dynamics of fungal succession of sequences (Smith et al. 1992). In general, genets can only be identified after isolating and further investigating the properties of each species. Somatic incompatibility is the technique traditionally used to resolve individual Basidiomycete genotypes (Dahlberg and Stenlid 1994), but it is not feasible for all ECM species (e.g., Russulaceae) because of its very limited growth in culture. Inter-simple sequence repeats (ISSRs) have therefore been used to characterize the genetic variation within fungal populations as it is a highly reproducible technique, which allows the detection of ECM fungal genets from sporocarps (Hantula et al. 1996; Anderson et al. 1998; Sawyer et al. 1999; Zhou et al. 1999). ISSR polymorphism is a useful tool to distinguish the otherwise morphologically indistinguishable individuals of fungi (Hantula et al.

1996). This technique was particularly appropriate in our study because the *Russula* genus contains a large number of species notorious for exhibiting high phenotypic plasticity (Miller and Buyck 2002). However, even when the individuals making up a population can be defined phenotypically or genotypically, their breeding behavior is rarely immediately obvious. In particular, the sampling of basidiocarps aboveground may not be an adequate estimate of the size, frequency, or spatial extension of genets belowground (Gardes and Bruns 1996; Jonsson et al. 1999a, b). In investigations of community structure above- and belowground for ECM fungi, the abundance of basidiocarps was not always indicative of the mycorrhizal morphotypes (Gardes and Bruns 1996; Dahlberg et al. 1997; Jonsson et al. 1999a). Because root tip or soil DNA extract analyses were not feasible under our field conditions, the information available on the distribution of basidiocarps was the only possible indicator of the presence and activity of individual mycelia.

The genet size of ECM fungal populations is assumed to vary with species identity, host forest age, and environmental conditions (Dahlberg and Stenlid 1995). For example, pioneer genera such as *Hebeloma* and *Laccaria* generally show numerous small (<3.5 m^2) and nonpersistent genets (Baar et al. 1994; Gryta et al. 1997). In contrast, fungi appearing late in succession, such as *Cortinarius rotundisporus*, spread primarily from hyphal networks, and their genets are large (up to 30 m) and temporally persistent (Sawyer et al. 1999). However, the overall picture is not that clear since certain genera such as *Suillus* may be found in disturbed areas and also occur in mature forests, thus following a mixed strategy (Dahlberg et al. 1997; Bonello et al. 1998). As a general principle, a high number of genotypes would be expected as a result of reproduction primarily by spores, whereas the formation of larger clones would be predicted if reproduction occurred primarily by mycelial expansion (Dahlberg and Stenlid 1990).

The Russulaceae are of particular interest for examination of the genetic structure and dynamics of late stage ECM populations. Despite the fact they are diverse and abundant in many types of forest ecosystems (Mason et al. 1987), the way they survive and spread in nature is still controversial. On the one hand, they are believed to be typical protagonists of the late stage field succession (Deacon and Fleming 1992; Keizer and Arnolds 1994) as they represent the majority of basidiocarps found in mature stands of temperate forest. Although little is known about the ability of Russulaceae to colonize tree seedlings in the field, laboratory studies indicate difficulties in germination from spores (Redecker et al. 2001). On the other hand, some studies have found small genets of *Russula* in late stage forests, suggesting that the role of sporulation in the life history of the Russulaceae growing on undisturbed forest may play a much more important role than previously recognized (Redecker et al. 2001; Bergemann and Miller 2002). Another interesting aspect of Russulaceae is that they are often dominant in tropical rainforests of Africa, Asia, and Madagascar (Buyck et al. 1996; Lee et al. 1997; Watling and Lee 1998). Once again, to our knowledge, genet distribution of ECM comes essentially from temperate ecosystems and virtually nothing is known for tropical rainforests.

4.2.2 Methods

In such a context, the aim of this study was to investigate the relative sizes of the individual genotype of *Russula* sp. *foetentinae* in a primary rainforest dominated by Dipterocarps species using ISSRs. To minimize the possibility of errors in assigning genotypes to genets (see Redecker et al. 2001), we quantified the probability of obtaining a given genotype by chance and calculated a Pairwise Jaccard's similarity among samples. The study site was located in a dense evergreen forest in the Kadamakal Reserve (12°30′N; 75°39′E). The annual rainfall reaches 5,200 mm with a marked dry season from December to March. The vegetation is dominated by *Vateria indica* L. (Diperocarpaceae), *Humboltia brunonis* Wall. (Fabaceae), *Myristica dactyloïdes* Gaertn. (Myristicaceae), and *Dipterocarpus indicus* Bedd. (Dipterocarpaceae). They represent more than 48% of the trees and 55% of the basal area (Pascal and Pélissier 1996) with pioneer species accounting for only 1.1% of the trees. Dominant ECM tree species include *V. indica* and *D. Indicus*. They represent 21% of the tree density (Pascal and Pélissier 1996). Basidiocarps of *Russula* sp. *foetentinae* were collected and mapped to the nearest 0.1 m within a study plot of 7,700 m^2 (110 m × 70 m), represented in Fig. 4.3.

DNA was extracted from each sample of basidiocarp. The molecular identification of each *Russula* sp. *foetentinae* sporocrap was confirmed using both nuclear and mitochondrial DNA fragments. The GenBank accession numbers for the sequences described here are DQ093423 and DQ093424. Finally, ISSR-PCR reactions were performed with three primers named ISSR1 (5′BDB (ACA)5), ISSR2 (5′DDB (CCA)5), and ISSR3 (5′DHB (CGA)5), where B is C, G, or T; D is A, G or T; H for a A, C or T (Hantula et al. 1996; for protocols see details in Rivière et al. 2005).

4.2.2.1 Spatial Randomness

To determine whether the pattern of basidiocarp distribution was random, aggregated, or regular, we used the first-order pair correlation function $G(r)$. This function gives the expected number of points at a distance r from an arbitrary point, divided by the intensity λ of the pattern (Diggle 1983; for details see Rivière et al. 2005)

4.2.2.2 Genotype Identification

In order to test for nonidentity due to somatic, miss-scored gels or reproducibility problems, we calculated Pairwise Jaccard's similarity (S_j) between genotypes on arcsin-transformed data using the following formula: $S_j = a/(a+b+c)<$; Where a is number of bands common to both genotypes and b and c the number of bands present in only one of the two genotypes. All statistical analyses were performed

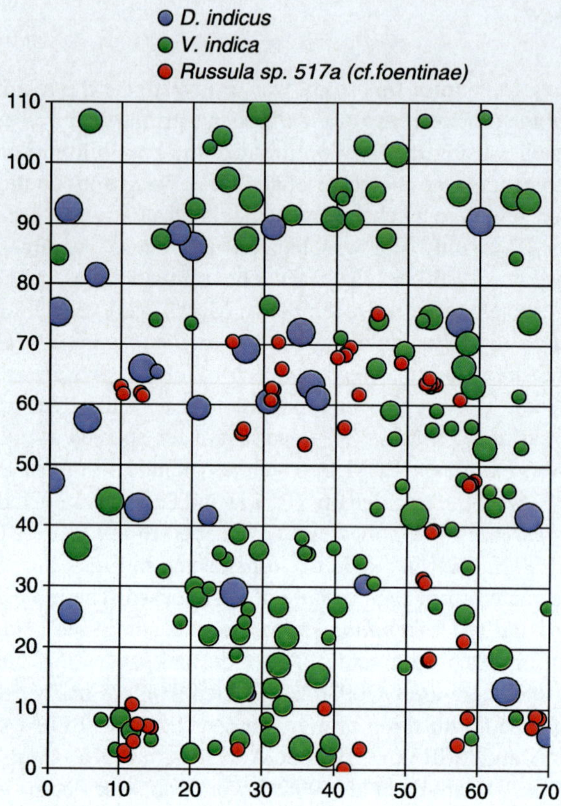

Fig. 4.3 Spatial distribution of *Russula* sp. *subfoetens* sporocarps, *Dipterocarpus indicus*, and *Vateria indica* within the study plot of 7,700 m^2 (110 m × 70 m). Tree diameters are indicated >200 cm; 100–200 cm, 50–100 cm; <50 cm

using R 2.0.1 (free software available online: http://www.r-project.org – The R Development Core Team). The histogram of the pairwise S_j was obtained using Grapher 4.00 (Golden software, Colorado).

4.2.3 Discussion

Forty five basidiocarps of *Russula* sp. *foetentinae* were mapped. Spatial analyses of the point pattern revealed a significant aggregation of the basidiocarps in a central zone of the study area, representing less than 30% of the total surface prospected. The first paired correlation function, $g(r)$, showed that 60% of all basidiocarps were located at a distance lower than 1 m from the nearest basidiocarp. To test the possibility of reproducibility errors in inferring genets from genotypes, we plotted all pairwise Jaccard similarity coefficients of genotypes for statistical outliers.

Our results showed a bell-shaped distribution, typical of randomly mating population (detail see Rivière et al. 2005). Identical genotypes were clearly separated from the right tail of the distribution. Cluster analyses on the 45 ISSR patterns revealed a total of 29 different genotypes with 18 genotypes represented by only one basidiocarp and 11 genotypes represented by two to three basidiocarps. Basidiocarps belonging to the same genet were not necessarily collected on the same day. The eleven detected genets could be separated into two size classes (Fig. 4.4). Six genets were large with a diameter ranging from 31 m (genet J) to 70 m (genet D). The five other genets (A, C, G, I, and K) were of smaller size with diameters ranging from 0.5 m (Genet I) to 5 m (Genet G). ISSRs revealed the presence of large *Russula* sp. *foetentinae* mycelial individuals in the mature Dipterocarps stand we studied. The largest individual comprises three sporocarps, two of which were situated 70 m from each other. We are aware that caution is needed in the interpretation of this

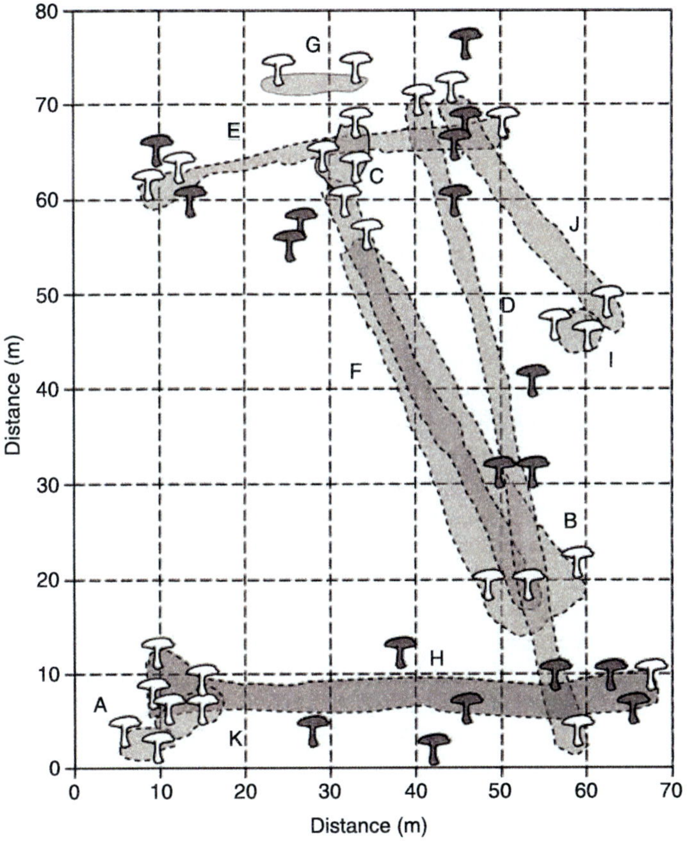

Fig. 4.4 Schematic relative positions and arbitrary sizes of *Russula subsect. foetentinae* genotypes plotted on a portion of the Uppangala study site (5,600 m^2). Each individual basidiocarp is identified with an *icon* (*open*, sporocarps belonging to a genet; *full*, individual basidiocarps). *Dotted lines* represent arbitrary surfaces of the separate genets

result. First, the size estimates are based only on the presence of aboveground fruiting bodies. Second, no connecting sporocarps were found over this distance, leaving the possibility of an error during handling of the specimens. Third, the action of fungivores or physical constraints of soil may have separated ECM mycelia into several genetically identical genotypes or ramets (Dahlberg and Stenlid 1995; Griffiths et al. 1996). It is however not possible from our data to assess the extent to which fragmentation has occurred within the mycelial individuals at the site. Despite all these potential limitations, we found five other large genets (>30 m) in the study site, which suggests the capacity of *Russula* sp. *foetentinae* individuals to spread over a long distance.

Russulaceae species are usually believed to spread via vegetative reproduction and to form relatively small genets; e.g., *Russula cremolicolor* (1.1 m^2, Redecker et al. 2001), *Russula brevipes* (3–18 m, Bergemann and Miller 2002), *Russula vinosa* (<1 m, Liang et al. 2004). Our data indicate genets that are much larger than previously described for other Russulaceae species, but the presence of large genets is not uncommon among ECM fungal species. Large genets have been found for *Suillus bovinus* and *S. variegatus* (40 m, Dahlberg and Stenlid 1995; Dahlberg 1997), *S. pungens* (40 m, Bonello et al. 1998), *Pisolithus tinctorius* (30 m, Anderson et al. 1998), and *Xerocomus chrysenteron* (110 m, Fiore-Donno and Martin 2001). Comparison with data from the literature is however limited as the size and extent of the mycelial phase can differ between genera and species or between different genets of the same species (Bonello et al. 1998; Redecker et al. 2001; Bergemann and Miller 2002). The presence of large genets may suggest that *Russula* sp. *foetentinae* can colonize by mycelial expansion and may be indicative of more mature mycelial systems that have grown from point sources of individual mating events over a long period (Dahlberg and Stenlid 1990). Because the spatial extent of genets has been correlated with age of host stands (Dahlberg and Stenlid 1994), our main hypothesis to explain the presence of large genets of *Russula* sp. *foetentinae* would be the absence of disturbance over a long period of time in the studied primary forest (>100 years, Loffeier 1989). However, the presence of smaller genets for 18 basidiocarps suggests that *Russula* sp. *foetentinae* could also spread via sexual reproduction of basidiocarps.

Field knowledge is lacking on the ecology of tropical ECM symbiosis, particularly in primary rainforest ecosystems. To date, most studies on ECM fungi in such ecosystems have consisted in species inventories (Buyck et al. 1996; Béreau et al. 1997; Lee et al. 1997). There is little information about life history strategies of tropical ECM fungi and only recently, Onguene and Kuyper (2001) revealed in a tropical forest of Cameroon that ECM fungal mycelium might form important networks acting as "nursery zones" for young trees. Our results give preliminary information on reproductive strategies of *Russula* sp. *foetentinae*, which may have important implications in terms of conservation of Asiatic primary rainforests. Dipterocarps are one of the most important timber species of tropical rain forest in Southeast Asia and are mostly ECM. Furthermore, they are mainly associated with Russulaceae and Amanitaceae species (Alexander and Högberg 1986; Watling and Lee 1998). The failure of dipterocarp regeneration in logged forests has been related to inadequate or inefficient ECM formations (Smits 1983). The latter have further been proved to enhance dipterocarps seedling growth both in nursery and in natural habitats (Lee and Alexander 1994; Lee et al. 1997).

This study represents a first step in our understanding of ECM fungal population dynamics in tropical primary forest ecosystems but only a snap shot of the genet distribution of *Russula* sp. *foetentinae*. Our results suggest that mature stands may shelter well-spread underground mycelium and may be crucial for durable interaction with the plant partner. Although the consistency of these results has to be confirmed on additional sites and over several years, they could be of particular importance in the light of current destruction of tropical forest or degradation into secondary stands. Further studies on the temporal persistence of these large genets and on the consequences of human-induced changes on the dynamics of fungal populations are urgently needed.

4.3 Conclusion

Nowadays, few field studies looking at ECM species composition and communities are conducted in tropical rainforests, mainly due to the sites' inaccessibility. So far, the only method to link an ECM root tip to a tree has been to physically trace it back to the stem. Due to molecular methods, it is now feasible to extract fungal and plant DNA from a single root tip, which allow scientists to unambiguously assign a fungal species to its specific plant host. Furthermore, molecular fingerprinting methods are routinely used to identify individual host tree from ECM root tips in addition to identifying the ECM fungus – from a single DNA extraction. Such efficient methods are of great help to investigate a new fine scale in the mycorrhizal ecology: it is now possible to describe the ECM communities between individual trees.

The first study indicates that there is still an unknown diversity of ECM fungi in tropical rainforests, including new species. It also points to the fact that further investigations are urgently required of both above- and belowground diversity over extended periods of time and over larger areas of forest biomes. The second objective was to evaluate the spatial distribution of individual mycelial systems in an undisturbed tropical rainforest. The genets identified reveal an underlying vast mycelium network. Due to unsustainable logging of tropical rainforests, the influence of destruction of such ECM network, upon subsequent forest restoration, must be seriously considered.

References

Alexander IJ, Högberg P (1986) Ectomycorrhizas of tropical angiospermous trees. New Phytol 102:541–549

Anderson IC, Chambers SM, Cairney JWG (1998) Use of molecular methods to estimate the size and distribution of mycelial individuals of the ectomycorrhizal basidiomycete *Pisolithus tinctorius*. Mycol Res 102:295–300

Andrade ACS, Queiroz MH, Hermes RAL, Oliveira VL (2000) Mycorrhizal status of some plants of the Araucaria forest and the Atlantic rainforest in Santa Catarina, Brazil. Mycorrhiza 10(3):131–136

Baar J, Ozinga WA, Kuyper TW (1994) Spatial distribution of *Laccaria bicolor* genets reflected by basidiocarps after removal of litter and humus layers in a *Pinus sylvestris* forest. Mycol Res 98:726–728

Berbee ML, Taylor JW (1993) Dating the evolutionary radiations of the true fungi. Can J Bot 71:1114–1127

Berbee ML, Taylor JW (2001) Fungal molecular evolution: gene trees and geologic time. In: McLaughlin DJ, McLaughlin RG, Lemke PA (eds) The Mycota VII. Part B. Systematics and evolution. Springer-Verlag, Berlin, pp 229–245

Béreau M, Gazel M, Garbaye J (1997) Mycorrhizal symbiosis in trees of the tropical rainforest of French Guiana. Can J Bot 75:711–716

Bergemann SE, Miller SL (2002) Size, distribution and persistence of genets in local populations of the late-stage ectomycorrhizal basidiomycete, *Russula brevipes*. New Phytol 156:313–320

Binder M, Hibbett DS (2002) Higher-level phylogenetic relationships of Homobasidiomycetes (Mushroom-forming fungi) inferred from four rDNA regions. Mol Phyl Evol 22:76–90

Bonello P, Bruns TD, Gardes M (1998) Genetic structure of a natural population of the ectomycorrhizal fungus *Suillus pungens*. New Phytol 138:533–542

Bruns TD, Szaro TM (1992) Rate and mode differences between nuclear and mitochondrial small-subunit rRNA genes in mushrooms. Mol Biol Evol 9:836–855

Bruns TD, Szaro TM, Gardes M, Cullings KW, Pan JJ, Taylor DL, Horton TR, Kretzer A, Garbelotto M, Li Y (1998) A sequence database for the identification of ectomycorrhizal Basidiomycetes by phylogenetic analysis. Mol Ecol 7(3):257–272

Bull JJ, Huelsenbeck PP, Cunningham CW, Swofford DL, Waddell PJ (1993) Partitioning and combining data in phylogenetic analysis. Sys Biol 42:384–397

Burnett J (2003) Fungal populations and species. Oxford University Press, Oxford, pp 1–368

Buyck B (1994a) *Russula* I (Russulaceae). Flore illustrée des champignons d'Afrique centrale 15:335–408

Buyck B (1994b) *Russula* II (Russulaceae). Flore illustrée des champignons d'Afrique centrale 16:411–542

Buyck B (1997) *Russula* III (Russulaceae). Flore illustrée des champignons d'Afrique centrale 17:545–597

Buyck B, Thoen D, Walting R (1996) Ectomycorrhizal fungi of the Guinea-Congo Region. Proc R Soc Edinb B 104:313–333

Dahlberg A (1997) Population ecology of *Suillus variegatus* in old Swedish Scots pine forests. Mycol Res 107:47–54

Dahlberg A, Stenlid J (1990) Population structure and dynamics in *Suillus bovinus* as indicated by spatial distribution of fungal clones. New Phytol 115:487–493

Dahlberg A, Stenlid J (1994) Size, distribution and biomass of genets in populations of *Suillus bovinus* (L.: Fr.) Roussel revealed by somatic incompatibility. New Phytol 128:225–234

Dahlberg A, Stenlid J (1995) Spatiotemporal patterns in ectomycorrhizal populations. Can J Bot 73(Suppl 1):1222–1230

Dahlberg A, Jonsson L, Nylund JE (1997) Species diversity and distribution of biomas above and below-ground among ectomycorrhizal fungi in an old-growth Norway spruce forest in South Sweden. Can J Bot 75:1323–1335

Deacon JW, Fleming LV (1992) Interactions of ectomycorrhizal fungi. In: Allen MF (ed) Mycorrhizal functionning: an integrative plant-fungal process. Chapman and Hall, New York, pp 249–300

Debaud JC, Marmeisse R, Gay G (1999) Intraspecific genetic variation and populations of ecomycorrhizal fungi. In: Varma AK, Hock B (eds) Mycorrhiza: structure, molecular biology and function. Springer-Verlag, Berlin, pp 75–110

Den Bakker HC, Zuccarello GC, Kuyper THW, Noordeloos ME (2004) Evolution and host specificity in the ectomycorrhizal genus *Leccinum*. New Phytol 163:201–215

Diggle PJ (1983) Statistical analysis of spatial point patterns. Academic Press, London, pp 1–444

Doyle JJ (1992) Gene trees and species trees: molecular systematics as one-character taxonomy. Sys Bot 17:144–163

Eberhardt U (2002) Molecular analyses of the agaricoid *Russulaceae*: correspondence with mycorrhizal and sporocarp features in the genus *Russula*. Mycol Prog 1(2):201–224

Eberhardt U, Verbeken A (2004) Sequestrate *Lactarius* species from tropical Africa: *L. angiocarpus* sp. nov. and *L. dolichocaulis* comb. nov. Mycol Res 108:1042–1052

Egger KN (1995) Molecular analysis of ectomycorrhizal fungal communities. Can J Bot 73:S1415–S1422

Erland S, Jonsson T, Mahmood S, Finlay RD (1999) Below-ground ectomycorrhizal community structure in tow Picea abies forests in southern Sweden. Scand J For Res 14:209–217

Finlay RD, Ek H, Odham G, Söderström B (1998) Mycelial uptake, translocation and assimilation of nitrogen from ^{15}N-labelled ammonium by *Pinus sylvestris* plants infected with four different ectomycorrhizal fungi. New Phytol 110:59–66

Fiore-Donno AM, Martin F (2001) Populations of ectomycorrhizal *Laccaria amesthystina* and *Xerocomus* spp. show contrasting colonization patterns in a mixed forest. New Phytol 152:533–542

Gardes M, Bruns TD (1993) ITS primers with enhanced specificity for Basidiomycetes – application for the identification of mycorrhizas and rusts. Mol Ecol 2:113–118

Gardes M, Bruns TD (1996) Community structure of ectomycorrhizal fungi in a *Pinus muricata* forest: above and below-ground views. Can J Bot 74:1572–1583

Gardes M, White TJ, Fortin J, Bruns TD, Taylor JW (1991) Identification of indigenous and introduced symbiotic fungi in ectomycorrhizas by amplification of nuclear and mitochondrial ribosomal DNA. Can J Bot 69:180–190

Griffiths RP, Bradshaw GA, Marks B, Lienkaemper GW (1996) Spatial distribution of ectomycorrhizal mats in coniferous forests of the Pacific Northwest, USA. Plant Soil 180:147–158

Grogan P, Baar J, Bruns TD (2000) Below-ground ectomycorrhizal community structure in a recently burned bishop pine forest. J Ecol 88:1051–1062

Gryta H, Bebaud JC, Effosse A, Gay G, Marmeisse R (1997) Fine scale structure of populations of the ectomycorrhizal fungus *Hebeloma cylindrosporum* in coastal sand dune ecosystems. Mol Ecol 6:353–364

Hantula J, Dusabenyagasani M, Hamelin RC (1996) Random Amplified Microsatellites (RAMS), a novel method for characterizing genetic variation within fungi. Eur J Path 26:159–166

Henkel TW, Terborgh J, Vilgalys RJ (2002) Ectomycorrhizal fungi and their leguminous hosts in the Pakaraima Mountains of Guyana. Mycol Res 106:515–531

Hibbett DS, Gilbert LB, Donoghue MJ (2000) Evolutionary instability of ectomycorrhizal symbioses in Basidiomycetes. Nature 407:506–508

Horton TR, Bruns TD (2001) The molecular revolution in ectomycorrhizal ecology: peeking into the black-box. Mol Ecol 10:1855–1871

Jonsson L, Dahlberg A, Nilsson M, Zackrisson O, Karen O (1999a) Ectomycorrhizal fungal communities in late-successional Swedish boreal forests, and their composition following wildfire. Mol Ecol 8:205–215

Jonsson L, Kokalj S, Finlay R, Erland S (1999b) Ectomycorrhizal community structure in a limed spruce forest. Mycol Res 103:501–508

Jonsson L, Dahlberg A, Brandrud TE (2000) Spatiotemporal distribution of an ectomycorrhizal community in an oligotrophic Swedish *Picea abies* forest subjected to experimental nitrogen addition: above and below-ground views. For Ecol Man 132:143–156

Keizer PJ, Arnolds E (1994) Succession of ectomycorrhizal fungi in roadside verges planted with common oak (*Quercus robur L.*) in Drenthe, The Netherlands. Mycorrhiza 4:147–159

Kendall JM (2007) Introduction to molecular analysis of ectomycorrhizal communities. Invited mini-review for the molecular based approaches to soil microbiology symposium, vol 71, pp 601–610

Koide RT, Xu B, Sharda J (2005) Contrasting below-ground views of an ectomycorrhizal fungal community. New Phytol 166:251–262

Kõljalg U, Dahlberg A, Taylor AFS, Larsson E, Hallenberg N, Stenlid J, Larsson KH, Fransson PM, Kåren O, Jonsson L (2000) Diversity and abundance of resupinate thelephoroid fungi as ectomycorrhizal symbionts in Swedish boreal forests. Mol Ecol 9:1985–1996

Kretzer AM, Dunham S, Molina R, Spatafora JW (2003) Microsatellite markers reveal the below ground distribution of genets in two species of *Rhizopogon* forming tuberculate ectomycorrhizas on Douglas fir. New Phytol 161:313–320

Landeweert R, Leeflang P, Kuyper TW, Hoffland E, Rosling A, Wernars K, Smit E (2003) Molecular identification of ectomycorrhizal mycelium in soil horizons. Appl Environ Microbiol 69:327–333

Lee SS, Alexander IJ (1994) The response of seedlings of two dipterocarp species to nutrient additions and ectomycorrhizal infection. Plant Soil 163:299–306

Lee SS, Alexander IJ, Watling R (1997) Ectomycorrhizas and putative ectomycorrhizal fungi of *Shorea leprosula* Miq. (Dipterocarpaceae). Mycorrhiza 7:63–81

Liang Yu, Guo Liang-dong, Ma Ke-ping (2004) Genetic structure of a population of the ectomycorrhizal fungus *Russula vinosa* in subtropical woodlands in southwest China. Mycorrhiza 14:235–240

Loffeier ME (1989) Sylviculture et sylvigénèse en forêt sempervirente du Coorg (sud-ouest de l'Inde). Travaux de la section Scientifique et Technique, Tome XXVI, Institut Français de Pondichéry, Inde.1–21

Mason PA, Last FT, Wilson J, Deacon JW, Fleming LV, Fox FM (1987) Fruiting and succession of ectomycorrhizal fungi. In: Pegg GF, Ayres PG (eds) Fungal infection of plant. Cambridge University Press, Cambridge, pp 253–268

Maury-Lechon G, Curtet L (1998) Biogeography and evolutionary systematics of Dipterocarpaceae. In: Appanah S, TurnbullU MJ (eds) A review of dipterocarps taxonomy, ecology and silviculture. CIFOR, Jakarta, pp 5–44

Miller SL, Buyck B (2002) Molecular phylogeny of the genus *Russula* in Europe with a comparison of modern infrageneric classifications. Mycol Res 106:259–276

Moncalvo JM, Drehmel D, Vilgalys R (2000) Variation in modes and rates of evolution in nuclear and mitochondrial ribosomal DNA in the mushroom genus *Amanita* (Agaricales, Basidiomycota): phylogenetic implications. Mol Phyl Evol 16:48–63

Natarajan K, Senthilarasu G, Kumaresan V, Rivière T (2005) Diversity in ectomycorrhizal fungi of a dipterocarp forest in Western Ghats. Curr Sci 88(12):1893–1895

Newbery DM, Alexander IJ, Rother LA (1997) Phosphorus dynamics in lowland African rainforest: the influence of ectomycorrhizal trees. Ecol Monogr 67:367–409

Onguene NA, Kuyper TW (2001) Mycorrhizal associations in the rainforest of South Cameroon. For Ecol Man 140:277–287

Pascal JP (1984) Les forêts denses humides sempervirentes des Ghâts Occidentaux de l'Inde. Tome XX, travaux de la section scientifique et technique, Institut Français de Pondichéry, Pondicherry, pp 1–365

Pascal JP, Pélissier R (1996) Structure and floristic composition of a tropical evergreen forest in south-west India. J Trop Ecol 12:191–211

Peay KG, Kennedy PG, Davies SJ, Tan S, Bruns TD (2010) Potential link between plant and fungal distributions in a dipterocarp rainforest: community and phylogenetic structure of tropical ectomycorrhizal fungi across a plant and soil écotone. New phytol 182(2):529–542

Peintner U, Moser MM, Thomas KA, Manimohan P (2003) First records of ectomycorrhizal *Cortinarius* species (Agaricales, Basidiomycetes) from tropical India and their phylogenetic position based on rDNA ITS sequences. Mycol Res 107(4):485–494

Pélissier R, Pascal JP, Houllier F, Laborde H (1998) Impact of selective logging on the dynamics of a low elevation dense moist evergreen forest in the Western Ghats (South India). For Ecol Man 105:107–119

Perez-Moreno J, Read DJ (2000) Mobilization and transfer of nutrients from litter to tree seedlings via the vegetative mycelium of ectomycorrhizal plants. New Phytol 145:301–309

Picone C (2000) Diversity and abundance of arbuscular-mycorrhizal fungus spores in tropical forest and pasture. Biotropica 32(4a):734–750

Redecker D, Szaro TM, Bowman RJ, Bruns TD (2001) Small genets of *Lactarius xanthogalactus*, *Russula cremoricolor* and *Amanita francheti* in late-stage ectomycorrhizal successions. Mol Ecol 10:1025–1034

Redhead JF (1980) Mycorrhiza in natural tropical forests. In: Mikola P (ed) Tropical mycorrhiza research. Clarendon Press, Oxford, pp 127–142

Rivière T (2004) Biodiversity, molecular ecology and phylogeography of tropical ectomycorrhizal symbiosis. PhD Université de Montpellier II, France, pp 1–256

Rivière T, Senthilarasug V, Kumaresan V, Natarajan K, Dreyfus B (2005) Spatial distribution of ectomycorrhizal Basidiomycete *Russula* sp. *foetentinae* populations in a primary Dipterocarp rainforest. Mycorrhiza 7:124–127

Rivière T, Senthilarasug V, Kumaresan V, Natarajan K, Dreyfus B (2007) Genetic diversity of ectomycorrhizal Basidiomycetes from African and Indian tropical rainforests. Mycorrhiza 17:415–428

Romagnesi H (1985) Les Russules d'Europe et d'Afrique du Nord. Reprint with supplement. J. Cramer, Lehre

Sanon KB, Bâ AM, Dexheimer J (1997) Mycorrhizal status of some fungi fruiting beneath indigenous trees in Burkina Faso. For Ecol Man 98:61–69

Sawyer NA, Chambers SM, Cairney JWG (1999) Molecular investigation of genet distribution and genetic variation of *Cortinarius rotundisporus* in eastern Australian sclerophyll forests. New Phytol 142:561–568

Shimono Y, Kato M, Takamatsu S (2004) Molecular phylogeny of *Russulaceae* (Basidiomycetes; Russulales) from the nucleotide sequences of nuclear large subunit rDNA. Mycoscience 45:303–316

Singer R (1986) The Agaricales in modern taxonomy, 4th edn. Koeltz Scientific Books, Koeningstein, pp 1–981

Singer R, Araujo IJS (1979) Litter decomposition and ectomycorrhiza in Amazonian forests: 1. A comparison of litter decomposition and ectomycorrhizal Basidiomycetes in latosol-terra firme rainforest and white podzol Campinarana. Acta Amaz 9:25–41

Smith SE, Read DJ (1997) Mycorrhizal symbiosis. Academic Press, San Diego, CA

Smith ML, Bruhn JN, Anderson JB (1992) The fungus *Armillaria bulbosa* is among the largest and oldest living organisms. Nature 356:428–431

Smits WTM (1983) Dipterocarps and mycorrhiza, an ecological adapation and a factor in forest regeneration. Flora Mal Bul 36:3926–3937

Smits WTM (1992) Mycorrhizal studies in dipterocarp forests in Indonesia. In: Read DJ, Lewis DH, Fitter AH, Alexander IJ (eds) Mycorrhizas in ecosystems. Cambridge University Press, Cambridge, pp 283–292

Thoen D, Bâ AM (1989) Ectomycorrhizas and putative ectomycorrhizal fungi of *Afzelia africana* Sm. and *Uapaca guineensis* Müll. Arg. in southern Senegal. New Phytol 113:549–559

Thomazini LI (1974) Mycorrhiza in plants of the "Cerrado". Plant Soil 41(3):707–711

Wall JD (2003) Estimating ancestral population sizes and divergence times. Genetics 163:395–404

Watling R, Lee SS (1998) Ectomycorrhizal fungi associated with members of the Dipterocarpaceae in Peninsular Malaysia. J Trop For Sci 10:421–430

Zhou Z, Miwa M, Hogetsu T (1999) Analysis of genetic structure of a *Suillus grevillei* population in a *Larix kaempferi* stand by polymorphism of inter-simple sequence repeat (ISSR). New Phytol 144:55–63

Chapter 5
The Molecular Ectomycorrhizal Fungus Essence in Association: A Review of Differentially Expressed Fungal Genes During Symbiosis Formation

Bartolomeu Acioli-Santos, Helder Elísio E. Vieira, Cláudia E.P. Lima, and Leonor C. Maia

5.1 Introduction

The formation of ectomycorrhiza is a complex process governed by biochemical and molecular interactions that start before the physical contact between potential partners. This involves intense gene expression control in both partners that leads to drastic morphological and physiological changes that are crucial to the development of mutualism and symbiotic harmony.

The purpose of this chapter was to review the main genes (Expressed Sequence Tags-ESTs, mRNAs or proteins) from ectomycorrhizal fungi (EMF) expressed in association with roots or under mycorrhizal stimulus. It was possible to group a large number of these genes/proteins into different physiological categories according to probable functions within symbiotic associations. For data collection, researchers looked at different EMF and host interactions, screened different experimental conditions and techniques (microarray, DDRT-PCR, subtractive suppression hybridization techniques, and other approaches). The data were analyzed and discussed under current microbiological views.

In total 389 ESTs/proteins were listed from different fungal genus and species, specifically in *Pisolithus*, *Paxillus*, *Tuber*, *Suillus* and *Laccaria*. Ectomycorrhizal genes were grouped in eight physiological categories as follows: cell growth and

B. Acioli-Santos (✉)
Departamento de Micologia, Universidade Federal de Pernambuco, Av. Prof. Nelson Chaves, s/n, Cidade Universitária, Recife, PE 50670-420, Brasil
and
Departamento de Virologia e Terapia Experimental-LaViTE, Centro de Pesquisas Aggeu Magalhães, Fundação Instituto Oswaldo Cruz, Recife, PE 50670-420, Brasil
e-mail: bartacioli@cpqam.fiocruz.br; bartacioli@yahoo.com

H.E.E. Vieira, C.E.P. Lima, and L.C. Maia
Departamento de Micologia, Universidade Federal de Pernambuco, Av. Prof. Nelson Chaves, s/n, Cidade Universitária, Recife, PE 50670-420, Brasil

M. Rai and A. Varma (eds.), *Diversity and Biotechnology of Ectomycorrhizae*, Soil Biology 25, DOI 10.1007/978-3-642-15196-5_5,

organization, morphogenesis, energy metabolism, protein synthesis and protein interaction, transcriptional and translation regulation, ion transport, amino acid and peptide transport, cell signaling and structural membrane proteins, RNA/DNA processing and cell defense and apoptosis.

5.2 Ectomycorrhizal Genes Differentially Expressed During Symbiosis Formation

5.2.1 Growth and Cell Organization Genes

Ectomycorrhiza formation requires intense cell activity that occurs both before and during physical contact between potential partners. The class of growth and cell organization genes encompassed 51 ESTs/proteins (Table 5.1).

The α-tubulin and actin proteins/ESTs were identified in *Suillus variegatus*, *Paxillus involutus, Tuber borchii* and *Pisolithus microcarpus* mycelia under association. Studies revealed the α-tubulin, actin and associated proteins (actin-binding proteins) were active in symbiotic tissues (Timonen and Peterson 2002). Also the centractin-like protein, which functions similar to the tubulin (Menotta et al. 2004), was isolated from *T. borchii–Tilia americana* mycelium at 30 days of development. Expression of centractin is similar to the increased expression of E-MAP 115 (microtubule-associated protein) in *Laccaria bicolor–Pinus resinosa* mycelium between 6 and 72 h of contact (Podila et al. 2002). This protein is necessary to the microtubule complex, cytoskeleton formation, and cell polarity maintenance in epithelial cells (Masson and Kreis 1993). Likewise, the *GAS 2*-homologue (observed in 30-day-old associated *T. borchii* mycelium) has direct participation in the reorganization of cytoskeleton and mammal cell exponential growth (Manzow et al. 1996). The enzyme apparatus that supports these cytoskeleton proteins is exemplified by the expression of Rho family mRNAs observed in *P. involutus–Betula pendula* (repressed at 25 days), *T. borchii–Tilia americana* (stimulated at 30 days) and *P. microcarpus–Eucalyptus globulus* (up-regulated at 12 days) mycelium. Rho family members are involved in molecular signaling processes prior to cell fission in yeast (Hirata et al. 1998; Imai et al. 2002).

The transcription of elongation factor 1α (translation elongation factor 1α) promotes ribosome/tRNA binding and regulates growing peptide fidelity. This molecule was observed in quick-growing cell cultures and is directly related to cell growth and proliferation in diverse eukaryotes (Condeelis 1995). Increased transcription of elongation factor 1α was observed in 4-day-old *Pisolithus tinctorius–Eucalyptus globulus* mycelium (Voiblet et al. 2001).

Positive expression of *Sur4*, *P300/CBP* and septin transcripts at 6–72 h in *L. bicolor–Pinus resinosa* mycelium produces proteins important for bipolar growth in yeast (Zahner et al. 1996), proliferation/cell differentiation (Chan and La Thangue 2001), and cytokinesis (Tasto et al. 2003). The 26S proteasome

Table 5.1 Some growth and cell organization genes differentially expressed in ectomycorrizal fungus

Gene	Ectomycorrhiza	Time/gene expression tendency	Reference
26 s Proteasome regulatory subunit mts4	*Tuber borchii–Tilia americana*	30d↑	Menotta et al. (2004)
Actin	*Suillus variegatus–Pinus contorta*	1–60d↑	Timonen et al. (1996)
Actin	*Paxillus involutus–Pinus contorta*	1–60d↑	Timonen et al. (1996)
Actin	*Paxillus involutus–Betula pendula*	2–8d↓	Le Quéré et al. (2005)
Actin	*Pisolithus microcarpus–Eucalyptus globulus*	4–7↓ 12↑ 21d↓	Duplessis et al. (2005)[a]
Actin binding protein	*Pisolithus microcarpus–Eucalyptus globulus*	4–7↓ 12↑ 21d↓	Duplessis et al. (2005)[a]
Actin cytoskeleton-assiociated	*Tuber borchii–Tilia americana*	30d↑	Menotta et al. (2004)
Actin-binding protein Sop2/arp2/Arp3	*Pisolithus tinctorius–Eucalypthus globulus*	4d↑	Voiblet et al. (2001)
Actin-like protein 3	*Pisolithus microcarpus–Eucalyptus globulus*	4–7↓ 12↑ 21d↓	Duplessis et al. (2005)[a]
Actin-organizing complex	*Paxillus involutus–Betula pendula*	2↓14d↓	Le Quéré et al. (2005)
Alpha tubulin	*Paxillus involutus–Betula pendula*	25d↑	Johansson et al. (2004)
Alpha tubulin	*Paxillus involutus–Betula pendula*	2–8d↓	Le Quéré et al. (2005)
Alpha tubulin	*Suillus variegatus–Pinus contorta*	1–60d↑	Timonen et al. (1996)
Alpha tubulin	*Paxillus involutus–Pinus contorta*	1–60d↑	Timonen et al. (1996)
Alpha tubulin A	*Paxillus involutus–Betula pendula*	2↓14d↓	Le Quéré et al. (2005)
Centractin-like protein	*Tuber borchii–Tilia americana*	30d↑	Menotta et al. (2004)
Chitin synthase I	*Laccaria bicolor–Pinus resinosa*	6–72h↑	Podila et al. (2002)
Cofilin	*Paxillus involutus–Betula pendula*	2–8d↓	Le Quéré et al. (2005)
Cornichon homolog	*Pisolithus microcarpus–Eucalyptus globulus*	4–7↓ 12↑ 21d↓	Duplessis et al. (2005)[a]
Cyclophilin	*Paxillus involutus–Betula pendula*	2–8↓14d↑	Le Quéré et al. (2005)
E-MAP 115	*Laccaria bicolor–Pinus resinosa*	6–72h↑	Podila et al. (2002)

(*continued*)

Table 5.1 (continued)

Gene	Ectomycorrhiza	Time/gene expression tendency	Reference
Endopolygalacturonase 1 (pectinase)	*Terfezia boudieri–Cistus incanus*	4w↑	Zaretsky et al. (2006)
Eukariotic initiation factor eif-3p66	*Pisolithus microcarpus–Eucalyptus globulus*	4–7↓ 12↑ 21d↓	Duplessis et al. (2005)[a]
GAS-2 homologue	*Tuber borchii–Tilia americana*	30d↑	Menotta et al. (2004)
Histidine-rich protein	*Tuber borchii–Tilia americana*	30d↑	Menotta et al. (2004)
Lbras	*Laccaria bicolor–Pinus resinosa*	48h↑	Sundaram et al. (2001)
Lis1 homolog	*Pisolithus microcarpus–Eucalyptus globulus*	4↓ 7↑ 12–21d↓	Duplessis et al. (2005)[a]
Matin-type switching protein	*Paxillus involutus–Betula pendula*	2–8↑14d↓	Le Quéré et al. (2005)
Methionine aminopept.	*Pisolithus microcarpus–Eucalyptus globulus*	4–7↓ 12↑ 21d↓	Duplessis et al. (2005)[a]
Mitotic checkpoint protein	*Pisolithus microcarpus–Eucalyptus globulus*	4↑ 7–21d↓	Duplessis et al. (2005)[a]
Myosin-IA	*Tuber borchii–Tilia americana*	30d↑	Menotta et al. (2004)
P300/CBP	*Laccaria bicolor–Pinus resinosa*	6–72h↑	Podila et al. (2002)
Rah1/Rad51	*Pisolithus tinctorius–Eucalypthus globulus*	4d↓	Voiblet et al. (2001)
Ras	*Laccaria bicolor–Pinus resinosa*	6–72h↑	Podila et al. (2002)
Ras related protein	*Laccaria bicolor–Pinus resinosa*	6–72h↑	Podila et al. (2002)
RAS related protein	*Pisolithus microcarpus–Eucalyptus globulus*	4↓ 7↑ 12–21d↓	Duplessis et al. (2005)[a]
Ras1p	*Pisolithus microcarpus–Eucalyptus globulus*	4–7↓ 12↑ 21d↓	Duplessis et al. (2005)[a]
Ras-related protein	*Paxillus involutus–Betula pendula*	25d↓	Johansson et al. (2004)
Rho 2 protein	*Paxillus involutus–Betula pendula*	25d↓	Johansson et al. (2004)
Rho gdp dissociation	*Pisolithus microcarpus–Eucalyptus globulus*	4–7↓ 12↑ 21d↓	Duplessis et al. (2005)[a]
Rho gdp dissociation	*Pisolithus microcarpus–Eucalyptus globulus*	4–7↓ 12↑ 21d↓	Duplessis et al. (2005)[a]

(*continued*)

Table 5.1 (continued)

Gene	Ectomycorrhiza	Time/gene expression tendency	Reference
Rho gdp dissociation inhibitor.	*Tuber borchii–Tilia americana*	30d↑	Menotta et al. (2004)
Ribonucleotide reductase	*Paxillus involutus–Betula pendula*	2–14d↑	Le Quéré et al. (2005)
Ribonucleotide reductase	*Tuber borchii–Tilia americana*	30d↑	Menotta et al. (2004)
Septin Spn3	*Laccaria bicolor–Pinus resinosa*	6–72h↑	Podila et al. (2002)
Sur4	*Laccaria bicolor–Pinus resinosa*	6–24h↑	Kim et al. (1999b)
Sur4	*Laccaria bicolor–Pinus resinosa*	6–72h↑	Podila et al. (2002)
Temperature sensitive suppressor Bem1/bud5	*Paxillus involutus–Betula pendula*	25d↓	Johansson et al. (2004)
Translation-elongation fator 1 alpha	*Pisolithus tinctorius–Eucalypthus globulus*	4d↑	Voiblet et al. (2001)
Vacuolar H-ATPase assembly	*Paxillus involutus–Betula pendula*	25d↓	Johansson et al. (2004)
YlmF	*Pisolithus microcarpus–Eucalyptus globulus*	4–7↓ 12↑ 21d↓	Duplessis et al. (2005)[a]

[a]Presented in additional material. In the time and expression tendency column, hyphen between numbers indicates interval time, (h) hours, (d) days, and (w) weeks. ↑ Indicates over expression while ↓ indicates down expression

regulatory subunit, *mts4*, is part of a proteolytic degradation complex and is related to cytoplasmic and nuclear protein degradation, and was found in abundance in the nuclear periphery during the yeast fission process (Wilkinson et al. 1998) and in *T. borchii–T. americana* fungal cells at 30 days of development. The presence of chitin synthase I transcripts in *L. bicolor–P. resinosa* may be related with chitin reposition during cell replication (Santos and Snyder 2005). Similarly, the pectinase activity observed in *Terfezia boudieri* mycelium (4 weeks after ectomycorrhizal stimulation with *Cistus incanus* (Zaretsky et al. 2006) could be related to roots tissue invasion during the early symbiosis process (Poinssot et al. 2003).

The observation of Ras family mRNAs in ectomycorrhizal mycelium is difficult to interpret, due to the diverse functions of this protein family in cells (Lammers 2004). The varied functions of these proteins may explain the lack of a specific pattern of regulation in different associations (Table 5.1).

Down-regulation of cell resistance and stress response elements, such as the temperature sensitive suppressor *Bem1/Bud5* (Bender and Pringle 1991) and *RAH1* (Chowdhury et al. 1992) were identified in some symbiotic mycelia (Table 5.1).

5.2.2 Morphogenesis

Only four ESTs have been reported in this category and all of them have increased expression in *L. bicolor–P. resinosa* at 6–72 h of development (Podila et al. 2002). As observed in Table 5.2, all citations are related to homeotic genes.

5.2.3 Energy and Metabolism

In this category, 96 ESTs were recognized that are related to metabolic processes in different symbiotic combinations (Table 5.3).

5.2.3.1 Carbohydrate Metabolism

In eukaryote cells, carbohydrate metabolism is one of the major bioenergetic pathways utilizing simple sugars through glycolysis, the tricarboxylic acid cycle (TCA) and oxidative phosphorylation. The hexose transporters carry different hexoses (including glucose) across cell membranes. ESTs of this transporter have been identified in *P. involutus–B. pendula* association (2–14 days) potentially indicating increased metabolic activity during the early physical contact stage (Le Quéré et al. 2005). Sorbitol dehydrogenase expression was repressed in *P. involutus* from the second to the eighth day of interaction. Together with the aldose reductase, this enzyme converts sorbitol (sugar alcohol) into fructose, allowing for further glucose production without ATP consumption. Sorbitol dehydrogenase down-regulation may indicate the fungus's preference for the glucose pathway.

Table 5.2 Some morphogenesis genes differentially expressed in ectomycorrizal fungus

Gene	Ectomycorrhiza	Time/gene expression tendency	Reference
Homeobox b4 gene	*Laccaria bicolor–Pinus resinosa*	6–72h↑	Podila et al. (2002)
Homeobox genes Hox 2.6	*Laccaria bicolor–Pinus resinosa*	6–72h↑	Podila et al. (2002)
Teashirt patterning gene	*Laccaria bicolor–Pinus resinosa*	6–72h↑	Podila et al. (2002)
You-too homeotic gene	*Laccaria bicolor–Pinus resinosa*	6–72h↑	Podila et al. (2002)

In the time and expression tendency column, hyphen between numbers indicates interval time, (h) hours. ↑ Indicates over expression while ↓ indicates down expression

Table 5.3 Some energy and metabolism genes differentially expressed in ectomycorrizal fungus

Gene	Ectomycorrhiza	Time/gene expression tendency	Reference
α-ketoglutarate sulfonate dioxygenase	*Tuber borchii–Tilia platyphyllos*	5m↑	Polidori et al. (2002)
β-keto thiolase 1	*Laccaria bicolor–Pinus resinosa*	6–72h↑	Podila et al. (2002)
3-GPD-glyceraldehyde 3 phosphate dehydrogenase	*Laccaria bicolor–Pinus resinosa*	6–72h↑	Podila et al. (2002)
Acetyl CoA acetyl transferase	*Laccaria bicolor–Pinus resinosa*	6–72h↑	Podila et al. (2002)
Acid phosphatase	*Pisolithus tinctorius–Eucalyptus urophylla*	4d↑	Lei et al. (1990)
Acyl CoA oxidase	*Laccaria bicolor–Pinus resinosa*	6–72h↑	Podila et al. (2002)
Adenylylsulfate kinase	*Paxillus involutus–Betula pendula*	2↓ 14d↓	Le Quéré et al. (2005)
Adrenodoxin	*Laccaria bicolor–Pinus resinosa*	6–24h↑	Kim et al. (1999b)
Aldehyde dehydrogenase	*Pisolithus microcarpus–Eucalyptus globulus*	4–7↓ 12↑ 21d↓	Duplessis et al. (2005)[a]
Alpha mannosidase	*Paxillus involutus–Betula pendula*	12w↑	Morel et al. (2005)
Alpha/beta-gliadin precursor	*Tuber borchii–Tilia americana*	30d↑	Menotta et al. (2004)
Alpha-L-rhamnosidase A precursor	*Tuber borchii–Tilia americana*	30d↑	Menotta et al. (2004)
Argininosuccinate lyase	*Paxillus involutus–Betula pendula*	2↓14d↓	Le Quéré et al. (2005)
Aryl-alcohol dehydrogenase	*Laccaria bicolor–Pinus resinosa*	6–72h↑	Podila et al. (2002)
Asparagine synthase Asn2p	*Tuber borchii–Tilia americana*	30d↑	Menotta et al. (2004)
Aspartate aminotransferase	*Pisolithus microcarpus–Eucalyptus globulus*	4↓ 7↑ 12–21d↓	Duplessis et al. (2005)[a]
Aspartic protease	*Tuber borchii–Tilia americana*	30d↑	Menotta et al. (2004)
ATP synthase subunit	*Pisolithus microcarpus–Eucalyptus globulus*	4–7↓ 12↑ 21d↓	Duplessis et al. (2005)[a]
ATP synthase subunit f	*Paxillus involutus–Betula pendula*	2↓ 4–8↑ 14d↓	Le Quéré et al. (2005)
ATPase F0 subunit 9	*Pisolithus microcarpus–Eucalyptus globulus*	4↓ 7↑ 12–21d↓	Duplessis et al. (2005)[a]
B2-aldehyde-forming enz.	*Paxillus involutus–Betula pendula*	2–8↓14d↑	Le Quéré et al. (2005)

(*continued*)

Table 5.3 (continued)

Gene	Ectomycorrhiza	Time/gene expression tendency	Reference
CG15406-PA	*Tuber borchii–Tilia americana*	30d↑	Menotta et al. (2004)
Choline-P-cytidylylt.	*Pisolithus microcarpus–Eucalyptus globulus*	4–7↓ 12↑ 21d↓	Duplessis et al. (2005)[a]
Chorismate synthase	*Paxillus involutus–Betula pendula*	25d↓	Johansson et al. (2004)
Citrate synthase	*Pisolithus microcarpus–Eucalyptus globulus*	4–7↓ 12↑ 21d↓	Duplessis et al. (2005)[a]
COX1-i1 protein	*Tuber borchii–Tilia americana*	30d↑	Menotta et al. (2004)
Cyanate lyase	*Paxillus involutus–Betula pendula*	2–8d↓	Le Quéré et al. (2005)
Cystahione beta synthase	*Laccaria bicolor–Pinus resinosa*	6–72h↑	Podila et al. (2002)
Cysteine proteinase	*Pisolithus microcarpus–Eucalyptus globulus*	4–12↓ 21d↑	Duplessis et al. (2005)[a]
Cyt ochrome c reductase	*Paxillus involutus–Betula pendula*	2–8↑ 14d↓	Le Quéré et al. (2005)
Cytochrome c oxidase	*Paxillus involutus–Betula pendula*	2–14d↑	Le Quéré et al. (2005)
Cytochrome c oxidase	*Pisolithus microcarpus–Eucalyptus globulus*	4–7↓ 12↑ 21d↓	Duplessis et al. (2005)[a]
Delta-1-pyrroline-5-carboxilate synthetase	*Pisolithus microcarpus–Eucalyptus globulus*	4–7↓ 12↑ 21d↓	Duplessis et al. (2005)[a]
Dihydroorotase	*Paxillus involutus–Betula pendula*	12w↑	Morel et al. (2005)
Elastinolytic metalloprotease	*Pisolithus tinctorius–Eucalyptus globulus*	4d↑	Voiblet et al. (2001)
Enolase	*Tuber borchii–Tilia platyphyllos*	5m↑	Polidori et al. (2002)
F1-ATP synthase delta	*Pisolithus microcarpus–Eucalyptus globulus*	4–7↓ 12↑ 21d↓	Duplessis et al. (2005)[a]
Ferrodoxin	*Laccaria bicolor–Pinus resinosa*	6–72h↑	Podila et al. (2002)
Glucoamylase-related protein	*Pisolithus microcarpus–Eucalyptus globulus*	4↓ 7↑ 12–21d↓	Duplessis et al. (2005)[a]
Glycogen branching enzyme	*Terfezia boudieri–Cistus incanus*	4w↑	Zaretsky et al. (2006)
Glycosyl hydrolase	*Pisolithus microcarpus–Eucalyptus globulus*	4–7↓ 12↑ 21d↓	Duplessis et al. (2005)[a]

(*continued*)

Table 5.3 (continued)

Gene	Ectomycorrhiza	Time/gene expression tendency	Reference
Hemolysin	*Pisolithus microcarpus–Eucalyptus globulus*	4–7↓ 12↑ 21d↓	Duplessis et al. (2005)[a]
Hexokinase	*Paxillus involutus–Betula pendula*	25d↑	Johansson et al. (2004)
Hexokinase	*Pisolithus microcarpus–Eucalyptus globulus*	4↓ 7↑ 12–21d↓	Duplessis et al. (2005)[a]
Hexose transporter	*Paxillus involutus–Betula pendula*	2–14d↑	Le Quéré et al. (2005)
Histidine kinase	*Pisolithus microcarpus–Eucalyptus globulus*	4–7↓ 12↑ 21d↓	Duplessis et al. (2005)[a]
Homoserine kinase Th.	*Pisolithus microcarpus–Eucalyptus globulus*	4–7↓ 12↑ 21d↓	Duplessis et al. (2005)[a]
Inorganic pyrophosphatase	*Tuber borchii–Tilia americana*	30d↑	Menotta et al. (2004)
Isoamyl alcohol oxidase	*Paxillus involutus–Betula pendula*	4–14d↑	Le Quéré et al. (2005)
Isocitrate dehydrogenase	*Pisolithus microcarpus–Eucalyptus globulus*	4↓ 7↑ 12–21d↓	Duplessis et al. (2005)[a]
Kynurenine 3-monoxygenase	*Pisolithus microcarpus–Eucalyptus globulus*	4–7↓ 12↑ 21d↓	Duplessis et al. (2005)[a]
Lactonohydrolase	*Laccaria bicolor–Pinus resinosa*	6–72h↑	Podila et al. (2002)
Malate dehydrogenase	*Pisolithus microcarpus–Eucalyptus globulus*	4–7↓ 12↑ 21d↓	Duplessis et al. (2005)[a]
Malate dehydrogenase Mitochondrial	*Pisolithus tinctorius–Eucalyptus globulus*	4d↑	Voiblet et al. (2001)
Malate synthase	*Laccaria bicolor–Pinus resinosa*	6–72h↑	Podila et al. (2002)
Mandelate racemase/ muconate lactonizing	*Pisolithus tinctorius–Castanea sativa*	12h↓	Acioli-Santos et al. (2008)
Methylamine dehydrogenase	*Pisolithus microcarpus–Eucalyptus globulus*	4–12↓ 21d↑	Duplessis et al. (2005)[a]
Mg^{2+} chelatase	*Laccaria bicolor–Pinus resinosa*	6–72h↑	Podila et al. (2002)
Mitochondrial rib. protein 15.5 kDa YmL31	*Paxillus involutus–Betula pendula*	25d↓	Johansson et al. (2004)
N. enase (ubiquinone)	*Pisolithus microcarpus–Eucalyptus globulus*	4↓ 7↑ 12–21d↓	Duplessis et al. (2005)[a]

(*continued*)

Table 5.3 (continued)

Gene	Ectomycorrhiza	Time/gene expression tendency	Reference
NAD-GDH	*Pisolithus microcarpus–Eucalyptus globulus*	4–7↓ 12↑ 21d↓	Duplessis et al. (2005)[a]
NADH dehydrogenase	*Pisolithus microcarpus–Eucalyptus globulus*	4–7↓ 12↑ 21d↓	Duplessis et al. (2005)[a]
NAD-malate dehydrogenase	*Pisolithus microcarpus–Eucalyptus globulus*	4↓ 7↑ 12–21d↓	Duplessis et al. (2005)[a]
NADP GDH	*Pisolithus microcarpus–Eucalyptus globulus*	4–7↓ 12↑ 21d↓	Duplessis et al. (2005)[a]
NADP-dep. Alc. Dehydrogenase	*Paxillus involutus–Betula pendula*	2–8d↓	Le Quéré et al. (2005)
Nar1p	*Tuber borchii–Tilia americana*	30d↑	Menotta et al. (2004)
Nitrilase Nit1	*Pisolithus microcarpus–Eucalyptus globulus*	4–7↓ 12↑ 21d↓	Duplessis et al. (2005)[a]
Nucleoside diphosphate kinase	*Tuber borchii–Tilia platyphyllos*	5m↑	Polidori et al. (2002)
O-acetil-L-homoserine sulfhydrilase	*Paxillus involutus–Betula pendula*	25d↑	Johansson et al. (2004)
P-1P-4-tetraphosphat	*Pisolithus microcarpus–Eucalyptus globulus*	4–7↓ 12↑ 21d↓	Duplessis et al. (2005)[a]
P-2-dehydro-3-deoxyheptonate aldolase	*Paxillus involutus–Betula pendula*	4–14d↑	Le Quéré et al. (2005)
PEP carboxykinase	*Laccaria bicolor–Pinus resinosa*	6–72h↑	Podila et al. (2002)
Peroxisomal glycolat.	*Pisolithus microcarpus–Eucalyptus globulus*	4–12↓ 21d↑	Duplessis et al. (2005)[a]
Pho88p	*Paxillus involutus–Betula pendula*	25d↓	Johansson et al. (2004)
Phosphorylcholine transferase Pct1p	*Paxillus involutus–Betula pendula*	25d↓	Johansson et al. (2004)
P-hydroxybenzoate po.	*Pisolithus microcarpus–Eucalyptus globulus*	4–7↓ 12↑ 21d↓	Duplessis et al. (2005)[a]
Proline iminopeptida. met.	*Pisolithus microcarpus–Eucalyptus globulus*	4–7↓ 12↑ 21d↓	Duplessis et al. (2005)[a]
Purine nucleotide-bi.	*Pisolithus microcarpus–Eucalyptus globulus*	4–7↓ 12↑ 21d↓	Duplessis et al. (2005)[a]
Putative diphthine synthase	*Paxillus involutus–Betula pendula*	25d↓	Johansson et al. (2004)

(*continued*)

Table 5.3 (continued)

Gene	Ectomycorrhiza	Time/gene expression tendency	Reference
Putative GDP-mannose pyrophosphorylase	*Tuber borchii–Tilia americana*	30d↑	Menotta et al. (2004)
Pyruvate dehydrogenase	*Pisolithus microcarpus–Eucalyptus globulus*	4–7↓ 12↑ 21d↓	Duplessis et al. (2005)[a]
Pyruvate kinase	*Pisolithus microcarpus–Eucalyptus globulus*	4↓ 7↑ 12–21d↓	Duplessis et al. (2005)[a]
Riboflavin synthase	*Paxillus involutus–Betula pendula*	2↓ 14d↓	Le Quéré et al. (2005)
Ring-box protein 1	*Pisolithus tinctorius–Eucalyptus globulus*	4d↓	Voiblet et al. (2001)
rRNA mitocondrial, Large sub	*Pisolithus tinctorius–Castanea sativa*	12h↓	Acioli-Santos et al. (2008)
S-adenosyl-L-homocysteine hydrolase	*Pisolithus tinctorius–Castanea sativa*	12h↑	Acioli-Santos et al. (2008)
S-adenosylmethionine	*Pisolithus microcarpus–Eucalyptus globulus*	4↓ 7↑ 12–21d↓	Duplessis et al. (2005)[a]
Shp1 protein phosphatase	*Pisolithus tinctorius–Eucalyptus globulus*	4d↑	Voiblet et al. (2001)
Sorbitol dehydrogenase	*Paxillus involutus–Betula pendula*	2–8d↓	Le Quéré et al. (2005)
Squalene monooxygenase	*Laccaria bicolor–Pinus resinosa*	6–72h↑	Podila et al. (2002)
Stearoyl-Coa desaturase	*Paxillus involutus–Betula pendula*	25d↓	Johansson et al. (2004)
Sterol 14 alpha demethylase	*Pisolithus tinctorius–Eucalyptus globulus*	4d↑	Voiblet et al. (2001)
Tyrosinase precursor	*Paxillus involutus–Betula pendula*	25d↑	Johansson et al. (2004)
Ubiquinol-cytochrome	*Pisolithus microcarpus–Eucalyptus globulus*	4–7↓ 12↑ 21d↓	Duplessis et al. (2005)[a]
UMP-CMP kinase	*Pisolithus microcarpus–Eucalyptus globulus*	4–7↓ 12↑ 21d↓	Duplessis et al. (2005)[a]
Vacuolar ATP synthase	*Pisolithus microcarpus–Eucalyptus globulus*	4–7↓ 12↑ 21d↓	Duplessis et al. (2005)[a]

[a]Presented in additional material. In the time and expression tendency column, hyphen between numbers indicates interval time, (h) hours, (d) days, (w) weeks, and (m) month. ↑ indicates over expression while ↓ indicates down expression

Hexokinase (the first enzyme of the glycolysis pathway) was up-regulated in *P. involutus–B. pendula* (25 days) and in *P. microcarpus–E. globulus* mycelium (only on the seventh day).

Positive enolase expression was found in 5-month-old *T. borchii–Tilia platyphyllos* mycelium (Polidori et al. 2002). This enzyme promotes the reversible conversion of 2-phosphoglycerate to phosphoenolpyruvate (PEP) (Babbitt et al. 1996). Pyruvate kinase, which is up-regulated in 7-day-old *P. microcarpus–E. globulus* fungal cells, transforms phosphoenolpyruvate into pyruvate. Between 6 and 72 h of *L. bicolor–P. resinosa* interaction (Podila et al. 2002) differentially expressed PEP carboxykinase ESTs were identified. PEP carboxykinase catalyzes the decarboxylation of GTP (guanidine triphosphate) and oxaloacetate into phosphoenolpyruvate, GDP (guanidine diphosphate) and CO_2.

The conversion of isocitrate into α-ketoglutarate is performed by isocitrate dehydrogenase within the TCA cycle. This enzyme was identified as up-regulated in 7-day-old *P. microcarpus* mycelium. Malate dehydrogenase promotes the malate–oxaloacetate conversion; the ESTs for this enzyme were determined to be up-regulated at fourth day of *P. tinctorius–E. globulus* association. Moreover, in *P. microcarpus,* the malate dehydrogenase expression was increased between 7 and 12 days of association with the same host. The malate synthase is indirectly linked to the Krebs cycle by the glyoxylate pathway (Smith et al. 2003). This enzyme was up-regulated in *L. bicolor–P. resinosa* mycelium (6–72 h).

In the respiratory chain, several enzymes were found differentially expressed in ectomycorrhizal association. The ubiquinol (Q oxidoreductase complex) represents the second complex of the respiratory chain and is the entry point of electrons from $FADH_2$ molecules. Transcripts of this ubiquinol were stimulated in *P. microcarpus* mycelium at 12 days of ectomycorrhizal development with *E. globulus*. Cytochrome c reductase is part of complex III. This enzyme promotes the electron transfer from coenzyme Q to a cytochrome c molecule (reduction reaction) and pumps protons out of the mitochondrial matrix (Berg et al. 2004). The cytochrome c oxidase is the last enzyme of the electron transport chain (complex IV). The cytochrome c oxidase receives four electrons from cytochrome c and transfers them to an oxygen molecule. The addition of four H^+ protons converts O_2 into two H_2O molecules (Berg et al. 2004). Expression of both enzymes, cytochrome c reductase and cytochrome c oxidase, was increased in *P. involutus–B. pendula* (2–14 days) and *E. globulus–P. microcarpus* associations (12 days, only cytochrome c oxidase). This expression is consistent with the ATP synthase subunit activation in both *P. involutus* and *P. microcarpus* mycelium. The ATP synthase complex performs the final ATP formation in the respiratory chain.

Interestingly in two different situations, mRNAs encoding structural elements of mitochondria were identified as "repressed" in symbiotic or free-living fungus. ESTs that encode a 15.5-kDa mitochondrial ribosomal protein YmL31 was isolated from 25-day-old *P. involutus–B. pendula* mycelium, while mRNAs encoding the large mitochondrial subunit, rRNA were observed in *P. tinctorius–C. sativa* preinfection mycelium (12 h) (Acioli-Santos et al. 2008).

5.2.3.2 Lipid Metabolism

In this category, we can exemplify the negative expressions of enzymes that act in fatty acids, sterols, and hormone synthesis in ectomycorrhizal fungus. Both the phosphorylcholine transferase, related to the glycerophospholipide and aminophosphate pathways, and stearoyl-CoA desaturase transcript, an enzyme involved in fatty acids synthesis (Ntambi et al. 1988), was down-regulated in *P. involutus* associated with *B. pendula* (25 days).

The ESTs encoding the sterol-14-alpha-demethylase, involved in sterol synthesis (Bellamine et al. 1999), was reported as up-regulated in 4-day-old *P. tinctorius–E. globulus* mycelium. Others lipid related enzyme transcripts were identified in *L. bicolor* associated with *P. resinosa*. Among these, were mRNA of adrenodoxin, a gene that produces a protein that acts as an electron carrier in the transformation of cholesterol to steroid hormones (Lambeth et al. 1979), the acetyl CoA acetyl transferase, involved in glycerophospholipide metabolism, squalene monooxygenase, important in sterol and terpene biosynthesis, and acetyl CoA oxidase, an enzyme involved in fatty acid oxidation (Emanuelsson et al. 2003). We found either of the transcripts encoding β-ketothiolase, a participant in fatty acid, pyruvate and ketone body metabolic pathways (Slater et al. 1998). The β-ketothiolase expression may indicate the fungal physiology preference towards the fatty acid catabolism to minimize energy waste. This enzyme was identified in *L. bicolor* mycelium with *P. resinosa* interaction between 6 and 72 h.

5.2.3.3 Amino Acid and Protein Metabolism

In this category, we found ESTs of O-acetyl-L-homoserine sulfhydrylase and S-adenosyl-L-homocysteine hydrolase, important enzymes in the methionine pathway (Yamagata 1989), and chorismate synthase, functionally related to production of aromatic amino acid compounds (phenylalanine, tyrosine and tryptophan) in bacteria, fungi and plants (Schmid and Amrhein 1995; Macheroux et al. 1999). Only chorismate synthase was repressed. S-adenosyl-L-homocysteine hydrolase was identified as over expressed in *P. tinctorius–C. sativa* preinfection mycelium (12 h) (Acioli-Santos et al. 2008). In the methionine pathway, the S-adenosyl-L-homocysteine hydrolase promotes S-adenosyl-L-homocysteine hydrolysis into adenosine and homocysteine. To the homocysteine, a methyl group is added to promote methionine formation (Kawalleck et al. 1992). In addition, the ESTs for cystathione β-synthase were found in *L. bicolor–P. resinosa* fungal cells (6–72 h), potentially indicating the involvement of methionine and cysteine metabolism in ectomycorrhizal associations (Bydlowiski et al. 1998). ESTs for α-ketoglutarate sulfonate dioxygenase were identified in mycelium of *T. borchii–T. platyphyllos* after 5 months of contact. In *Saccharomyces cerevisiae*, this enzyme can act in the alternative sulfonated compound pathway (e.g., taurine and isotianato as a sulfur source) when cells are grown in the absence of sulfate (Hogan et al. 1999).

The α-mannosidase is involved in glycoprotein synthesis and has been observed in *Aspergillus nidulans* (Eades et al. 1998) and *Amanita muscaria* (Kong 1995). Transcripts of α-mannosidase were stimulated in *P. involutus* associated with *B. pendula* during 12 weeks of development.

5.2.3.4 Nucleic Acids Metabolism

The dihydroorotase mRNA, involved in the pentose and pyrimidine pathways (Porter et al. 2004), was identified in 12-week-old *P. involutus–B. pendula* mycelium. The nucleoside diphosphate kinase catalyzes, in turn, the nucleoside diphosphate phosphorylation to nucleoside triphosphate; this is a reversible reaction regulating the nucleotides supplied to the cell (Hasunuma et al. 2003). The nucleoside diphosphate kinase was reported in *T. borchii–T. platyphyllos* association at 5 months of development.

5.2.4 Protein Synthesis and Interaction, Transcriptional and Translational Regulation

A total of 104 ESTs were grouped in this category, 57 representing genes associated with the ribosomal machinery, i.e., they encode for ribosomal subunit proteins (Table 5.4). We observed that the 19 ribosomal genes found in the *P. involutus–B. pendula* fungal cells were repressed in 2 to 8 day old symbiosis (Le Quéré et al. 2005), while eight of them were stimulated later (Johansson et al. 2004; Morel et al. 2005). In the *P. microcarpus–E. globulus* ectomycorrhiza 15 ribosomal genes were stimulated at 12 days of development. Interestingly, ribosomal *P. tinctorius* ESTs were also repressed at the first 12 h of interaction (preinfection) with *C. sativa* (Acioli-Santos et al. 2008).

5.2.5 Ions, Amino Acids and Peptides Transports

36 ESTs/genes were categorized into four main groups in this category delineated as vesicular transport, traffic mitochondrial molecules, ion channels and transporters and protein transport (Table 5.5). The vesicular transport genes are involved in aspects of cellular vesicle formation for transport and secretion. As for example, coatomer zeta subunit, *COPII* and *NIPSNAPI*, are genes that produce proteins directly related to the vesicle formation and solutes secretory transport in the endoplasmic reticulum and Golgi complex (Kuge et al. 1993; Campbell and Scheckman 1997; Seroussi et al. 1998), respectively; although the function of *NIPSNAPI* is still under speculation. The ESTs for *COPII* and *NIPSNAPI* were

Table 5.4 Some protein synthesis and interaction, transcriptional and translational regulation genes differentially expressed in ectomycorrizal fungus

Gene	Ectomycorrhiza	Time/gene expression tendency	Reference
Bystin	*Paxillus involutus–Betula pendula*	25d↓	Johansson et al. (2004)
Calmodulin A	*Paxillus involutus–Betula pendula*	2–8d↓	Le Quéré et al. (2005)
Clathrin adapter protein	*Laccaria bicolor–Pinus resinosa*	6–72h↑	Podila et al. (2002)
COP9 subunit 7[a]	*Pisolithus microcarpus–Eucalyptus globulus*	4↑ 7–21d↓	Duplessis et al. (2005)[a]
Deoxyhypusine synthase	*Laccaria bicolor–Pinus resinosa*	6–72h↑	Podila et al. (2002)
Dipeptidyl-peptidase IV	*Terfezia boudieri–Cistus incanus*	4w↑	Zaretsky et al. (2006)
Disulfide isomerase	*Laccaria bicolor–Pinus resinosa*	6–72h↑	Podila et al. (2002)
Elongation factor EF	*Paxillus involutus–Betula pendula*	2–8d↓	Le Quéré et al. (2005)
Elongation factor EF-1 gamma	*Paxillus involutus–Betula pendula*	2–8↓14d↑	Le Quéré et al. (2005)
Elongation factor EF1a	*Laccaria bicolor–Pinus resinosa*	6–72h↑	Podila et al. (2002)
Eukariotic initiation factor eIF4 A	*Pisolithus tinctorius–Eucalyptus globulus*	4d↑	Voiblet et al. (2001)
Eukariotic initiation factor eIF-5	*Pisolithus microcarpus–Eucalyptus globulus*	4–7↓ 12↑ 21d↓	Duplessis et al. (2005)[a]
Eukariotic initiation factor eIF-6	*Paxillus involutus–Betula pendula*	2–8d↓	Le Quéré et al. (2005)
GPI anchor transamidase	*Pisolithus microcarpus–Eucalyptus globulus*	4–7↓ 12↑ 21d↓	Duplessis et al. (2005)[a]
Isopeptidase t	*Pisolithus microcarpus–Eucalyptus globulus*	4–7↓ 12↑ 21d↓	Duplessis et al. (2005)[a]
Kexin protein	*Pisolithus tinctorius–Eucalyptus globulus*	4d↑	Voiblet et al. (2001)
Peptidyl-prolyl cis/	*Pisolithus microcarpus–Eucalyptus globulus*	4–7↓ 12↑ 21d↓	Duplessis et al. (2005)[a]
Problabe translation release fator Erf3	*Paxillus involutus–Betula pendula*	12S↑	Morel et al. (2005)
Proteasome component	*Pisolithus microcarpus–Eucalyptus globulus*	4–7↓ 12↑ 21d↓	Duplessis et al. (2005)[a]
Proteasome p44.5 26S	*Pisolithus microcarpus–Eucalyptus globulus*	4–7↓ 12↑ 21d↓	Duplessis et al. (2005)[a]

(*continued*)

Table 5.4 (continued)

Gene	Ectomycorrhiza	Time/gene expression tendency	Reference
Proteasome subunit a	*Pisolithus microcarpus–Eucalyptus globulus*	4–7↓ 12↑ 21d↓	Duplessis et al. (2005)[a]
PWP2 GTP protein	*Pisolithus tinctorius–Eucalyptus globulus*	4d↑	Voiblet et al. (2001)
PWP2 periodic trypto..	*Pisolithus microcarpus–Eucalyptus globulus*	4–7↓ 12↑ 21d↓	Duplessis et al. (2005)[a]
RE63412p	*Paxillus involutus–Betula pendula*	2–8d↓	Le Quéré et al. (2005)
Rex p27 protein	*Pisolithus microcarpus–Eucalyptus globulus*	4–7↓ 12↑ 21d↓	Duplessis et al. (2005)[a]
Ribophorin II	*Paxillus involutus–Betula pendula*	25d↓	Johansson et al. (2004)
Ribosomal protein S2	*Paxillus involutus–Betula pendula*	2–8d↓	Le Quéré et al. (2005)
Ribosomal Acid protein P2	*Paxillus involutus–Betula pendula*	25d↑	Johansson et al. (2004)
Ribosomal protein	*Pisolithus microcarpus–Eucalyptus globulus*	4↑ 7–21d↓	Duplessis et al. (2005)[a]
Ribosomal protein 136	*Paxillus involutus–Betula pendula*	2–8d↓	Le Quéré et al. (2005)
Ribosomal protein 22	*Paxillus involutus–Betula pendula*	2–8d↓	Le Quéré et al. (2005)
Ribosomal protein 4 S1	*Pisolithus microcarpus–Eucalyptus globulus*	4↑ 7–21d↓	Duplessis et al. (2005)[a]
Ribosomal protein 4.S1	*Pisolithus microcarpus–Eucalyptus globulus*	4–7↓ 12↑ 21d↓	Duplessis et al. (2005)[a]
Ribosomal protein 4.S1	*Pisolithus microcarpus–Eucalyptus globulus*	4–7↓ 12↑ 21d↓	Duplessis et al. (2005)[a]
Ribosomal protein 4.S2	*Pisolithus microcarpus–Eucalyptus globulus*	4–7↓ 12↑ 21d↓	Duplessis et al. (2005)[a]
Ribosomal protein 4.S9	*Pisolithus microcarpus–Eucalyptus globulus*	4–7↓ 12↑ 21d↓	Duplessis et al. (2005)[a]
Ribosomal protein 40s	*Paxillus involutus–Betula pendula*	25d↑	Johansson et al. (2004)
Ribosomal protein 40s	*Laccaria bicolor–Pinus resinosa*	6–72h↑	Podila et al. (2002)
Ribosomal protein 40S	*Pisolithus microcarpus–Eucalyptus globulus*	4–7↓ 12–21d↑	Duplessis et al. (2005)[a]

(*continued*)

Table 5.4 (continued)

Gene	Ectomycorrhiza	Time/gene expression tendency	Reference
Ribosomal protein 40S	*Pisolithus microcarpus–Eucalyptus globulus*	4–7↓ 12–21d↑	Duplessis et al. (2005)[a]
Ribosomal protein 40s S10	*Paxillus involutus–Betula pendula*	12w↑	Morel et al. (2005)
Ribosomal protein 40s S16	*Paxillus involutus–Betula pendula*	2–8d↓	Le Quéré et al. (2005)
Ribosomal protein 50S	*Pisolithus microcarpus–Eucalyptus globulus*	4–12↓ 21d↑	Duplessis et al. (2005)[a]
Ribosomal protein 6. 1	*Pisolithus microcarpus–Eucalyptus globulus*	4–7↓ 12d↑ 21d↓	Duplessis et al. (2005)[a]
Ribosomal protein 6. L15	*Pisolithus microcarpus–Eucalyptus globulus*	4↓ 7↑ 12–21d↓	Duplessis et al. (2005)[a]
Ribosomal protein 6..L1	*Pisolithus microcarpus–Eucalyptus globulus*	4–7↓ 12↑ 21d↓	Duplessis et al. (2005)[a]
Ribosomal protein 6.L2	*Pisolithus microcarpus–Eucalyptus globulus*	4–7↓ 12↑ 21d↓	Duplessis et al. (2005)[a]
Ribosomal protein 60S	*Pisolithus microcarpus–Eucalyptus globulus*	4↓ 7↑ 12–21d↓	Duplessis et al. (2005)[a]
Ribosomal protein 60S	*Pisolithus microcarpus–Eucalyptus globulus*	4–7↓ 12↑ 21d↓	Duplessis et al. (2005)[a]
Ribosomal protein E L3	*Pisolithus microcarpus–Eucalyptus globulus*	4↑ 7–21d↓	Duplessis et al. (2005)[a]
Ribosomal protein E. L2	*Pisolithus microcarpus–Eucalyptus globulus*	4↓ 7↑ 12–21d↓	Duplessis et al. (2005)[a]
Ribosomal protein E. L4	*Pisolithus microcarpus–Eucalyptus globulus*	4↓ 7↑ 12–21d↓	Duplessis et al. (2005)[a]
Ribosomal protein E.S15	*Pisolithus microcarpus–Eucalyptus globulus*	4–7↓ 12↑ 21d↓	Duplessis et al. (2005)[a]
Ribosomal protein L10	*Paxillus involutus–Betula pendula*	2–8d↓	Le Quéré et al. (2005)
Ribosomal protein L10	*Pisolithus microcarpus–Eucalyptus globulus*	4–7↓ 12↑ 21d↓	Duplessis et al. (2005)[a]
Ribosomal protein L11	*Paxillus involutus–Betula pendula*	25d↑	Johansson et al. (2004)
Ribosomal protein L12	*Paxillus involutus–Betula pendula*	2–8d↓	Le Quéré et al. (2005)

(*continued*)

Table 5.4 (continued)

Gene	Ectomycorrhiza	Time/gene expression tendency	Reference
Ribosomal protein L13A	*Paxillus involutus–Betula pendula*	25d↑	Johansson et al. (2004)
Ribosomal protein L15	*Paxillus involutus–Betula pendula*	2–8d↓	Le Quéré et al. (2005)
Ribosomal protein L18	*Paxillus involutus–Betula pendula*	2–8d↓	Le Quéré et al. (2005)
Ribosomal protein L23a	*Paxillus involutus–Betula pendula*	2–8d↓	Le Quéré et al. (2005)
Ribosomal protein L28	*Paxillus involutus–Betula pendula*	2–8d↓	Le Quéré et al. (2005)
Ribosomal protein L31	*Paxillus involutus–Betula pendula*	2–8d↓	Le Quéré et al. (2005)
Ribosomal protein L32	*Pisolithus microcarpus–Eucalyptus globulus*	4–7↓ 12d↑ 21d↓	Duplessis et al. (2005)[a]
Ribosomal protein L37	*Paxillus involutus–Betula pendula*	2–8d↓	Le Quéré et al. (2005)
Ribosomal protein L38	*Paxillus involutus–Betula pendula*	2–8d↓	Le Quéré et al. (2005)
Ribosomal protein L39	*Paxillus involutus–Betula pendula*	2–8d↓	Le Quéré et al. (2005)
Ribosomal protein L41	*Paxillus involutus–Betula pendula*	25d↑	Johansson et al. (2004)
Ribosomal protein L41	*Paxillus involutus–Betula pendula*	2–8↓ 14d↑	Le Quéré et al. (2005)
Ribosomal protein r.L32	*Pisolithus microcarpus–Eucalyptus globulus*	4–7↓ 12↑ 21d↓	Duplessis et al. (2005)[a]
Ribosomal protein r.S25	*Pisolithus microcarpus–Eucalyptus globulus*	4–7↓ 12↑ 21d↓	Duplessis et al. (2005)[a]
Ribosomal protein rs6/l7a	*Pisolithus microcarpus–Eucalyptus globulus*	4↓ 7↑ 12–21d↓	Duplessis et al. (2005)[a]
Ribosomal protein S11 homologue	*Paxillus involutus–Betula pendula*	25d↑	Johansson et al. (2004)
Ribosomal protein S12	*Paxillus involutus–Betula pendula*	2–8d↓	Le Quéré et al. (2005)
Ribosomal protein S14	*Paxillus involutus–Betula pendula*	2–8↓ 14d↑	Le Quéré et al. (2005)
Ribosomal protein S19	*Paxillus involutus–Betula pendula*	2–8d↓	Le Quéré et al. (2005)
Ribosomal protein S2	*Pisolithus microcarpus–Eucalyptus globulus*	4↑ 7–21d↓	Duplessis et al. (2005)[a]
Ribosomal protein S28	*Paxillus involutus–Betula pendula*	12w↑	Morel et al. (2005)
Ribosomal protein S29A	*Paxillus involutus–Betula pendula*	2–14d↑	Le Quéré et al. (2005)

(*continued*)

Table 5.4 (continued)

Gene	Ectomycorrhiza	Time/gene expression tendency	Reference
Ribosomal protein S31	*Paxillus involutus–Betula pendula*	2–8d↓	Le Quéré et al. (2005)
Ribosomal RNA 18S	*Pisolithus tinctorius–Castanea sativa*	12h↓	Acioli-Santos et al. (2008)
Ribosomal RNA 28S	*Pisolithus tinctorius–Castanea sativa*	12h↓	Acioli-Santos et al. (2008)
Ribosomal RNA 5S	*Pisolithus tinctorius–Castanea sativa*	12h↓	Acioli-Santos et al. (2008)
Serine/threonine protein	*Pisolithus microcarpus–Eucalyptus globulus*	4–7↓ 12↑ 21d↓	Duplessis et al. (2005)[a]
Shp1 protein	*Pisolithus microcarpus–Eucalyptus globulus*	4–7↓ 12↑ 21d↓	Duplessis et al. (2005)[a]
Subtilisin–like serine protease	*Paxillus involutus–Betula pendula*	12w↑	Morel et al. (2005)
SUG1 26S protein reg sub	*Laccaria bicolor–Pinus resinosa*	6–72h↑	Podila et al. (2002)
Threonyl-tRNA synthe.	*Pisolithus microcarpus–Eucalyptus globulus*	4↓ 7↑ 12–21d↓	Duplessis et al. (2005)[a]
Transcription elongation factor tef1-beta	*Pisolithus microcarpus–Eucalyptus globulus*	4↑ 7–21d↓	Duplessis et al. (2005)[a]
Transcription elongation factor tef1-g	*Pisolithus microcarpus–Eucalyptus globulus*	4–7↓ 12↑ 21d↓	Duplessis et al. (2005)[a]
Transcription elongation factor tef1-g	*Pisolithus microcarpus–Eucalyptus globulus*	4–7↓ 12–21d↑	Duplessis et al. (2005)[a]
Transcription elongation factor tef1-g	*Pisolithus microcarpus–Eucalyptus globulus*	4–12↓ 21d↑	Duplessis et al. (2005)[a]
Transcription factor	*Pisolithus microcarpus–Eucalyptus globulus*	4↓ 7↑ 12–21d↓	Duplessis et al. (2005)[a]
Transcription initiation	*Pisolithus microcarpus–Eucalyptus globulus*	4–7↓ 12↑ 21d↓	Duplessis et al. (2005)[a]
Translation initiation fator 3 Sui1p	*Paxillus involutus–Betula pendula*	25d↓	Johansson et al. (2004)
Translation initiation SUI1	*Paxillus involutus–Betula pendula*	2–8↓ 14d↑	Le Quéré et al. (2005)
Ubiquitin	*Paxillus involutus–Betula pendula*	2–8d↓	Le Quéré et al. (2005)
Ubiquitin / ribosomal S27a	*Paxillus involutus–Betula pendula*	2–8d↓	Le Quéré et al. (2005)

(*continued*)

Table 5.4 (continued)

Gene	Ectomycorrhiza	Time/gene expression tendency	Reference
Ubiquitin C-terminal	*Pisolithus microcarpus–Eucalyptus globulus*	4–7↓ 12↑ 21d↓	Duplessis et al. (2005)[a]
Ubiquitin like protein	*Paxillus involutus–Betula pendula*	12S↑	Morel et al. (2005)
Ubiquitin-conjugatin	*Pisolithus microcarpus–Eucalyptus globulus*	4–7↓ 12↑ 21d↓	Duplessis et al. (2005)[a]
Ubiquitin-conjugating enz.	*Paxillus involutus–Betula pendula*	2–8d↓	Le Quéré et al. (2005)
WD40 domain	*Terfezia boudieri–Cistus incanus*	4S↑	Zaretsky et al. (2006)
WD-repeat protein	*Pisolithus microcarpus–Eucalyptus globulus*	4–7↓ 12↑ 21d↓	Duplessis et al. (2005)[a]

[a]Presented in additional material. In the time and expression tendency column, hyphen between numbers indicates interval time, (h) hours, (d) days, (w) weeks, and (m) month. ↑ Indicates over expression while ↓ indicates down expression

down-regulated in *P. involutus–B. pendula* (25 days) and *P. tinctorius–Eucalyptus globulus* (4 days) mycelium, respectively. ESTs encoding for the endosomal Vps protein complex subunit were isolated from *Terfezia boudieri–Cistus incanus* fungal cells. Vps proteins act in particle transport from the Golgi complex to the cytoplasm vacuole and have been identified in *S. cerevisiae* cells (Tatsumi et al. 2007). In addition, the *Lb-AUT-7* gene is related to the cell mechanisms of phagocytosis (Podila et al. 2002).

Mitochondrial trafficking genes identified encode for proteins found in the mitochondrial inner membrane, which promote nuclear protein transport to mitochondria (Rehling et al. 2003; Palmieri 1994), and the mitochondrial carrier protein, which participates in phosphate anion transport (Palmieri 1994). The ADP/ATP carrier protein, also a mitochondrial carrier protein, performs translocation of ADP/ATP to the mitochondrial membranes, and here the protein catalyzes the reduction reaction of internal ADP to ATP by translocation of a radical phosphate. Interestingly, the ADP/ATP carried gene was repressed in 25-day-old *P. involutus* and *B. pendula* mycelium. The ADP/ATP carrier protein was also observed in *P. tinctorius* associated with *E. globulus*, both as repressed in the early and in the late development stages (i.e., at 4 and 21 days of ectomycorrhizal development) (Table 5.5).

Ion channel proteins were represented by calcium channels (specifically, Al voltage dependent calcium channels) and acid-sensitive K^+ channels. The detection of a putative Ca^{2+} transporting ATPase molecule as a calcium carrier was also reported. Genes for all of these were up-regulated in *L. bicolor* mycelium associated with *P. resinosa* (Kim et al. 1999b; Podila et al. 2002). The plasma

Table 5.5 Some ion, amino acids and peptide transport genes differentially expressed in ectomycorrizal fungus

Gene	Ectomycorrhiza	Time/gene expression tendency	Reference
β-importin	*Laccaria bicolor–Pinus resinosa*	6–72h↑	Podila et al. (2002)
A1 voltage dependente calcium Channel	*Laccaria bicolor–Pinus resinosa*	6–72h↑	Podila et al. (2002)
Acid sensitive K^+ channel	*Laccaria bicolor–Pinus resinosa*	6–72h↑	Podila et al. (2002)
Adenine nucleotide translocation	*Paxillus involutus–Betula pendula*	25d↓	Johansson et al. (2004)
ADP/ATP carrier protein	*Paxillus involutus–Betula pendula*	25d↓	Johansson et al. (2004)
ADP-ATP carrier protein	*Pisolithus microcarpus–Eucalyptus globulus*	4–7↓ 12↑ 21d↓	Duplessis et al. (2005)[a]
BIP protein	*Laccaria bicolor–Pinus resinosa*	6–72h↑	Podila et al. (2002)
Brefeldin A resistant.	*Pisolithus microcarpus–Eucalyptus globulus*	4–7↓ 12↑ 21d↓	Duplessis et al. (2005)[a]
Calcium channel	*Laccaria bicolor–Pinus resinosa*	6–24h↑	Kim et al. (1999b)
Endosomal Vps protein complex subunit	*Terfezia boudieri–Cistus incanus*	4w↑	Zaretsky et al. (2006)
Glutamate transporter	*Laccaria bicolor–Pinus resinosa*	6–72h↑	Podila et al. (2002)
Golgi memb. sorting protein	*Paxillus involutus–Betula pendula*	12w↑	Morel et al. (2005)
Huntingtin interacti.	*Pisolithus microcarpus–Eucalyptus globulus*	4–7↓ 12↑ 21d↓	Duplessis et al. (2005)[a]
Import protein Tim9p	*Paxillus involutus–Betula pendula*	2–14d↑	Le Quéré et al. (2005)
Lb-AUT7	*Laccaria bicolor–Pinus resinosa*	6–72h↑	Podila et al. (2002)
Lb-AUT7-autophagocitose	*Laccaria bicolor–Pinus resinosa*	6–24h↑	Kim et al. (1999a)
Mir1p	*Pisolithus microcarpus–Eucalyptus globulus*	4–7↓ 12↑ 21d↓	Duplessis et al. (2005)[a]
Mitochondrial carrier protein	*Paxillus involutus–Betula pendula*	25d↓	Johansson et al. (2004)
Mitochondrial Carrier YHM1	*Paxillus involutus–Betula pendula*	2–8↓ 14d↑	Le Quéré et al. (2005)
Mitochondrial inner memb. Translocase subunit Tim17 homologue	*Paxillus involutus–Betula pendula*	25d↓	Johansson et al. (2004)

(*continued*)

Table 5.5 (continued)

Gene	Ectomycorrhiza	Time/gene expression tendency	Reference
MSF Transporter	*Paxillus involutus–Betula pendula*	25d↓	Johansson et al. (2004)
NIPSNAPI	*Pisolithus tinctorius–Eucalyptus globulus*	4d↓	Voiblet et al. (2001)
Nuclear Transport factor 2	*Paxillus involutus–Betula pendula*	2–8d↓	Le Quéré et al. (2005)
Outer membrane protein	*Pisolithus microcarpus–Eucalyptus globulus*	4–7↓ 12↑ 21d↓	Duplessis et al. (2005)[a]
Oxaloacetate transport protein	*Paxillus involutus–Betula pendula*	2–8↑ 14d↓	Le Quéré et al. (2005)
Peptide transport protein	*Laccaria bicolor–Pinus resinosa*	6–72h↑	Podila et al. (2002)
Plasma membrane $H^{(+)}$-ATPase	*Paxillus involutus–Betula pendula*	2d↓	Le Quéré et al. (2005)
Probable mitochondrial carrier protein	*Paxillus involutus–Betula pendula*	25d↓	Johansson et al. (2004)
Putaive Coatomer zeta subunit	*Paxillus involutus–Betula pendula*	25d↑	Johansson et al. (2004)
Putaive syntaxin	*Paxillus involutus–Betula pendula*	25d↑	Johansson et al. (2004)
Putative Ca^{2+} transporting ATPase	*Laccaria bicolor–Pinus resinosa*	6–72h↑	Podila et al. (2002)
Putative component of COPII-coated vesicle	*Paxillus involutus–Betula pendula*	25d↓	Johansson et al. (2004)
Ran interactin giant nucleopore protein	*Laccaria bicolor–Pinus resinosa*	6–72h↑	Podila et al. (2002)
Ryanodine receptor	*Laccaria bicolor–Pinus resinosa*	6–24h↑	Kim et al. (1999b)
Urea active transport protein	*Paxillus involutus–Betula pendula*	12w↑	Morel et al. (2005)
Zn transporter	*Laccaria bicolor–Pinus resinosa*	6–72h↑	Podila et al. (2002)

[a]Presented in additional material. In the time and expression tendency column, hyphen between numbers indicates interval time, (h) hours, (d) days, (w) weeks, and (m) month. ↑ Indicates over expression while ↓ indicates down expression

membrane $H^{(+)}$-ATPase is involved in inorganic phosphorus transport especially in low availability conditions (Shen et al. 2006). ESTs for these genes were repressed in a 2-day-old *P. involutus–B. pendula* mycelium.

In the protein transport class of altered genes, the peptide transporter (stimulated between 6 and 72 h of *L. bicolor* and *P. involutus* mycelium), the nuclear transport factor 2 (repressed between 2 and 8 days of *Paxillus involutus–Betula pendula* association) and ß-importin gene were found. The ß-importin gene encodes for a protein that translocates histones from the cytoplasm to the nucleus through the nuclear channel (Jakel et al. 1999); and here it was observed as up-regulated in *L. bicolor–P. resinosa* mycelium. The Bip protein, which is related to increased cellular protein excretion via cellular vesicle, was up-regulated in *L. bicolor–P. resinosa* mycelium (Punt et al. 1998).

5.2.6 Signaling and Structural Membrane Proteins

Two groups of molecules are in evidence in this category: the SRAPs (symbiosis-regulated acidic polypeptides) and hydrophobins (Table 5.6). SRAPs are commonly observed to be stimulated in *P. tinctorius–E. globulus* ectomycorrhiza formation and considered to be a strong association marker for this interaction (Voiblet et al. 2001). However, *P. microcarpus–E. globulus* mycelium demonstrated several ESTs that encode for SRAPs that were suppressed in early ectomycorrhizal stages (Duplessis et al. 2005).

Hydrophobins are small molecular weight peptides that are moderately hydrophobic proteins containing eight conserved cysteine residues. They are found in the hypha cell wall or are secreted by hyphae (Kershaw and Talbot 1998; Wösten and Vocht 2000; Tagu et al. 2001). Several hydrophobins have been reported in ectomycorrhiza and many of them are stimulated in fully developed or advanced stages of symbiosis (Table 5.6). However, recently, a few different hydrophobin transcripts (hydrophobin 2 and 3) were determined to be repressed during *P. tinctorius – C. sativa* preinfections mycelium indicating a difference in hydrophobin gene regulation during the early interaction stages (Acioli-Santos et al. 2008).

5.2.7 DNA/RNA Processing

Thirteen ESTs were related to DNA replication and nucleic acid maintenance/processing. Interestingly, most of these genes were repressed in early ectomycorrhizal association (Table 5.7). Maintenance and processing of DNA genes mostly showed increased expression in the intermediate ectomycorrhiza formation stage. Helicases, important enzymes during DNA replication, were up-regulated in *P. microcarpus* associated with *E. globulus* at 12 days of association. The coiled coil domain is related to DNA α-helix folding, and works as a gene expression controller element (Yu 2002); this was observed in *P. involutus* associated with *B. pendula* (up-regulated between 4 and 8 days).

Table 5.6 Some signaling and structural membrane protein genes differentially expressed in ectomycorrizal fungus

Gene	Ectomycorrhiza	Time/gene expression tendency	Reference
β-transducin	*Laccaria bicolor–Pinus resinosa*	6–72h↑	Podila et al. (2002)
CFTR	*Laccaria bicolor–Pinus resinosa*	6–24 h↑	Kim et al. (1999b)
Cornichon homolog	*Pisolithus microcarpus–Eucalyptus globulus*	4–7↓ 12↑ 21d↓	Duplessis et al. (2005)[a]
Extensin like plant protein	*Amanita muscaria–Picea abies*	2m↑	Nehls et al. (1999)
FUN34/ATO protein	*Pisolithus microcarpus–Eucalyptus globulus*	4–12↓ 21d↑	Duplessis et al. (2005)[a]
GNS1/SUR4 family protein	*Paxillus involutus–Betula pendula*	2–14d↑	Le Quéré et al. (2005)
GPR/FUN34 family protein	*Paxillus involutus–Betula pendula*	2–8d↓	Le Quéré et al. (2005)
GTP bindin protein	*Paxillus involutus–Betula pendula*	2–8d↓	Le Quéré et al. (2005)
Het-c	*Paxillus involutus–Betula pendula*	2↓14d↓	Le Quéré et al. (2005)
Hydorphobin hydPt-8	*Pisolithus tinctorius–Eucalyptus globulus*	4d↑	Voiblet et al. (2001)
HydPt-1 prom	*Pisolithus microcarpus–Eucalyptus globulus*	4–7↓ 12↑ 21d↓	Duplessis et al. (2005)[a]
Hydrofobin hydPt-1	*Pisolithus tinctorius–Eucalyptus globulus*	0.5–7d↑	Tagu et al. (1996)
Hydrofobin hydPt-2	*Pisolithus tinctorius–Eucalyptus globulus*	0.5–7d↑	Tagu et al. (1996)
Hydrofobin hydPt-5	*Pisolithus tinctorius–Eucalyptus globulus*	4d↑	Voiblet et al. (2001)
Hydrofobin hydPt-6	*Pisolithus tinctorius–Eucalyptus globulus*	4d↑	Voiblet et al. (2001)
Hydrophobin-3 precursor	*Paxillus involutus–Betula pendula*	25d↑	Johansson et al. (2004)
Hydrophobin 1	*Paxillus involutus–Betula pendula*	4–14d↑	Le Quéré et al. (2005)
Hydrophobin hydPt-4	*Pisolithus tinctorius–Eucalyptus globulus*	4d↑	Voiblet et al. (2001)
Hydrophobin-2	*Pisolithus tinctorius–Castanea sativa*	12h↓	Acioli-Santos et al. (2008)

(*continued*)

Table 5.6 (continued)

Gene	Ectomycorrhiza	Time/gene expression tendency	Reference
Hydrophobin-2	*Pisolithus microcarpus–Eucalyptus globulus*	4↑ 7–21d↓	Duplessis et al. (2005)[a]
Hydrophobin-2	*Pisolithus microcarpus–Eucalyptus globulus*	4↓ 7↑ 12–21d↓	Duplessis et al. (2005)[a]
Hydrophobin-3	*Pisolithus tinctorius–Castanea sativa*	12h↓	Acioli-Santos et al. (2008)
Hydrophobin-3	*Pisolithus microcarpus–Eucalyptus globulus*	4↑ 7–21d↓	Duplessis et al. (2005)[a]
Hydrophobin-3	*Pisolithus microcarpus–Eucalyptus globulus*	4↓ 7↑ 12–21d↓	Duplessis et al. (2005)[a]
Hydrophobin-3	*Pisolithus microcarpus–Eucalyptus globulus*	4–12↓ 21d↑	Duplessis et al. (2005)[a]
IRS1 like protein	*Laccaria bicolor–Pinus resinosa*	6–72h↑	Podila et al. (2002)
LZK protein kinase	*Laccaria bicolor–Pinus resinosa*	6–72h↑	Podila et al. (2002)
Malignant T cell amplified protein	*Paxillus involutus–Betula pendula*	2↓14d↓	Le Quéré et al. (2005)
Probable membrane protein YOL 063e	*Laccaria bicolor–Pinus resinosa*	6–72h↑	Podila et al. (2002)
Profilaggrin	*Laccaria bicolor–Pinus resinosa*	6–72h↑	Podila et al. (2002)
Receptor kinase like protein	*Laccaria bicolor–Pinus resinosa*	6–72h↑	Podila et al. (2002)
S receptor kinase	*Laccaria bicolor–Pinus resinosa*	6–72h↑	Podila et al. (2002)
S.ar ribonucleoprotein	*Pisolithus microcarpus–Eucalyptus globulus*	4–7↓ 12↑ 21d↓	Duplessis et al. (2005)[a]
Small ribonucleoprotein	*Paxillus involutus–Betula pendula*	2–14d↑	Le Quéré et al. (2005)
Snod Prot1	*Pisolithus microcarpus–Eucalyptus globulus*	4–7↓ 12–21d↑	Duplessis et al. (2005)[a]
Spil-GTP-biding protein	*Paxillus involutus–Betula pendula*	2–8d↓	Le Quéré et al. (2005)
SRAP17	*Pisolithus tinctorius–Eucalyptus globulus*	4d↑	Voiblet et al. (2001)
SRAP17	*Pisolithus microcarpus–Eucalyptus globulus*	4↑ 7–21d↓	Duplessis et al. (2005)[a]

(*continued*)

Table 5.6 (continued)

Gene	Ectomycorrhiza	Time/gene expression tendency	Reference
SRAP17	*Pisolithus microcarpus–Eucalyptus globulus*	4–7↓ 12–21d↑	Duplessis et al. (2005)[a]
SRAP17	*Pisolithus microcarpus–Eucalyptus globulus*	4–12↓ 21d↑	Duplessis et al. (2005)[a]
SRAP17	*Pisolithus microcarpus–Eucalyptus globulus*	4–7↓ 12↑ 21d↓	Duplessis et al. (2005)[a]
SRAP32 Type I	*Pisolithus microcarpus–Eucalyptus globulus*	4↑ 7–21d↓	Duplessis et al. (2005)[a]
SRAP32 Type II	*Pisolithus microcarpus–Eucalyptus globulus*	4–7↓ 12↑ 21d↓	Duplessis et al. (2005)[a]
SRAP32-1	*Pisolithus tinctorius–Eucalyptus globulus*	4d↑	Voiblet et al. (2001)
SRAP32-3	*Pisolithus tinctorius–Eucalyptus globulus*	4d↑	Voiblet et al. (2001)
Steroid binding protein	*Paxillus involutus–Betula pendula*	2↓14d↓	Le Quéré et al. (2005)
Transmembrane FUN 34 protein	*Pisolithus tinctorius–Eucalyptus globulus*	4d↑	Voiblet et al. (2001)

[a]Presented in additional material. In the time and expression tendency column, hyphen between numbers indicates interval time, (h) hours, (d) days, (w) weeks, and (m) month. ↑ Indicates over expression while ↓ indicates down expression

In RNA processing, five ESTs showed repressed expression during short interaction periods. These ESTs include transcripts that encode for RNA binding motif protein, PolyA-binding protein, pre-rRNA processing, mRNA maturase bl2 and U6 snRNA-associated Sm-like protein Lsm8. The mRNA maturase bl2 participates in mRNA splicing (Szczepaneck and Lazowska 1996) and U6 snRNA-associated/Sm-like protein Lsm8 is a protein within the mRNA decay enzyme complex (He and Parker 2000).

5.2.8 Cell Defense and Apoptosis

Table 5.8 shows some oxidative defense genes observed in EMF. Brain selenoprotein, glutathione peroxidase and thiol-specific antioxidant are enzymes that act to protect organisms against reactive oxygen species (Moon et al. 1994, Koga et al. 1998). These genes were observed in *L. bicolor* and *P. involutus* mycelium.

Table 5.7 Some DNA/RNA processing genes differentially expressed in ectomycorrizal fungus

Gene	Ectomycorrhiza	Time/gene expression tendency	Reference
Coiled-coil protein	*Paxillus involutus–Betula pendula*	2↓ 4–8↑ 14d↓	Le Quéré et al. (2005)
GATA zinc finger protein	*Pisolithus microcarpus–Eucalyptus globulus*	4↓ 7↑ 12–21d↓	Duplessis et al. (2005)[a]
Helicase homolog	*Pisolithus microcarpus–Eucalyptus globulus*	4–7↓ 12↑ 21d↓	Duplessis et al. (2005)[a]
Histone	*Pisolithus microcarpus–Eucalyptus globulus*	4–7↓ 12↑ 21d↓	Duplessis et al. (2005)[a]
Histone gene complex	*Pisolithus microcarpus–Eucalyptus globulus*	4–7↓ 12↑ 21d↓	Duplessis et al. (2005)[a]
mRNA maturase bl2	*Pisolithus tinctorius–Eucalyptus globulus*	4d↓	Voiblet et al. (2001)
PolyA-binding protein	*Pisolithus microcarpus–Eucalyptus globulus*	4–7↓ 12↑ 21d↓	Duplessis et al. (2005)[a]
Pre-rRNA processing	*Pisolithus microcarpus–Eucalyptus globulus*	4–7↓ 12↑ 21d↓	Duplessis et al. (2005)[a]
Rad51	*Pisolithus microcarpus–Eucalyptus globulus*	4–7↓ 12↑ 21d↓	Duplessis et al. (2005)[a]
RNA binding motif protein	*Pisolithus microcarpus–Eucalyptus globulus*	4–7↓ 12↑ 21d↓	Duplessis et al. (2005)[a]
Sin3 associated poly.	*Pisolithus microcarpus–Eucalyptus globulus*	4–7↓ 12↑ 21d↓	Duplessis et al. (2005)[a]
ssDNA biding protein	*Laccaria bicolor–Pinus resinosa*	6–72h↑	Podila et al. (2002)
U6 snRNA-associated Sm-like protein Lsm8	*Pisolithus tinctorius–Eucalyptus globulus*	4d↓	Voiblet et al. (2001)

[a]Presented in additional material. In the time and expression tendency column, hyphen between numbers indicates interval time, (h) hours, (d) days, (w) weeks, and (m) month. ↑ Indicates over expression while ↓ indicates down expression

The metallothionein mRNA identified in *P. involutus* and *P. microcarpus* mycelium and the arsenite ATPase mRNA both have increased expression in advanced stages of development in their respective ectomycorrhizal systems. These genes encode for enzymes important for toxic metal tolerance (Haerslev et al. 1995) and resistance to arsenic (Silver et al. 1989), respectively. Aldose reductase acts on the metabolism of carbohydrates (such as fructose, mannose,

Table 5.8 Some cell defense and appopitosis genes differentially expressed in ectomycorrizal fungus

Gene	Ectomycorrhiza	Time/gene expression tendency	Reference
ABC transporter MSBA	*Pisolithus microcarpus–Eucalyptus globulus*	4↑ 7–21d↓	Duplessis et al. (2005)[a]
Aldose reductase	*Paxillus involutus–Betula pendula*	12w↑	Morel et al. (2005)
Arsenite-translocating ATPase	*Paxillus involutus–Betula pendula*	12w↑	Morel et al. (2005)
Bap31 protein	*Pisolithus microcarpus–Eucalyptus globulus*	4–7↓ 12↑ 21d↓	Duplessis et al. (2005)[a]
Brain seleno protein	*Laccaria bicolor–Pinus resinosa*	6–72h↑	Podila et al. (2002)
Calcineurin B subunit	*Pisolithus microcarpus–Eucalyptus globulus*	4–7↓ 12↑ 21d↓	Duplessis et al. (2005)[a]
Cycloheximide resist.	*Pisolithus microcarpus–Eucalyptus globulus*	4–7↓ 12–21d↑	Duplessis et al. (2005)[a]
Cytochrome P450	*Paxillus involutus–Betula pendula*	25d↓	Johansson et al. (2004)
Cytochrome P450	*Paxillus involutus–Betula pendula*	2–8↑ 14d↓	Le Quéré et al. (2005)
Cytochrome P450	*Tuber borchii–Tilia americana*	30d↑	Menotta et al. (2004)
Dna J protein	*Pisolithus microcarpus–Eucalyptus globulus*	4–7↓ 12↑ 21d↓	Duplessis et al. (2005)[a]
Dna J Protein	*Laccaria bicolor–Pinus resinosa*	6–72h↑	Podila et al. (2002)
Glutathione peroxidase	*Laccaria bicolor–Pinus resinosa*	6–72h↑	Podila et al. (2002)
Glutathione S-transferase III	*Laccaria bicolor–Pinus resinosa*	6–72h↑	Podila et al. (2002)
Glutathione-S-transferase	*Paxillus involutus–Betula pendula*	25d↓	Johansson et al. (2004)
Glutathione-S-transferase	*Paxillus involutus–Betula pendula*	2–8d↓	Le Quéré et al. (2005)
Glutathione-S-transferase	*Paxillus involutus–Betula pendula*	2–14d↑	Le Quéré et al. (2005)
HSP 104	*Pisolithus tinctorius–Castanea sativa*	12h↓	Acioli-Santos et al. (2008)
HSP 70	*Paxillus involutus–Betula pendula*	2–8d↓	Le Quéré et al. (2005)
HSP 70-related protein HSS1	*Pisolithus tinctorius–Castanea sativa*	12h↓	Acioli-Santos et al. (2008)
HSP 90 BB, 1 beta isoform	*Pisolithus tinctorius–Castanea sativa*	12h↓	Acioli-Santos et al. (2008)

(*continued*)

Table 5.8 (continued)

Gene	Ectomycorrhiza	Time/gene expression tendency	Reference
HSP 10 mitochondrial	*Pisolithus microcarpus–Eucalyptus globulus*	4–7↓ 12↑ 21d↓	Duplessis et al. (2005)[a]
LEA protein homolog	*Pisolithus microcarpus–Eucalyptus globulus*	4–7↓ 12↑ 21d↓	Duplessis et al. (2005)[a]
Macrolide-binding protein	*Pisolithus microcarpus–Eucalyptus globulus*	4–7↓ 12–21d↑	Duplessis et al. (2005)[a]
Metallothionein	*Paxillus involutus–Betula pendula*	25d↑	Johansson et al. (2004)
Metallothionein	*Pisolithus microcarpus–Eucalyptus globulus*	4–12↓ 21d↑	Duplessis et al. (2005)[a]
PAPS reductase	*Paxillus involutus–Betula pendula*	2↓ 14d↓	Le Quéré et al. (2005)
PCAR apoptotic related	*Laccaria bicolor–Pinus resinosa*	6–72h↑	Podila et al. (2002)
Rehydrin-like protein	*Paxillus involutus–Betula pendula*	25d↑	Johansson et al. (2004)
Related to trichodiene oxygenase Cytochrome P450	*Tuber borchii–Tilia americana*	30d↑	Menotta et al. (2004)
TBP like Cyp450	*Laccaria bicolor–Pinus resinosa*	6–72h↑	Podila et al. (2002)
Thiol-spec. antioxidant protein	*Paxillus involutus–Betula pendula*	2–8↓ 14d↑	Le Quéré et al. (2005)
Thioredoxin	*Paxillus involutus–Betula pendula*	2–8d↓	Le Quéré et al. (2005)
Thioredoxin reductase	*Pisolithus tinctorius–Eucalyptus globulus*	4d↑	Voiblet et al. (2001)
Thioredoxin reductase	*Pisolithus microcarpus–Eucalyptus globulus*	4–7↓ 12↑ 21d↓	Duplessis et al. (2005)[a]
Tpo3 MFS multidrug efflux transporter	*Paxillus involutus–Betula pendula*	12w↑	Morel et al. (2005)
Txl protein	*Pisolithus microcarpus–Eucalyptus globulus*	4↓ 7↑ 12–21d↓	Duplessis et al. (2005)[a]
WSC4 homolog	*Pisolithus microcarpus–Eucalyptus globulus*	4–7↓ 12↑ 21d↓	Duplessis et al. (2005)[a]

[a]Presented in additional material. In the time and expression tendency column, hyphen between numbers indicates interval time, (h) hours, (d) days, (w) weeks, and (m) month. ↑ Indicates over expression while ↓ indicates down expression

and galactose) and, in yeast, is related to several stress response mechanisms such as NaCl, H_2O_2 and carbon unavailability (Aguilera and Prieto 2001). Aldose reductase ESTs were identified as up-regulated in 12-week-old *P. involutus–B. pendula* mycelium (Morel et al. 2005).

Glutathione S transferase acts to detoxify the cell (Hayes and Pulford 1995) and was identified in ectomycorrhiza mycelium. Podila et al. (2002) observed up regulation of Glutathione S transferase III in *L. bicolor* associated with *P. resinosa* between 6 and 72 h of contact. However, Johansson et al. (2004) identified the same gene repressed in mycelium from *P. involutus–B. pendula* symbiosis at 25 days of association.

The cytochrome P450 family acts broadly in many oxidative functions including defense mechanisms and stress resistance (Chapple 1998). Cytochrome P450 ESTs were identified in different ectomycorrhizal combinations (Table 5.8). Rehydrin ESTs were observed overexpressed after 25 days in *P. involutus–B. pendula* mycelium potentially relating to dehydration resistance (Velten and Oliver 2001). The HSP (heat shock proteins) are stress cellular response proteins with the major cellular HSPs including HSP60, HSP70 and HSP90 (molecular weights of 60 kD, 70 kD and 90 kD, respectively). These proteins play an essential chaperone role facilitating intracellular transport and peptide interactions (Gupta 1995). Five HSP ESTs were observed as differentially expressed in mycelium associated (or in the process of association) with different host plants. In *P. tinctorius–C. sativa* system (Acioli-Santos et al. 2008); the mycelium showed three HSP transcripts down regulated after 12 h (HSP90BB 1 beta isoform, HSP70 HSS1-related protein, and HSP104) in response to mycorrhizal stimulus. The HSP70 EST was also observed repressed in *P. involutus–B. pendula* fungal cells (2–8 days). Finally, the HSP10 was repressed in *P. microcarpus–E. globulus* during early and then late symbiosis development. These data suggest a HSP repression trend in short interaction/formation intervals.

The PCAR apoptotic gene was observed in *L. bicolor–P. resinosa* fungal cells (Podila et al. 2002) and may be related to the hyphal apoptosis processes because apoptotic genes have been identified in yeast (Madeo et al. 2002).

5.3 Conclusions

Symbiotic ectomycorrhizal final tissue results from roots and fungal cells interaction under temporal/spatial regulation by both partner genetic programs. This process involves environmental sensing and cell–cell communication that culminates in the construction of new symbiotic structures. The reviewed data indicates the use of new genetic programs during different ectomycorrhizal phases. Thus, related genes change expression levels during formation and maintenance of symbiotic relationships; however, all of them are basally expressed in most symbiotic conditions. There seems to be a tendency for gene expression suppression (down-regulation) during short association periods or even the preinfection interaction. This tendency can be observed in protein synthesis genes, carbon metabolism genes and membrane proteins like SRAPs and hydrophobins, which are proteins believed to be important agents for ectomycorrhiza formation. If these findings are confirmed, the research focus for association genetic markers should be directed toward studying shorter interaction intervals to identify potentials key elements of ectomycorrhizal development.

However, in several experiments not discussed in this chapter, genes of unknown function that could be potential participants in ectomycorrhizal associations (Podila et al. 2002; Menotta et al. 2004; Johansson et al. 2004; Acioli-Santos et al. 2008; Martin et al. 2008) were observed. This fact highlights a knowledge gap concerning fungal origin sequences in symbiosis and the urgent emerging need to encourage future studies in molecular ectomycorrhizal associations.

References

Acioli-Santos B, Sebastiana M, Pessoa F, Sousa L, Figueiredo A, Fortes AM, Balde A, Maia LC, Pais MS (2008) Fungal transcript pattern during the preinfection stage (12 h) of ectomycorrhiza formed between *Pisolithus tinctorius* and *Castanea sativa* roots, identified using cDNA microarrays. Curr Microbiol 57:620–625

Aguilera J, Prieto JA (2001) The *Saccharomyces cerevisiae* aldose reductase is implied in the metabolism of methylglyoxal in response to stress conditions. Curr Genet 39:273–283

Babbitt PC, Hasson MS, Wedekind JE, Palmer DJ, Barrett WC, Reed GH, Rayment I, Ringe D, Kenyon GL, Gerlt JA (1996) The enolase superfamily: a general strategy for enzyme-catalyzed abstraction of the r-protons of carboxylic acids. Biochemistry 35:16489–16501

Bellamine A, Mangla AT, Nes WD, Waterman MR (1999) Characterization and catalytic properties of the sterol 14 α-demethylase from *Mycobacterium tuberculosis*. Biochemistry 96:8937–8942

Bender A, Pringle JR (1991) Use of a screen for synthetic lethal and multicopy suppressed mutants to identify two new genes involved in morphogenesis in *Saccharomyces cerevisiae*. Mol Cell Biol 11:1295–1305

Berg JM, Tymoczko JL, Stryer L (2004) Bioquímica 5ª Ed. Guanabara Koogan S.A., Rio de Janeiro, p p1059

Bydlowiski P, Magnannelli AC, Chamone DAF (1998) Hyper-homocistenemia e doenças vaso-oclusivas. Arquivos Brasileiros de Cardiologia 71:69–76

Campbell JL, Scheckman R (1997) Selective packaging of cargo molecules into endoplasmic reticulum-derived COPII vesicles. Proc Natl Acad Sci 94:837–842

Chan HM, La Thangue NB (2001) P300/CBP proteins: HATs for transcriptional bridges and scaffolds. J Cell Sci 114:2363–2373

Chapple C (1998) Molecular-genetics analysis of plant cytochrome P450-dependent monoxigenases. Annu Rev Plant Physiol Plant Mol Biol 49:311–343

Chowdhury S, Smith KW, Gustin MC (1992) Osmotic stress and the yeast cytoskeleton: phenotype-specific suppression of an actin mutation. J Cell Biol 118:561–571

Condeelis J (1995) Elongation factor 1α, translation and the cytoskeleton. Trends Biochem Sci 20:169–170

Duplessis S, Courty P, Tagu D, Martim F (2005) Transcript patterns associated with ectomycorrhiza development in *Eucalyptus globulus* and *Pisolithus microcarpus*. New Phytol 165:599–611

Eades CJ, Gilbert A, Goodman CD, Hintz WE (1998) Identification and analysis of a class 2 alpha-mannosidase from *Aspergillus nidulans*. Glycobiology 8:17–33

Emanuelsson O, Elofsson A, Von Heijne G, Cristobal S (2003) In silico prediction of the peroxisomal proteome in fungi, plants and animals. J Mol Biol 330:443–456

Gupta RS (1995) Phylogenetic analysis of 90kD heat shock family of protein sequences and an examination of the relationship among animals, plants and fungi species. Mol Biol Evol 12:1063–1073

Haerslev T, Jacobson GK, Zedeler K (1995) The prognostic significance of immunohisto-chemically detectable metallothionein in primary breast carcinomas. Acta Pathol Microbiol Immunol Scand 103:279–285

Hasunuma K, Yabe N, Yoshida Y, Ogura Y, Hamada T (2003) Putative functions of nucleoside diphosphate kinase in plants and fungi. J Bioenerg Biomembr 35:57–65

Hayes JD, Pulford DJ (1995) The glutathione s-transferase supergene family: regulation of GST and the contribution of the isoenzymes to cancer chemoprotection and drug resistance. Crit Rev Biochem Mol Biol 30:445–600

He W, Parker R (2000) Functions of Lsm proteins in mRNA degradation and splicing. Curr Opin Cell Biol 12:346–350

Hirata D, Nakano K, Fukui M, Takenaka H, Miyakawa T, Mabuchi I (1998) Genes that cause aberrant cell morphology by overexpression in fission yeast: a role of a small GTP-binding protein Rho2 in cell morphogenesis. J Cell Sci 111:149–159

Hogan DA, Auchtung TA, Hausinger RP (1999) Cloning and characterization of a sulfonate/ a-ketoglutarate dioxygenase from *Saccharomyces cerevisiae*. J Bacteriol 181:5876–5879

Imai K, Kijima T, Noda Y, Sutoh K, Yoda K, Adachi H (2002) A Rho GDP-dissociation inhibitor is involved in cytokinesis of Dictyostelium. Biochem Biophys Res Commun 296:305–312

Jakel S, Albig W, Kutay U, Bischoff FR, Schwamborn K, Doenecke D, Gorlich D (1999) The importin ß/importin 7 heterodimer is a functional nuclear import receptor for histone H1. Eur Mol Biol Organ J 18:2411–2423

Johansson T, Le Quéré A, Ahren D, Söderström B, Erlandsson R, Lundeberg J, Uhlén M, Tunlid A (2004) Transcriptional responses of *Paxillus involutus* and Betula pendula during formation of ectomycorrhizal root tissue. Mol Plant Microbe Interact 17:202–205

Kawalleck P, Plesh G, Hahlbrock K, Somssich IE (1992) Induction by fungal elicitor of S-adenosyl-L-methionine synthetase and S-adenosyl-L-homocysteine hydrolase mRNAs in cultured cell and leaves of *Petroselinum crispum*. Proc Natl Acad Sci 89:4713–4717

Kershaw MJ, Talbot NJ (1998) Hydrophobins and repellents: proteins with fundamental roles in fungal morphogenesis. Fungal Genet Biol 23:18–33

Kim S, Bernreuther D, Thumm M, Podila G (1999a) *LB.AUT7*, a novel symbisis-regulated gene from an ectomycorrhizal fungus, *Laccaria bicolor*, is functionally related to vesicular transport and autophagocytosis. J Bacteriol 181:1963–1967

Kim S, Shivanand T, Podila G (1999b) Cloning and identification of symbiosis-regulated genes from the ectomycorrhizal *Laccaria bicolor*. Mycol Res 103:168–172

Koga M, Tanaka H, Yomogida K, Tsuchida J, Uchida K, Kitamura M, Sakoda S, Matsumiya K, Okuyama A, Nishimune Y (1998) Expression of selenoprotein-p messenger ribonucleic acid in the rat testis. Biol Reprod 58:261–265

Kong FK (1995) Influence of copper, manganese and pH on the growth and several enzyme activities in mycorrhizal fungus *Amanita muscaria*. Hemosphere 30:199–207

Kuge O, Hara-Kuge S, Orci L, Ravazzola M, Amherdt M, Tanigawa G, Wieland FT, Rothman JE (1993) ζ-COP, a subunit of coatomer, is required for COP-coated vesicle assembly. J Cell Biol 123:1727–1734

Lambeth JD, Seybert DW, Kamin H (1979) Ionic effects on adrenal steroidogenic electron transport. The role of adrenodoxin as an electron shuttle. J Biol Chem 254:7255–7264

Lammers PJ (2004) Symbiotic signalling: new functions for familiar proteins. New Phytol 161:324–326

Le Quéré A, Wright DP, Söderström B, Tunlid A, Johansson T (2005) Global patterns of gene regulation associated with the development of ectomycorrhiza between Birch (*Betula pendula* Roth.) and *Paxillus involutus*. Mol Plant Microbe Interact 18:659–673

Lei J, Lapayerie F, Malajczuk N, Dexheimer J (1990) Infectivity of pine and eucalypt isolates of *Pisolithus tinctorius* (Pers.) Coker & Couch on toots of *Eucalyptus urophylla* S. T. Blake *in vitro*. II. Ultrastructural and biochemical changes at the early stage of mycorrhizal formation. New Phytol 116:115–122

Macheroux P, Schmid J, Amrhein N, Schaller A (1999) A unique reaction in a common pathway: mechanism and function of chorismate synthase in the shikimate pathway. Planta 207:325–334

Madeo F, Herker E, Maldener C, Wissing S, LäChelt S, Herlan M, Fehr M, Lauber K, Sigrist SJ, Wesselborg S, Fröhlich KU (2002) A caspase-related protease regulates apoptosis in yeast. Mol Cell 9:911–917

Manzow S, Brancolini C, Marks F, Richter KH (1996) Expression of growth arrest-specific (Gas) genes in murine keratinocytes: Gas2 is specifically regulated. Exp Cell Res 224:200–203

Martin F, Aerts A, Ahrén D, Brun A, Danchin EGJ, Duchaussoy F, Gibon J, Kohler A, Lindquist E, Pereda V, Salamov A, Shapiro HJ, Wuyts J, Blaudez D, Buée M, Brokstein P, Canbäck B, Cohen D, Courty PE, Coutinho PM, Delaruelle C, Detter JC, Deveau A, Difazio S, Duplessis S, Tachet LF, Lucic E, Klett PF, Fourrey C, Feussner I, Gay G, Grimwood J, Hoegger PJ, Jain P, Kilaru S, Labbé J, Lin YC, Legué V, Le Tacon F, Marmeisse R, Melayah D, Montanini B, Muratet M, Nehls U, Hirzel HN, Le Secq MPO, Peter M, Quesneville H, Rajashekar B, Reich M, Rouhier N, Schmutz J, Yin T, Chalot M, Henrissat B, Kües U, Lucas S, Peer YV, Podila GK, Polle A, Pukkila PJ, Richardson PM, Rouzé P, Sanders IR, Stajich JE, Tunlid A, Tuskan G, Grigoriev IV (2008) The genome of *Laccaria bicolor* provides insights into mycorrhizal symbiosis. Nature 452:88–92

Masson D, Kreis TE (1993) Identification and molecular characterization of E-MAP -115, a novel microtubule-associated protein predominantly expressed in epithelial cells. J Cell Biol 123: 357–371

Menotta M, Amicucci A, Sisti D, Gioacchini AM, Stocchi V (2004) Differential gene expression during pre-symbiotic interaction between *Tuber borchii* Vittad. and *Tilia Americana* L. Curr Genet 46:158–165

Moon BY, Ho ZC, Sue GR, Chock PB, Earl RS (1994) On the protective mechanism of the thiol-specific antioxidant enzyme against the oxidative damage of biomacromolecules. J Biol Chem 269:1621–1626

Morel M, Jacob C, Kohler A, Johansson T, Martin F, Charlot M, Brun A (2005) Indentification of genes differentially expressed in extraradical mycelium and ectomycorrizal roots during *Paxillus involutus–Petula pendula* ectomycorrhizal symbiosis. Appl Environ Microbiol 71: 382–391

Nehls U, Mikolajewski S, Ecke M, Hampp R (1999) Identification and expression analysis of two fungal cDNA regulated by ectomycorrhiza and fruit body formation. New Phytol 144: 195–202

Ntambi JM, Buhrow SA, Kaestner KH, Christy RJ, Sibley E, Kelly TJ Jr, Lane MD (1988) Differentiation-induced gene expression in 3T3-Ll preadipocytes. Characterization of a differentially expressed gene encoding steaoryl-coa desaturase. J Biol Chem 263:17291–17300

Palmieri F (1994) Mitochondrial carrier proteins. FEBS Lett 346:48–54

Podila GK, Zheng J, Balasubramanian S, Sundaram S, Hiremath S, Brand JH, Hymes MJ (2002) Fungal gene expression in early symbiotic interactions between *Laccaria bicolor* and red pine. Plant Soil 244:117–128

Poinssot B, Vandelle E, Bentéjac M, Adrian M, Levis C, Brygoo Y, Garin J, Sicilia F, Thévenot PC, Pugin A (2003) The Endopolygalacturonase 1 from *Botrytis cinerea* activates grapevine defense reactions unrelated to its enzymatic activity. Mol Plant Microbe Interact 16:553–564

Polidori E, Agostini D, Zeppa S, Potenza L, Palma F, Sisti D, Stocchi V (2002) Identification of differentially expressed cDNA clones in *Tilia platiphyllos-Tuber borchii* ectomycorrhizae using a differential screening approach. Mol Genet Genomics 266:858–864

Porter TN, Li Y, Raushel FM (2004) Mechanism of the dihydroorotase reaction. Biochemistry 43:16285–16292

Punt PJ, Van Gemeren IA, Drint-Kuijvenhoven J, Hessing JGM, Van Muijlwijk-Harteveld GM, Beijersbergen A, Verrips CT, Van Den Hondel CAMJJ (1998) Analysis of the role of the gene *bipA,* encoding the major endoplasmic reticulum chaperone protein the secretion of homologous and heterologous proteins in black *Aspergilli.* Appl Microbiol Biotechnol 50:447–454

Rehling P, Model K, Brandner K, Kovermann P, Sickmann A, Meyer HE, Kühlbrandt W, Wagner R, Truscott KN, Pfanner N (2003) Protein insertion into the mitochondrial inner membrane by a twin-pore translocase. Science 299:1747–1751

Santos B, Snyder M (2005) Targeting of chitin synthase 3 to polarized growth sites in yeast requires Chs5p and Myo2p. J Cell Biol 136:95–110

Schmid J, Amrhein N (1995) Molecular organization of the shikimate pathway in higher plants. Phytochemistry 39:737–749

Seroussi F, Pan HQ, Kedra D, Roe BA, Dumanski JP (1998) Caracterization of human NIPSNAP1 gene from 22q12: a member of novel gene family. Gene 212:13–20

Shen H, Chen J, Wang Z, Yang C, Sasaki T, Yamamoto Y, Matsumoto H, Yan X (2006) Root plasma membrane H1-ATPase is involved in the adaptation of soybean to phosphorus starvation. J Exp Bot 57:1353–1362

Silver S, Misra TK, Laddaga RA (1989) DNA sequence analysis of bacterial toxic heavy metal resistances. Biol Trace Elem Res 21:145–163

Slater S, Houmiel KL, Tran M, Mitski TA, Taylor NB, Padgette SR, Gruys KJ (1998) Multiple b-ketothiolases mediate poly(b-hydroxyalkanoate) copolymer synthesis in *Ralstonia eutropha.* J Bacteriol 180:1979–1987

Smith CV, Huang C, Miczak A, Russell DG, Sacchettini JC, Bentrup KH (2003) Biochemical and structural studies of malate synthase from *Mycobacterium tuberculosis.* J Biol Chem 278: 1735–1743

Sundaram S, Kim SJ, Suzuki H, Mcquattie CJ, Hiremath ST, Podila G (2001) Isolation and characterization of a symbiosis-regulated *ras* from the ectomycorrhizal fungus *Laccaria bicolor.* Mol Plant Microbe Interact 14:618–628

Szczepaneck T, Lazowska J (1996) Replacement of two non-adjacent amino acid in the *S. serevisiae* bi2 intron-incoded RNA maturase is sufficient to gain a homing- endonuclease activity. Eur Mol Biol Organ J 15:3758–3767

Tagu D, Nasse B, Martin F (1996) Cloning and caracterization of hydrophobins-enconding cDNAs from the ectomycorrhizal Basidiomycete *Pisolithus tinctorius.* Gene 168:93–97

Tagu D, De Bellis R, Balestrini R, De Vries OMH, Piccolli G, Stocchi V, Bonfante P, Martin F (2001) Immuno localization of hydrophobin HYDPt-1 from de ectomycorrhizal basidiomycete *Pisolithus tinctorius* during colonization of *Eucalyptus globulus* roots. New Phytol 149: 127–135

Tasto JJ, Morrell JL, Gould KL (2003) An anillin homologue, Mid2p, acts during fission yeast cytokinesis to organize the septin ring and promote cell separation. J Cell Biol 160:1093–1103

Tatsumi A, Shoji J, Kikuma T, Arioka M, Kitamoto K (2007) Aggregation of endosomal-vacuolar compartments in the Aovps24-deleted strain in the filamentous fungus *Aspergillus oryzae.* Biochem Biophys Res Commun 362:474–479

Timonen S, Peterson L (2002) Cytoskeletons in mycorrhizal symbioss. Plant Soil 244:199–210

Timonen S, Söderström B, Raudaskoski M (1996) Dynamics of cytosckeletal proteins in developing pine ectomycorriza. Mycorrhiza 6:423–429

Velten J, Oliver MJ (2001) Tr288, a rehydrin with a dehydrin twist. Plant Mol Biol 45:713–722

Voiblet C, Duplessis S, Encelot L, Martin F (2001) Identifications of symbiosis-reguleted genes in *Eucalyptus globulus-Pisolitus tinctorius* ectomycorrhiza by differential hybridization of arrayed cDNAs. Plant J 25:181–191

Wilkinson CRM, Wallace M, Morphew M, Perry P, Allshire R, Javerzat JP, Mcintosh JR, Gordon C (1998) Localization of the 26S proteasome during mitosis and meiosis in fission yeast. EMBO J 17:6465–6476

Wösten HAB, Vocht ML (2000) Hydrophobins, the fungal coat unravelled. Biochim Biophys Acta 1469:79–86

Yamagata S (1989) Roles of O-acetyl-L-homoserine sulfhydrylases in micro-organisms. Biochimie 71:1125–1143

Yu YB (2002) Coiled-coils: stability, specificity, and drug delivery potential. Adv Drug Deliv Rev 54:1113–1129

Zahner HA, Harkins JE, Pringle JR (1996) Genetic analysis of the bipolar pattern of bud site selection in the yeast *Saccharomyces cerevisiae*. Mol Cell Biol 16:1857–1870

Zaretsky M, Sitrit Y, Mills D, Roth-Bejerano N, Kagan-Zur V (2006) Differential expression of fungal genes at preinfection and mycorrhiza establishment between *Terfezia boudieri* isolates and *Cistus incanus* hairy root clones. New Phytol 171:837–846

Chapter 6
Agrobacterium tumefaciens-Mediated Transformation of Ectomycorrhizal Fungi

Minna J. Kemppainen, Maria C. Alvarez Crespo, and Alejandro G. Pardo

6.1 Introduction

Introduction of homologous or heterologous genes into filamentous fungi requires DNA-mediated transformation systems. The breakthrough of efficient fungal transformation took place with the yeast *Saccharomyces cerevisiae* (Hinnen et al. 1978). This advance rapidly led to the development of DNA-transformation techniques for ascomycete filamentous model species such as *Neurospora crassa* and *Aspergillus nidulans* (Case et al. 1979; Ballance et al 1983; Tilburn et al. 1983; Yelton et al. 1984) which later on were used for genetic transformation of basidiomycete dimorphic pathogenic species such as *Ustilago maydis* and *Cryptococcus neoformans* (Wang et al. 1988; Edman and Kwon-Chung 1990) and filamentous saprotrophic basidiomycetes such as *Schizophyllum commune*, *Phanerochaete crysosporum*, *Coprinus* spp. and *Agaricus bisporus* (Binninger et al. 1987; Specht et al. 1988; Alic et al. 1989; van de Rhee et al. 1996). Most of these methodologies are based on permeabilization of cell membranes with polyethylene glycol (PEG) or electroporation, or a combination of both (Punt et al. 1987; Richey et al. 1989; Ward et al. 1989; Chakraborty and Kapoor 1990; Kuwano et al. 2008) and generally require preparation of fungal protoplasts. Introduction of DNA can be accompanied with restriction enzyme-mediated integration (REMI) for increasing transformation frequency (Kahmann and Basse 1999). Also direct introduction of DNA into intact cells by particle bombardment has been used for transforming filamentous fungi (Lorito et al. 1993; Christiansen et al. 1995). Transformation of filamentous fungi is traditionally carried out with linear or circular DNA that integrates into the fungal genome by nonhomologous or homologous integration mechanisms. The frequency

---

M.J. Kemppainen, M.C. Alvarez Crespo, and A.G. Pardo (✉)
Laboratorio de Micología Molecular, Departamento de Ciencia y Tecnología, Universidad Nacional de Quilmes, y Consejo Nacional de Investigaciones Científicas y Tecnicas (CONICET), Roque Sáenz Peña 352, (B1876BXD) Bernal, Provincia de Buenos Aires, Argentina
e-mail: apardo@unq.edu.ar

M. Rai and A. Varma (eds.), *Diversity and Biotechnology of Ectomycorrhizae*, Soil Biology 25, DOI 10.1007/978-3-642-15196-5_6,

of homologous vs. nonhomologous integration varies and it is highly dependent on the fungal species. The massive genetic research over the last 30 years on model filamentous ascomycetes has led to the availability of several integrative transformation vectors and genetic tools for this group of fungi. Also auto-replicative DNA elements are available for transformation of some fungal species such as *A. nidulans* and *U. maydis* (Tsukuda et al. 1988; Aleksenko 1994). Transformants selection of filamentous fungi is performed either with nutritional markers, requiring auxotrophic strains, or with dominant antibiotic markers. The most common antibiotic markers used for fungal transformation are hygromycin B, phelomycin, bialophos, sulfonylurea, nourseothricin, carboxin, blasticidin S and benomyl (Weld et al. 2006).

The transformation efficiency of filamentous fungi by traditional transformation techniques is often limited by the researcher's capacity to produce the necessary quantity of viable protoplasts. This is rather challenging and it is carried out by enzymatic degradation of the fungal cell wall. Optimizing the cell wall degradation protocol can be complicated by differences in the physiological state of the starting fungal material and batch variations in commercially available enzyme preparations. Successful protoplast preparation also generally requires young homogenous hyphal material. This is preferably obtained by freshly germinating sexual or asexual spores, oidia, or the yeast-phase of dimorphic filamentous fungi. If these are not produced by the fungus under study, vegetative mycelium can also be used for protoplast production. This is however not an optimal material due to wide physiological variation between different parts of the mycelium. Also PEG is known to cause fusion of protoplasts in filamentous fungi and fungal cells used to prepare protoplasts of *A. nidulans* usually contain multiple nuclei (Fiddy and Trinci 1976; Kuwano et al. 2008).

6.2 Genetic Modification of ECM Fungi with Traditional Transformation Methods

Basidiomycete ECM fungi have turned out to be especially difficult to modify via traditional genetic transformation methods. This is partly due to the fact that spore production cannot be obtained under laboratory conditions and production of viable protoplast needed for many transformation techniques has been difficult and requires working with rather slowly growing vegetative mycelia. On the other hand, regeneration of protoplast is extremely difficult for many ECM fungi, presenting another obstacle for genetic transformation. Furthermore, many antibiotic markers used for ascomycetes cannot be used in basidiomycetes due to natural resistance. Despite these difficulties some ECM fungi such as *Hebeloma cylindrosporum* and *Laccaria laccata* have been transformed via protoplast (Barret et al. 1990; Marmeisse et al. 1992). Particle bombardment mediated transformation, avoiding tedious protoplast production, has been reported for *Paxillus involutus*, *Pisolithus tinctorius* and *Suillus* spp. (Bills et al. 1995; Rodríguez-Tovar et al. 2005; Sunagawa et al. 2007) and also for *Laccaria bicolor* (Bills et al. 1999). However, the transformation efficiencies of ECM fungi have been rather low and obtaining transformants has been irregular and time consuming. Therefore, these original reports have not led to the efficient and widespread use of genetic fungal transformation in mycorrhizal research. On the other hand, the lack of functional genetic tools adapted to filamentous basidiomycetes (transformation plasmids adapted to ECM fungi, availability of selection markers, etc.) partly due to the low number of researchers dedicated to the subject has slowed down the advances in genetic transformation of ECM fungi.

6.3 *Agrobacterium*-Mediated Transformation of Fungi

Agrobacterium tumefaciens and *A. rhizogenes* are Gram-negative bacteria which cause crown gall tumors and hairy root disease in their natural hosts, dicotyledonous plants. These diseases are caused by the transfer of plant hormone biosynthesis genes and genes responsible for production of opine from bacteria to the plant. These genes are located in Ti-(tumor inducing) and Ri (root inducing)-plasmids and transferred as single-stranded T-DNA (ssT-DNA or T-strand), via a type IV secretion system to the host cell where then double-stranded T-DNA (dsT-DNA or T-DNA) incorporates in the host genome (Ream 2009). Opines produced by the plant in tumor tissue are used by the bacteria as C and N sources. For a long time *Agrobacterium* was thought to be the only example of horizontal interkingdom gene transfer in nature but also some other bacteria such as *Rhizobium* sp. *NGR234*, *Sinorhizobium meliloti* and *Mesorhizobium loti* were demonstrated to be capable of genetically transforming different plant tissues and plant species (Broothaerts et al. 2005).

The natural capacity of *Agrobacterium* to transfer genes to intact plant cells has been exploited by plant researchers to genetically modify both dicotyledonous and monocotyledonous plants for decades. *Agrobacterium*-mediated transformation (AMT) had already become a routine tool in plant sciences in the 1980s. The finding that the natural oncogenes in the T-DNA could be replaced by sequences of interest, and that these were equally transferred to the host plant for trans-genesis revolutionized plant research. Development of binary vector systems, where the T-DNA and virulence genes in a large size Ti-plasmid (~200 kb) were separated into two different replicons, an easily handled T-DNA containing binary vector and a virulence Ti-helper plasmid, also greatly increased the use of *Agrobacterium* in plant transformation (de Framond et al. 1983; Hoekema et al. 1983).

In 1995 Bundock et al. published the first report on AMT of *Saccharomyces cerevisiae* demonstrating that *A. tumefacies* is capable of transferring T-DNA not only to its natural hosts, plants, but also to fungi. This finding initiated a true revolution in fungal research. Demonstration of AMT of filamentous ascomycetes (de Groot et al. 1998) has led to the widespread use of *Agrobacterium* as a tool for fungal transformation. Several ascomycetes, basidiomycetes and also zygomycetes have been shown to be susceptible to this transformation methodology (Michielse et al. 2005b) and new fungal species are joining this group with increasing speed. The great benefit of AMT of fungi is that there is no need for complicated protoplast preparation. Intact fungal cells can be transformed via *Agrobacterium* and these cells can originate from different sources available: spores, germlings, vegetative mycelium or even fruiting body tissue. Many fungal species, before recalcitrant to or presenting very low transformation efficiencies with traditional gene transformation methods, have been efficiently modified via AMT. These also include several filamentous basidiomycetes. Despite the wide success of AMT on fungi not all species are transformed with this technique. Furthermore, the fungal susceptibility to AMT has been show to vary between different isolates of the same species (Covert et al. 2001; Sullivan et al. 2002; Fitzgerald et al. 2003). The precise reasons for this variation are not known but strain-specific variations on cell wall structures could be involved. Also several studies have shown that *Agrobacterium* strain/binary vector combination used is a fundamental factor for fungal transformation efficiency (Vijn and Govers 2003; Park and Kim 2004).

6.4 AMT of ECM Fungi

Successful *Agrobacterium* transformation of the homobasidiomycete *Agaricus bisporus* with hygromycin B resistance (de Groot et al. 1998; Chen et al. 2000) gave the first insight that AMT could also be used for genetic transformation of basidiomycete ECM fungi. This transformation method was soon proved functional by transforming vegetative mycelia of the ECM fungi *Suillus bovinus*, *Paxillus involutus* and *Hebeloma cylindrosporum* with the *Streptoalloteicus hindustanus* phleomycin resistance gene (*Sh BLE*) and the *Escherichia coli* hygromycin resistance gene (*hph*) as selection markers (Pardo et al. 2002; Hanif et al. 2002; Combier et al. 2003). Other AMT-modified ECM fungi are *Pisolithus tinctorius* (Pardo et al. 2005; Rodríguez-Tovar et al. 2005) and *Laccaria bicolor* (Kemppainen et al. 2005). *Tuber borchii*, the model ascomycete ECM fungus, has also been transformed with *Agrobacterium*. This transformation, however, did not lead to stable genomic integrations of the transgenes (Grimaldi et al. 2005). A similar situation has been shown in ECM basidiomycetes with phleomycin resistance as selection marker (Pardo et al. 2002, 2005). Also the ECM fungus *Amanita muscaria* has not been successfully transformed with *Agrobacterium* (Nehls, personal communication). It is clear that susceptibility of EMC fungi to AMT shows species variation and setting-up an AMT-based transformation system for a new species requires optimization.

Figure 6.1 shows a comparison between different transformation methods of fungi.

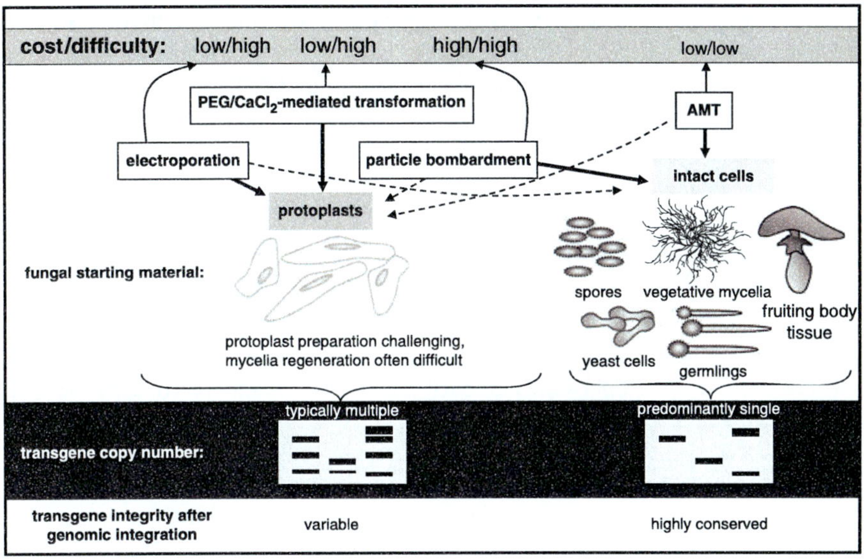

Fig. 6.1 A comparison of traditional genetic transformation methods and AMT of fungi. Most of the traditional methods require tedious preparation of protoplast, which is not the case for AMT. Also the number of AMT integrated transgenes is generally low and their integrity highly conserved making analysis of transformed strains simple. The cost/difficulty comparison includes both the price of reagents and special equipments needed, and the overall efficiency of each fungal transformation method

6.5 AMT-Mechanisms: T-DNA Transfer to the Host Cell

Agrobacterium natural hosts are plants but it can also transfer the T-DNA element to other eukaryotic cells such as fungi and also, under laboratory conditions to, mammalian cells (Kunik et al. 2001). The T-DNA transfer has been exhaustively studied in plant cells (McCullen and Binns 2006; Citovsky et al. 2007; Dafny-Yelin et al. 2008; Gelvin 2009; Ream 2009). How horizontal gene transfer between *Agrobacterium* and other nonplant hosts is carried out and which are the fundamental processes shared between all the hosts is not yet known (Lacroix et al. 2006). AMT is already a well established and a routine transformation method for many fungal species. Despite its popular use the knowledge on AMT of fungi at the molecular level is minimal and should receive more research interests in future.

The T-DNA transfer between bacteria and the plant host is a multistep process and it can be divided into three fundamental phases: (1) pretransfer bacterial processes, (2) T-DNA transfer and (3) host-cell events leading to T-DNA integration in the host genome. Steps 1 and 2 are mostly controlled by the bacterial protein-machinery but step 3 depends both on imported bacterial proteins and host cellular mechanisms. *Agrobacterium*-infection also activates several plant-host defense mechanisms such as production of reactive oxygen species, systemic acquired resistance processes and RNA silencing during early stages of the interaction but these are attenuated afterwards probably by direct repression of defense genes by the bacterium. Even though several factors participating in T-DNA mobilization and integration are known, the full understanding of AMT on plant cells is still incomplete and under active research.

6.5.1 Pretransfer Bacterial Processes Leading to T-DNA Mobilization

These events depend on the activity of bacterial chromosomal virulence genes (*chv*) and tumor-inducing Ti plasmid virulence genes (*vir*) which control host recognition, attachment and T-DNA production.

Host cell recognition starts by sensing the vicinity of a wounded plant. Wounded plant cells release phenolic compounds that are detected by the *Agrobacterium* two-component regulatory system of VirA and VirG and this induction leads to activation of other *vir* genes responsible for T-DNA mobilization. Acetosyringone (3,5-Dimethoxyacetophenone) a commercially available phenolic molecule, is a potent inducer of *Agrobacterium* and is used for inducing bacteria for transforming nonplant hosts or increasing plant transformation efficiency. The inner-membrane protein VirA, a kinase present as a dimer, detects the phenolic compound and this leads to its autophosphorylation. This phosphorylation is further transferred to VirG, a transcriptional activator protein with DNA-binding capacity. The activated VirG causes transcriptional activation of itself and other *vir*-genes in the Ti plasmid [such as *virB* operon genes (1–11), *virD1*, *virD2*, *virD4*, *virE1*, *virE2*, *virE3*]. The chromosomally encoded ChvE protein, present in the periplasmic space further increases the *vir*-gene induction level by interacting with VirA in the presence of aldose monosaccharides, low pH and low phosphorus conditions. Activity of virD1 and virD2 proteins leads to excision and liberation of a single-stranded DNA molecule, T-strand, from the T-region of the Ti plasmid. This region is surrounded by 24-bp border repeats called left- and right border (LB and RB respectively) forming the T-DNA. The VirD1-VirD2 heterodimer nicks the T-DNA at the border repeats and VirD2 covalently binds to the 5′ end of it. The VirD2/T-strand is further mobilized towards the bacterial membrane for horizontal transfer.

6.5.2 T-Strand Transfer

VirB proteins (1–11) and virD4 are structural compounds of a specialized trans-membrane type IV secretion system (T4SS) apparatus with a T-pilus. These generally localize to cell poles of *Agrobacterium*. The T4SS is used for transferring VirD2/T-strand to the host cell. Also several virulence proteins (VirE2, VirE3, VirF and VirD5) are transported via T4SS to the host. VirD4 of the transporter has ATP binding domains and it has been shown to interact with transported virulence proteins and T-strand indicating that VirD4 physically mediates and, most likely also provides energy for their transport. The export of VirE2 needs a chaperon protein VirE1 that avoids VirE2 to form filamentous aggregates with itself.

Attachment of *Agrobacterium* cells to the host cell is fundamental for T-strand transfer. The nature of how bacteria interact with plant host cell wall is not clear. Bacterial chromosomal genes seem to be needed for this step. Several putative bacterial receptors and host proteins are postulated to be involved, such as victronectin-like protein, rhicadhesin protein, a cellulose synthatase-like protein, and several VirB2-interacting proteins. The attachment of *Agrobacterium* to the host cell is not Ti-plasmid dependent, ruling out attachment via T-pilus. On the other hand, chromosomal genes *chvA*, *chvB* and *pscA*(*exoC*), involved in synthesis and/or localization of β 1-2-glucan are required even though the role of this compound in attachment is not clear.

Exactly how the VirD2/T-strand conjugate and Vir-proteins are transferred to the host cell cytoplasm is not yet understood. In T4SS-directed plasmid conjugations between bacteria the role of the pilus seems to be pulling the cells close together forming the so called mating junction. This is characterized in Gram-negative bacteria by fusion of outer membranes and there is no evidence of pilus in the sites of genetic material transfer. Whether similar membrane unions are taking place between *Agrobacterium* and the recipient eukaryotic cell is not known.

6.5.3 Host Cell Events Leading to T-DNA Integration in the Host Genome

Once within the host cell VirD2/T-strand conjugate is believed to be covered with VirE2 to form the T-complex. VirE2 is believed to protect the T-strand from nuclease attacks because in the absence of this protein T-DNA integrations are often truncated at their left ends. Recent data is, however, indicating that the covering of T-strand by VirE2 would be taking place in the nucleus and not in the cytoplasm (Ream 2009). VirD2 has a nuclear localization signal and T-strand is moved, probably by an active dynein motor mechanism towards host cell nucleus. Host proteins would be involved in this intracellular transport. Also VirE2 promotes localization of T-strand to the nucleus. VirD5 might function in plants to aid nuclear localization of VirE2.

Plant proteins VIP1 and α-importin proteins are involved in nuclear import of VirE2. VirE3 can mimic the function of VIP1 and probably participate in the nuclear import of VirE2. VirD2-T-DNA conjugate is most probably transported across the nuclear membrane by an α-importin 1 protein. Nuclear localization signal of VirD2 is fundamental for T-strand entrance to the nucleus.

Once inside the nucleus the T-complex may be directed to chromatin with the assistance of plant VIP2 protein for integration. Also VIP1 has been proposed to play role in T-DNA integration by interacting with histone H2A. VIP2 seems to be required for stable, but not transient transformation of plants. A targeted proteolysis of VirE2 takes place with the assistance of nuclear localizing VirF protein and prior to integration in host genome the ssT-DNA (T-strand) is completed to the ds-form (T-DNA) by host DNA repair mechanisms. Later on DNA breaks are produced in the host genome and the dsT-DNA is ligated to these breaks. Complete understanding of these integration events is still missing. Most likely double-stranded breaks are involved.

Integration events are conducted by host proteins and the T-DNA can be integrated via nonhomologous end joining (NHEJ) or homologous recombination (HR) mechanisms depending on the predominant host mechanism of DNA integration. HR can take place if T-DNA shares homology with the recipient genome. In plants T-DNA integration occurs predominantly via NHEJ while in other nonplant hosts also HR-integration can take place.

6.6 Molecular Mechanism of AMT in Fungi

AMT of fungi depends on acetosyringone induction of *Agrobacterium* indicating participation of *vir*-genes and T4SS in T-strand transfer to a nonplant host (de Groot et al. 1998). The bacterial pretransfer steps are therefore most probably identical during transformation of plant and nonplant hosts. However, the attachment, T-strand transfer and host-cell directed events of T-DNA integration show some differences between different eukaryotic hosts. Most information for these differences comes from AMT of *S. cerevisieae* and *Aspergillus awamori* with *Agrobacterium* strains mutated in different virulence genes (Bundock et al. 1995; Piers et al. 1996; Michielse et al. 2004a, b). A different requirement for VirE protein has been reported. While lack of this protein almost inhibits transformation of plants, fungal transformation efficiency is only reduced. Also transformation of yeast with *Agrobacterium* mutated in *chvA*, *chvB* and *exoC* is as efficient as with a wild type strain. Activity of these chromosomally encoded proteins is shown to be fundamental for attachment and transformation of plant cells but does not seem to play an important role in AMT of fungi. Interestingly, ChvA and ChvB are needed for attachment to human cells (Kunik et al. 2001) demonstrating that variation exist in requirements of AMT on different nonplant hosts. Yeast and human cells also lack plant type VIP1 protein which is involved in nuclear import of VirE2. However, bacterial origin VirE3 can mimic the VIP1 function and could partly explain the functionality of AMT in these nonplant hosts.

The T-DNA integration step is conducted mainly by the host-cell protein machinery. Therefore, it shows significant difference between plant and nonplant hosts and is also dependent on the nucleotide sequence of the T-DNA. While in plants T-DNA, even in the presence of long flanks homologous to plant genome, integrates via a nonhomologous end-joining mechanism, in nonplant species, such as fungi, both homologous (HR) and nonhomologous recombination (NHR) mechanisms can participate in T-DNA integration. Which cellular mechanism is dominant depends on the organism, length of homologous sequence in the T-DNA and the degree of homology. The main molecular knowledge of T-DNA integration comes from *S. cerevisiae* where both NHEJ and HR take place. Several DNA-repair genes have been demonstrated to be essential or participate in HR and NHR of T-DNA (van Attikum et al. 2001; van Attikum and Hooykaas 2003). In yeast the Ku70 encoding gene is determinant for NHR and Rad52 for HR, and double mutants of these genes are completely blocked in T-DNA integration (van Attikum and Hooykaas 2003). The information on HR in yeast has been exploited by plant researchers to improve low HR rates presented in these organisms. The heterologous expression of another yeast DNA-repair protein, Rad54 involved in HR, in plant cells has resulted in increased homologous T-DNA integration in this system (Shaked et al. 2005).

6.7 Use of AMT in Fungal Genetics

The capacity of *Agrobacterium* to transfer genetic material to fungi has been increasingly used for genetic modification of this group of eukaryotes (Michielse et al. 2005b; Weld et al. 2006). It has proven to be a powerful tool for filamentous fungi overcoming problems rising from traditional DNA based transformation techniques such as the need for protoplast preparation. AMT has also resulted in higher transformation efficiencies in many fungal species for which other transformation methods already existed resulting in less complex transgene integration patterns. Traditional transformation methods, such as PEG mediated transformation, electroporation and particle bombardment, generally produce a high percentage of multiple and often tandem transgene integrations. This has complicated the interpretation of produced fungal phenotypes. AMT on the other hand results in predominantly single integration of the T-DNA in the fungal genome and this integration, in absence of homologous sequences, occurs randomly enough for making random gene tagging possible. Sequencing of fungal genomes has initiated a true interest for functional genomics research in filamentous fungi. This requires the optimization of high-throughput transformation and gene disruption tools such as AMT.

6.8 AMT-Mediated Random Gene Tagging

Traditional fungal transformation methods generally produce multiple transgene integration events. The transgene copy number can be reduced by incorporating a REMI step but the use of restriction enzymes can lead to genomic reorganizations and phenotypes not linked to the integration site of the transgene. The number of T-DNAs integrated in the fungal genome via AMT can also vary. The number of integrations can be influenced by factors such as cocultivation time, concentration of acetosyringone or bacterial/fungal cell ratio during transformation. Despite the species and transformation protocol a high percentage (50–80%) of *Agrobacterium* transformed fungal strains are reported to carry a single T-DNA integration (Michielse et al. 2005b). This is a much higher percentage than with AMT of plants which generally produces multilocus and/or multicopy integrations. Even though small deletions in the border sequences of the integrated T-DNAs are common in fungi, as in transformed plants, the T-DNAs of known sequence are usually quite complete making integration site analysis by PCR-based methods such as TAIL-PCR or inverted PCR possible. Due to this predominant single transgene copy integration nature, AMT has been the preferred methodology for producing random insertional fungal libraries of *Magnaporthe grisea*, *Cryptococcus neoformans*, *Leptosphaeria maculans* and *Colletotrichum higginsianum* (Tsuji et al. 2003; Idnurm et al. 2004; Walton et al. 2005; Betts et al. 2007; Blaise et al. 2007; Jeon et al. 2007; Li et al. 2007; Talhinhas et al. 2008; Huser et al. 2009). These random gene disruption collections have been used for successful identification of novel gene functions involved in fungal pathogenesis.

The AMT random tagging methodology can also be used for detecting promoters or increase transcription of contiguous genes. These can be done by enhancer trap constructions where a minimal promoter is linked to a reporter gene, with a promoterless reporter gene construct or placing a strong promoter close to the border of the T-DNA. These approaches have been used in fungi with traditional transformation methods but not yet with AMT. There is no reason to believe that these genetic modification techniques are not compatible with AMT. Also incorporation of autoreplicative elements in transferred T-DNAs has been shown to enhance transformation efficiency in *S. cerevisiae*. In this case the T-DNA does not need to integrate in the recipient genome (Bundock et al. 1995). Even though autoreplicative sequences are available for some filamentous fungi, these have not been used in combination with AMT.

6.9 Is Random T-DNA Tagging of Fungal Genome Really at Random?

The randomness of T-DNA integration in the recipient genome has been a matter of ongoing scientific debate which has direct implications on the use of AMT in saturated mutagenesis. Several plant studies have indicated a bias towards transcriptionally active sites, especially in regulatory regions (Koncz et al. 1989; Brunaud et al. 2002; Alonso et al. 2003; Sallaud et al. 2004). The transcriptional status and the open chromatin structure have been proposed to determine this T-DNA attraction to gene-rich zones. However, recent studies on plants are indicating that this observed bias has been an artifact of T-DNA libraries generated under selection pressure. Analysis of plant T-DNA target sites under selection-free conditions indicates that the integration profile of T-DNAs in the plant genome is more random than previously believed (Kim et al. 2007). Recovery of transgenic lines under selection pressure requires a strong expression of the selection marker carried by the T-DNA. This is more probable in transcriptionally gene-rich zones than in heterochromatin integrations where transcription of the selection marker can be hindered by the epigenetic status of the integration site (Francis and Spiker 2005).

Similar biased T-DNA integration has been reported for filamentous fungi. A preference of T-DNA to integrate in fungal promoters and gene rich regions has been reported for *M. grisea* and *C. neoformans* (Walton et al. 2005; Choi et al. 2007; Li et al. 2007; Meng et al. 2007). The T-DNA tagged analyzed fungal strains in these studies were all originated from selective conditions. Therefore, it is highly probable that T-DNA integration is also more random in fungi as now postulated for plants. From the point of view of use of AMT for random mutagenesis and transgene expression this bias towards gene rich zone is however more a benefit than a problem. Higher percentage of tagging of active genes can be expected with this genetic transformation method. Studies carried out in our group on gene transfer from *Agrobacterium* and recovery of the transgene by plasmid rescue show that T-DNA integration in the genome of the ectomycorrhizal model fungus *L. bicolor* occurs mainly in genes and as single copy. These indicate that T-DNA insertion by AMT would be very useful for high-throughput tagged mutagenesis in ectomycorrhizal fungi (Kemppainen et al. 2008).

6.10 Targeted Gene Disruption by AMT

DNA integration into the genome via HR is highly rare in plants but more common in fungi. Therefore, the integration of T-DNA harboring homology with the fungal genome can also occur via HR. This has made possible the use of AMT not only for random gene tagging but for targeted gene disruption. In gene targeting, a selection marker cassette in the T-DNA is surrounded by DNA flanks homologous to the target gene (or a zone around it). The successful integration of T-DNA via HR results in interruption of the target gene by the selection cassette. This approach has been used for gene inactivation already in several ascomycete species such as *S. cerevisiae*, *Monilinia fructicola*, *Verticillium dahliae*, *Aspergillus fumigatus*, *A. awamori*, *Metarhizium anisopliae* and *Mycosphaerella graminicola* (Zwiers and de Waard 2001; van Attikum and Hooykaas 2003; Michielse et al. 2005a; Rauyaree et al. 2005; Sugui et al. 2005; Lee and Bostock 2006; Staats et al. 2007). However, the frequency of HR events is a fungal species specific character and it is also highly dependent on the length of the homologous flanks used. While HR is the dominant pathway in *S. cerevisiae* and just 100 bp homology flanks can be used for targeting genes in this yeast, in filamentous fungi generally at least 1 kb flanks or even longer homologous sequences are needed for gene disruption (Wilson et al. 2002; Michielse et al. 2005b). Even in the presence of long homologous sequences HR rates in filamentous fungi are variable and generally low. For example in wild-type strains of *Acremonium chrysogenum*, *Aspergillus nidulans*, *N. crassa*, *Sordaria macrospora* and *M. grisea* HR rates are as low as 0.1–7% (Chaveroche et al. 2000; Liu et al. 2001; Talbot and Foster 2001; Colot et al. 2006; Pöggeler and Kück 2006). AMT when compared to other traditional genetic transformation methods has been shown to increase this generally low HR rate both in yeast and filamentous fungi (Bundock et al. 1999; Michielse et al. 2005a, b). Also the degree of dependency on the length of homologous sequence flanks is lower with AMT (Michielse et al. 2005a, b). This increased HR rate is proposed to depend on the ssT-DNA nature or the active host protein assisted localization into the nucleus. Even though AMT has been shown to increase HR, the difficulties in gene inactivation of filamentous fungi and in basidiomycete fungi in particular have led to the generation of mutant strains affected in their NHEJ-pathway. NHEJ is shown to depend on the activity of Ku70-Ku80 heterodimer that binds to broken DNA ends. These Ku70 and Ku80 proteins were also identified in fungi (Hefferin and Tomkinson 2005). Mutations in these genes result in inactivation of NHEJ and leaves HR as the dominant pathway for DNA integration. HR efficiency of 70–100% has been reported for ku70 or ku80 mutants of *Aspergillus oryzae*, *A. fumigatus*, *A. nidulans*, *N. crassa*, *S. macrospora*, *C. neoformans* and *M. grisea* (Ninomiya et al. 2004; da Silva Ferreira et al. 2006; Goins et al. 2006; Nayak et al. 2006; Pöggeler and Kück 2006; Takahashi et al. 2006; Villalba et al. 2008).

6.11 *Laccaria bicolor* Represents a Special Challenge for AMT as a Reverse Genetic Tool

The lack of an efficient transformation method for the genome-sequenced ECM fungus *L. bicolor* (Martin et al. 2008) was a major obstacle in ECM research. Even though this species was reported to be transformed via particle bombardment (Bills et al. 1999) and its closely related species *L. laccata* via PEG mediated protoplast transformation (Barret et al. 1990) both of these traditional genetic transformation techniques are time consuming and complicated, show low transformation efficiency and reproducibility, are of high costs and generally result in multiple transgene copies in the recipient genome. AMT on the other hand overcomes these limitations but not all fungal species are as susceptible to this method or do not produce stable integrations for still unknown reasons. We developed an efficient AMT system for *L. bicolor* (Kemppainen et al. 2005). However, *Laccaria* shows some special features that may or will complicate the use of AMT or any other genetic transformation method for reverse functional genetic studies in ECM. Unlike filamentous ascomycetes the symbiotic phase of the fungus is the dikaryon where two compatible nuclei coexist in each hyphal compartment. The life-cycle and basidiospore production cannot be completed without ECM formation with the plant host. Furthermore, fruiting body production is unpredictable and requires a long time and therefore cannot be achieved under axenic conditions. The dikaryotic nature of *Laccaria* generates a technical obstacle for the use of AMT as a random gene tagging tool for searching gene functions relevant in ECM. An interruption of a gene copy in one of the nuclei may not be enough for altering gene expression and thus produce functional phenotypes. Also a gene interruption may lead to increased transcription from the intact allele in the untransformed nucleus. This compensation may result in unaltered gene expression level in tagged dikaryons. Tagging of *Laccaria* monokaryons could however be used for studying general fungal gene function and especially for identifying genes involved in control of dikaryotization.

The same dilemma of the dikaryon also affects the use of AMT for targeted inactivation of ECM-regulated genes. Targeted disruption of these genes should be carried out independently in two compatible monokaryons which could form functional double knock-out dikaryotic strains by further crossing. Homobasidiomycetes usually show extremely low HR rates. Working with two *Laccaria* monokaryotic strains for producing knock-out dikaryons can be predicted to be extremely

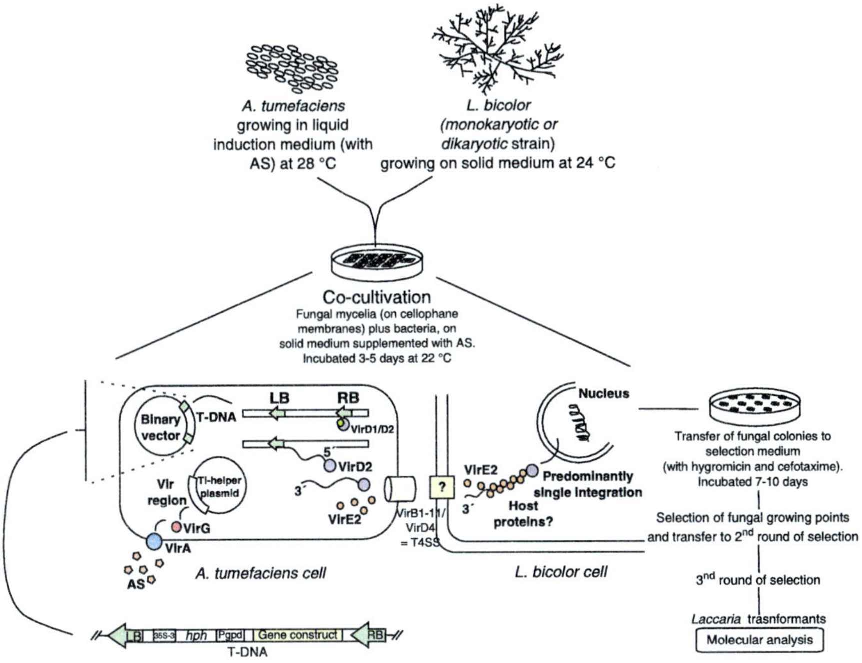

Fig. 6.2 Schematic representation of *Laccaria bicolor* transformation mediated by *Agrobacterium tumefaciens*. 35S-3′: cauliflower mosaic virus 35S terminator. AS: acetosyringone. Cefotaxime: cephalosporin antibiotic which inhibits the final transpeptidation step of peptidoglycan synthesis in bacterial cell walls. Used for killing *Agrobacterium* cells. Gene construct: different gen constructs can be transferred via *Agrobacterium* and integrated in *Laccaria* genome (i.e., plasmid rescue T-DNAs, RNAi-triggers, heterologous genes of interest, reporters, etc.). *hph: hph* gene of *E. coli* coding for an aminocyclitol phosphotransferase that confers resistance to hygromicin B (fungal selection marker) and structurally related antibiotics. *LB* T-DNA left border of pCAMBIA1300. *Pgpd* glyceraldehide-3-phosphate dehydrogenase promoter of *Agaricus bisporus*. *RB* T-DNA right border of pCAMBIA1300. *T4SS* type IV secretion system. For details on *Agrobacterium*-mediated transformation of *Laccaria* see Kemppainen et al. (2005, 2008)

challenging and time consuming if not impossible. The use of the RNA-silencing technology could however represent an alternative and more straightforward approach for studying ECM-regulated *Laccaria* genes (Kemppainen et al. 2009; Kemppainen and Pardo 2010; this book, Chap. 9). A diagram showing the use of *Agrobacterium* for genetic transformation of *Laccaria* is presented in Fig. 6.2.

6.12 Conclusion

Despite the great ecological and economic importance of ECM the current comprehension of host-fungus recognition, establishment and functions of ECM is still rather limited. During the last ten years the use of novel molecular methods such as EST-libraries and cDNA micro– and macro arrays has dramatically increased the knowledge on genetic regulation underlying the ECM interaction. More importantly, the decision of the Department of Energy Joint Genome Institute (JGI) to sequence the genomes of micobionts of the first genome-sequenced tree, poplar, took the mycorrhizal research towards the genomic era. As a result the full genomic sequence of the basidiomycete ECM fungus *Laccaria bicolor* strain S238N-H82 was resolved in collaboration between JGI and the *Laccaria* Genome Consortium which joined the efforts of several ECM research laboratories around the world. The *Laccaria* genome sequence is currently been used for genome-wide expression profiles at different stages of ECM development. This approach is generating an immense amount of valuable data. However, the current ECM research faces a serious technical obstacle. Further studies for resolving the biological relevance of the identified symbiosis-regulated genes depend on the availability of reverse genetic tools, among them an efficient genetic transformation technique. We established a high throughput AMT based on hygromycin B resistance for *Laccaria* dikaryotic and monokaryotic strains. This gene transfer from *Agrobacterium* and recovery of the transgene by plasmid rescue showed that T-DNA integration in the genome of *L. bicolor* occurs mainly in genes and as single copy. These indicate that T-DNA insertion by AMT would be very useful for high throughput tagged mutagenesis in ectomycorrhizal fungi. AMT has opened a new era for genetic studies in mycorrhizal research and its use in combination with the RNA silencing technology will certainly help in elucidating the function of symbiosis-regulated genes in ECM development.

References

Aleksenko AY (1994) Cointegration of transforming DNAs in *Aspergillus nidulans*: a model using autonomously-replicating plasmids. Curr Genet 26:352–358

Alic M, Kornegay JR, Pribnow D, Gold MH (1989) Transformation by complementation of an adenine auxotroph of the lignin-degrading basidiomycete *Phanerochaete chrysosporium*. Appl Environ Micobiol 55:406–411

Alonso JM, Stepanova AN, Leisse TJ, Kim CJ, Chen H, Shinn P, Stevenson DK, Zimmerman J, Barajas P, Cheuk R, Gadrinab C, Heller C, Jeske A, Koesema E, Meyers CC, Parker H, Prednis L, Ansari Y, Choy N, Deen H, Geralt M, Hazari N, Hom E, Karnes M, Mulholland C, Ndubaku R, Schmidt I, Guzman P, Aguilar-Henonin L, Schmid M, Weigel D, Carter DE, Marchand T, Risseeuw E, Brogden D, Zeko A, Crosby WL, Berry CC, Ecker JR (2003) Genome wide insertional mutagenesis of *Arabidopsis thaliana*. Science 301:653–657

Ballance DJ, Buxton FP, Turner G (1983) Transformation of *Aspergillus nidulans* by the orotidine-5-phosphate decarboxylase gene of *Neurospora crassa*. Biochem Biophys Res Commun 112:284–289

Barret V, Dixon RK, Lemke PA (1990) Genetic transformation of a mycorrhizal fungus. Appl Microbiol Biotechnol 33:313–316

Betts MF, Tucker SL, Galadima N, Meng Y, Patel G, Li L, Donofrio N, Floyd A, Nolin S, Brown D, Mandel MA, Mitchell TK, Xu JR, Dean RA, Farman ML, Orbach MJ (2007) Development of a high throughput transformation system for insertional mutagenesis in *Magnaporthe oryzae*. Fungal Genet Biol 44:1035–1049

Bills SN, Richter DL, Podila GK (1995) Genetic transformation of the ectomycorrhizal fungus *Paxillus involutus* by particle bombardment. Mycol Res 99:557–561

Bills S, Podila G, Hiremath S (1999) Genetic engineering of a fungus *Laccaria bicolor* for use as a biological control agent. Mycologia 91:237–242

Binninger DM, Skrzynia C, Pukkila PJ, Casselton LA (1987) DNA-mediated transformation of the basidiomycte *Coprinus cinereus*. EMBO J 6:835–840

Blaise F, Remy E, Meyer M, Zhou L, Narcy JP, Roux J, Balesdent MH, Rouxel T (2007) A critical assessment of *Agrobacterium tumefaciens*-mediated transformation as a tool for pathogenicity gene discovery in the phytopathogenic fungus *Leptosphaeria maculans*. Fungal Genet Biol 44:123–138

Broothaerts W, Mitchell HJ, Weir B, Kaines S, Smith LM, Yang W, Mayer JE, Roa-Rodríguez C, Jefferson RA (2005) Gene transfer to plants by diverse species of bacteria. Nature 433:583–584

Brunaud V, Balzergue S, Dubreucq B, Aubourg S, Samson F, Chauvin S, Bechtold N, Cruaud C, DeRose R, Pelletier G, Lepiniec L, Caboche M, Lecharny A (2002) T-DNA integration into the *Arabidopsis* genome depends on sequences of pre-insertion sites. EMBO Rep 12:1152–1157

Bundock P, Dulk-Ras A, Beijersbergen A, Hooykaas PJ (1995) Trans-kingdom T-DNA transfer from *Agrobacterium tumefaciens* to *Saccharomyces cerevisiae*. EMBO J 14:3206–3214

Bundock P, Mroczek K, Winkler AA, Steensma HY, Hooykaas PJ (1999) T-DNA from *Agrobacterium tumefaciens* as an efficient tool for gene targeting in *Kluyveromyces lactis*. Mol Gen Genet 261:115–121

Case ME, Schweizer M, Kushner SR, Giles NH (1979) Efficient transformation of *Neurospora crassa* by utilizing hybrid plasmid DNA. Proc Natl Acad Sci USA 76:5259–5263

Chakraborty BN, Kapoor M (1990) Transformation of filamentous fungi by electroporation. Nucleic Acids Res 18:6737

Chaveroche MK, Ghigo JM, d'Enfert C (2000) A rapid method for efficient gene replacement in the filamentous fungus *Aspergillus nidulans*. Nucleic Acid Res 28:E97

Chen X, Stone M, Schlagnhaufer C, Romaine CP (2000) A fruiting body tissue method for efficient *Agrobacterium*-mediated transformation of *Agaricus bisporus*. Appl Environ Microbiol 66:4510–4513

Choi J, Park J, Jeon J, Chi MH, Goh J, Yoo SY, Park J, Jung K, Kim H, Park SY, Rho HS, Kim S, Kim BR, Han SS, Kang S, Lee YH (2007) Genome-wide analysis of T-DNA integration into the chromosomes of *Magnaporthe oryzae*. Mol Microbiol 66:371–382

Christiansen SK, Knudsen S, Giese H (1995) Biolistic transformation of the obligate plant pathogenic fungus, *Erysiphe graminis f.sp. hordei*. Curr Genet 29:100–102

Citovsky V, Kozlovsky SV, Lacroix B, Zaltsman A, Dafny-Yelin M, Vyas S, Tovkach A, Tzfira T (2007) Biological systems of the host cell involved in *Agrobacterium* infection. Cell Microbiol 9:9–20

Colot HV, Parks G, Turner GE, Ringelberg C, Crew CM, Litvinkova L, Weiss RL, Borkovich K, Dunlap JC (2006) A high-throughput gene knockout procedure for *Neurospora crassa* reveals functions for multiple transcription factors. Proc Natl Acad Sci USA 103:10353–10357

Combier JP, Melayah D, Raffier C, Gay G, Marmeisse R (2003) *Agrobacterium tumefaciens*-mediated transformation as a tool for insertional mutagenesis in the symbiotic ectomycorrhizal fungus *Hebeloma cylindrosporum*. FEMS Microbiol Lett 220:141–148

Covert SF, Kapoor P, Lee M, Briley A, Nairn CJ (2001) *Agrobacterium*-mediated transformation of *Fusarium circinatum*. Mycol Res 105:259–264

da Silva Ferreira ME, Kress MR, Savoldi M, Goldman MH, Hartl A, Heinekamp T, Brakhage AA, Goldman GH (2006) The akuB (KU80) mutant deficient for nonhomologous end joining is a powerful tool for analyzing pathogenicity in *Aspergillus fumigatus*. Eukaryot Cell 5:207–211

Dafny-Yelin M, Levy A, Tzfira T (2008) The ongoing saga of *Agrobacterium*–host interactions. Trends Plant Sci 13:102–105

de Framond AJ, Barton KA, Chilton MD (1983) Mini-Ti: a new vector strategy for plant genetic engineering. Biotechnol 5:262–269

de Groot MJ, Bundock P, Hooykaas PJ, Beijersbergen AG (1998) *Agrobacterium tumefaciens*-mediated transformation of filamentous fungi. Nat Biotechnol 16:839–842

Edman JC, Kwon-Chung KJ (1990) Isolation of the URA5 gene from *Cryptococcus neoformans var. neoformans* and its use as a selective marker for transformation. Mol Cell 10:4538–4544

Fiddy C, Trinci AP (1976) Mitosis, septation, branching and the duplication cycle in *Aspergillus nidulans*. J Gen Microbiol 97:169–184

Fitzgerald AM, Mudge AM, Gleave AP, Plummer KM (2003) *Agrobacterium* and PEG-mediated transformation of the phytopathogen *Venturia inaequalis*. Mycol Res 107:810–813

Francis KE, Spiker S (2005) Identification of *Arabidopsis thaliana* transformants without selection reveals a high occurrence of silenced T-DNA integrations. Plant J 41:464–477

Gelvin SB (2009) *Agrobacterium* in the genomics age. Plant Physiol 150:1665–1676

Goins CL, Gerik KJ, Lodge JK (2006) Improvements to gene deletion in the fungal pathogen *Cryptococcus neoformans*: absence of Ku proteins increases homologous recombination, and co-transformation of independent DNA molecules allows rapid complementation of deletion phenotypes. Fungal Genet Biol 43:531–544

Grimaldi B, de Raaf MA, Filetici P, Ottonello S, Ballario P (2005) *Agrobacterium*-mediated gene transfer and enhanced green fluorescent protein visualization in the mycorrhizal ascomycete *Tuber borchii*: a first step towards truffle genetics. Curr Genet 48:69–74

Hanif M, Pardo AG, Gorfer M, Raudaskoski M (2002) T-DNA transfer and integration in the ectomycorrhizal fungus *Suillus bovinus* using hygromycin B as a selectable marker. Curr Genet 41:183–188

Hefferin ML, Tomkinson AE (2005) Mechanism of DNA doublestrand break repair by non-homologous end joining. DNA Repair 4:639–648

Hinnen A, Hicks JB, Fink GR (1978) Transformation of yeast. Proc Natl Acad Sci USA 75:1929–1933

Hoekema A, Hirsch PR, Hooykaas PJJ, Schilperoort RA (1983) A binary plant vector strategy based on separation of vir- and T-region of the *Agrobacterium tumefaciens* Ti-plasmid. Nature 303:179–180

Huser A, Takahara H, Schmalenbach W, O'Connell R (2009) Discovery of pathogenicity genes in the crucifer anthracnose fungus *Colletotrichum higginsianum*, using random insertional mutagenesis. Mol Plant Microb Interact 22:143–156

Idnurm A, Reedy JL, Nussbaum JC, Heitman J (2004) *Cryptococcus neoformans* virulence gene discovery through insertional mutagenesis. Eukaryot Cell 3:420–429

Jeon J, Park S-Y, Chi M-H, Choi J, Park J, Rho H-S, Kim S, Goh J, Yoo S, Choi J, Park J-Y, Yi M, Yang S, Kwon M-J, Han S-S, Kim BR, Khang CH, Park B, Lim S-E, Jung K, King S, Karunakaran M, Oh H-S, Kim H, Kim S, Park J, Kang S, Choi W-B, Kang S, Lee Y-H

(2007) Genome-wide functional analysis of pathogenicity genes in the rice blast fungus. Nat Genet 39:561–565

Kahmann R, Basse C (1999) REMI (Restriction enzyme mediated integration) and its impact on the isolation of pathogenicity genes in fungi attacking plants. Eur J Plant Pathol 105:221–229

Kemppainen M, Pardo AG (2010) pHg/pSILBAγ vector system for efficient gene silencing in homobasidiomycetes: optimization of ihpRNA-triggering in the mycorrhizal fungus *Laccaria bicolor*. Microb Biotechnol 3:178–200

Kemppainen M, Circosta A, Tagu D, Martin F, Pardo AG (2005) *Agrobacterium*-mediated transformation of the ectomycorrhizal symbiont *Laccaria bicolor* S238N. Mycorrhiza 16:19–22

Kemppainen M, Duplessin S, Martin F, Pardo AG (2008) T-DNA insertion, plasmid rescue and integration analysis in the model mycorrhizal fungus *Laccaria bicolor*. Microb Biotechnol 1:259–269

Kemppainen M, Duplessin S, Martin F, Pardo AG (2009) RNA silencing in the model mycorrhizal fungus *Laccaria bicolor*: gene knock-down of nitrate reductase results in inhibition of symbiosis with *Populus*. Environ Microbiol 11:1878–1896

Kim SI, Veena HJ, Gelcin SB (2007) Genome-wide analysis of *Agrobacterium* T-DNA integration sites in the *Arabidopsis* genome generated under non-selective conditions. Plant J 51:779–791

Koncz C, Martini N, Mayerhofer R, Koncz-Kalman Z, Korber H, Redei GP, Schell J (1989) High-frequency T-DNA-mediated gene tagging in plants. Proc Natl Acad Sci USA 86:8467–8471

Kunik T, Tzfira T, Kapulnik Y, Gafni Y, Dingwall C, Citovsky V (2001) Genetic transformation of HeLa cells by *Agrobacterium*. Proc Natl Acad Sci USA 98:1871–1876

Kuwano T, Shirataki C, Itoh Y (2008) Comparison between polyethyleneglycol- and polyethyleneimine-mediated transformation of *Aspergillus nidulans*. Curr Genet 54:95–103

Lacroix B, Tzfira T, Vainstein A, Citovsky V (2006) A case of promiscuity: *Agrobacterium*'s endless hunt for new partners. Trends Genet 22:29–37

Lee MH, Bostock RM (2006) *Agrobacterium* T-DNA-mediated integration and gene replacement in the brown rot pathogen *Monilinia fructicola*. Curr Genet 49:309–322

Li G, Zhou W, Liu G, Zheng F (2007) Characterization of T-DNA insertion patterns in the genome of the rice blast fungus *Magnaporthe oryzae*. Curr Genet 51:233–243

Liu G, Casqueiro J, Bañuelos O, Cardoza RE, Gutiérrez S, Martín JF (2001) Targeted inactivation of the *mecB* gene, encoding cystathione-gamma-lyase, shows that the reverse transsulfuration pathway is required for high-level cephalosporin biosíntesis in *Acremonium chysogenum* C10 but not for methionine induction of the cephalosporin genes. J Bacteriol 183:1765–1772

Lorito M, Hayes CK, Di Pietro A, Harman GE (1993) Biolistic transformation of *Trichoderma harzianum* and *Gliocladium virens* using plasmid and genomic DNA. Curr Genet 24:349–356

Marmeisse R, Gay G, Debaud JC, Casselton LA (1992) Genetic transformation of the symbiotic fungus *Hebeloma cylindrosporum*. Curr Genet 22:41–45

Martin F, Aerts A, Ahren D, Brun A, Danchin EGJ, Duchaussoy F, et al (2008) The genome of *Laccaria bicolor* provides insights into mycorrhizal symbiosis. Nature 452:88–92

McCullen CA, Binns AN (2006) *Agrobacterium tumefaciens* and plant cell interactions and activities required for interkingdom macromolecular transfer. Annu Rev Cell Dev Biol 22:101–127

Meng Y, Patel G, Heist M, Betts MF, Tucker SL, Galadima N, Donofrio NM, Brown D, Mitchell TK, Li L, Xu JR, Orbach M, Thon M, Dean RA, Farman ML (2007) A systematic analysis of T-DNA insertion events in *Magnaporthe oryzae*. Fungal Genet Biol 44:1050–1064

Michielse CB, Ram AFJ, Hooykaas PJJ, van den Hondel CAMJJ (2004a) *Agrobacterium*-mediated transformation of *Aspergillus awamori* in the absence of full length VirD2, VirC2 or VirE2 leads to insertion of aberrant T-DNA structures. J Bacteriol 186:2038–2045

Michielse CB, Ram AFJ, Hooykaas PJJ, van den Hondel CAMJJ (2004b) Role of bacterial virulence proteins in *Agrobacterium* mediated transformation of *Aspergillus awamori*. Fungal Genet Biol 45:571–578

Michielse CB, Hooykaas PJJ, van den Hondel CAMJJ, Ram AFJ (2005b) *Agrobacterium*-mediated transformation as a tool for functional genomics in fungi. Curr Genet 48:1–17

Nayak T, Szewczyk E, Oakley CE, Osmani A, Ukil L, Murray SL, Hynes MJ, Osmani SA, Oakley BR (2006) A versatile and efficient gene-targeting system for *Aspergillus nidulans*. Genetics 172:1557–1566

Ninomiya Y, Suzuki K, Ishii C, Inoue H (2004) Highly efficient gene replacements in *Neurospora* strains deficient for non-homologous end joining. Proc Natl Acad Sci USA 101:12248–12253

Pardo AG, Hanif M, Raudaskoski M, Gorfer M (2002) Genetic transformation of ectomycorrhizal fungi mediated by *Agrobacterium tumefaciens*. Mycol Res 106:132–137

Pardo AG, Kemppainen M, Valdemoros D, Duplessis S, Martin F, Tagu D (2005) T-DNA transfer from *Agrobacterium tumefaciens* to the ectomycorrhizal fungus *Pisolithus microcarpus*. Rev Argent Microbiol 37:69–72

Park SM, Kim DK (2004) Transformation of a filamentous fungus *Cryphonectria parasitica* using *Agrobacterium tumefaciens*. Biotechnol Bioprocess Eng 9:217–222

Piers KL, Heath JD, Liang X, Stephens KM, Nester EW (1996) *Agrobacterium tumefaciens*-mediated transformation of yeast. Proc Natl Acad Sci USA 93:1613–1618

Pöggeler S, Kück U (2006) Highly efficient generation of signal transduction knockout mutants using a fungal strain deficient in the mammalian *ku70* ortholog. Gene 378:1–10

Punt PJ, Oliver RP, Dingemanse MA, Pouwels PH, van den Hondel CA (1987) Transformation of *Aspergillus* based on the hygromycin B resistance marker from *Escherichia coli*. Gene 56:117–124

Rauyaree P, Ospina-Giraldo MD, Kang S, Bhat RG, Subbarao KV, Grant SJ, Dobinson KF (2005) Mutations in VMK1, a mitogen activated protein kinase gene, affect microsclerotia formation and pathogenicity in *Vericillium dahliae*. Curr Genet 48:109–116

Ream W (2009) *Agrobacterium tumefaciens* and *A. rhizogenes* use different proteins to transport bacterial DNA into the plant cell nucleus. Microb Biotechnol 2:416–427

Richey MG, Marek ET, Schardl CL, Smith DA (1989) Transformation of filamentous fungi with plasmid DNA by electroporation. Phytopathology 79:844–847

Rodríguez-Tovar AV, Ruiz-Medrano R, Herrera-Martínez A, Barrera-Figueroa BE, Hidalgo-Lara ME, Reyes-Márquez BE, Cabrera-Ponce JL, Valdés M, Xoconostle-Cázares B (2005) Stable genetic transformation of the ectomycorrhizal fungus *Pisolithus tinctorius*. J Microbiol Methods 63:45–54

Sallaud C, Gay C, Larmande P, Bès M, Piffanelli P, Piégu B, Droc G, Regad F, Bourgeois E, Meynard D, Périn C, Sabau X, Ghesquière A, Glaszmann JC, Delseny M, Guiderdoni E (2004) High throughput T-DNA insertion mutagenesis in rice: a first step towards in silico reverse genetics. Plant J 39:450–464

Shaked H, Melamed-Bessudo C, Levy AA (2005) High-frequency gene targeting in *Arabidopsis* plants expressing the yeast *RAD54* gene. Proc Natl Acad Sci USA 102:12265–12269

Specht CA, Muñoz-Rivas A, Novotny CP, Ullrich RC (1988) Transformation of *Schizophyllum commune*; and análisis of parameters for improving transformation frequencies. Exp Mycol 12:357–366

Staats CC, Junges A, Fitarellim M, Fulaneto MC, Henning Vainstein M, Schrank A (2007) Gene inactivation mediated by *Agrobacterium tumefaciens* in the filamentous fungi *Metarhizium anisopliae*. Appl Microbiol Biotechnol 76:945–950

Sugui JA, Chan YC, Kwon-Chung KJ (2005) *Agrobacterium tumefaciens*-mediated transformation of *Aspergillus fumigatus*: and efficient tool for insertional mutagenesis and targeted gene disruption. Appl Environ Microbiol 71:1798–1802

Sullivan TD, Rooney PJ, Klein BS (2002) *Agrobacterium tumefaciens* integrates transfer DNA into single chromosomal sites of dimorphic fungi and yields homokaryotic progeny from multinucleate yeast. Eukaryot Cell 1:895–905

Sunagawa M, Murata H, Miyazaki Y, Nakamura M (2007) Transformation of the mycorrhizal basidiomycetes. *Suillus grevillei* and *S. bovinus*, by particle bombardment. Biosci Biotechnol Biochem 71:47–50

Takahashi T, Masuda T, Koyama Y (2006) Enhanced gene targeting frequency in *ku70* and *ku80* disruption mutants of *Aspergillus soj*ae and *Aspergillus oryzae*. Mol Genet Genomics 275:460–470

Talbot NJ, Foster AJ (2001) Genetics and genomics of the rice blast fungus *Magnaporthe grisea*: developing an experimental model for understanding fungal diseases of cereals. Adv Bot Res 34:263–287

Talhinhas P, Muthumeenakshi S, Neves-Martins J, Oliveira H, Sreenivasaprasad S (2008) *Agrobacterium*-mediated transformation and insertional mutagenesis in *Colletotrichum acutatum* for investigating varied pathogenicity lifestyles. Mol Biotechnol 39:57–67

Tilburn J, Scazzocchio C, Taylor GG, Zabicky-Zissman JH, Lockington RA, Davies RW (1983) Transformation by integration in *Aspergillus nidulans*. Gene 26:205–221

Tsuji G, Fujii S, Fujihara N, Hirose C, Tsuge S, Shiraishi T, Kubo Y (2003) *Agrobacterium tumefaciens*-mediated transformation for random insertional mutagenesis in *Colletotrichum lagenarium*. J Gen Plant Pathol 69:230–239

Tsukuda T, Carleton S, Fotheringham S, Holloman WK (1988) Isolation and characterization of an autonomously replicating sequence from *Ustilago maydis*. Mol Cell Biol 8:3703–3709

van Attikum H, Hooykaas PJ (2003) Genetic requirements for the targeted integration of *Agrobacterium* T-DNA in *Saccharomyces cerevisiae*. Nucleic Acid Res 31:826–832

van Attikum H, Bundock P, Hooykaas PJ (2001) Non-homologous end-joining proteins are required for *Agrobacterium* T-DNA integration. EMBO J 20:6550–6558

van de Rhee MD, Graça PM, Huizing HJ, Mooibroek H (1996) Transformation of the cultivated mushroom, *Agaricus bisporus*, to hygromycin B resistance. Mol Gen Genet 250: 252–258

Vijn I, Govers F (2003) *Agrobacterium tumefaciens* mediated transformation of the oomycete plant pathogen *Phytophthora infestans*. Mol Plant Pathol 4:459–467

Villalba F, Collemare J, Landraud P, Lambou K, Brozek V, Cirer B, Morin D, Bruel C, Beffa R, Lebrun MH (2008) Improved gene targeting in *Magnaporthe grisea* by inactivation of *MgKU80* required for non-homologous end joining. Fungal Genet Biol 45:68–75

Walton FJ, Idnurm A, Heitman J (2005) Novel functions required for melanization of the human pathogen *Cryptococcus neoformans*. Mol Microbiol 57:1381–1386

Wang J, Holden DW, Leong SA (1988) Gene transfer system for the phytopathogenic fungus *Ustilago maydis*. Proc Natl Acad Sci USA 85:865–869

Ward M, Kodama KH, Wilson LJ (1989) Transformation of *Aspergillus awamori* and *A. niger* by electroporation. Exp Mycol 13:289–293

Weld RJ, Lady CC, Ridgway HJ (2006) *Agrobacterium*-mediated transformation of *Sclerotinia sclerotiorum*. J Microbiol Methods 65:202–207

Wilson LM, Idnurm A, Howlett BJ (2002) Characterization of a gene (*sp1*) encoding a secreted protein from *Leptosphaeria maculans*, the blackleg pathogen of *Brassica napus*. Mol Plant Pathol 3:487–493

Yelton MM, Hamer JE, Timberlake WE (1984) Transformation of *Aspergillus nidulans* by using a trpC plasmid. Proc Natl Acad Sci USA 81:1470–1474

Zwiers LH, De Waard MA (2001) Efficient *Agrobacterium tumefaciens*-mediated gene disruption in the phytopathogen *Mycosphaerella graminicola*. Curr Genet 39:388–393

Chapter 7
Biotechnological Processes Used in Controlled Ectomycorrhizal Practices

Par Duponnois Robin, Bâ Amadou, Mousain Daniel, Galiana Antoine, Baudoin Ezékiel, Dreyfus Bernard, and Prin Yves

7.1 Introduction

Excessive industrial exploitation, clearing for industrial purposes and collection of firewood, has led to a dramatic deforestation during recent decades in Mediterranean and tropical areas (Piéri 1991). One of the critical issues of deforestation is the acceleration of soil degradation and desertification processes that involved a loss or reduction of major physico-chemical and biological soil properties (Requena et al. 2001). This lack or scarcity of plant cover largely contributed to soil erosion increase and consequently, to the decrease of soil fertility and soil microbial activities (Garcia et al. 1997). Numerous studies have reported that in such conditions, indigenous inoculum levels of mycorrhizal fungi were significantly reduced (Duponnois et al. 2001; Azcon-Aguilar et al. 2003). The mycorrhizal symbiotic

P.D. Robin (✉)
IRD, UMR 113 CIRAD/INRA/IRD/SUP-AGRO/UM2, Laboratoire des Symbioses Tropicales et Méditerranéennes (LSTM), TA A-82/J, Campus International de Baillarguet, Montpellier, Cedex 5, France
IRD, Laboratoire Commun de Microbiologie IRD/ISRA/UCAD, Centre de Recherche de Bel Air, BP 1386 Dakar, Sénégal
e-mail: Robin.Duponnois@ird.fr

B. Amadou, B. Ezékiel and D. Bernard
IRD, UMR 113 CIRAD/INRA/IRD/SUP-AGRO/UM2, Laboratoire des Symbioses Tropicales et Méditerranéennes (LSTM), TA A-82/J, Campus International de Baillarguet, Montpellier, Cedex 5, France

M. Daniel
INRA, UMR 113 CIRAD/INRA/IRD/SUP-AGRO/UM2, Laboratoire des Symbioses Tropicales et Méditerranéennes (LSTM), TA A-82/J, Campus International de Baillarguet, Montpellier, Cedex 5, France

G. Antoine and P. Yves
IRD, UMR 113 CIRAD/INRA/IRD/SUP-AGRO/UM2, Laboratoire des Symbioses Tropicales et Méditerranéennes (LSTM), TA A-82/J, Campus International de Baillarguet, Montpellier, Cedex 5, France

M. Rai and A. Varma (eds.), *Diversity and Biotechnology of Ectomycorrhizae*, Soil Biology 25, DOI 10.1007/978-3-642-15196-5_7,

process mobilizes and transports nutrients to roots, improves soil aggregation in eroded soils, and reduces water stress (Smith and Read 1997).

Mycorrhizal fungi are ubiquitous components of most ecosystems throughout the world and are considered key ecological factors in governing the cycles of major plant nutrients and in sustaining the vegetation cover (van der Hejden et al. 1998; Requena et al. 2001; Schreiner et al. 2003). Two major forms of mycorrhizas are usually recognized: the arbuscular mycorrhizas (AM) and the ectomycorrhizas (ECM). AM symbiosis is the most widespread mycorrhizal association type with plants that have true roots, i.e., pteridophytes, gymnosperms, and angiosperms (Read et al. 2000). These fungal symbionts affect about 80–90% land plants in natural, agricultural, and forest ecosystems (Brundrett 2002). ECMs affect trees and shrubs, gymnosperms (Pinaceae), and angiosperms and are usually the result of the association of Homobasidiomycetes with about 20 families of mainly woody plants (Smith and Read 1997). These woody species are associated with a larger (compared to the AM symbiosis) diversity of fungi, comprising 4,000 to 6,000 species, mainly Basidiomycetes and Ascomycetes (Allen et al. 1995; Valentine et al. 2004). The distribution of ectomycorrhizal fungi is not uniform in terms of their presence, abundance, or community composition. This has important implication for reforestation in areas with no native ectomycorrhizal vegetation (Marx 1991) or where ectomycorrhizal fungi are absent resulting from natural (Terwilliger and Pastor 1999) or anthropogenic disturbance (Jones et al. 2003).

The lack of mycorrhizal fungi on root systems is a leading cause of poor plant establishment and growth in a variety of forest landscapes. Numerous studies have shown that specific ectomycorrhizal fungi are able to improve the survival and early growth of various tree species in the field (Castellano and Molina 1989; Kropp and Langlois 1990; Marx et al. 1991; Castellano 1996; Roldan et al. 1996; Garbaye and Churin 1997; Duponnois et al. 2005, 2007). Since mycorrhizal association is estimated to occur in 95% of native undisturbed vegetation, whereas it occurs in less than 1% of vegetation from disturbed sites, mycorrhizal symbiosis has to be reestablished in order to benefit from the mycorrhizal effect on plant growth. Hence, this topic can be achieved by inoculating seedlings before they are transplanted to disturbed sites. However, different fungal associations do not lead to the same benefit to the host plant. A significant variability in response linked with the nature of the fungal–plant association has been reported (Guehl et al. 1990; Bâ et al. 2002; Duponnois and Plenchette 2003). These variations generally result from different factors such as the degree of host–fungus compatibility, mycorrhizal dependency of the host, fungal effectiveness in relation to biotic and abiotic site conditions, and the abundance and effectiveness of indigenous fungi (Garbaye 1988). In order to ensure the performance of afforestation, it is necessary that nurseries produce tree seedlings associated with mycorrhizal fungi that are ecologically compatible with the tree species and the planting sites. According to these conditions that have to be taken in account, different methods of controlled inoculation have been identified to optimize the effect on plant growth.

The main objective of this chapter was to describe some methods required to produce large quantities of efficient fungal inoculums and to present some tree growth data resulting from the use of controlled ectomycorrhization in nursery and field conditions in tropical and Mediterranean areas.

7.2 Criteria to Adopt an Inoculum Formulation

The main criteria that must be taken into account can be summarized as follows:

- The positive impact of the selected ectomycorrhizal fungi on the growth and the survival of tree species targeted for plantation.
- The viability of the fungal propagules which has to be maintained in storage to ensure the fungal efficiency on plant growth after inoculation.
- Ectomycorrhizal fungal inoculums have to be cost-effective to produce.

7.3 Spore-Based Fungal Inoculum

Spores collected from fruiting bodies can serve as a natural source of inoculum. This inoculation practice has been widely used in forest nurseries (Castellano 1994). However, this inoculation technique is more effective with fungal species that produce abundant sporocarps with large spore numbers (i.e., *Pisolithus* and *Scleroderma*).

7.3.1 Formulation of Spore Inoculums and Effect on Plant Growth

There are numerous formulations of spore inoculums. In a first step, sporocarps are collected, carefully identified, and kept in paper bags. Then, they are brushed free of adhering soil. These kinds of inoculum are frequently based on dry spores obtained by air-drying fruit bodies at temperatures below 35°C. Then dried sporocarps are crushed in plastic bags and sieved through 200–500 μm sieve. Then dried spores can be used in different formulations described as follows:

- Spore powder can be mixed with sterilized fine sand (1:100, w/w). Then containers are inoculated by mixing the soil substrate with this fungal mixture or by adding a small quantity of the fungal inoculum at the base of seedlings.
- Spore powder can be included into an inert carrier (i.e., clay) to form pellets which are applied at the base of seedlings (de la Cruz et al. 1990). According to Turjaman et al. (2005), in an experiment conducted to determine the effect of ectomycorrhizal fungi on the growth of dipterocarp species in peat soils, the sporocarps were crushed manually in plastic bags to ensure minimal loss of

spores and cross-contamination between fungi (de la Cruz et al. 1990). Crushed sporocarps and clay were mixed and formed into pellets with a ratio of 1:100 (w/w). Inoculation was performed out 10 days after seed germination. One hole was made in each pot, and a tablet (0.4 g) of fungal inoculum was applied to a potted seedling 1 cm below the soil surface, at the proximity of the root.

– Another process of ectomycorrhizal spore inoculation is realized by coating seeds with a mixture of spores and a binding agent such as clay (Marx et al. 1984).

The beneficial effects of ectomycorrhizal spore inoculation on the plant growth in nursery conditions have been frequently reported (Table 7.1). Several disadvantages have also been recorded with this type of fungal inoculum:

– Difficulties to collect large quantities of fruit bodies with some ectomycorrhizal fungal species.
– Low efficiency of the fungal inoculum due to slow germination or low spore viability ; this inoculation process is easy to apply and is an efficient method for storage and transport of spores for some mycorrhizal fungi such as *Pisolithus* or *Scleroderma*.

7.4 Mycelium-Based Fungal Inoculums

These inoculum forms require the isolation of fungal symbionts in axenic conditions in order to obtain a purified living material of a single fungal strain. These pure cultures are often obtained from fruit bodies of ectomycorrhizal fungi collected in the field but they can also be performed by using mycorrhizal roots, sclerotia, rhizomorphs, and spores (Molina and Palmer 1982). In this chapter, the fungal isolation from fruiting bodies will be described.

7.4.1 Fungal Isolation from Fruit Bodies

This procedure is usually considered as the most successful method to obtain pure fungal culture. Sporocarps are brushed free of adhering soil and fractured carefully in a laminar flow hood. A small amount of tissue is then removed with a fine forceps and placed on a nutrient medium agar. As growth requirements vary between ectomycorrhizal fungi, different nutrient media can be used (Table 7.2). The fungal cultures are incubated at 25°C in the dark and subcultured until all the contaminating microorganisms are eliminated. Fungal cultures are usually maintained in Petri dishes over a required nutrient agar medium and at 20–25°C. Fresh subcultures are made every 6–12 weeks depending on the fungal growth rates.

Table 7.1 Effect of ectomycorrhizal spore inoculation on plant growth in nursery conditions

References	Fungal strain	Plant species	Inoculum formulation	Effects on plant growth parameters (%) and ectomycorrhizal colonization			
				Height	Shoot biomass	Root biomass	Ecto. colonization (%)
Turjaman et al. (2005)	*Pisolithus arhizus*	*Shorea pinanga*	Ectomycorrhizal spore pellets	+46.1[a]	+66.7	nd[b]	87
Turjaman et al. (2005)	*Scleroderma* sp.	*S. pinanga*	Ectomycorrhizal spore pellets	+41.3	+60.5	nd	86
Aggangan et al. (2010)	*P. tinctorius*	*Acacia mangium*	Ectomycorrhizal spore tablets	nd	+29.1	+40.7	52
Rincon et al. (2007)	*Rhizopogon roseolus*	*Pinus halepensis*	Spore suspension	−30.1	nd	nd	48
Rincon et al. (2007)	*Suillus collinitus*	*Pinus halepensis*	Spore suspension	+27.8	nd	nd	75
Chen et al. (2006)	*Scleroderma albidum*	*Eucalyptus globulus*	Spore suspension	+32.1	+19.7	+42.5	nd
Chen et al. (2006)	*S. areolatum*	*E. globulus*	Spore suspension	+17.5	+9.1	+33.5	nd
Chen et al. (2006)	*S. cepa*	*E. globulus*	Spore suspension	+30.8	+8.8	+2.0	nd
Chen et al. (2006)	*S. albidum*	*E. urophylla*	Spore suspension	−3.6	+3.1	−0.7	nd
Chen et al. (2006)	*S. areolatum*	*E. urophylla*	Spore suspension	+7.2	+4.9	+4.0	nd
Chen et al. (2006)	*S. cepa*	*E. urophylla*	Spore suspension	+6.3	+13.0	+1.4	nd
Torres and Honrubia (1994)	*P. tinctorius*	*P. halepensis*	Spore suspension	+37.1	+44.9	+41.7	55.4
Torres and Honrubia (1994)	*R. roseolus*	*P. halepensis*	Spore suspension	+38.3	+44.6	+28.6	39.5
Torres and Honrubia (1994)	*Suillus collinitus*	*P. halepensis*	Spore suspension	+35.1	+28.2	+30.6	28.9

[a](mean value of ectomycorrhizal plants – mean value of the nonectomycorrhizal plants) $\times$ 100)/(mean value of the ectomycorrhizal plants)
[b]nd: not determined

Table 7.2 Composition of some nutrient media commonly used for the isolation and culture of ectomycorrhizal fungi (from Brundrett et al. 1996)

Components	Nutrient media		
	MMN[a]	Pachlewski[b]	FDA[c]
Mineral nutrients (mg l^{-1})			
$(NH_4)_2HPO_4$	250		
NH_4Cl			500
$C_4H_{12}N_2O_6$[d]		500	
KH_2PO_4	500	1000	500
$MgSO_4 \cdot 7H_2O$	150	500	500
$CaCl_2 \cdot 2H_2O$	50	50	
NaCl	25		
Fe EDTA	20	20	
H_3BO_3		2.8	
$MnCl_2 \cdot 2H_2O$		3.0	
$ZnSO_4 \cdot 7H_2O$		2.3	
$CuCl_2 \cdot 2H_2O$		0.63	
$Na_2Mo_4 \cdot 2H_2O$		0.27	
Carbohydrate source (g l^{-1})			
Maltose		5	
Glucose	10	20	20
Malt extract	3		
Vitamins (μg l^{-1})			
Thiamine HCl	0.1	0.1	
Agar (g l^{-1})	20	20	20
pH			
Adjusted pH to	5.8	5.4	5.0

[a]MMN: Modified Menin Norkrans medium (Marx 1969)
[b]Pachlewski medium (Pachlewski and Pachlewski 1974)
[c]Ferry and Das (1968)
[d]Ammonium tartrate

7.4.2 *Formulation of Mycelium-Based Inoculums and Effect on Plant Growth*

These procedures are only suitable with some fungal species that are able to grow in culture for the economical production of fungal inoculums. Two main supports are usually used to produce fungal inoculums: the multiplication of the fungal strain on a peat/vermiculite substrate and the production of an ectomycorrhizal inoculum entrapped in a hydrogel (i.e., calcium alginate gel).

7.4.2.1 Vermiculite–Peat Inoculums (Marx and Bryan 1975)

Glass jars (1.6 l) containing 1.3 l expanded vermiculite–sphagnum peat mixture (4:5–1:5, v:v, pH 5.5) are autoclaved (120°C, 20 min). Another ratio of vermiculite–peat can be used (2:3–1:3) but the peat can release some substances which could be toxic for the fungal development. For this reason, the first ratio mixture is usually used. Then the mixture is moistened to field capacity with 600 ml modified as a liquid nutrient medium (see Table 7.2). The jars are blocked with lids with a 1-cm diameter hole. This hole is fitted with a 4-cm long tube filled with cotton wool. The jars are then autoclaved a second time (120°C for 20 min). After cooling, about eight mycelial plugs are laid on top of the substrate. Mycelium grows down into the substrate, which is completely colonized after 6–10 weeks at 25°C depending on the fungal growth rate. For faster growth, jars can be filled with a smaller quantity of substrate and shaken after mycelia have colonized a few centimeters: in this manner, mycelium is evenly distributed throughout the substrate and incubation time is shortened. This inoculum can be stored at 4°C for up to 6 months.

7.4.2.2 Encapsulation of Hyphal Fragments Within Beads of Alginate Gel (Mauperin et al. 1987)

The process of including fungal mycelium in polymeric gels (especially calcium alginate) has been previously described by Dommergues et al. (1979) and Le Tacon et al. (1983, 1985). The inoculum prepared in this manner is more efficient than the vermiculite–peat formulation (Mortier et al. 1988) because the mycelium is protected in the gel from physical stresses (e.g., water stress) and from competitor microorganisms. With this process, it is also possible to accurately determine the weight of mycelium or the number of living propagules contained in the inoculum. Different methods have been tested to measure the quantity of mycelium in the vermiculite–peat inoculum:ergosterol assay (Martin et al. 1990); chitin assay (Vignon et al. 1986), but none of them gave reliable results (Mortier et al. 1988) because of the peat which interferes with colorimetric measurements. Using fungal strains that are able to grow from fragmented hyphae, a mycelial suspension is produced in Erlenmeyer flasks or in fermentor filled with the required liquid nutrient medium. The mycelium is then washed in tap water in order to remove residual nutrients, homogenized in a Waring blender for about 10 s, and resuspended in distilled water. This kind of fungal inoculum is quantified by measuring the fungal dry weight per milliliter or by counting living propagules (determination of colony-forming units by spreading 1 ml of suspension on a 6-cm Petri dish with nutrient agar). Then the hyphal fragment suspension is mixed (1:1, v:v) with distilled water containing 20 g l^{-1} sodium alginate and 50 g l^{-1} autoclaved dry powdered sphagnum peat. The final solution is pumped throughout a pipe with 2-mm holes. The drops fall into a 100 g l^{-1} $CaCl_2$ solution and form beads of reticulated calcium alginate gel (Mauperin et al. 1987). The beads are kept in $CaCl_2$ for 24 h at room temperature in order to ensure complete reticulation. Then they are washed with tap water to eliminate NaCl and $CaCl_2$ and stored in air-tight containers at 4°C in order to prevent drying. This type of inoculum can be kept up to 9 months in these conditions. The beads are prepared with 1–2 g mycelium (dry weight) per liter of final solution (Mortier et al. 1988).

7.4.2.3 Controlled Ectomycorrhization in Glasshouse, Nursery, and Field Conditions

The aims of these procedures are to screen fungal isolates for their compatibility with host plants and to determine plant growth promotion in order to select the best fungus to be inoculated in nurseries. This topic will be mainly illustrated by some data resulting from controlled ectomycorrhization trials with Australian *Acacia* species and tropical fungal strains.

Screening for Efficient Fungal Isolates in Controlled Conditions

Seedlings can be planted in soil (disinfected or not after autoclaving) or nondisinfected vermiculite–peat mixture (1:1, v:v). Several kinds of containers can be used for growing seedlings in the glasshouse depending on the aim of the experiment. Then the ectomycorrhizal inoculation was performed by mixing the fungal inoculum (10/1; v/v) with the soil. The treatment without fungus (control) received an autoclaved mixture of vermiculite and peat moss moistened in nutrient medium at the same rate. Then the pots were arranged in a randomized complete block design. Numerous studies have reported the beneficial effects on the growth and nutrient status of tree seedlings in controlled conditions with the ultimate goal of selecting the most promising and potential mycorrhizal inoculants for large-scale inoculation programs in degraded lands in the tropics and Mediterranean areas (Table 7.3). Although several methods of inoculation can be used in practice, only some species of the genera *Laccaria*, *Hebeloma*, *Paxillus*, *Pisolithus*, *Scleroderma*, *Suillus*, or *Rhizopogon* have been routinely used (Castellano 1996).

Controlled Inoculation in Field Conditions

Numerous studies have shown that controlled mycorrhizal symbiosis can significantly improve the growth of tree species in degraded soils. However, most of these experiments have been performed in controlled glasshouse conditions, hence data on the sustainability of the positive effect of mycorrhiza on plant growth in field conditions are lacking, especially in Sahelian areas. Controlled ectomycorrhization experiments have been recently set up in Senegal using an Australian *Acacia* species, *A. holosericea*, and an ectomycorrhizal fungus, *Pisolithus albus* strain IR100 (Duponnois et al. 2007). A positive effect of mycorrhizal inoculation on *A. holosericea* development has been recorded in all of the experiments (Fig. 7.1). These results suggest that without suitable symbionts, this species has poor ability to scavenge for P under P-limiting conditions (Dommergues et al. 1999) and clearly shows that the controlled mycorrhization of *A. holosericea* could be a beneficial tool in improving the survival and productivity of *Acacia* species and consequently for the reafforestaion of Sahelian regions. However, in order to evaluate the possible economic advantage of ectomycorrhizal inoculation, it is essential to carry out field experiments in a range of situations representative of major plantations.

Table 7.3 Growth of some Australian *Acacia* species inoculated with different ectomycorrhizal fungal strains (peat–vermiculite formulation) after 4 months culturing in glasshouse conditions

Fungal species	*Acacia* species	Effects on plant growth parameters (%) and ectomycorrhizal colonization			References
		Shoot biomass	Root biomass	Ectomycorrhizal colonization (%)	
Pisolithus albus IR100	*A. auriculiformis*	+42.1[a]	+38.6	45.2	Duponnois and Plenchette (2003)
P. albus IR100	*A. mangium*	+35.9	+44.1	20.1	Duponnois and Plenchette (2003)
P. albus IR100	*A. platycarpa*	+43.1	+9.8	31.6	Duponnois and Plenchette (2003)
Pisolithus sp. SL2	*A. holosericea*	+56.7	+10.6	48.3	Duponnois and Plenchette (2003)
P. albus COI007	*A. holosericea*	+55.6	+25.3	43.8	Duponnois and Plenchette (2003)
P. albus COI024	*A. holosericea*	+50.4	+17.1	10.8	Duponnois and Plenchette (2003)
Pisolithus sp. COI032	*A. holosericea*	+54.9	+12.6	15.0	Duponnois and Plenchette (2003)
P. albus IR100	*A. holosericea*	+57.1	+48.9	25.2	Duponnois and Plenchette (2003)
P. tinctorius GEMAS	*A. holosericea*	+57.8	+14.1	43.1	Duponnois and Plenchette (2003)
Scleroderma dictyosporum IR109	*A. holosericea*	+52.9	+21.0	53.4	Duponnois and Plenchette (2003)
S. verrucosum IR500	*A. holosericea*	+64.4	+14.1	13.8	Duponnois and Plenchette (2003)
P. tinctorius GEMAS	*A. crassicarpa*	+77.1	+52.3	49.4	Lesueur and Duponnois (2005)
P. albus COI024	*A. mangium*	+54.5	+52.1	41.7	Duponnois et al. (2002)
Scleroderma sp. IR408	*A. holosericea*	+82.1	+89.6	13.8	Duponnois et al. (2006)
S. dictyosporum IR412	*A. holosericea*	+72.9	+82.6	12.5	Duponnois et al. (2006)

[a](mean value of ectomycorrhizal plants – mean value of the nonectomycorrhizal plants) × 100)/(mean value of the ectomycorrhizal plants)

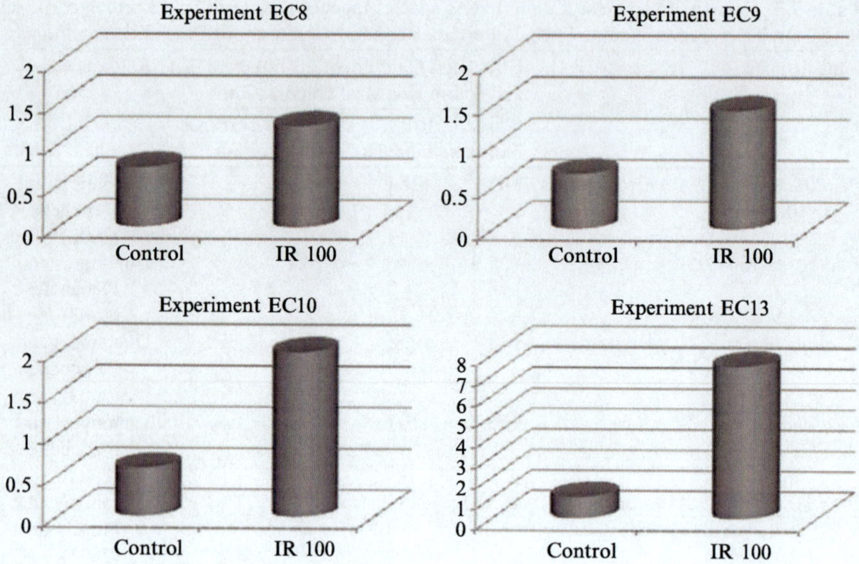

Fig. 7.1 Field performance of *A. holosericea* trees (expressed in tons of wood biomass per ha with 1,200 planted trees per ha) inoculated or not with *Pisolithus albus* strain IR100 in field trials conducted in Senegal after 18 month plantation (from Duponnois et al. 2007)

7.5 Conclusion

Numerous studies have reported the benefits that result from the use of fungal inoculants, which can improve the development of tree seedlings in glasshouse and nursery conditions. In the same way, a lot of technical procedures have been elaborated to produce cost-effective mycorrhizal inoculum. Unfortunately, mycorrhizal inoculation remains underexploited in nursery cultural practices although the management of tree mycorrhizal status could have very important implications for tropical reforestation programs in the developing countries. The results presented in this chapter suggest that inoculation with ectomycorrhizal fungi can improve the early growth of the main tree species in tropical and Mediterranean forests and that this technique will accelerate the rehabilitation of degraded forests.

References

Aggangan NS, Moon HK, Han SH (2010) Growth response of *Acacia mangium* Willd. seedlings to arbuscular mycorrhizal fungi and four isolates of the ectomycorrhizal fungal *Pisolithus tinctorius* (Pers.) Coker and Couch. New For 39:215–230

Allen EB, Allen MF, Helm DJ, Trappe JM, Molina R, Rincon E (1995) Patterns and regulation of mycorrhizal plant and fungal diversity. Plant Soil 170:47–62

Azcon-Aguilar C, Palenzuela J, Roldan A, Bautista S, Vallejo R, Barea JM (2003) Analysis of the mycorrhizal potential in the rhizosphere of representative plant species from desertification-threatened Mediterranean shrublands. Appl Soil Ecol 14:165–175

Bâ AM, Sanon KB, Duponnois R (2002) Influence of ectomycorrhizal inoculation on *Afzelia quanzensis* Welw. Seedlings in a nutrient-deficient soil. For Ecol Manage 161:215–219

Brundrett MC (2002) Coevolution of roots and mycorrhizas of land plants. New Phytol 154:275–304

Brundrett M, Bougher N, Dell B, Grove T, Malajczuk N (1996) Working with mycorrhizas in Forestry and Agriculture. ACIAR Monograph, 373p

Castellano MA (1994) Current status of outplanting studies using ectomycorrhiza-inoculated forest trees. In: Pfleger F, Linderman B (eds) A reappraisal of mycorrhizae in plant health. The American Phytopathological Society, St Paul, pp 261–281

Castellano MA (1996) Outplanting performance of mycorrhizal inoculated seedlings. In: Mukerji KG (ed) Concepts in mycorrhizal research. Kluwer Academic Publishers, Dordrecht, The Netherlands, pp 223–301

Castellano MA, Molina R (1989) Mycorrhizae. In: Landis TD (ed) The biological component: nursery pest and mycorrhizae manual. Agric. Handbook 674, vol 5. USDA Forest Service, Washington, DC, pp 101–167

Chen YL, Dell B, Malajczuk N (2006) Effect of *Scleroderma* spore density and age on mycorrhiza formation and growth of containerized *Eucalyptus globulus* and *E. urophylla* seedlings. New For 31:453–467

de la Cruz RE, Lorilla EB, Aggrangan NS (1990) Ectomycorrhizal tablets for *Eucalyptus* species. In: Werner D, Muller P (eds) Fast growing trees and nitrogen fixing trees. Gustav Fisher Verlag, Stuttgard, p 371

Dommergues R, Diem HG, Divies C (1979) Microbiological process for controlling the productivity of cultivated plants. US Patent No 4.155.737, 22 May 1979

Dommergues YR, Duhoux E, Diem HG (1999) Les arbres fixateurs d'azote, Editions Espaces 34. ORSTOM, FAO, Montpellier, 501pp

Duponnois R, Plenchette C (2003) A mycorrhiza helper bacterium (MHB) enhances ectomycorrhizal and endomycorrhizal symbiosis of Australian *Acacia* species. Mycorrhiza 13:85–91

Duponnois R, Plenchette C, Thioulouse J, Cadet P (2001) The mycorrhizal soil infectivity and arbuscular mycorrhizal fungal spore communities in soils of different aged fallows in Senegal. Appl Soil Ecol 17:239–251

Duponnois R, Founoune H, Lesueur D (2002) Influence of the controlled dual ectomycorrhizal and rhizobial symbiosis on the growth of *Acacia mangium* provenances, the indigenous symbiotic microflora and the structure of plant parasitic nematode communities. Geoderma 109:85–102

Duponnois R, Founoune H, Masse D, Pontanier R (2005) Inoculation of *Acacia holosericea* with ectomycorrhizal fungi in a semi-arid site in Senegal: growth response and influences on the mycorrhizal soil infectivity after 2 years plantation. For Ecol Manag 207:351–362

Duponnois R, Assigbetse K, Ramanankierana H, Kisa M, Thioulouse J, Lepage M (2006) Litter-forager termite mounds enhance the ectomycorrhizal symbiosis between *Acacia holosericea* A. Cunn. Ex G. Don and Scleroderma dictyosporum isolates. FEMS Microbiol Ecol 56:292–303

Duponnois R, Plenchette C, Prin Y, Ducousso M, Kisa M, Bâ AM, Galiana A (2007) Use of mycorrhizal inoculation to improve reafforestation process with Australian *Acacia* in Sahelian ecozones. Ecol Eng 29:105–112

Ferry BW, Das M (1968) Carbon nutrition of some mycorrhizal *Boletus* species. Trans Br Mycol Soc 51:795–798

Garbaye J (1988) Les plantations forestières tropicales: un champ d'application privilégié pour la mycorhization contrôlée. Revue Bois et Forêts des Tropiques 216:23–34

Garbaye J, Churin JL (1997) Growth stimulation of young oak plantations inoculated with the ectomycorrhizal fungus *Paxillus involutus* with special reference to summer drought. For Ecol Manag 98:221–228

Garcia C, Roldan A, Hernandez T (1997) Changes in microbial activity after abandonment of cultivation in a semi-arid Mediterranean environment. J Environ Qual 26:285–291

Guehl JM, Mousain D, Falconnet G, Gruez J (1990) Growth, carbon dioxide assimilation capacity and water-use efficiency of *Pinus pinea* L. seedlings inoculated with different ectomycorrhizal fungi. Ann Sci For 47:91–100

Jones MD, Durall DM, Cairney JWG (2003) Ectomycorrhizal fungal communities in young forest stands regenerating after clearcut logging. New Phytol 157:399–422

Kropp BR, Langlois CG (1990) Ectomycorrhizae in reforestation. Can J For Res 20:438–451

Le Tacon F, Jung G, Michelot P, Mugnier J (1983) Efficacité en pépinière forestière d'un inoculum de champignon ectomycorhizien produit en fermenteur et inclus dans une matrice de polymères. Ann Sci For 40:165–176

Le Tacon F, Jung G, Mugnier J, Michelot P, Mauperin C (1985) Efficiency in a forest nursery of an ectomycorhizal fungus inoculums produced in a fermentor and entrapped in polymeric gels. Can J Bot 63:1664–1668

Lesueur D, Duponnois R (2005) Relations between rhizobial nodulation and root colonization of *Acacia crassicarpa* provenances by an arbuscular mycorrhizal fungus, *Glomus intraradices* Schenk and Smith or an ectomycorrhizal fungus, *Pisolithus tinctorius* Coker & Couch. Ann For Sci 62:467–474

Martin F, Delaruelle C, Hilbert JL (1990) An improved ergosterol assay to estimate the fungal biomass in ectomycorrhizas. Mycol Res 94:1069–1074

Marx DH (1969) The influence of ectotrophic mycorrhizal fungi on the resistance of pine roots to pathogenic infections. I. Antagonism of mycorrhizal fungi to root pathogenic fungi and soil bacteria. Phytopathology 59:153–163

Marx DH (1991) The practical significance of ectomycorrhizae in forest establishment. In: Ecophysiology of ectomycorrhizae of forest trees, Marcus Wallenberg Foundation symposia proceedings, vol 7, pp 54–90

Marx DH, Bryan WC (1975) Growth and ectomycorrhizal development of Loblolly pine seedlings in fumigated soil infested with the fungal symbiont *Pisolithus tinctorius*. For Sci 21:242–254

Marx DH, Jarl K, Ruehle JL, Bell W (1984) Development of *Pisolithus tinctorius* ectomycorrhizae on pine seedlings using basidio-encapsulated seed. For Sci 30:897–907

Marx DH, Ruehle JL, Cordell CE (1991) Methods for studying nursery and field response of trees to specific ectomycorrhiza. In: Norris JR, Read DJ, Varma AK (eds) Methods in microbiology, vol 23, Techniques for the study of mycorrhiza. Academic Press, London, pp 383–411

Mauperin C, Mortier F, Garbaye J, Le Tacon F, Carr G (1987) Viability of an ectomycorrhizal inoculum produced in a liquid medium and entrapped in a calcium alginate gel. Can J Bot 65:2326–2329

Molina R, Palmer JG (1982) Isolation, maintenance and pure culture manipulation of ectomycorrhizal fungi. In: Schenck NC (ed) Methods and principles of mycorrhizal research. The American Phytopathological Society, St Paul, pp 115–129

Mortier F, Le Tacon F, Garbaye J (1988) Effect of inoculum type and inoculation dose on ectomycorrhizal development, root necrosis and growth of Douglas fir seedlings inoculated with *Laccaria laccata* in a nursery. Ann Sci For 45:301–310

Pachlewski R, Pachlewski J (1974) Studies on symbiotic properties of mycorrhizal fungi of pine (*Pinus silvestris* L.) with the aid of the method of mycorrhizal synthesis in pure culture on agar. Forest Research Institute, Warsaw, p 228

Piéri C (1991) Les bases agronomiques de l'amélioration et du maintien de la fertilité des terres de savanes au sud Sahara. In: Savanes d'Afrique, terre fertile? Actes des Rencontres Internationales, 10–14 Dec 1990, Montpellier, France, pp 43–74

Read DJ, Duckett JG, Francis R, Ligrone R, Russell A (2000) Symbiotic fungal associations in "lower" land plants. Philos Trans R Soc Lond B Biol Sci 355:815–830

Requena N, Perez-Solis E, Azcon-Aguilar C, Jeffries P, Barea JM (2001) Management of indigenous plant–microbe symbioses aids restoration of desertified ecosystems. Appl Environ Microbiol 67:495–498

Rincon A, de Felipe MR, Fernandez-Pascual M (2007) Inoculation of *Pinus halepensis* Mill. with selected ectomycorrhizal fungi improves seedling establishment 2 years after planting in a degraded gypsum soil. Mycorrhiza 18:23–32

Roldan A, Querejeta I, Albadalejo J, Castillo V (1996) Growth response of *Pinus halepensis* to inoculation with *Pisolithus arhizus* in a terraced rangeland with urban refuse. Plant Soil 179:35–43

Schreiner RP, Mihara KL, McDaniel KL, Bethlenfalvay GJ (2003) Mycorrhizal fungi influence plant and soil functions and interactions. Plant Soil 188:199–209

Smith SE, Read DJ (1997) Mycorrhizal symbiosis, 2nd edn. Academic Press, UK, p 605

Terwilliger J, Pastor J (1999) Small mammals, ectomycorrhizae, and conifer succession in beaver meadows. Oikos 85:83–94

Torres P, Honrubia M (1994) Inoculation of containerized *Pinus halepensis* (Miller) seedlings with basidiospores of *Pisolithus arhizus* (Pers) Rauschert, *Rhizopogon roseolus* (Corda) and *Suillus collinitus* (Fr) O Kuntze. Ann Sci For 51:521–528

Turjaman M, Tamai Y, Segah H, Limin SH, Cha JY, Osaki M, Tawaraya K (2005) Inoculation with the ectomycorrhizal fungi *Pisolithus arhizus* and *Scleroderma* sp. improves early growth of *Shorea pinanga* nursery seedlings. New For 30:67–73

Valentine LL, Fieldler TL, Hart AA, Petersen CA, Berninghausen HK, Southworth D (2004) Diversity of ectomycorrhizas associated with *Quercus garryana* in southern Oregon. Can J Bot 82:123–135

van der Hejden MGA, Klironomos JN, Ursic M, Moutoglis P, Streitwolf-Engel R, Boller T, Wiemken A, Sanders IR (1998) Mycorrhizal fungal diversity determines plant biodiversity ecosystem variability and productivity. Nature 396:69–72

Vignon C, Plassard C, Moussain D, Salsac L (1986) Assay of fungal chitin and estimation of mycorrhizal infection. Physiol Vég 24:201–207

Chapter 8
Signaling in Ectomycorrhizal Symbiosis Establishment

Paula Baptista, Rui Manuel Tavares, and Teresa Lino-Neto

8.1 Introduction

A multiplicity of strategies has been developed by plants for their survival in a wide range of biotic and abiotic stimuli. One successful approach is the development of mutualistic associations, such as ectomycorrhizas (ECM) formed with soil-borne fungi from Basidiomycota, Ascomycota, and Zygomycota. In forest ecosystems, the ECM association is indeed the predominant form of mycorrhiza. The plant species involved are diverse and belong mainly to Fagaceae, Betulaceae, Pinaceae, and Dipterocarpaceae families (Smith and Read 2008). This symbiosis is characterized by the presence of a fungal sheath around the fine root tips (the mantle) and by the hyphae colonization of the intercellular space between root cells, the so-called Hartig-net. From the mantle, hyphae spread over into the surrounding substrate, increasing the root surface area. This represents an effective way to enhance nutrient, carbon, and water exchanges between the fungus and the plant roots. While the plant takes advantage from the improved nutrient uptake, the fungus receives photosynthetic carbohydrates from plant (Smith and Read 2008). Other advantages for the host plant include improved growth, enhanced water acquisition and use, as well as enhanced soil stability (Smith and Read 2008; Futai et al. 2008; Finlay 2008). In addition, plants displaying ECMs also exhibit increased tolerance to pathogens (Marx 1972), heavy metals (Brunner and Frey 2000), or drought (Osonubi et al. 1991; Davies et al. 1996; Alvarez et al. 2009).

The ECM symbiosis results from the interaction between roots and fungi in a time-regulated sequence of highly coordinated events. In this process, the

P. Baptista (✉)
CIMO/School of Agriculture, Polytechnic Institute of Bragança, Campus de Santa Apolónia, Apartado 1172, 5301-855 Bragança, Portugal
e-mail: pbaptista@ipb.pt

R.M. Tavares and T. Lino-Neto
Centre for Biodiversity Functional and Integrative Genomics (BioFIG), Plant Functional Biology Centre, University of Minho, Campus de Gualtar, 4710-057 Braga, Portugal
e-mail: tavares@bio.uminho.pt; tlneto@bio.uminho.pt

M. Rai and A. Varma (eds.), *Diversity and Biotechnology of Ectomycorrhizae*,
Soil Biology 25, DOI 10.1007/978-3-642-15196-5_8,

mycobiont has the ability to recognize and escape the host defense surveillance, being able to become associated with host roots and establish bidirectional nutrient transfer. Exchange of signals between partners should occur not only to determine their compatibility, but also to prepare them for further physiological and morphological changes for the formation of the symbiotic organ (Martin et al. 2001; Voiblet et al. 2001; Podila 2002). Although representing an active research area, little is known about the nature of the signaling molecules, and only few genes and proteins associated to ectomycorrhizal symbiosis have been identified. The mechanistic details of root–fungi perception and recognition are also basically unknown or scarcely defined. Here, we review the recent results shedding light on the nature of the molecular signals that determine ECM development.

8.2 Ectomycorrhiza Formation

The ECM development usually includes four stages: preinfection, initiation, differentiation, and functioning (Table 8.1) (Martin and Tagu 1995; Le Quéré et al. 2005). In the preinfection stage, hyphae perceive the presence of host roots and the potential partner recognition occurs. In the initiation stage, hyphae start to grow and reach the root surface establishing physical contact. After attachment onto epidermal cells, hyphae proliferate in a series of layers that surround the roots and will differentiate into the mantle. Their penetration between epidermal (angiosperms) and also cortical (gymnosperms) root cells allows the differentiation of the Hartig-net.

Table 8.1 Time course developmental stages of ectomycorrhizal formation between *Pinus pinaster* and *Pisolithus tinctorius* promoted under in vitro conditions

Preinfection	Initiation	Differentiation	Functioning
0–12 h	12–24 h	2–5 days	5–15 days
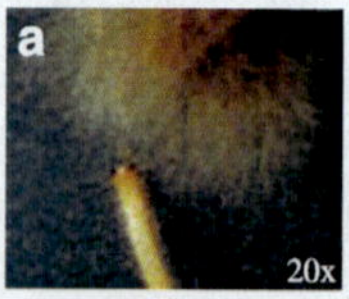	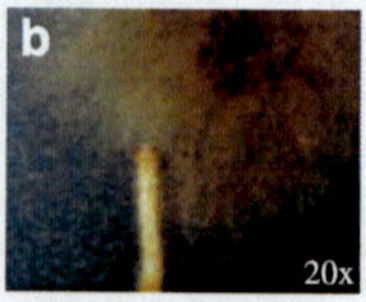	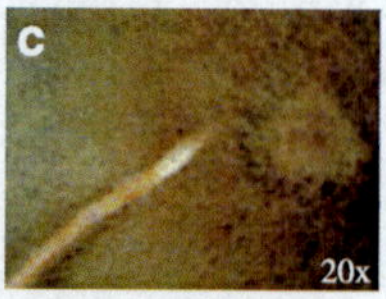	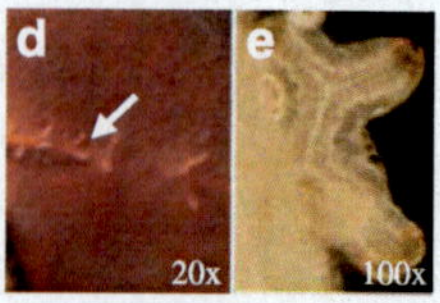
Recognition between symbionts	Hyphae contact with the root Fungal attachment to the root epidermis	Hyphae growth around the roots (mantle formation) Hyphae penetration and growth between epidermal and cortical root cells (Hartig-net formation)	Well-developed mantle is tightly adherent to epidermal cells Bidirectional translocation of nutrients between partners

The time required for attaining each stage and the principal events that occur are represented. Figures a–d display the morphological aspects of root–fungus interaction and figure e depicts a cross-section from a 15-day-old mycorrhiza. The arrow indicates an emerging lateral root from *P. pinaster*

Finally, in the functioning stage, the bidirectional transfer of nutrients occurs between both symbionts.

The time required for each stage of ECM development is variable and dependent on the genetic compatibility between plant and fungi, as well on abiotic factors (Smith and Read 2008). In addition, the method used for ECM establishment (axenic or nonaxenic culture) could influence this process. The development of a functional ECM association under axenic conditions occurs within 2–3 weeks. For example, the development of *Paxillus involutus–Betula pendula* (Le Quéré et al. 2005) and *Pisolithus microcarpus–Eucalyptus globulus* (Duplessis et al. 2005) ECM associations occurs within 21 days. When ectomycorrhizal development was established between *Pisolithus tinctorius* and *Pinus pinaster* under in vitro conditions, a mature mycorrhiza was observed 15 days postcontact (Table 8.1). In this case, the preinfection and hyphae adherence onto root surface occurred within the first 24 h. After 2 days of contact, the mantle started to be formed at the apical region of the main root axis. After 5–15 days of contact, newly emerging lateral roots were produced and became completely enveloped by the fungal mantle. At the same time the Hartig-net became evident.

In diverse ECM associations, differential gene expression studies revealed changes in the transcript levels of genes from both the fungus and host plant during the interaction, resulting in the identification of several symbiosis-related (SR) genes (Voiblet et al. 2001; Johansson et al. 2004; Duplessis et al. 2005; Le Quéré et al. 2005; Acioli-Santos et al. 2008; Sebastiana et al. 2009). The identified genes are involved in many cellular functions like fungal cell division and proliferation, differentiation and signaling, synthesis of plant cell wall and extracellular matrices, plant defense or stress responses, and primary metabolism (e.g., glycolysis, amino acid biosynthesis, and transporter activity) (Martin et al. 2007). These results suggest that ECM formation is a highly dynamic process, in which plant and fungi are always receiving and sending signals and both are exposed to high stress conditions (Martin et al. 2001).

8.3 Rhizospheric Signals Emitted During the Preinfection

The exchange of signals between the symbiotic partners starts long before the occurrence of any physical contact, suggesting the involvement of diffusible elicitors in the early stages of ECM establishment (Fig. 8.1). Accordingly, previous to any physical contact with the symbiotic host *Tilia Americana*, 58 genes of the ectomycorrhizal fungus *Tuber borchii* were found to be differentially expressed (Menotta et al. 2004). It was also observed that fungal genes encoding putative signaling pathway components, such as *ras* (Martin et al. 2001; Sundaram et al. 2001) or *PF6.2* (Kim et al. 1999), were induced prior to fungal contact with host roots.

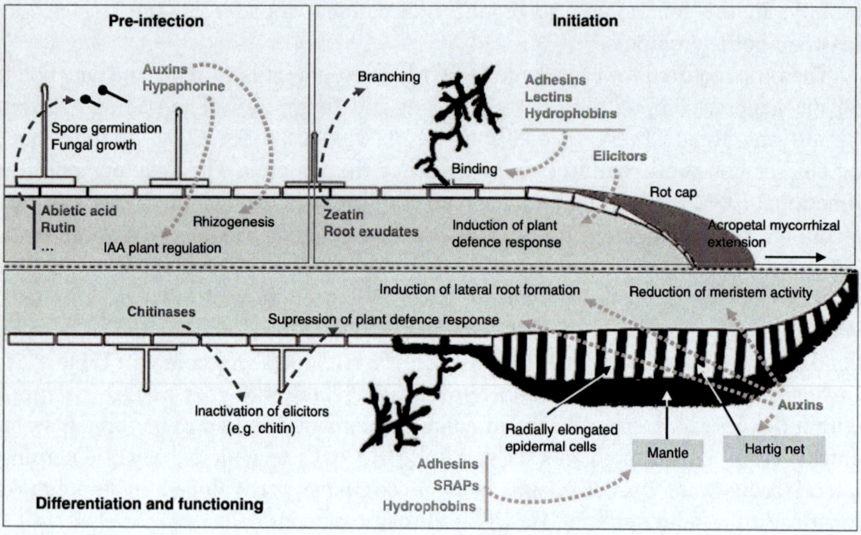

Fig. 8.1 Signals required for the ectomycorrhizal development. Root and fungal exudates are the source of specific signals during the preinfection stage that lead to the perception and recognition of compatible partners. The morphological changes that occur for forming the symbiotic structures (in *gray boxes*) and the main signals that induce them are outlined. Signal molecules produced by the plant are represented in *dark gray bold* and the developmental processes in which they are involved are indicated by *dashed gray arrows*. Fungal signals are represented in *light gray bold* and the processes controlled by them pointed out by *dotted arrows*

In the preinfection stage, hyphae sense the presence of a host root. Colonizer hyphae may result from a germinated spore or from a mycelium already involved in the development of other mycorrhizas (Tagu et al. 2002). When originating from spore germination, the hyphae seem to sense the root system by the nutritional composition changes that occur in the rhizosphere. The root nitrogen uptake and secretion of sugars from the host plant turn the surroundings of the roots depleted of nitrogen and enriched in carbon. These alterations could represent trophic signals for fungus perception of the roots presence. However, as these trophic signals are not specific, they could also be used by saprophytic or pathogenic fungi to recognize the presence of roots (Tagu et al. 2002).

Root and fungal exudates seem to be the source of more specific signals for the perception and recognition of a compatible association. The plant secretion of certain signal molecules into the rhizosphere seems to promote spore germination and hyphal growth (Barker et al. 1998; Barker and Tagu 2000). Root exudates and extracts from Scotch pine (*Pinus sylvestris*) enhance spore germination of several ectomycorrhiza-forming *Suillus* species (*S. granulatus*, *S. grevillei*, *S. luteus* and *S. variegatus*) (Fries et al. 1987). The germination-induction compound was identified as abietic acid, which was described to confer selective advantage to *Suillus* spp. against other fungal species present in the same rhizosphere. In root exudates of *Eucalyptus globulus* ssp. *bicostata* the presence of flavonols, namely the rutin, was also reported to promote the growth of several strains of *Pisolithus tinctorius*

(Lagrange et al. 2001). In addition, during the initial stages of *Suillus tridentinus* colonization of *in vitro Larix decidua* roots, an increase in phenylpropanoids (mainly flavonoids), benzoic, and cinnamic acids suggests their participation as signals in ECM symbiosis (Weiss et al. 1997). However, even after the detection of their effects, the chemical identification of inducing fungal growth molecules was not always achieved. Although the diffusible molecules released by eucalyptus roots had been described as chemoattractants towards *Pisolithus tinctorius* and *Paxillus involutus* mycelia, their chemical identification is still lacking (Horan and Chilvers 1990).

The ability of root extracts to regulate ectomycorrhizal fungal growth was also confirmed during the early stages of *Pisolithus tinctorius–Castanea sativa* association (Baptista et al. 2007). In this study, *P. tinctorius* growth was evaluated after 8 and 17 days of culture in the presence of crude extracts from *P. tinctorius*–elicited *C. sativa* roots, prepared during the early stages of fungal contact (0–48 h) (Fig. 8.2). In the first 8 days of culture, root-extracts inhibited *P. tinctorius* growth, mainly after 3–9 h and 15 h of interaction (Fig. 8.2a). The enhanced fungal growth in control (in the absence of root extracts) suggests that compounds from the elicited plants could reduce the fungal growth. After 17 days of culture, the inhibitory effect was not observed and an inducing effect became evident (Fig. 8.2b). These results suggest that elicited plant could transiently restrain the mycelium growth but the fungus is able to overcome the inhibitory effect. After degradation/inactivation of the inhibitory compound(s), root extracts promote fungal growth as previously described. Nevertheless, growth-stimulating activity of root extracts was still reduced after 3, 9, and 15 h of root interaction with *P. tinctorius*. As the H_2O_2 production was coincident with the inhibitory profile, both phenomena may be related (Baptista et al. 2007).

The ectomycorrhizal fungi also produce signaling molecules that play a key role in the early stages of ECM development (Barker and Tagu 2000; Martin et al. 2001). Several authors have attributed to the fungal auxin an important role as a signaling molecule operating in the initial stages of the mycorrhiza formation (Gay et al. 1994; Nehls et al. 1998, Martin et al. 1999; Rincón et al. 2001). Pine roots inoculated with mutant strains of *Hebeloma cylindrosporum* overproducing indole-3-acetic acid (IAA) develop more ectomycorrhizal roots than when inoculated with wild-type mycelium (Gay et al. 1994). The stimulation of lateral roots by IAA was suggested to create new targets for further fungal colonization, supporting the idea that fungal IAA controls major anatomical features of pine ECM. The effect of auxins on rhizogenesis can then be interpreted as a preparation of the root system for a more efficient colonization by the mycelium (Barker et al. 1998). Accordingly, the addition of an inhibitor of auxin transport (2,3,5-triiodobenzoic acid, TIBA) blocked the colonization of the Norway spruce root by *Laccaria bicolor*, restricting the hyphal growth between cortical cells and limiting the Hartig-net formation (Karabaghli-Degron et al. 1998). The levels of IAA produced by the fungus also seem to control the production of elicitors/signal molecules by the host plant (Mensen et al. 1998; Rincón et al. 2001).

The fungal production of hypaphorine, another indolic compound, also appears to influence the ECM establishment (Béguiristain et al. 1995; Béguiristain and

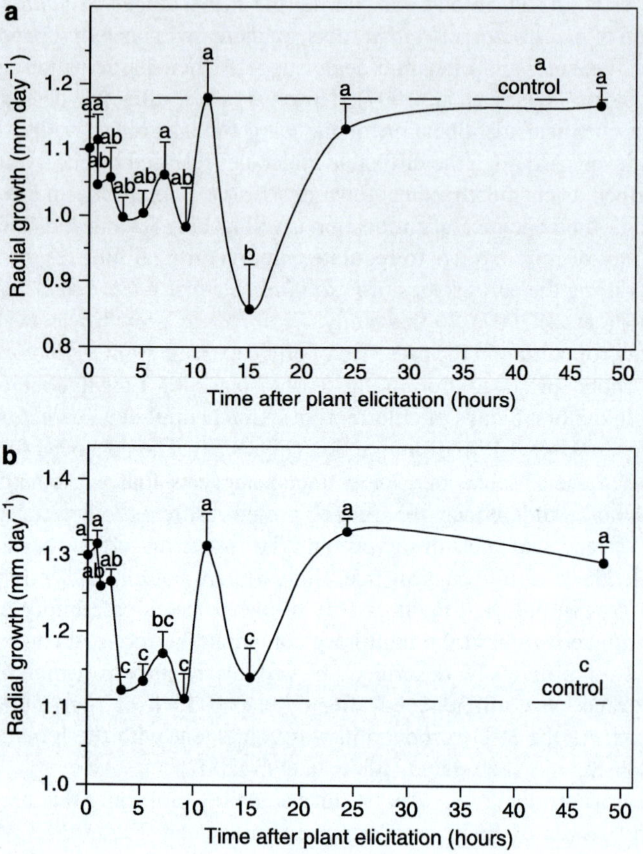

Fig. 8.2 Radial growth of *Pisolithus tinctorius* after 8 days (**a**) and 17 days (**b**) of culture on agar MMN media, in the presence of crude extracts obtained from elicited *Castanea sativa* roots with *P. tinctorius* (from 0 to 48 h). Protein extraction buffer was used as control. Results are presented as means $\pm$ SE ($n = 5$). The letters above the columns indicate significant difference at $p < 0.001$

Lapeyrie 1997; Ditengou et al. 2000). This fungal alkaloid, present in *P. tinctorius* mycelium axenically grown (Béguiristain et al. 1995), is highly produced during ECM formation (Béguiristain and Lapeyrie 1997). In *Eucalyptus globulus* roots inoculated with *P. tinctorius*, the fungal hypaphorine levels increased three- to five-fold when compared to the pure mycelium culture (Béguiristain and Lapeyrie 1997). The host plant appears to stimulate the production of hypaphorine in the fungus, which in turn seems to regulate the activity and levels of plant auxins (Nehls et al. 1998; Jambois et al. 2005). The regulation of auxin levels is further supported by the identification of several ECM-regulated genes in host tissues involved in auxin metabolism (Voiblet et al. 2001; Johansson et al. 2004; Duplessis et al. 2005; Le Quéré et al. 2005). According to Ditengou and Lapeyrie (2000), the hypaphorine could participate in plant IAA regulation by (1) regulating IAA transport;

(2) inducing a cellular IAA-specific detoxification mechanism; (3) competing with IAA for some receptor; or (4) interfering with IAA signal transduction. Fungal hypaphorine seems also to display a morphogenetic effect on *Eucalyptus globulus* roots by reducing root hair elongation (Béguiristain and Lapeyrie 1997; Ditengou et al. 2000; Ditengou and Lapeyrie 2000; Ditengou et al. 2003). The reduction in the root hair growth is indeed a marker feature of ectomycorrhizal roots.

8.4 Proteins Involved in Fungal Attachment to the Root Host Surface and Formation of Symbiotic Interface

The ectomycorrhizas are structurally distinguished by the presence of the mantle and the Hartig-net, both involving the occurrence of physical contact between the fungus and root surface (cell-to-cell contact). Soon after the initial contact, fungal cells swell and adhere firmly to the root surface, which in turn induces rapid modifications on fungal cell morphology. The fungal cells become enlarged and intensively branched. These alterations are accompanied by the increase in nuclear division (Martin and Tagu 1995). The signal or mechanism responsible for triggering such alterations is still not full elucidated. Nutrient limitation in the vicinity of root system, root exudates, and the topographic sensing of root surface by fungus could contribute to these morphologic alterations (Barker et al. 1998; Tagu et al. 2002). The fungal growth pattern as well as the hyphal branching seems also to be controlled by the host roots through the production of cytokinins, namely zeatin (Martin et al. 2001; Tagu et al. 2002). These early morphologic modifications seem to constitute a reliable indicator of a compatible interaction between symbionts (Peterson and Bonfante 1994).

During the process of ectomycorrhizal fungi adhesion to host roots, distinct protein classes displaying cell adhesion properties are secreted by the fungi (Fig. 8.1). Ultrastructural observations of fungus–root interface in the early stages of colonization revealed the presence of fibrillar material rich in glycoproteins (Lei et al. 1990a, b; Martin et al. 1999; Rincón et al. 2001). These proteins (adhesins) are described as protein complexes able to recognize and bind to receptors located in plant root cell surface (Martin et al. 1999; Gross et al. 2004). Adhesins are derived from fungi cell wall and expand towards the root surface. Besides playing a role in the adhesion of hyphae to root surface, these proteins are also important in mycelium aggregation during the mantle formation at the final stages of mycorrhization (Martin and Tagu 1995).

The lectins seem also to be important in the fungi–root adhesion, not only in the early stages by playing a similar role to adhesins, but also at the final stages of ECM establishment. Receptor sites for the lectin produced by the specific symbiont of spruce, *Lactarius deterrimus*, were found in the roots of *Picea abies* (Giollant et al. 1993). In this study, the lectin was mainly bounded to young tissues, such as root hairs and tips of lateral roots. By contrast, no binding to the spruce root was observed for the lectin isolated from *Lactarius deliciosus*, a symbiont of the pine. This result suggests a role of the fungal lectin in the recognition and partner specificity during the early stages of ECM development (Giollant et al. 1993). However, the differential gene expression analysis performed during *Betula pendula* and *Paxillus involutus* ECM development revealed a repression of fungal lectin-encoding genes during the early stages of interaction and their induction 14 days after inoculation (Le Quéré et al. 2005).

Search for differentially expressed genes during ECM interactions has contributed for the identification of additional gene products that could play a role in fungi–host root adhesion. In *Pisolithus–Eucalyptus* ectomycorrhizal symbiosis, genes encoding hydrophobins are up-regulated, suggesting the implication of the gene products in fungi–cell adhesion (Tagu et al. 1996; Duplessis et al. 2001; Voiblet et al. 2001; Peter et al. 2003; Duplessis et al. 2005). Hydrophobins exhibit a conserved spacing of eight cystein residues which can form disulfide bonds between themselves (Linder et al. 2005). These hydrophobic proteins are secreted by the fungus into the medium but can also be found at the fungi cell wall. In this case, they exhibit a hydrophilic domain through which they are linked to the hydrophilic cell wall and a hydrophobic domain that could be exposed to a hydrophobic external medium. Thus, these proteins have the ability to self-assemble into an amphipathic membrane at a hydrophilic/hydrophobic interface. The hydrophilic side of the amphipathic membrane orients and attaches to the fungal cell wall, while the hydrophobic side becomes exposed to the hydrophobic environment, such as the air or the hydrophobic surface of a host. In this way, the aerial hyphae and spores become hydrophobic, whereas hyphae grown over a hydrophobic substrate become attached to themselves (Wösten 2001). As a result, hydrophobins seem to play a dual role in ECM, not only in promoting the hyphae adhesion to the host surface but also contributing to hyphae aggregation (Kershaw and Talbot 1998; Wösten 2001).

Hydrophobins are encoded by multigene families (Linder et al. 2005). For example, in *Pisolithus tinctorius* seven different genes have already been identified (*hydPt-1* to *hydPt-6* and *hydPt-8*), although the corresponding gene products display minor structural differences (Tagu et al. 1996; Duplessis et al. 2001, 2005; Voiblet et al. 2001). As suggested by Linder et al. (2005), the presence of multiple hydrophobin genes in an organism may be significant in two different ways. They could complement each other by having differential expression in distinct developmental stages or as a response to environmental conditions. Alternatively, they could fulfill different functional roles as a result of the minor structural differences they exhibit.

Hydrophobin-like proteins have already been reported at least in 20 fungal species belonging to ascomycetes, basidiomycetes, and zygomycetes (Wessels 1996). According to their structure, these proteins are described as being involved in several fungal developmental processes, such as emergence of aerial hyphae, hyphae aggregation during fruiting bodies (carpophores) development, sporulation and spore dissemination, as well as during plant or insect infections caused by pathogenic fungi (Wessels 1996; Kershaw and Talbot 1998; Wösten 2001; Linder et al. 2005). In this case, hydrophobins role seems to be essential since the pathogen must attach to the hydrophobic surface of the host before penetration and infection can occur (Wessels 1996; Wösten 2001). Besides their role in the adhesion of fungi to the host root during ectomycorrhizal formation, hydrophobins have also been described as being involved in the fungal recognition of the host plant, as well as in the fungal specificity displayed by the host plant (Mankel et al. 2002).

During *Pisolithus tinctorius* ECM development, the expression analysis of hydrophobin genes suggests the involvement of these proteins in the morphogenetic events related to the fungus attachment to root surfaces, but not in the late stages of fungal mantle or Hartig-net formation (Martin et al. 1999). In *Pisolithus–Eucalyptus* association, hydrophobin gene expression was induced during the early steps of root colonization, being repressed after 4 days of contact (Tagu et al. 1996; Martin et al. 1999). During the first 12 h of *Castanea sativa* root–*P. tinctorius* contact, a strong downregulation of two hydrophobin genes (*HydPt-2* and *HydPt-3*) was detected (Acioli-Santos et al. 2008). Other differential gene expression studies performed in distinct ECM interactions describe different results concerning the expression of hydrophobin genes (Mankel et al. 2002; Johansson et al. 2004; Le Quéré et al. 2005). However, even when the hydrophobin gene expression is transient, the gene product could be detected in late developmental stages. For example, in the previously referred *Pisolithus–Eucalyptus* association, the hydrophobins were immunologically detected in hyphae forming the mantle and Hartig-net (Tagu et al. 2001). The expression levels of hydrophobin genes could differ depending on fungal species or host plants, or even be influenced by the presence of certain nutrients in the substrate, suggesting a complex regulation of hydrophobin gene expression during ECM development (Martin et al. 1999).

Another family of cell adhesion proteins was identified in *P. tinctorius* after detection of gene overexpression in the early stages of the *Eucalyptus globulus–P. tinctorius* interaction (Laurent et al. 1999). The identified protein family – *symbiosis-regulated acidic polypeptides* (SRAP) – comprises at least six different isoforms of 31–32 kDa and is characterized by the presence of Arg-Gly-Asp tripeptide (RGD) motif (Martin et al. 1999). This motif is commonly associated to cell adhesion in animals (Critchley et al. 1999) and plants (Mellersh and Heath 2001; Meinhardt et al. 2002). Cellular localization of a 32-kDa SRAP ($SRAP_{32}$) during ECM development showed a preferential accumulation of this protein in fungal cell walls, fungus–root interfaces, and in penetrating hyphae forming the Hartig-net (Laurent et al. 1999). The suggestion that SRAP family could be involved in the hyphae aggregation for the fungal mantle formation and for the Hartig-net development is corroborated by differential expression studies during ECM development. After 4 days of *E. globulus–P. tinctorius* interaction, an increase of $SRAP_{32}$ transcript levels (4.1-fold) was detected, as well as the expression levels (4.7-fold) of a new family member of low molecular weight ($SRAP_{17}$) (Voiblet et al. 2001). The participation of SRAP family members in the ectomycorrhizal development seems to decrease during the final stages of the interaction. Genes encoding $SRAP_{32}$ and $SRAP_{17}$ are repressed after 7–12 days of *E. globulus–Pisolithus microcarpus* interaction (Duplessis et al. 2005).

8.5 Ectomycorrhizal Morphogenesis

The development of a functional ECM involves large morphological alterations in both partners. The major morphogenetic adaptations that occur in the symbiotic fungus includes: the hyphae aggregation for the formation of the mantle enveloping the root surface, and the hyphae penetration and growth between epidermal and cortical root cells forming the Hartig-net (Smith and Read 2008). Development of a mature mantle proceeds through a programmed and highly coordinated series of events. After attachment onto root epidermal cells, hyphae multiply to form a series of layers that will differentiate into the mature mantle (Martin et al. 2001). The mantle morphology is dependent on both the plant and fungus genomes. Indeed, the identification of mycobiont species of field-collected ectomycorrhizas is often performed by observation of mantle morphology (Peterson and Bonfante 1994). The host plant seems to control the mantle formation by stimulating hyphal growth and exerting a morphogenetic effect leading to compact hyphal mantle development (Martin et al. 2001). In addition, several abiotic factors, such as nutrients and oxygen availability, as well as the presence of a physical support, are described to play a role in mantle formation (Peterson and Bonfante 1994; Martin et al. 1999).

The hyphae that enter the intercellular spaces of the root will develop the Hartig-net. The fungus penetration and growth within the apoplastic space is mainly accomplished by a mechanic process, but the production of lytic enzymes could also be involved in the digestion of host cell walls (Cairney and Burke 1994). Once inside the root, hyphae could only grow between epidermal cells (in the majority of angiosperms) or could also penetrate between cortical cells up to the endodermis (typical in gymnosperms) (Smith and Read 2008). The restriction of fungal ingress into the cortical tissues seems to be due to the fungus inability to degrade the suberin and lignin that are present in the endodermal cell walls (Barker et al. 1998). As the same fungal species could only colonize the epidermal layer or also the cortical layer, depending on the host plant, the hyphal growth was suggested to be under the host plant control (Barker et al. 1998).

The progression of ectomycorrhizal fungus within intercellular spaces of root epidermal cells promotes morphologic alterations in the majority of angiosperm roots (Fig. 8.1). The physical contact with the fungus promotes a radial elongation of the cell roots, thus changing their orientation of growth and shape (Peterson and Bonfante 1994). Several studies have showed that these modifications are related to cytoskeleton rearrangements (Timonen et al. 1993). Changes in the amount of actin and tubulin proteins in root and/or fungal cells were observed during ECM formation, suggesting its involvement in the ECM morphogenesis (Timonen et al. 1993; Carnero Diaz et al. 1996; Timonen et al. 1996; Johansson et al. 2004). In agreement, fungal genes encoding proteins associated with actin and microtubules were found to be differentially regulated in distinct ECM associations (Voiblet et al. 2001;

Timonen and Peterson 2002; Menotta et al. 2004; Le Quéré et al. 2005). However, it still remains to be clarified if changes in cytoskeletal gene expression are the cause or the consequence of mycorrhizal root morphogenesis (Barker et al. 1998).

In contrast to what is observed in the initial mycorrhization stages, after fungal root colonization there is a reduction of plant meristem activity, as well as early formation of lateral roots. In the case of Pinaceae ectomycorrhiza, dichotomous branching of short roots occurs, resulting occasionally in the formation of coralloid structures (Peterson and Bonfante 1994). These morphological modifications were suggested to be caused in part by fungal auxins, which have been also related to the Hartig-net formation. Several studies have demonstrated that fungal auxins are morphogen factors in ECM development, playing a key role in inducing the fungal proliferation alterations (Gay et al. 1994; Martin and Tagu 1995; Karabaghli-Degron et al. 1998; Mensen et al. 1998; Rincón et al. 2001). The fungal production of auxins in roots completely surrounded by ectomycorrhizal hyphae could alter the internal plant auxin balance, and consequently promote typical ectomycorrhizal root morphogenesis (Barker and Tagu 2000). Inhibition of auxin transporters has shown that these morphogenetic effects, like the lateral root dichotomy, are dependent on auxin concentration and distribution in the root meristem (Karabaghli-Degron et al. 1998; Kaska et al. 1999). However, to our knowledge, the clear effect of ectomycorrhizal fungi on concentration and/or distribution of auxin in the roots still remains to be elucidated.

Differential gene expression studies have shown that the morphological and physiological changes observed throughout ECM development are accompanied by changes in gene expression in both partners (Heller et al. 2008). The number of genes and the amplitude of their expression vary in time but seem to be more significant at the early stages of development, soon after plant–fungus contact. Accordingly, gene expression studies revealed that the gene overexpression levels were more evident during the first 8 days of interaction being repressed in advanced stages of mycorrhization (Duplessis et al. 2005; Le Quéré et al. 2005). In addition, it was shown that the number of fungal differentially expressed genes is much higher than the number of plant host genes.

In all the ECM associations studied up to now, ECM-specific genes were not yet identified. Therefore, the ontogenic and metabolic programs that lead to the development of symbiosis seem to be driven by the differential expression of preexisting transcription factors and/or transduction pathways, rather than by the expression of symbiosis-specific gene arrays (Martin et al. 2007). However, a more detailed study encompassing all differentially expressed genes is needed in order to clarify this question. Many up- or downregulated genes during ECM development have not shown similarity to known sequences (Voiblet et al. 2001; Podila et al. 2002; Krüger et al. 2004). Some of these genes may represent unidentified fungal mycorrhiza-specific genes (Heller et al. 2008), or be unique to a particular ectomycorrhizal fungus, or even represent very rare transcripts that have not been previously identified and/or characterized (Podila et al. 2002).

8.6 Induction of Plant Defense Responses by Ectomycorrhizal Fungi

When infected by pathogens, plants can activate a multitude of defense mechanisms (Dixon and Lamb 1990; Baker and Orlandi 1995; Doke et al. 1996; Mehdy et al. 1996; Lamb and Dixon 1997). The recognition of pathogen molecules (elicitors) by specific plant cell receptors could lead to the activation of kinases and phosphorylation cascades, involving phosphatases, G proteins, and ionic fluxes. These events induce a typical pattern of plant defense responses that includes (1) the production of reactive oxygen species (ROS) in the so-called oxidative burst; (2) the hypersensitive response; (3) the induction of defense response genes or pathogen-related genes (PR genes); (4) the reinforcement of cell wall, including the deposition of callose, lignin, or hydroxiproline-rich glycoproteins; (5) synthesis of jasmonate, ethylene, and/or salicylic acid; (6) production of antimicrobial secondary metabolites, such as phytoalexins; and (7) production of hydrolytic enzymes, such as chitinases and glucanases. Plants could also induce systemic-acquired resistance and produce signals to other plants.

As observed for plant–pathogen interactions, the plant inoculation with ectomycorrhizal fungi comprises the induction/suppression of mechanisms associated to plant defense responses (Fig. 8.1). These mechanisms seem to play a key role in determining the symbiotic compatibility, as well as in ectomycorrhizal development and functioning (Martin et al. 2001). Suspension-cultured cells of *Picea abies* subjected to cell wall elicitors from the ectomycorrhizal fungi *Amanita muscaria* (Schwacke and Hager 1992) and *Hebeloma crustuliniforme* (Schwacke and Hager 1992; Salzer et al. 1996) exhibit biochemical resistance responses similar to those described for incompatible pathogen attack. Soon after elicitation, an efflux of Cl^- and K^+ takes place, followed by an influx of Ca^{2+} into plant cells (Schwacke and Hager 1992; Salzer et al. 1996). Four minutes after elicitation, protein phosphorylation and dephosphorylation (Salzer et al. 1996), alkalinization of the medium, and transient accumulation of ROS, such as hydrogen peroxide (H_2O_2) (Schwacke and Hager 1992; Salzer et al. 1996), can be detected. The activity of antioxidative enzymes, such as superoxide dismutase (SOD) and catalase (CAT), seems to play a role in regulating ROS levels in ECM interactions (Schwacke and Hager 1992). During the first 48 h of contact between ectomycorrhizal fungus *Pisolithus tinctorius* and *Castanea sativa* roots, three peaks of H_2O_2 production were observed, the first two synchronized with $O_2^{\bullet-}$ bursts (Baptista et al. 2007). The accumulation of H_2O_2 was presumed to be the result of a regulated time-dependent stimulation of ROS-generating systems and decrease of ROS-scavenging enzyme activities. Accordingly, the first H_2O_2 production peak coincided with an increase in SOD activity, while CAT activity seemed to be implicated in subsequent H_2O_2 scavenging (Baptista et al. 2007).

The similarity of signal transduction pathways in incompatible plant–pathogen interactions and ECM symbiosis suggests that the fungus produces elicitors, at the early stages of the symbiotic association that will induce typical plant defense responses (Hebe et al. 1999). Ultrastructural observations revealed that, when *Pinus nigra* was inoculated with *Suillus collinitus*, a defense reaction was activated in the host plant with the formation of wall thickenings containing β-1,3-glucans, a callose constituent (Bonfante et al. 1998). In addition, Norway spruce seedlings inoculated with the ectomycorrhizal fungus *Pisolithus tinctorius* displayed increased guaiacol peroxidase activity in the roots, which is involved in cell-wall strengthening (Matevž and Regvar 2008). The accumulation of phenolics in plant cell wall observed in distinct ECM associations was also suggested to restrain fungal progression during the formation of ECM structures, thus limiting the Hartig-net formation (Weiss et al. 1997, 1999; Feugey et al. 1999). During the initial fungal progression into apoplastic root regions to form the Hartig-net, a transient increase in the activity of the phenylpropanoid pathway enzyme phenylalanine ammonia lyase was observed, together with the increase in transcripts of PR-encoding genes (Feugey et al. 1999). Differential gene expression studies during ectomycorrhiza formation have also confirmed the transient increase in plant defense/stress responses which has been interpreted as an initial reaction of plant to restrict fungal growth (Voiblet et al. 2001; Johansson et al. 2004; Morel et al. 2005; Le Quéré et al. 2005; Duplessis et al. 2005, Sebastiana et al. 2009).

In contrast to what has been reported for incompatible plant–pathogen interactions, in ectomycorrhizal symbiosis a rapid decline in plant defense responses seems to occur (Salzer et al. 1996). In *P. abies* this response suppression seems to be related to the production of chitinases by the plant that inactivates elicitors, like chitin fragments produced by the ectomycorrhizal fungus (Salzer et al. 1997a, b). These results suggest that elicitor inactivation could be a prerequisite for a compatible interaction between plant and ectomycorrhizal fungus. Furthermore, the stimulation of chitinase production in host plant could contribute to hyphae growth within root intercellular spaces, promoting the formation of the ECM structures (Sauter and Hager 1989). This hypothesis is supported by the increase in ECM root number in the presence of high chitinase amounts (Albrecht et al. 1994). The auxins produced by the ectomycorrhizal fungus during the colonization process also seem to play a role in declining the defense capacity of plant cells (Mensen et al. 1998).

The induction of plant defense responses during the early stages of mycorrhization is not always followed by its suppression at later stages. In the co-cultures of *P. abies* callus and *Lactarius deterrimus* or *Suillus variegatus* the development of typical hypersensitive response reactions was observed. Initially the spruce cells become brownish and finally the whole callus became necrotic (Sirrenberg et al. 1995). Therefore, plant defense responses induced by ectomycorrhizal fungi seem to be diverse and often contradictory in what concerns their attenuation during the later stages of the symbiosis. According to Martin et al. (1999), depending on the host and ectomycorrhizal fungus, different biochemical and molecular mechanisms could exist, thus explaining the differences in the reported results. The environmental factors occurring during plant–fungus interaction could also be decisive for the defense response displayed by host plant.

8.7 Conclusions

The establishment of an ectomycorrhizal symbiosis is a complex process triggered by signals produced by both partners (Fig. 8.1). The ECM development involves five main events: hyphae survival in the rhizosphere; hyphae attachment to the host roots; invasion of root tissues by fungi; plant–fungus coordinated formation of symbiotic structures; and bilateral transfer of assimilates. During the preinfection, diffusible signal molecules are exchanged between the host plant and the fungus. The recognition of these signals by both partners determines the compatibility of symbiosis. Although the nature of these signaling molecules is scarcely known, diffusible molecules (such as abietic acid or rutin) are released by host plant and would chemoattract, promote spore germination, or regulate ectomycorrhizal fungal growth. The fungus also takes part in this signaling process through the secretion of auxins and hypaphorin. Besides their signaling role, these compounds have also a direct effect on rhizogenesis and ECM structure formation.

Once mutually recognized, root and fungus establish physical contact. During the initiation, fungal proteins (such as adhesins, lectins, or hydrophobins) promote fungus adhesion to host roots. The initial contact induces a defense response in host plant that seems to restrain the fungal progression into the plant tissues. After the attenuation/suppression of host surveillance mechanisms, drastic morphological and physiological modifications occur in both fungal and root cells for forming the symbiotic structures. Fungal tissues differentiate into the mantle and the Hartig-net, whereas root cells change their shape and orientation, and root tips proliferate and develop lateral roots. These auxin-regulated plant developmental processes are induced by the mycobiont through the action of secreted auxins.

The ontogenesis of a functional ECM symbiosis requires a finely regulated crosstalk between plant and fungi. Despite the contribution given by gene expression studies, the molecular information on ECM establishment is still scarce. A few genes and proteins associated to ECM symbiosis have been recognized and a small number of signaling molecules chemically identified. Further studies are needed to understand the formation and functioning of ECM symbiosis.

References

Acioli-Santos B, Sebastiana M, Pessoa F, Sousa L, Figueiredo A, Fortes AM, Baldé A, Maia LC, Pais MS (2008) Fungal transcript pattern during the preinfection stage (12 h) of ectomycorrhiza formed between *Pisolithus tinctorius* and *Castanea sativa* roots, identified using cDNA microarrays. Curr Microbiol 57:620–625

Albrecht C, Asselin A, Piché Y, Lapeyrie F (1994) Chitinase activities are induced in *Eucalyptus globulus* roots by ectomycorrhizal or pathogenic fungi, during early colonization. Physiol Plant 91:104–110

Alvarez M, Huygens D, Fernandez C, Gacitúa Y, Olivares E, Saavedra I, Alberdi M, Valenzuela E (2009) Effect of ectomycorrhizal colonization and drought on reactive oxygen species metabolism of Nothofagus dombeyi roots. Tree Physiol 29:1047–1057

Baker CJ, Orlandi EW (1995) Active oxygen in plant pathogenesis. Annu Rev Phytopathol 33: 299–321

Baptista P, Martins A, Pais MS, Tavares RM, Lino-Neto T (2007) Involvement of reactive oxygen species during early stages of ectomycorrhiza establishment between *Castanea sativa* and *Pisolithus tinctorius*. Mycorrhiza 17:185–193

Barker SJ, Tagu D (2000) The roles of auxins and cytokinins in mycorrhizal symbioses. J Plant Growth Regul 19:144–154

Barker SJ, Tagu D, Delp G (1998) Regulation of root and fungal morphogenesis in mycorrhizal symbioses. Plant Physiol 116:1201–1207

Béguiristain T, Lapeyrie F (1997) Host plant stimulates hypaphorine accumulation in *Pisolithus tinctorius* hyphae during ectomycorrhizal infection while excreted fungal hypaphorine controls root hair development. New Phytol 136:525–532

Béguiristain T, Côte R, Rubini P, Jay-Allemand C, Lapeyrie F (1995) Hypaphorine accumulation in the hyphae of the ectomycorrhizal fungus *Pisolithus tinctorius*. Phytochemistry 40: 1089–1091

Bonfante P, Balestrini R, Martino E, Perotto S, Plassard C, Mousain D (1998) Morphological analysis of early contacts between pine roots and two ectomycorrhizal *Suillus* strains. Mycorrhiza 8:1–10

Brunner I, Frey B (2000) Detection and localization of aluminum and heavy metals in ectomycorrhizal Norway spruce seedlings. Environ Pollut 108:121–128

Cairney J, Burke R (1994) Fungal enzymes degrading plant cell walls: their possible significance in the ectomycorrhizal symbiosis. Mycol Res 98:1345–1356

Carnero Diaz E, Martin F, Tagu D (1996) Eucalypt alpha-tubulin: cDNA cloning and increased level of transcripts in ectomycorrhizal root system. Plant Mol Biol 31:905–910

Critchley DR, Holt MR, Barry ST, Priddle H, Hemmings L, Norman J (1999) Integrin-mediated cell adhesion: the cytoskeletal connection. Biochem Soc Symp 65:6579–6599

Ditengou F, Lapeyrie F (2000) Hypaphorine from the ectomycorrhizal fungus *Pisolithus tinctorius* counteracts activities of indole-3-acetic acid and ethylene but not synthetic auxins in eucalypt seedlings. MPMI 13:151–158

Ditengou F, Béguiristain T, Lapeyrie F (2000) Root hair elongation is inhibited by hypaphorine, the indole alkaloid from the ectomycorrhizal fungus *Pisolithus tinctorius*, and restored by indole-3-acetic acid. Planta 211:722–728

Ditengou F, Raudaskoski M, Lapeyrie F (2003) Hypaphorine, an indole-3-acetic acid antagonist delivered by the ectomycorrhizal fungus *Pisolithus tinctorius*, induces reorganisation of actin and the microtubule cytoskeleton in *Eucalyptus globulus* ssp. *bicostata* root hairs. Planta 218:217–225

Dixon RA, Lamb CJ (1990) Molecular communication in interactions between plants and microbial pathogens. Annu Rev Plant Physiol Plant Mol Biol 41:339–367

Doke N, Miura Y, Sanchez LM, Park H-J, Noritake T, Yoshioka H, Kawakita K (1996) The oxidative burst protects plants against pathogen attack: mechanism and role as an emergency signal for plant bio-defence – a review. Gene 179:45–51

Duplessis S, Sorin C, Voiblet C, Palin B, Martin F, Tagu D (2001) Cloning and expression analysis of a new hydrophobin cDNA from the ectomycorrhizal basidiomycete *Pisolithus*. Curr Genet 39:335–339

Duplessis S, Courty P-E, Tagu D, Martin F (2005) Transcript patterns associated with ectomycorrhiza development in *Eucalyptus globulus* and *Pisolithus microcarpus*. New Phytol 165: 599–611

Feugey L, Strullu D-G, Poupard P, Simoneau P (1999) Induced defence responses limit Hartig-net formation in ectomycorrhizal birch roots. New Phytol 144:541–547

Finlay RD (2008) Ecological aspects of mycorrhizal symbiosis: with special emphasis on the functional diversity of interactions involving the extraradical mycelium. J Exp Bot 59:1115–1126

Fries N, Serck-Hanssen K, Hall-Dimberg L, Theander O (1987) Abietic acid, an activator of basidiospore germination in ectomycorrhizal species of the genus *Suillus* (Boletaceae). Exp Mycol 11:360–363

Davies FT Jr, Svenson SE, Cole JC, Phavaphutanon L, Duray SA, Olalde-Portugal V, Meier CE, Bo SH (1996) Non-nutritional stress acclimation of mycorrhizal woody plants exposed to drought. Tree Physiol 16:985–993

Futai K, Taniguchi T, Kataoka R (2008) Ectomycorrhizae and their importance in forest ecosystems. In: Mycorrhizae: sustainable agriculture and forestry. Springer, Netherlands, pp 241–285

Gay G, Normand L, Marmeisse R, Sotta B, Debaud JC (1994) Auxin overproducer mutants of *Hebeloma cylindrosporum Romagnesi* have increased mycorrhizal activity. New Phytol 128:645–657

Giollant M, Guillot J, Damez M, Dusser M, Didier P, Didier E (1993) Characterization of a lectin from *Lactarius deterrimus*. Research on the possible involvement of the fungal lectin in recognition between mushroom and spruce during the early stages of mycorrhizae formation. Plant Physiol 101:513–522

Gross E, Casagrande LIT, Caetano FH (2004) Ultrastructural study of ectomycorrhizas on *Pinus caribaea Morelet*. var. hondurensis Barr. & Golf. seedlings. Acta Botanica Brasilica 18:1–7

Hebe G, Hager A, Salzer P (1999) Initial signalling processes induced by elicitors of ectomycorrhiza-forming fungi in spruce cells can also be triggered by G-protein-activating mastoparan and protein phosphatase-inhibiting cantharidin. Planta 207:418–425

Heller G, Adomas A, Li G, Osborne J, van Zyl L, Sederoff R, Finlay R, Stenlid J, Asiegbu F (2008) Transcriptional analysis of *Pinus sylvestris* roots challenged with the ectomycorrhizal fungus *Laccaria bicolour*. BMC Plant Biol 8:19

Horan DP, Chilvers GA (1990) Chemotropism: the key to ectomycorrhizal formation? New Phytol 116:297–301

Jambois A, Dauphin A, Kawano T, Ditengou FA, Bouteau F, Legue V, Lapeyrie F (2005) Competitive antagonism between IAA and indole alkaloid hypaphorine must contribute to regulate ontogenesis. Physiol Plant 123:120–129

Johansson T, Le Quéré A, Ahren D, Söderström B, Erlandsson R, Lundeberg J, Uhlén M, Tunlid A (2004) Transcriptional responses of *Paxillus involutus* and *Betula pendula* during formation of ectomycorrhizal root tissue. MPMI 17:202–215

Karabaghli-Degron C, Sotta B, Bonnet M, Gay G, Le Tacon F (1998) The auxin transport inhibitor 2,3,5-triidobenzoic acid (TIBA) inhibits the stimulation of in vitro lateral root formation and the colonization of the tap-root cortex of Norway spruce (Picea abies) seedlings by the ectomycorrhizal fungus *Laccaria bicolor*. New Phytol 140:723–733

Kaska DD, Myllylä R, J.B C (1999) Auxin transport inhibitors act through ethylene to regulate dichotomous branching of lateral root meristems in pine. New Phytol 142:49–58

Kershaw MJ, Talbot NJ (1998) Hydrophobins and repellents: proteins with fundamental roles in fungal morphogenesis. Fungal Genet Biol 23:18–33

Kim S-J, Hiremath ST, Podila GK (1999) Cloning and identification of symbiosis-regulated genes from the ectomycorrhizal *Laccaria bicolor*. Mycol Res 103:168–172

Krüger A, Peskan-Berghöfer T, Frettinger P, Herrmann S, Buscot F, Oelmüller R (2004) Identification of premycorrhiza-related plant genes in the association between *Quercus robur* and *Piloderma croceum*. New Phytol 163:149–157

Lagrange H, Jay-Allemand C, Lapeyrie F (2001) Rutin, the phenolglycoside from Eucalyptus root exudates, stimulates *Pisolithus* hyphal growth at picomolar concentrations. New Phytol 149:349–355

Lamb C, Dixon RA (1997) The oxidative burst in plant disease resistance. Annu Rev Plant Physiol Plant Mol Biol 48:251–275

Laurent P, Voiblet C, Tagu D, Carvalho D, Nehls U, De Bellis R, Balestrini R, Bauw G, Bonfante P, Martin F (1999) A novel class of ectomycorrhiza-regulated cell wall polypeptides in *Pisolithus tinctorius*. MPMI 12:862–871

Le Quéré A, Wright DP, Söderström B, Tunlid A, Johansson T (2005) Global patterns of gene regulation associated with the development of ectomycorrhiza between birch (*Betula pendula* Roth.) and *Paxillus involutus* (Batsch) Fr. MPMI 18:659–673

Lei J, Ding H, Lapeyrie F, Piché Y, Malajczuk N, Dexheimer J (1990a) Ectomycorrhizal formation on the roots of *Eucalyptus globulus* and *Pinus caribaea* with two isolates of *Pisolithus tinctorius*: structural and cytochemical observations. In: Nardon P, Gianinazzi-Pearson V, Greiner AM, Margulis L, Smith DC (eds) Endocytobiology, vol IV. INRA, Paris, France, pp 123–126

Lei J, Lapeyrie F, Malajczuk N, Dexheimer J (1990b) Infectivity of pine and eucalypt isolates of *Pisolithus tinctorius* (Pers.) Coker & Couch on roots of *Eucalyptus urophylla* S.T. Blake in vitro. II. Ultrastructural and biochemical changes at the early stage of mycorrhiza formation. New Phytol 116:115–122

Linder MB, Szilvay GR, Nakari-Setälä TN, Penttilä ME (2005) Hydrophobins: the protein-amphiphiles of filamentous fungi. FEMS Microbiol Rev 29:877–896

Mankel A, Krause K, Kothe E (2002) Identification of a hydrophobin gene that is developmentally regulated in the ectomycorrhizal fungus *Tricholoma terreum*. Appl Environ Microbiol 68:1408–1413

Martin F, Tagu D (1995) Ectomycorrhiza development: a molecular perspective. In: Varma A, Hock B (eds) Mycorrhiza: structure, function, molecular biology and biotechnology. Springer, Berlin, pp 30–58

Martin F, Laurent P, de Carvalho D, Voiblet C, Balestrini R, Bonfante P, Tagu D (1999) Cell wall proteins of the ectomycorrhizal basidiomycete *Pisolithus tinctorius*: identification, function, and expression in symbiosis. Fungal Genet Biol 27:161–174

Martin F, Duplessis S, Ditengou F, Lagrange H, Voiblet C, Lapeyrie F (2001) Developmental cross talking in the ectomycorrhizal symbiosis: signals and communication genes. New Phytol 151:145–154

Martin F, Kohler A, Duplessis S (2007) Living in harmony in the wood underground: ectomycorrhizal genomics. Curr Opin Plant Biol 10:204–210

Marx DH (1972) Ectomycorrhizae as biological deterrents to pathogenic root infections. Annu Rev Phytopathol 10:429–454

Matevž L, Regvar M (2008) Early defence reactions in Norway spruce seedlings inoculated with the mycorrhizal fungus *Pisolithus tinctorius* (Persoon) Coker & Couch and the pathogen *Heterobasidion annosum*. Trees Struct Funct 22:861–868

Mehdy M, Sharma YK, Sathasivan K, Bays NW (1996) The role of activated oxygen species in plant disease resistance. Physiol Plant 98:365–374

Meinhardt SW, Cheng W, Kwon SY, Donohue CM, Rasmussen JB (2002) Role of the Arginyl-Glycyl-Aspartic motif in the action of Ptr ToxA produced by *Pyrenophora tritici-repentis*. Plant Physiol 130:1545–1551

Mellersh DG, Heath MC (2001) Plasma membrane–cell wall adhesion is required for expression of plant defence responses during fungal penetration. Plant Cell 13:413–424

Menotta M, Amicucci A, Sisti D, Gioacchini AM, Stocchi V (2004) Differential gene expression during pre-symbiotic interaction between *Tuber borchii* Vittad. and *Tilia americana* L. Curr Genet 46:158–165

Mensen R, Hager A, Salzer P (1998) Elicitor-induced changes of wall-bound and secreted peroxidase activities in suspension-cultured spruce (*Picea abies*) cells are attenuated by auxins. Physiol Plant 102:539–546

Morel M, Jacob C, Kohler A, Johansson T, Martin F, Chalot M, Brun A (2005) Identification of genes differentially expressed in extraradical mycelium and ectomycorrhizal roots during *Paxillus involutus-Betula pendula* ectomycorrhizal symbiosis. Appl Environ Microbiol 71:382–391

Nehls U, Béguiristain T, Ditengou F, Lapeyrie F, Martin F (1998) The expression of a symbiosis-regulated gene in eucalypt roots is regulated by auxins and hypaphorine, the tryptophan betaine of the ectomycorrhizal basidiomycete *Pisolithus tinctorius*. Planta 207:296–302

Osonubi O, Mulongoy K, Awotoye OO, Atayese MO, Okali DUU (1991) Effects of ectomycorrhizal and vesicular-arbuscular mycorrhizal fungi on drought tolerance of four leguminous woody seedlings. Plant Soil 136:131–143

Peter M, Courty P-E, Kohler A, Delaruelle C, Martin D, Tagu D, Frey-Klett P, Duplessis S, Chalot M, Podila G, Martin F (2003) Analysis of expressed sequence tags from the ectomycorrhizal basidiomycetes *Laccaria bicolor* and *Pisolithus microcarpus*. New Phytol 159:117–129

Peterson RL, Bonfante P (1994) Comparative structure of vesicular-arbuscular mycorrhizas and ectomycorrhizas. Plant Soil 159:79–88

Podila GK (2002) Signalling in mycorrhizal symbioses – elegant mutants lead the way. New Phytol 154:541–545

Podila GK, Zheng J, Balasubramanian S, Sundaram S, Hiremath S, Brand JH, Hymes MJ (2002) Fungal gene expression in early symbiotic interactions between *Laccaria bicolor* and red pine. Plant Soil 244:117–128

Rincón A, Gérard J, Dexheimer J, Le Tacon F (2001) Effect of an auxin transport inhibitor on aggregation and attachment process during ectomycorrhiza formation between *Laccaria bicolor* S238N and *Picea abies*. Can J Bot 79:1152–1160

Salzer P, Hebe G, Reith A, Zitterell-Haid B, Stransky H, Gaschler K, Hager A (1996) Rapid reactions of spruce cells to elicitors released from the ectomycorrhizal fungus *Hebeloma crustuliniforme*, and inactivation of these elicitors by extracellular spruce cell enzymes. Planta 198:118–126

Salzer P, Hebe G, Hager A (1997a) Cleavage of chitinous elicitors from the ectomycorrhizal fungus *Hebeloma crustuliniforme* by host chitinases prevents induction of K^+ and Cl^- release, extracellular alkalinization and H_2O_2 synthesis of *Picea abies* cells. Planta 203:470–479

Salzer P, Hübner B, Sirrenberg A, Hager A (1997b) Differential effect of purified spruce chitinases and beta-1,3-glucanases on the activity of elicitors from ectomycorrhizal fungi. Plant Physiol 114:957–968

Sauter M, Hager A (1989) The mycorrhizal fungus *Amanita muscaria* induces chitinase activity in roots and in suspension-cultured cells of its host *Picea abies*. Planta 179:61–66

Schwacke R, Hager A (1992) Fungal elicitors induce a transient release of active oxygen species from cultured spruce cells that is dependent on Ca^{2+} and protein-kinase activity. Planta 187:136–141

Sebastiana M, Figueiredo A, Acioli B, Sousa L, Pessoa F, Baldé A, Pais MS (2009) Identification of plant genes involved on the initial contact between ectomycorrhizal symbionts (*Castanea sativa* – European chestnut and *Pisolithus tinctorius*). Eur J Soil Biol 45:275–282

Sirrenberg A, Salzer P, Hager A (1995) Induction of mycorrhiza-like structures and defence reactions in dual cultures of spruce callus and ectomycorrhizal fungi. New Phytol 130:149–156

Smith SE, Read DJ (2008) Mycorrhizal symbiosis, 3rd edn. Academic Press, London

Sundaram S, Kim SJ, Suzuki H, Mcquattie CJ, Hiremath ST, Podila GK (2001) Isolation and characterization of a symbiosis-regulated *ras* from the ectomycorrhizal fungus *Laccaria bicolor*. MPMI 14:618–628

Tagu D, Nasse B, Martin F (1996) Cloning and characterization of hydrophobins-enconding cDNAs from the ectomycorrhizal basidiomycete *Pisolithus tinctorius*. Gene 168:93–97

Tagu D, De Bellis R, Balestrini R, De Vries OMH, Piccoli G, Stocchi V, Bonfante P, Martin F (2001) Immunolocalization of hydrophobin HydPt-1 from the ectomycorrhizal basidiomycete *Pisolithus tinctorius* during colonization of *Eucalyptus globulus* roots. New Phytol 149: 127–135

Tagu D, Lapeyrie F, Martin F (2002) The ectomycorrhizal symbiosis: genetics and development. Plant Soil 244:97–105

Timonen S, Peterson RL (2002) Cytoskeleton in mycorrhizal symbiosis. Plant Soil 244:199–210

Timonen S, Finlay RD, Söderström B, Raudaskoski M (1993) Identification of cytoskeletal components in pine ectomycorrhizas. New Phytol 124:83–92

Timonen S, Söderström B, Raudaskoski M (1996) Dynamics of cytoskeletal proteins in developing pine ectomycorrhiza. Mycorrhiza 5:423–429

Voiblet C, Duplessis S, Encelot N, Martin F (2001) Identification of symbiosis-regulated genes in *Eucalyptus globulus-Pisolithus tinctorius* ectomycorrhiza by differential hybridization of arrayed cDNAs. Plant J 25:181–191

Weiss M, Mikolajewski S, Peipp H, Schmitt U, Schmidt J, Wray V, Strack D (1997) Tissue-specific and development-dependent accumulation of phenylpropanoids in Larch Mycorrhizas. Plant Physiol 114:15–27

Weiss M, Schmidt J, Neumann D, Wray V, Christ R, Strack D (1999) Phenylpropanoids in mycorrhizas of the Pinaceae. Planta 208:491–502

Wessels JGH (1996) Fungal hydrophobins: proteins that function at an interface. Trends Plant Sci 1:9–15

Wösten HAB (2001) Hydrophobins: multipurpose proteins. Annu Rev Microbiol 55:625–646

Chapter 9
RNA Silencing in Ectomycorrhizal Fungi

Minna J. Kemppainen and Alejandro G. Pardo

9.1 Introduction

RNA was for a long time believed to be a rather passive molecule. The only function was to transfer coding information from genomic DNA to the protein translation machinery. However, during the last 10 years this role of RNA has gone through a radical transformation which has in turn revolutionized the way the eukaryotic cells are understood to function today. Several types of small nonprotein encoding RNAs are shown to be highly important as controllers of gene expression and genome stability in plants, animals and fungi. These RNA molecules originate predominantly from double-stranded RNA precursors (dsRNA) and they can affect the gene transcript level both transcriptionally and post-transcriptionally. They also modify and control the epigenetic state of nuclear DNA by participating in the formation and maintenance of heterochromatin. These RNA-dependent eukaryotic regulatory networks are now referred to as RNA interference (RNAi). RNAi is believed to have evolved as a protection mechanism against invading nucleic acids such as viruses that would have later become involved in general genetic control of cells. The cellular mechanisms in which RNAi is currently known to participate are multiple and the small-RNA type is organism specific. However, they all seem to share a highly controlled processing of larger RNAs into small effector RNAs which via sequence specific binding target homologous RNAs (or probably also DNA). The outcome of RNAi pathways is usually a reduction of gene expression and this phenomenon is therefore also called RNA silencing.

The finding that an artificial introduction of genetic components (DNA or RNA) can trigger cell's RNAi and lead to targeted modification of gene activity has made the RNA-research one of the most blooming fields of investigation in biosciences. Besides its use as a basic tool for studying gene function, RNAi has numerous

M.J. Kemppainen and A.G. Pardo (✉)
Laboratorio de Micología Molecular, Departamento de Ciencia y Tecnología, Universidad Nacional de Quilmes, y Consejo Nacional de Investigaciones Científicas y Tecnicas (CONICET), Roque Sáenz Peña 352, (B1876BXD) Bernal, Provincia de Buenos Aires, Argentina
e-mail: apardo@unq.edu.ar

M. Rai and A. Varma (eds.), *Diversity and Biotechnology of Ectomycorrhizae*, Soil Biology 25, DOI 10.1007/978-3-642-15196-5_9,

biotechnological and therapeutic applications. The discovery of the RNAi mechanism in 1998 (Fire et al. 1998) was awarded with the Nobel Prize for Medicine or Physiology in 2006. The cellular mechanisms connected to and controlled by RNAi are under massive scientific research. Even though many aspects of RNAi are still poorly understood the basic steps of various pathways have already been established. With increasing information it is however becoming evident that previous classifications of RNAi dependent mechanisms have been oversimplifications. Different small RNA dependent pathways are now known to work in a simultaneous and interconnected manner and new RNAi regulatory networks are also identified with increasing speed.

9.2 RNAi Mechanisms

Several gene inactivation phenomena described separately such as transgene-induced cosuppression of endogenes (Jorgensen 1990) and virus-induced gene silencing (VIGS) (Ruiz et al. 1998) in plants as well as quelling in *Neurospora crassa* (equivalent to cosuppression in plants) (Romano and Macino 1992), are now known to be mediated by small 20–30 nt cellular RNAs originating from dsRNA. The RNAi mechanism was first described in the model nematode *Caenorhabditidis elegans* (Fire et al. 1998). It is now known to be a conserved eukaryotic pathway which affects genes and genomes by controlling chromatin structure, chromosome segregation, protection against invading viruses, transposon inactivation, gene transcription, RNA processing, RNA stability and translation in plants, fungi and animals (Carthew and Sontheimer 2009; Moazed 2009).

RNAi pathways are shown to act in cells both in cytosolic and nuclear environments and even though highly variable in their actions and outcomes they all rely on the sequence complementary of the small regulatory RNAs to their target transcripts. Three main classes of cellular small regulatory RNAs have been described this far: small interfering RNAs (siRNAs), micro-RNAs (miRNAs) and PIWI-interacting RNAs (piRNAs).

The siRNAs act at several regulatory levels and via variable RNAi target mechanisms in plants, animals and fungi while the miRNA-dependent translational silencing pathway is detected both in animals and plants but it seems to be missing in the fungal kingdom. The biogenesis of siRNAs and miRNAs requires dsRNA precursors and strictly depends on activities of two RNAi proteins: ribonuclease (RNase) Dicer (Bernstein et al. 2001; Carmell and Hannon 2004) and Argonaute (Hammond et al. 2001; Hutvagner and Simard 2008). The piRNAs are the newest and less understood class of regulatory small RNAs. They silence transposons in animal germline cells and their biosynthesis is not dsRNA dependent and neither does it need Dicer activity. They however interact with PIWI-clade Argonaute proteins (Girard et al. 2006). This classification of small regulatory RNA is based on the biogenesis and mechanisms of action and is to some extent artificial though some biogenesis classes can perform different organism-specific target activities.

Many of the proteins involved in RNAi are present as multigene families in eukaryotes. The homologous proteins are demonstrated to have different RNAi pathway specificities and/or several pathways can be also interconnected by their protein or RNA components. The facts that many of the already known protein components are still uncharacterized and many of the detected RNA-dependent cellular mechanisms are not fully resolved, indicate that there are certainly other important regulatory roles for RNAs to be identified in future.

Also while the basic RNAi mechanisms are shown to be conserved, each eukaryotic group has had its own evolutionary diversification resulting in distinctive characteristics of RNA silencing pathways. Because siRNAs are the only class of small regulatory RNAs described in fungi till today their biogenesis and target actions are discussed in more detail. The use of RNAi as a genetic tool for fungi also depends on the artificial introduction of this type of small regulatory RNAs. The biogenesis and action of another important group, miRNAs, which are responsible for gene regulation of at least 30% of human genes (Lewis et al. 2005) are also presented as their target functions may overlap with siRNAs.

9.3 siRNA-Dependent RNA Silencing

9.3.1 siRNAs

The majority of information on basic RNAi mechanism originates from studies on siRNAs (Carthew and Sontheimer 2009). siRNAs are double stranded small regulatory RNAs of 21–25 nt which are processed from dsRNA precursors in plants, animals, fungi and ciliates cells. Several siRNA-mediated RNA silencing pathways have been detected and they act both in cytosol and nucleus. In cytosol siRNAs mediate post-transcriptional gene silencing (PTGS) predominantly via target mRNA degradation but they can also cause translational arrest. In the nucleus siRNAs have been connected to chromatin-dependent gene silencing (CDGS) both by transcriptional gene silencing (TGS) and cotranscriptional gene silencing (CTGS) mechanisms (Moazed 2009).

SiRNAs were first detected in plants where they were believed to be of exogenous origin and degradation products of invading viral, transgene or transposon RNA (Mello and Conte 2004). However, several classes of siRNAs homologous to centromeric and other natural genome repetitive sequences were detected indicating that siRNAs were processed in cells also from endogenous RNA sources in plants (Lippman and Martienssen 2004). This led to the classification of exo-siRNAs and endo-siRNAs. Endo-siRNA production has an obligate nuclear step while exo-siRNA production can take place directly in cytosol. Besides plants various classes of endo-siRNAs have also been detected in animals and fungi proving that nuclear DNA is a natural source of siRNAs in different eukaryote kingdoms (Golden et al. 2008; Moazed 2009).

9.3.2 Sources of dsRNA in Cells

Long dsRNA precursors of siRNAs can be a result of direct invasion of dsRNAs into cells such as viral RNA. Also exogenously introduced single-stranded RNAs (ssRNA) can be turned into dsRNA in cytosol by RNA dependent RNA polymerases (RdRP), enzymes present in plants and fungi but absent in insects and vertebrates. Also some noncoding or aberrant ssRNAs (such as mRNAs lacking proper processing signals) originating from nuclear transcription can be detected as a template for RdRPs directly in the nucleus or in cytosol. dsRNA molecules can also be a direct result of nuclear transcription without RdRP activity. Antisense transcription can result in dsRNAs by binding with sense transcripts. Also transcripts of repetitive sequences, inverted repeats, and transposons can produce dsRNAs (Moazed 2009).

Once dsRNA is present in the cell, either in the nucleus or in cytoplasm, it is detected by the RNAi machinery. The cytosolic production of siRNAs and PTGS cascade is best characterized in *Drosophila*, plant and human cells, while nuclear siRNA activities involved in CDGS and heterochromatin control is profoundly studied in the model fission yeast *Shizosaccharomyces pombe*. Today the artificial triggering of cytosolic siRNA-dependent PTGS pathway is by far the most common approach in altering gene expression by the RNA silencing technology in filamentous fungi.

9.3.3 Cytosolic siRNA-Mediated PTGS

When dsRNA is detected in cytosol it initiates a multistep RNAi cascade that is characterized by (1) digestion into double-stranded 21–25 nt primary-siRNAs by the RNase III domain of the Dicer protein, (2) loading of one of the siRNA strands (guide strand) into a multiprotein siRNA-induced silencing complex (siRISC) including Argonaute protein, (3) sequence specific targeting of siRISC to a homologous ssRNA by guide-siRNA base paring and destruction of the target by the slicing activity of RNase H-like fold of Argonaute, (4) dissociation of the siRISC from the cut target RNA and targeting a new mRNA.

Dicer activity on dsRNAs is characterized by 2 nt 3′overhangs leaving a 5′ monophosphate in siRNA-ends. Binding of Dicer to dsRNA is also assisted by proteins with dsRNA-binding domains (Tomari and Zamora 2005). If an organism has several Dicer-like proteins these are generally specialized to process a certain type of dsRNA templates (siRNA precursors or miRNA precursors respectively). The role of Dicer as a siRNA-cutter can also be extended to RISC assembly as especially Dicer-associated dsRBPs are shown to be important in the early steps of RISC-formation (so called pre-RISC).

Argonaute (AGO) is a core silencing protein and many organism have several AGO proteins specialized for acting in different RNAi pathways. While *S. pombe* has only one Argonaute, five, eight and twenty-seven paralogs exist in *Drosophila*, humans and *C. elegans* (Carthew and Sontheimer 2009). All Argonautes do not have endonucleolytic capacity and these might be important in RNAi pathways not leading to direct target RNA destruction. The election of the guide-siRNA strand in siRISC was for a long time a mystery but now it is known that the less thermodynamically stable 5′-terminus is chosen as a guide. siRNAs from different sources such as viral siRNAs and transgene siRNAs in plants and endo- and exo-siRNAs in *Drosophila* have different requirements for RISC-assembly depending on different dsRBD (dsRNA binding domain) proteins (Golden et al. 2008). The precise mechanisms leading to such a precise discrimination of different origin siRNAs is not yet known. The slicing activity of Argonaute requires a perfect match in base paring between the siRNA-guide strand and the target RNA. The AGO slicing generates 5′-monophosphate and 3′-hydroxyl termini in the RNA fragments produced. These are attacked by cellular exonucleases and degraded further (Orban and Izaurralde 2005). RISC is demonstrated to localize in cytoplasm in special structures, the processing bodies (P bodies), which are implicated in the regulation of mRNA translation, storage, and degradation (Jagannath and Wood 2009).

Eukaryote genomes generally encode for several Dicer and Argonaute proteins. Some isoforms have been linked to separate RNAi pathways such as virus defense, chromatin modifications, miRNA processing or siRNA synthesis while some have overlapping functions and are functional in several pathways. How different isoforms discriminate their dsRNA-targets and how they are engaged in specific RNAi pathways is still poorly understood and under vigorous scientific investigation.

9.4 miRNA-Mediated Translational Arrest

If the base pairing with the target RNA is not perfect enough or the RISC contains an Argonaute protein without slicing activity the direct degradation cannot take place. However, these siRISCs can theoretically still act on target RNAs via a second PTGS mechanism characteristic of another class of small regulatory RNAs, miRNAs.

miRNAs were originally detected in viruses as sequences that affected expression of specific host mRNAs (Pfeffer et al. 2004). Later on endogenous miRNAs of 21–25 nt were described in plants and animals and they are now known to play a fundamental role in gene expression regulation in these organisms. The majority of miRNAs originate from nuclear transcriptional units of nonprotein encoding RNAs and they are processed from 65–70 nt sequence that forms a ~33 bp mismatch stem-loop precursor (pri-miRNAs) in nucleus by Dicer (plants) or the specific Dicer-like protein Drosha (animals) to pre-miRNAs (Kim 2005). The pre-miRNAs with a stem-loop structure are further processed in nucleus (plants) or in cytosol (animals) to a mature ~22 bp miRNAs by Dicer and the mature guide-strand of miRNA is incorporated into miRISC and targeted to a homologous RNA in cytosol. The animal miRNAs generally target gene transcripts at their 3′UTR (untranslated region) and this base paring is imperfect. Plant miRNAs however show nearly perfect paring and they target coding sequences. In the case that the Argonaute slicing activity is not compromised and the target binding is perfect or nearly perfect, miRISC act in a similar manner as siRISC and degrades the target. This often happens in plants where miRNAs can act in a siRNA manner. However, an imperfect match, typical of animal miRNAs, inhibits slicing and leads to prolonged binding of miRISC to the target and translational repression of the mRNA. The mechanisms behind the translational repression are not yet clear but it might involve various pathways that affect translation and direct mRNA to degradation by deadenylation (Petersen et al. 2006; Wu et al. 2006).

9.5 RdRP Amplification of the Silencing Trigger

Organisms which have RdRP activity can amplify the primary siRNA trigger and generate the so-called secondary siRNA cascade. This amplification of the original effector molecule in plants, animals and fungi is a key for understanding how a small amount of dsRNA can result in a strong and persistent RNA silencing effect in cells. The amplification of the silencing signal is also behind the systemic spreading of locally initiated RNAi in multicellular organisms. Furthermore, it explains how the silencing signal can spread its effect and result in Argonaute-mediated slicing of the whole target mRNA, and not only the RNA site sharing sequence homology, with the primary siRNAs. The latter is known as transitive silencing and it can have important implications for the use of RNA silencing as a reverse genetic tool in organisms with RdRP activity.

The production of secondary siRNAs depends on RdRPs and this process is shown to differ between plants and *C. elegans* (Petersen and Albrechtsen 2005; Axtell et al. 2006; Ruby et al. 2006; Sijen et al. 2007; Pak and Fire 2007). In *C. elegans* transitive silencing spreads toward the 5′ but in plants to both 5′ and 3′ of target mRNAs. In plants secondary siRNAs are also efficiently produced when the target mRNA is already cut by Argonaute and these cleaved molecules are then recognized by RdRP that generates more dsRNA molecules. These new dsRNAs are further digested by Dicer into secondary siRNAs which can target homologous RNAs. In *C. elegans* the primary siRNAs can directly lead to RdRP activity on target mRNA and produce 22–23 nt dsRNA which are later cut by a still uncharacterized endonuclease activity into secondary siRNAs. The secondary siRNA pools in plants and *C. elegans* differ also in their 5′-end due their different origins. The plant secondary siRNA carry 5′-monophosphates while C. *elegans* siRNAs have 5′-di or triphosphates. 5′-monophosphate cleavage is a typical Dicer cut siRNA character. The secondary siRNAs in *C. elegans* differ also in a detected strand-bias. They are predominantly antisense to the target RNA. They are also in phase respect to each other. The first secondary siRNA carries a sequence starting close to the primary siRNA and the following one cover the next 22 bases and so on. This dsRNA synthesis in the nematode seems not to be primed by the primary siRNAs and how these exact 22 nt-long secondary siRNA synthesis are controlled and how they are cleaved is not yet understood. The characteristics and extent of secondary siRNA cascade in different fungal taxa has been rather poorly studied. However, this cascade must be widespread as many fungi posses RdRPs.

The RdRP-dependent amplification of the original silencing signal is behind the systemic silencing phenomenon observed in multicellular organisms. Introduction of a silencing trigger in one part of an organism with RdRP activity can lead to spreading of silencing to the whole plant, nematode or fungal colony. The molecular identity of the cell-to-cell mobile signal in RNAi is however not yet clear but it is believed to be siRNA (Kalantidis et al. 2008). As there is no secondary siRNA amplification in mammals or in *Drosophila* like in plants, fungi and nematodes (Cogoni and Macino 1999; Dalmay et al. 2000; Sijen et al. 2001) the direct introduction of synthetic siRNAs to these organisms can cause only transient silencing effect.

Besides the systemic silencing, another important aspect of transitive silencing is that in these organisms the RNA silencing effect may not be strictly primary trigger specific. Secondary cascade siRNAs may act on other mRNAs that share homologous sequences with the original target outside the primary silencing trigger binding site. These are called off-target effects and cellular silencing pathways responsible for them can vary according to the grade of homology between the secondary siRNAs and the RNA targets. Perfect matches can produce Argonaute-mediated slicing and the imperfect base pairing can lead to miRNA-type translational arrest and RNA decay. The majority of transitive silencing linked off-target effects are believed to rise via the latter mechanism due to the predominantly imperfect base paring nature of the secondary siRNAs. Interestingly, no plant or animal type miRNAs have been detected in fungi and whether the imperfectly paired secondary siRNAs in these eukaryotes can act via translational arrest and mRNA decay is not clear.

However, these silencing off-target effects can also be a benefit. Transitive silencing makes possible more efficient simultaneous silencing of homologous members in multigene families. This can be highly useful in polyploid plants. However, species specific differences for transitive silencing and efficiency depending whether the target gene is an endo- or a transgene have been reported (Miki et al. 2005; Bleys et al. 2006a, b). The molecular mechanisms leading to such sequence discrimination are not yet understood but as plants and fungi generally have a numerous set of RdRPs this suggest a specialized activity of different isoforms on specific RNA substrates. Studies on some of the *Arabidopsis* six RdRPs have in fact revealed the specificity of independent enzymes, some participating in sense transgene-triggered RNA silencing but not in RNA silencing activated by inverted repeated transgenes or RNA viruses while others are responsible for viral defense (Yu et al. 2003).

9.6 siRISC Can Operate on Target mRNAs in Nuclear Environment Also

siRNA-dependent slicing of target mRNAs was believed to occur only in cytosol. However, recent data strongly indicates that Argonaute-mediated slicing takes place also in the nuclear environment. Nuclear RISC (nRISC) has been detected in human cells and it is loaded with siRNAs in cytosol. This nRISC is later imported to the nucleus where it targets homologous mRNAs either by slicing or directing them to other decay pathways (Ohrt et al. 2008; Weinmann et al. 2009). The nRISC appears to be a conserved part of the eukaryotic silencing machinery. A nuclear RNAi–specific Argonaute protein has also been described in *C. elegans*. Its localization into the nucleus is siRNA-binding dependent and this binding takes place in cytosol (Guang et al. 2008).

The existence of cytosol-loaded nRISC is now explaining some long known silencing phenomena such as transcriptional inactivation of host genes via VIGS. This nuclear effect was detected in plants when infected with cytosolic dsRNA viruses (Wassenegger et al. 1994). Equal to siRNAs, other types of small regulatory RNAs are also now proposed to function both in cytosolic and nuclear environments. Besides the direct post-transcriptional silencing effect on target mRNAs in cytosol and also in nucleus these small RNAs can also be involved in epigenetic modification of homologous DNA sequences and control gene expression at the transcriptional level.

9.7 Epigenetic Effects Associated with RNAi

RNAi is connected to epigenetic modifications of the host genome. These include direct DNA methylation and/or histone modifications depending on the organism. It is increasingly being recognized that all types of small regulatory RNAs have some epigenetic nuclear effects and that the basic mechanisms responsible for them seem to be conserved within the eukaryotic domain (Djupedal and Ekwall 2009; Verdel et al. 2009). These chromatin modifications are complex, long lasting, even heritable, and they can affect gene expression as well as be implicated in different diseases. Epigenetics along with RNAi are the current hot spots of biological research.

Majority of RNAi-mediated DNA modifications can be seen as auto-defense responses which secure genome integrity by inactivating transposable elements and imbedding repetitive sequences into heterochromatin. Also some genes can be under RNAi-dependent epigenetic control. siRNA-mediated heterochromatin control and the histone methylation of the pericentromeric repeats has been exhaustively studied in the fission yeast *S. pombe*. Heterochromatin, a traditionally transcriptionally inactive part of the genome is now, paradoxically, shown to depend on active transcription for its formation and maintenance (Bühler 2009).

A nascent-transcript model has been proposed where siRNAs, originating from transcription of genome repetitive elements and produced via RdRP and Dicer activities in nucleus, are incorporated into RITS (RNA-induced transcriptional silencing complex) which binds to nascent homologous RNA transcripts at their site of synthesis. This RITS-binding serves as a mark for histone modifying proteins (*S. pombe* does not have direct DNA methylation response) which initiate heterochromatin formation and maintain it in repetitive genomic loci (Moazed 2009). Equivalent mechanisms based on active transcription and nuclear siRNA–RNA binding are proposed to control epigenetic modifications present in natural or introduced repetitive sequences in plants, fungi and animals. These effects are called CTGS.

Heterochromatin formation can reduce transcription of genes embedded in the region. Such a heterochromatization leading to TGS can be a result of RNAi mechanisms and to be mediated via siRNAs. If dsRNA precursor in the cell carries a sequence homologous to a gene promoter, siRNAs processed from such a precursor can be engaged in a RNAi pathway targeting homologous DNA in nucleus. This results in the inhibition of the target gene transcription and this reduced transcription correlates with an altered epigenetic status (increased DNA and histone methylation) of the targeted promoter sequences. TGS has been demonstrated in plants, fungi, insects and also in mammalian cells (Weinberg 2006; Verdel et al. 2009). The function of siRNAs in TGS is not yet totally resolved but they seem to target homologous promoter sequences by RNA–RNA, and not via direct RNA–DNA binding, probably according to the nascent transcript model. The production of primary siRNAs involved in TGS can, however, also be cytosolic and not only nuclear as postulated for *S. pombe* heterochromatin formation. A cytosolic nRISC formation and nuclear import may therefore take place before homologous DNA sequence targeting in nucleus. Besides siRNAs, both plant and animal miRNAs are also now known to have the capacity to target, not only mRNAs in cytosol, but to act on homologous DNA sequences in nucleus (Gonzalez et al. 2008; Kim et al. 2008). This targeting, probably via RNA–RNA interactions, causes epigenetic modification and can result in TGS. Epigenetic effects linked to RNAi are highly complex and a field of massive scientific research.

A scheme showing different RNA silencing pathways demonstrated to be active in eukaryotic cells is presented in Fig. 9.1.

9.8 Use of RNAi as a Genetic Tool

RNAi technology has offered an exceptional tool for functional genomic studies. RNAi can be launched in different organisms by artificial introduction of RNAi inducing RNA (or DNA) triggers. Because RNAi results in reduced mRNA levels but does not completely abolish gene function it has also made possible to study genes whose knock-outs would be lethal. Due to the cytosolic nature of PTGS it offers a direct and fast approach to alter gene expression in dikaryotic, diploid and

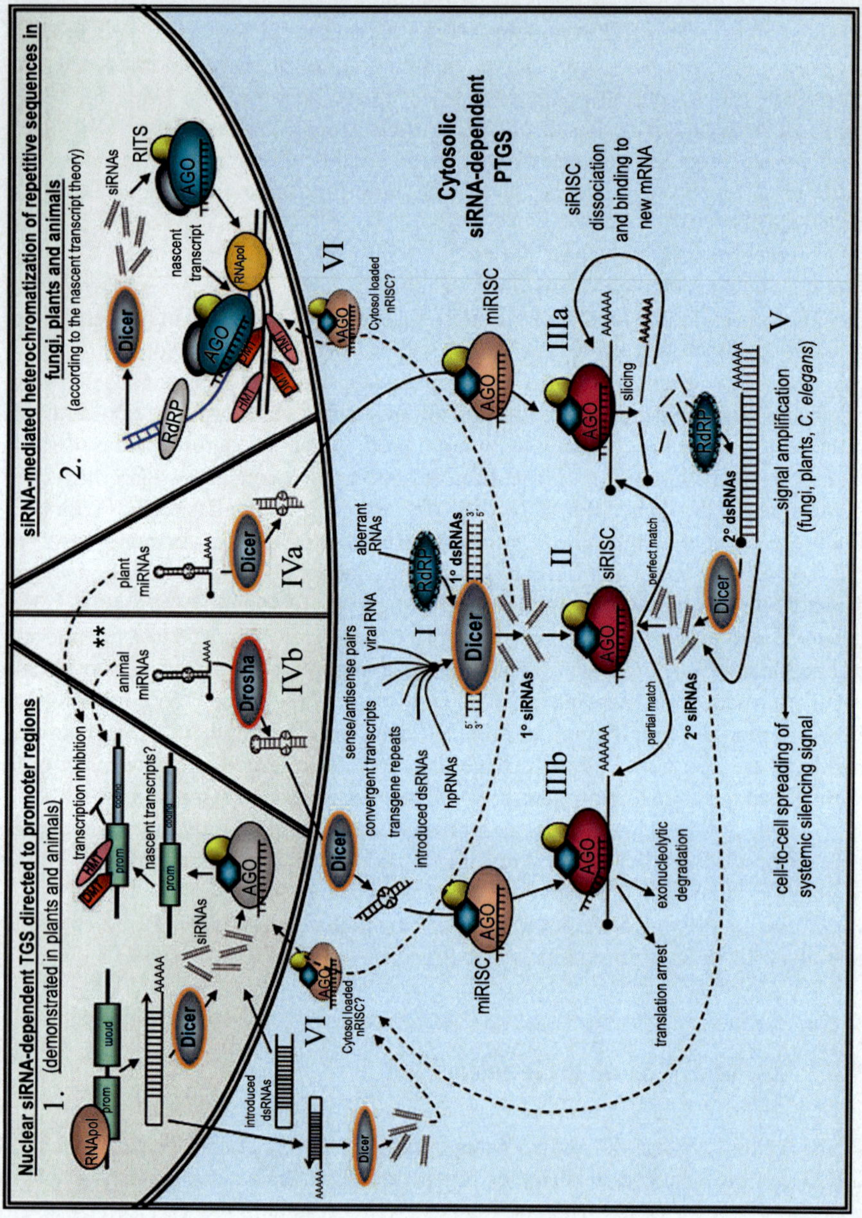

Nuclear siRNA-dependent TGS directed to promoter regions
1. (demonstrated in plants and animals)
siRNA-mediated heterochromatization of repetitive sequences in fungi, plants and animals
2. (according to the nascent transcript theory)
RNApol
prom
transcription inhibition
nascent transcripts?
siRNAs
Dicer
AGO
introduced dsRNAs
animal miRNAs
plant miRNAs
Drosha
IVb
IVa
VI
Cytosol loaded nRISC?
RdRP
nascent transcript
RITS
Cytosolic siRNA-dependent PTGS
sense/antisense pairs
viral RNA
convergent transcripts
transgene repeats
introduced dsRNAs
hpRNAs
aberrant RNAs
1° dsRNAs
1° siRNAs
I
II
siRISC
miRISC
IIIa
IIIb
slicing
perfect match
partial match
2° siRNAs
2° dsRNAs
siRISC dissociation and binding to new mRNA
exonucleolytic degradation
translation arrest
V
signal amplification (fungi, plants, C. elegans)
cell-to-cell spreading of systemic silencing signal

Fig. 9.1 Schematic representation of different conserved siRNA- and miRNA-dependent cytosolic and nuclear RNA silencing pathways demonstrated to be active in eukaryotic cells. The figure shows the key processing steps leading to RNAi-dependent post-transcriptional (PTGS), transcriptional (TGS) gene silencing and siRNA-dependent detection and epigenetic modification of repetitive genomic sequences. Not all the protein components demonstrated to participate in each step of the different pathways have been illustrated. The figure presents the cytosolic PTGS pathway. However similar mechanisms are now known to act also in nuclear environment. **I-V:** siRNA- and miRNA-mediated cytosolic PTGS. **I** Different dsRNAs of endo- or exogenous origin or aberrant RNAs turned into dsRNAs via RdRP-activity (in plants, fungi and *C. elegans*) can be recognized by ribonuclease protein Dicer. Dicer's RNase activity leads to cutting of dsRNA into 21–24 bp dsRNAs, the so called primary (1°) cascade siRNAs. **II** One strand of the 1° cascade siRNAs is loaded into a multiprotein complex (siRISC) which contains a core RNAi protein, Argonaute (AGO). SiRISC can bind to homologous mRNA sequences by siRNA base paring. **III** The degree of sequence homology of siRISC binding to mRNA can result in two silencing outcomes: **IIIa** A perfect mach permits slicing activity of Argonaute and the mRNA is cut. The siRISC dissociates from the cut mRNA and returns to act on new homologous mRNAs. This siRISC-mediated slicing leads to reduced cytosolic concentration of target mRNA and gene silencing. Also plant miRNAs produced by nuclear transcription and Dicer activity **(IVa)** share this siRISC-pathway as they typically act via perfect match of miRISC producing target mRNA slicing. **IIIb** In the case of imperfect match with the target (or when associated with AGO without slicing activity) Argonaute slicing of mRNA cannot occur. The PTGS however takes place as this type of siRISC binding results in translational arrest and exonucleolytic degradation of the target mRNA. Animal miRNAs which are produced via nuclear transcription and further processed by the nuclear Dicer-like protein Drosha and cytosolic Dicer **(IVb)** regulate their target mRNAs in cytosol via the imperfect match strategy. The majority of off-target effects caused by artificially initiated silencing in different eukaryotes are proposed to be a result of this RNAi pathway as siRNA binding to homologous non-target mRNAs can be expected to be predominantly imperfect. **V** In organisms with RdRP activity (plants, fungi, *C. elegans*) the initial silencing trigger can be further amplified resulting in the so-called transitive silencing. This is characterized by the production of a secondary (2°) siRNA cascade, not homologous to the initial Dicer-cut dsRNA-silencing trigger. The siRISC-cut mRNAs (or intact mRNAs) can act as templates for RdRP-catalyzed dsRNA synthesis which are processed into siRNAs and further incorporated into RISC. The synthesis of these 2° siRNAs is demonstrated to show some fundamental differences between plants and *C. elegans*. While in plants siRNAs are a product of Dicer cut of dsRNAs in *C. elegans* they are produced by RdRP using the mRNA as template toward the 5′-end but no Dicer-slicing takes place. Transitive silencing potentially increases the off-target effects as the silencing is not strictly primary dsRNA trigger-dependent. However, this RdRP-activity and silencing amplification also explains the long lasting and systemic silencing signal in the case of local or transitive dsRNA triggering in these organism. The degree of transitive silencing is organism-dependent but it also shows target gene specific behavior. In plants endogenes seem to act as weaker templates for transitive silencing than heterologous transgenes. **VI:** The recent demonstrations of cytosolic-loading of nuclear targeting siRISC (nRISC) confirms that cytosol-produced or introduced siRNAs can also potentially act in nuclear environment either directly on homologous target mRNAs by Argonaute-mediated slicing and transcript decay or by mediating epigenetic effects on DNA sequences. NRISC is a link with potential to connect the cytosolic and nuclear RNAi pathways. Both primary and secondary siRNAs might participate in nRISC. The presence of nRISC in fungi has not been demonstrated yet. **1 and 2:** siRNA-mediated nuclear RNAi can result in TGS and heterochromatization of repetitive genomic sequences by DNA methylation and/or histone modifications. **1.** siRNAs with homologous sequences to gene promoters can initiate nuclear Dicer/AGO-dependent RNAi pathway leading to direct DNA methylation and/or histone modifications in these genomic sequences. As a result the transcriptional activity from the targeted promoters is reduced causing transcriptional gene silencing (TGS). The siRNAs linked to TGS can be a result of nuclear transcription, artificially-introduced to nucleus, or of cytosolic origin and entering to nucleus via mechanisms such as nRISC. How these siRNAs can polyploid organisms. RNAi has been successfully used for studying specific gene functions in a wide range of eukaryotes (Travella and Keller 2009). Today RNAi research is mainly dedicated to the use of artificial miRNAs instead of siRNAs due to their higher target specificity (Ossowski et al. 2008; Carthew and Sontheimer 2009). However, miRNA applications are functional in organisms which posses this specific silencing pathway, these not including fungi.

The number of full genomic sequences is growing almost on a daily basis and RNAi technology has made possible, with a relative low investment, to look for the function of practically any gene of interest. Studies at full genome scale have already been conducted in *C. elegans*. Of nematode's approximately 19,000 genes 86% have been screened for their function with the RNAi feeding library technology (Kamath et al. 2003). Of the genes analyzed 1,722 produced identified phenotypes proving RNAi a highly efficient tool for reverse genetics in this organism. RNAi has also been used for several genome wide studies in *Drosophila* (Boutros et al. 2004; Chen et al. 2008; Cronin et al. 2009; Mummery-Widmer et al. 2009). Similar kinds of studies with the RNAi technology are now initiated in plants (Ossowski et al. 2008). Moreover, RNAi is massively used for screening gene functions is mammalian cells. These studies are headed for resolving the genetic background of different diseases and developing RNA therapy drugs (Gobeil et al. 2008; Zhou et al. 2008; Krishnan et al. 2008; Castanotto and Rossi 2009).

Fig. 9.1 (continued) initiate de novo methylation in homologous genomic DNA sites is not yet well understood but it seems to depend on nascent RNA-siRNA binding similarly to RNAi-dependent maintenance of heterochromatin in pericentromeric repetitive sequences of *S. pombe* (see below) **2.** Nuclear siRNAs are also shown to participate in heterochromatization in different eukaryotes. siRNAs can target both natural and transgenic genomic repetitions resulting in target DNA methylation and/or histone modifications and this process involves Dicer, AGO and RdRP proteins. The RNAi-dependent maintenance of cellular heterochromatin has been most thoroughly studied in *S. pombe* where a nascent transcript model has been proposed: RNA transcripts constantly produced from these repetitive genomic loci are responsible for the physical co-localization of siRNAs-protein complexes (RITS) to these sites and also for the further production of the homologous siRNAs. The RNA-RNA binding of RITS with the nascent transcripts serves as a flag for nuclear histone modifying enzymes to maintain the heterochromatic state of these sites. Also these same nascent transcripts are detected by RdRP to produce dsRNAs which are processed by Dicer into siRNAs and loaded into RITS. Even though characterized in *S. pombe* the involvement of RNAi and nascent RNA transcripts in heterochromatization of genomic sequence repetitions seem to be a conserved cellular mechanism among different eukaryotes. To which extent these nuclear siRNAs can initiate de novo chromatin modification however is not yet clear. Also the participation of cytosol processed siRNAs in this nuclear RNAi–pathway and in initiation of epigenetic modifications cannot be excluded. *hpRNA* hairpin RNA; *dsRNA* double-stranded RNA; *siRNA* small interfering RNA; *miRNA* micro-RNA; *RdRP* RNA dependent RNA polymerase; *AGO* Argonaute protein; *siRISC* siRNA-induced silencing complex; *miRISC* miRNA-induced silencing complex; *nRISC* nuclear siRISC; *RITS* RNA-induced transcriptional silencing complex; *DMT* DNA methylation; *HMT* histone methylation; *RNAPol* RNA polymerase; *Prom* gene promoter region

9.9 RNA Silencing in Fungi

In fungi transgene induced endogene cosuppression was first reported in the filamentous ascomycete *N. crassa* where the phenomenon was called quelling (Romano and Macino 1992). Now it is known that quelling is a RNAi gene inactivation process dependent on multiple transgene arrays. Further studies on quelling have resulted in the discovery of many of the underlying mechanisms of RNAi in eukaryotes and turned this fungus as one of the model organisms in RNAi research (Fulci and Macino 2007). Also in the fission yeast *S. pombe* studies about RNAi-mediated heterochromatin formation have resolved how nonprotein coding natural small RNAs control genome integrity in eukaryotes (Moazed 2009; Bühler 2009). Later on, the RNA silencing pathway has been shown to exist in many other yeast and filamentous fungal species. Small nonprotein coding RNAs have been shown to target homologous mRNAs by Dicer- and Argonaute- dependent slicing in fungi, similar to other eukaryotes. RNAi has also been linked to cell defense against viruses, stabilization of transposons, control of heterochromatin formation and some programmed DNA elimination processes. Natural nonprotein coding small RNAs originating from transposons and repetitive genome sequences have recently been isolated in *M. oryzae*, *N. crassa,* and *A. fumigatus* demonstrating that RNAi pathways are involved in the control of genome integrity in filamentous fungi as well as in *S. pombe* (Murata et al. 2007; Jöchl et al. 2008; Cecere and Cogoni 2009). However, no regulatory miRNAs have been reported yet and the absence of this RNAi pathway could be characteristic of the fungal kingdom (Nakayashiki and Nguyen 2008).

9.10 RNA Silencing Protein Machinery in Fungi

With growing genomic sequence information from diverse fungi, comparative studies on evolution and diversification of RNA silencing proteins in this group of eukaryotes have become possible. A phylogenetic analysis of fungal species covering, ascomycetes, basidiomycetes and zygomycetes was carried out by Nakayashiki and Kadotani (2006). These authors used sequences coding for Dicer, RdRP and Argonaute proteins in order to establish the evolutionary history of diversification of RNA silencing pathways between and within different fungal taxa and also between fungal, plant (*Arabidopsis*) and animal (*Drosophila*) kingdoms. The existence of RNAi protein machinery has been demonstrated in ascomycetes, basidiomycetes and zygomycetes. However, different fungal taxa show an unexpected wide diversification in their RNAi protein repertoire. Some parts of the fungal RNAi machinery have gone through phases of expansion while a complete or partial loss of the pathway has happened independently in different nonphylogenetically related species. It seems that the whole *Saccharomyces* complex, including *S. cerevisiae* and *Candida* spp. has lost all Dicer and RdRP proteins and the same has happened in the basidiomycete *Ustilago maydis*. However, in the closely related species *U. hordei* the dsRNA-triggered RNAi pathway is functional (Laurie et al. 2008). Similarly different orthologous proteins such as RdRP-1, 2 and 3, present in different fungal phyla, cluster together indicating that these duplication events are ancestral and have occurred before fungal diversification. *Schizosaccharomyces pombe,* which has just one of each RNAi proteins, would thus have lost paralogous proteins during evolution.

A special RNAi gene expansion is detected among basidiomycetes where at least three distinct classes of Dicer and RdRP proteins and two of Argonaute-like proteins have been detected. This gene expansion has especially affected Argonaute and RdRP proteins in homobasidiomycetes such as *Phanerochaete chrysosporium* and *Coprinus cinereus*. The former has seven AGOs and nine RdRPs and the latter eight AGOs and seven RdRPs. Moreover, both of these basidiomycetes have three Dicer genes while filamentous ascomycetes usually have two. *Cryptococcus neoformans*, a basidiomycete yeast, seem to have evolved distinctly and its RNA silencing proteins show characteristics not observed in other fungal proteins. For example the *Cryptococcus* Dicer proteins lack the typical DEAD/DEAH box helicase, a feature that has also been reported in functional *Tetrahymena* Dicer-like protein (Dcl1) involved in RNA silencing related pathway of internal elimination sequences (IES) (Mochizuki and Gorovsky 2005). Also a comparison of RNAi protein machinery of seven *Aspergillus* species has revealed that significant variation exists between them indicating both gene duplication and truncation events (Hammond et al. 2008b).

These phylogenetic studies highlight some fundamental aspects of fungal RNAi. Firstly, different fungal taxa have highly different RNAi protein repertoire and secondly, some fungal species have lost all or are impaired in some RNAi pathways. Also closely related species can significantly differ in their RNAi capacity. The evolutionary forces driving eukaryotic RNA silencing gene change are unclear and in fungi in general, RNA silencing gene evolution appears to be more complex than in any other type of eukaryotes. Due to this high variation, RNA silencing and its functionality must be demonstrated in each fungal species of interest.

9.11 Viral Origin of RNAi in Fungi

RNAi mechanisms are activated in eukaryotes during viral infections. The evolutionary origin of RNAi is believed to be linked to the protection of cells against invading nucleic acids. Also both plant DNA and RNA viruses encode for suppressor proteins which inactivate plant RNAi mechanisms supporting this hypothesis further (Moissiard and Voinnet 2006; Díaz-Pendón and Ding 2008). Similar mechanisms are active also during viral infection in animal cells (Gitlin and Andino 2003; Li et al. 2004; Berry et al. 2009).

The first indications of the virus defense origin of RNAi pathways in fungi come from a recent work on the chestnut blight fungus *Cryphonectria parasitica*. Virulence attenuating hypoviruses of the species, *Cryphonectria* hypovirus 1 (CHV1), encode a papain-like protease, p29, that shares similarities with the potyvirus-encoded suppressor of RNA silencing HC-Pro. Expression of this protein in the fungus was shown to suppress hpRNA (hairpin RNA) induced RNA silencing of the reporter GFP gene (Segers et al. 2006). The direct link between RNA silencing and virus defense was further demonstrated by disruption of two Dicer-like genes of *Cryphonectria*, *dcl1* and *dcl2*. While *dcl1* disruption did not cause an obvious phenotype, *dcl2* or *dcl1/dcl2* double mutants were highly susceptible to hypovirus CHV1-EP13 infection (Segers et al. 2007) The virus-derived small RNAs (vsRNAs) accumulation, their dependency on DCL-2 activity and virus induced up-regulation of Dicer expression was also recently reported in *Cryphonectria* (Zhang et al. 2008). The activation of fungal RNAi during viral infection or when triggered with long dsRNAs has been demonstrated in *N. crassa* and *A. nidulans* as well (Choudhary et al. 2007; Hammond et al. 2008a).

9.12 RNA Silencing in Different Fungi

Description of the RNAi pathways and their efficient artificial triggering in the filamentous ascomycetes *N. crassa*, *M. oryzae* (syn. *M. grisea*) and the fission yeast *S. pombe* (Cogoni et al. 1996; Raponi and Arndt 2003; Kadotani *et al.* 2003) has led to the increasing use of RNAi as a reverse genetic tool in fungi. Especially in filamentous fungi, where gene knock-out experiments are often very challenging, RNAi has offered a novel and fast way for modifying gene expression. In last years siRNA-dependent silencing has been demonstrated and used for genetic studies in several ascomycete species such as *M. oryzae*, *Colletotrichum lagenarium*, *Venturia inaequalis*, *Cladosporum fulvum*, *S. pombe*, *Aspergillus fumigatus*, *A. nidulans*, *A. flavus*, *A parasiticus*, *Fusarium graminearum*, *Histoplasma capsulatum*, *Sclerotinia sclerotiorum*, *Ophiostoma floccosum*, *O. piceae*, *Coniothyrium minitans*, *Penicillium expansum*, *Acremonium chrysogenum* (syn. *Cephalosporium acremonium*), *Trichoderma harzianum* and in *N. crassa* (Hamada and Spanu 1998; Kadotani et al. 2003; Raponi and Arndt 2003; Schramke and Allshire 2003; Fitzgerald et al. 2004; Goldoni et al. 2004; Mouyna et al. 2004; Sigova et al. 2004; McDonald et al. 2005; Nakayashiki et al. 2005; Rappleye et al. 2004; Cardoza et al. 2006; Tanguay et al. 2006; Erental et al. 2007; Gong et al. 2007; Janus et al. 2007; Schümann and Hertweck 2007). RNA silencing has been also reported in the zygomycete *Mucor circinelloides* (Nicolás et al. 2003) and the basidiomycetes *Schizophyllum commune*, *Cryptococcus neoformans*, *Coprinus cinereus*, *P. chrysosporium* and *Moniliophthora perniciosa* (Schuurs et al. 1997; Gorlach et al. 2002; Liu et al. 2002; Namekawa et al. 2005; de Jong et al. 2006; Wälti et al. 2006; Matityahu et al. 2008; Caribé Dos Santos et al. 2009).

9.13 How RNAi Is Experimentally Initiated in Fungi

In fungi RNAi is generally launched by similar techniques used by plant researchers. These are characterized by a stable cellular production of RNAi trigger which results in a stable and not transient RNAi phenotype. This is achieved by transforming the fungus with a DNA construct which via host cell's transcriptional machinery produces dsRNAs and later siRNAs homologous to the target mRNA predominantly via PTGS. All fungal studies this far have used gene coding sequences for targeting homologous or heterologous mRNAs. Whether the triggered reduction of the target mRNA has been purely due to PTGS or TGS mechanisms have also contributed to the silencing phenotypes, has however not been verified in the majority of these studies.

The RNAi triggering genetic DNA elements introduced in fungi can be integrative or auto-replicative and the dsRNA in cells can be produced from different RNA precursors. These can directly have a dsRNA structure or form dsRNA by binding to cellular sense-transcripts. Fungal as well as plant cells tolerate long dsRNA molecules unlike vertebrate and mammalian cells where longer than 30 nt dsRNA launches strong interferon response (Stark et al. 1998). Therefore, long dsRNA hairpins (hpRNAs), antisense RNAs and expression of sense and antisense RNAs from convergent promoters can be used for triggering RNA silencing in fungi. However, the most potent and widely used RNAi triggers in fungi are promoter expressed hpRNAs with homology arms of around 200–500 bp.

An efficient RNA silencing/transformation vector system adapted to fungi however did not exist until Nakayashiki et al. (2005) published pSilent-1. This cloning vector was developed for an easy PCR based cloning of intronic spacer hairpin RNAs (ihpRNAs) to be expressed in ascomycetes. pSilent-1 was also designed to function for direct plasmid transformation and carries a hygromycin resistance cassette for antibiotic selection of the transformants. This cloning/transformation vector has made RNA silencing studies in ascomycetes easy to perform and has been successfully used, with or without modifications, in *M. oryzae*, *Colletotrichum lagenarium*, *Bipolaris oryzae* and *Acremonium chrysogenum* (Nakayashiki et al. 2005; Janus et al. 2007; Moriwaki et al. 2007). Now the second generation of silencing vectors for direct hpRNA or convergent promoter RNA expression is being designed. Further modifications of the pSilent-1 vector, pSilent-Dual1 (Nguyen et al. 2008), pTroya (Shafran et al. 2008), and pFANTAi4 (Krajaejun et al. 2007) are adapted for high-throughput functional genomic studies in ascomycetes, the latter two also incorporating the GATEWAY technology (Invitrogen) in fungal silencing vectors. Also other hpRNA cloning vectors such as pFIRD1, designed for *A. niger* and utilizing recombination in vitro (Oliveira et al. 2008), and an inducible promoter system for RNAi in *A. nidulans* (Barton and Prade 2008), have been released.

RNAi is already established as quite a standard genetic tool for working with ascomycetes (Nakayashiki and Nguyen 2008). While the number RNA silencing publications in ascomycetes is rapidly increasing, studies conducted in basidiomycetes are still few. This low number of RNA silencing publications partly reflects the complications with DNA-transformation techniques but most of all the lack of molecular genetic RNAi tools adapted for this group of fungi.

In *C. elegans* an exogenous exposure of the organism to small dsRNA molecules launches RNAi. A similar approach of a direct RNA triggers has not traditionally been used for fungal RNAi. Nevertheless, in a recent work on *A. nidulans* the cocultivation of the fungus with siRNAs targeted to the fungal key polyamine biosynthesis gene ornithine decarboxylase (ODC) was shown to launch RNAi. As a result target mRNA and cellular polyamine concentrations levels in the fungus were reduced (Khatri and Rajam 2007). This has been the first demonstration of uptake of siRNA by germlines from the culture medium and RNAi triggering in fungi. Also direct incubation of *A. niger* protoplast with siRNAs has been shown to launch transient RNAi (Barnes et al. 2008). These discoveries open the possibility of using siRNAs for new applications such as antifungal drug development. Also a direct electroporation of long dsRNAs have been recently shown to trigger long lasting RNAi response in the basidiomycete *Moniliophthora perniciosa* (Caribé Dos Santos et al. 2009). These reports strongly propose that, equal to animal research, dsRNAs as directly introduced silencing triggers could be used in fungal research as well.

Different DNA and RNA triggers used for initiating PTGS in fungi and the proposed cellular RNAi pathways involved are summarized in Fig. 9.2. Also, the possible epigenetic effects linked to silencing triggering are illustrated.

9.14 Simultaneous Silencing of Genes with a Single Trigger in Fungi

A singular silencing trigger can lead to simultaneous silencing of several targets. Genes belonging to multigene families sharing homologous sequences can theoretically be targeted with a single trigger. Also chimeric constructs carrying target endogene and reporter gene sequences can lead to simultaneous silencing of both. Such approach can be used for facilitating the screening of strongly silenced strains as silencing of the marker and the target endogene generally coincides. An efficient simultaneous silencing of two genes has been reported by a chimeric hairpin construct in *Venturia inaequalis*. Transformation with a hairpin carrying both reporter *egfp* and endogenous gene sequences resulted in reduced expression of both genes (Fitzgerald et al. 2004). A similar approach has been used for cosilencing a dsRED-maker gene in *Acremonium chrysogenum* (Janus et al. 2007), *ade2* in *Cryptococcus neoformans* (Liu et al. 2002) and *egfp* in *M. oryzae* (Nguyen et al. 2008).

RNA silencing of two fungal endogenes by a single hairpin has been demonstrated in *Coprinus cinereus*. Simultaneous silencing of genes encoding isogalectins (*cgl1* and *cgl2*) sharing only 87% identity was produced by dsRNA carrying a *cgl2* sequence (Wälti et al. 2006). These results indicate that RNAi serves also for

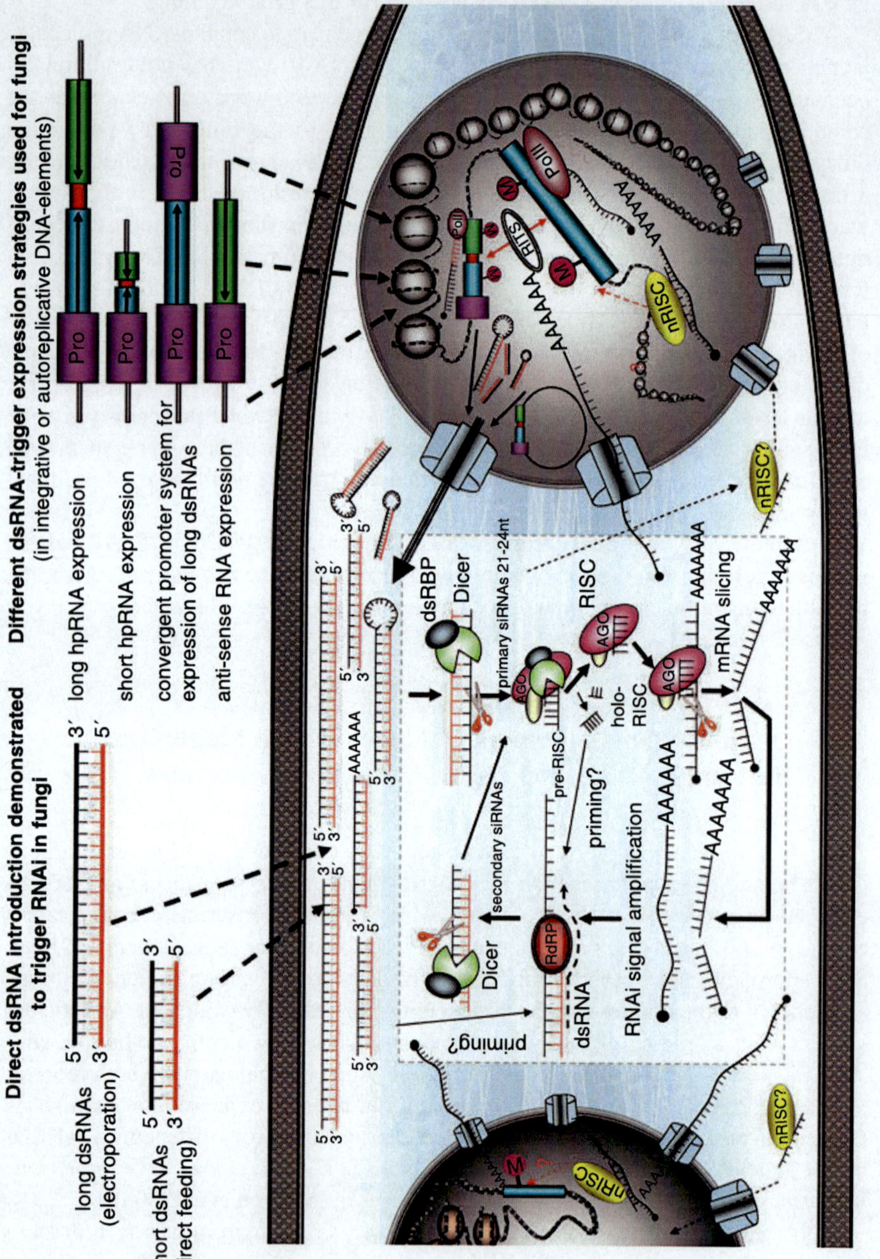
Direct dsRNA introduction demonstrated to trigger RNAi in fungi
long dsRNAs (electroporation)
short dsRNAs (direct feeding)
Different dsRNA-trigger expression strategies used for fungi (in integrative or autoreplicative DNA-elements)
long hpRNA expression
short hpRNA expression
convergent promoter system for expression of long dsRNAs
anti-sense RNA expression
Pro
dsRBP
Dicer
primary siRNAs 21-24nt
RISC
AGO
holo-RISC
pre-RISC
priming?
mRNA slicing
RNAi signal amplification
secondary siRNAs
RdRP
dsRNA
RITS
PolII
nRISC
nRISC?
AAAAAA

silencing gene families in fungi. The minimum sequence homology needed for efficient cosilencing is not yet determined for fungi and may vary between species. While simultaneous silencing can be a benefit, it is also the biggest drawback of the RNA silencing technology. Unwanted off-target effects originating from short sequence homologies are possible especially when long dsRNAs are used as primary RNAi triggers.

These off-target effects were recently studied by Nguyen et al. in *M. oryzae*. Genes sharing sequence similarity with an *E*-value of only about e^{-3} or e^{-4} were sometimes silenced at similar levels and only 13 nt perfect homology between genes was enough for simultaneous silencing of an unwanted target (Nguyen et al. 2008) Also in HeLa cells only a 7 nt sequence complementary between the unwanted target and a siRNA or a short-hairpin RNA trigger has been reported to launch off-site effects (Jackson et al. 2006). Special attention should thus be paid in future for designing RNAi triggers. It seems that RNA silencing is not as sequence specific as it was previously believed and probable off-target effects cannot be completely avoided. These can however be reduced by using more discriminatory shorter RNA triggers. In *Coprinopsis cinerea* (syn. *Coprinus cinereus*) as short as 19 bp dsRNA trigger can initiate a specific RNA silencing response (Costa et al. 2008).

As fungi possess RdRPs the silencing off-target effects are not necessarily linked only to the primary silencing trigger. The degree of transitive silencing and possible off-target effects of secondary cascade siRNAs has however not been profoundly

Fig. 9.2 The most common triggers for artificially initiated RNAi in fungi are variable integrative or auto-replicative promoter-directed DNA elements. Their introduction to cells results in direct dsRNA production by transcription (hpRNAs), dsRNA formation by base paring of produced sense and antisense transcripts (convergent promoter expression), or their form dsRNAs by paring with target mRNAs (antisense expression). Recently also direct dsRNA introduction by feeding or electroporation has been demonstrated to launch RNAi in fungi. dsRNAs originating from these different introduced triggers are recognized in cytosol by the fungal RNAi pathway leading to PTGS. This involves dsRNA Dicer cutting, formation of 21–24 nt primary siRNAs and RISC assembly. Dicer binding to dsRNA is assisted by dsRNA binding proteins (dsRBP) and RISC assembly initiates with pre-RISC formation, leads to discard of one of the siRNA strands and results in formation of mature holo-RISC. RISC targets mRNAs homologous to siRNAs. A perfect base pairing leads to slicing of target by Argonaute (AGO) and reduction of mRNA concentration (gene knock-down). However, imperfect mRNA–RISC pairing which leads to translational arrest or mRNA decay in plants and animals might also occur in fungi even though it has not yet been demonstrated. Cut mRNAs (or intact mRNAs) are potential targets for signal amplification by RdRP. These new dsRNAs can be detected by Dicer and processed into secondary siRNAs. The RNAi signal amplification step is not well characterized in filamentous fungi. Short dsRNAs or siRNAs may act as primers at this step. Neither is the extent of this transitive silencing effect clear in the fungal kingdom in general. RNAi is linked to epigenetic marks in different eukaryotes and also in fungi. These nuclear effects can potentially affect the silencing triggering locus and/or the target endogene by DNA methylations and histone modifications via mechanisms such as RNA-induced transcriptional silencing complex RITS. If these epigenetic modifications can also be internuclear, reaching untransformed nuclei in multinucleate fungal cells is not clear. A cytosol-loaded nuclear siRISC (nRISC), detected in animals and apparently also present in plants, could mediate such sequence specific effects between different nuclei through fungal cytoplasm studied yet in this group of eukaryotes. Transitive silencing has been demonstrated in the zygomycete *Mucor circinelloides* (Nicolás et al. 2003) but its existence in filamentous ascomycetes is not clear. At least no silencing spreading has been detected in *A. nidulans* toward the 3′ end on target mRNA transcripts (Barton and Prade 2008). The fungal RNAi protein machineries are however highly variable between different fungal taxa and the biological role of especially the RdRP-gene family expansion in filamentous homobasidiomycetes (Nakayashiki and Kadotani 2006) is intriguing. The possible role of these proteins in spreading the silencing effect should be investigated in future. Even though miRNAs have not been detected in fungi the possibility that secondary siRNAs could act via miRNA-type translational arrest on homologous mRNAs is not an excluded possibility.

9.15 Use of RNAi as a Reverse Genetic Tool in Fungi

RNA silencing does not completely abolish the target gene expression but its reduction has been shown to result in phenotypes equivalent or close to null-mutants. The percentage of functional phenotypes is generally much higher than with gene knock-out making gene function studies remarkably faster to perform. The power of this approach on modifying endogene expression has already been demonstrated in several studies where strongly silenced fungal strains have clarified the role of target genes in processes such as biosynthetic pathways, meiosis, asexual sporulation or fungal virulence control (Rappleye et al. 2004; McDonald et al. 2005; Namekawa et al. 2005; Cardoza et al. 2006; Bohse and Woods 2007; Nguyen et al. 2008; Nicolás et al. 2008; Cooper and Woods 2009; Panepinto et al. 2009). RNA silencing can be also used for confirming gene knock-out phenotypes obtained by other means (Gong et al. 2007). The true RNAi era on fungal research has started with the release of the pSilent-1 cloning vector (Nakayashiki et al. 2005) and therefore numerous reports on the RNAi technology in resolving biological fungal functions are expected to be released in the following years.

Although it is already successfully used in reverse genetics of several species, the basic cellular mechanisms behind RNAi in filamentous fungi are still poorly understood and studied. The detected high species level variability in RNAi protein machinery and the differences especially between ascomycetes and basidiomycetes suggest that the true flexibility and limitations of this gene knock-down technique must be evaluated on a species basis.

9.16 Gene Silencing in Ectomycorrhizal Fungi

Laccaria bicolor is a homobasidiomycete and ectomycorrhizal (ECM) fungus which forms symbiosis with several boreal and temperate forest trees such as birch, pine and poplar. *Laccaria*, the first symbiotic fungus with its genome sequenced (Martin et al. 2008), is also susceptible to genetic modification via *Agrobacterium*-mediated transformation (AMT) (Kemppainen et al. 2005, 2008; this book, Chap. 6). The availability of the full genomic sequence and the possibility to genetically modify this fungus has rapidly turned *Laccaria* into a true ECM model species. However, gene knock-outs have turned out to be difficult to obtain in this fungus. As the symbiotic phase of *Laccaria* is dikaryotic and interrupting one of the gene copies does not necessarily produce strong haploinsufficiency phenotypes (Kemppainen et al. 2008) RNA silencing represented an alternative and highly attractive way for altering expression of *Laccaria* genes involved in ECM. Gene knock-down had not been used before for studying plant–fungus symbiotic interactions. Neither had the functionality of RNA silencing pathway been demonstrated in ECM fungi.

A preliminary study of *Laccaria* full genome sequence revealed that it encodes for a minimum set of genes with predicted protein products shown to be essential for RNA silencing. Two putative Dicer-like, and several Argonaute and RNA-dependent RNA polymerase (RdRP) proteins were detected (Kemppainen et al. 2009). Especially high number of putative RdRPs (6) and Argonautes (6) in *Laccaria* reflects the homobasidiomycete specific gene expansion of the RNA silencing machinery (Nakayashiki and Kadotani 2006). The bioinformatical evidence of the existence of *Laccaria* RNAi machinery suggested the presence of a functional siRNA-dependent silencing pathway in this fungus. This was assayed with nitrate reductase as a test endogene target.

Laccaria nitrate reductase encoding gene (*Lbnr*) can be considered an optimal test target for RNA silencing. This is a single genomic gene and not a member of a multigene-family, thus reducing possible silencing off-target effects. Silencing of *Lbnr* would also allow a fast growth phenotype screening on nitrate medium. Moreover, as nitrate metabolism is generally repressed in fungi and other eukaryotes by primary N sources, growth of *Lbnr*-silenced strains would be expected to be unaffected on ammonium. As N is one of the main mobile nutrients in ECM, fungal N metabolism genes are therefore of special interest.

As with *Laccaria* the vast majority of ectomycorrhizal fungi studied this far can support growth on nitrate indicating that this metabolic trait is evolutionarily conserved (Nygren et al. 2008). The true role of nitrate as N source for ectomycorrhizal fungi in forest soil is however under current debate. Even though traditionally considered of minor importance for total N pool in forest soil other studies are proposing that this inorganic N form may have marked temporal and spatial concentration gradients and be more available for mycorrhizal fungi than previously believed (Stark and Hart 1997; Laverman et al. 2000). Moreover, the activity and differential expression of fungal and plant nitrate utilization genes has been connected to functional ectomycorrhizal symbiosis and enhanced host plant N nutrition. The nitrate reductases of plant hosts become down-regulated both in endo- and ectomycorrhizal roots and this repression has been directly linked to the activity of fungal N metabolism in ectomycorrhiza (Kaldorf et al. 1998; Guescini et al. 2003; Bailly et al. 2007). This proposes that the mycorrhized hosts can be highly dependent on the N uptake and utilization pathway of the fungal partner under nitrate feeding. The fungal response to symbiosis can however vary between different ectomycorrhizal species. While in the basidiomycete *H. cylindrosporum* nitrate reductase expression is lower in mycorrhizal structures than in extraradical mycelium, in ectomycorrhiza formed by the ascomycete *T. borchii* fungal nitrate utilization genes are induced in symbiotic structures (Bailly et al. 2007; Guescini et al. 2003, 2007, 2009). Despite the observed species specific variation, increasing data suggest that fungal nitrate metabolism genes can play an important role in the establishment and/or the function of ectomycorrhizal interactions in nature. Therefore, knocking-down of *Lbnr* could also offer information on how this mutualistic association is regulated.

The functionality of the *Laccaria* RNA silencing pathway was tested by transforming the dikaryotic fungus with a promoter-directed inverted repeated sequence of a partial coding sequence of *Lbnr* (Kemppainen et al. 2009). RNA silencing was accomplished in *L. bicolor* by AMT (Kemppainen et al. 2005, 2008; this book, Chap. 6). Promoter-directed expression of dsRNA resulted in fungal transgenic strains strongly affected in growth with nitrate as N source (Kemppainen et al. 2009). The phenotype correlated with a clear reduction of the target gene mRNA level and this effect was not caused by homologous recombination of the T-DNA in the nitrate reductase locus. Transformation with the hairpin sequence resulted in specific but moderate CpG methylation of both the silencing triggering transgene construct and the nitrate reductase encoding gene demonstrating that epigenetic modifications accompany RNA silencing in *Laccaria* like in other eukaryotes. The methylation in the target gene was restricted to the silencing trigger sequence and did not represent the entire genomic DNA in the dikaryon suggesting that the epigenetic changes accompanying RNA silencing affected only the transformed nucleus. This strongly proposes that the *Lbnr* silencing phenotypes were predominantly a result of siRNA-dependent cytosolic PTGS mechanisms (Kemppainen et al. 2009). Moreover, the silencing strength variation (SSV) between different transformed strains did not correlate with the transgene copy number but it seems to be linked to the nature of genomic integration sites of transgenes. Integrations in euchromatin zones, especially within the coding sequence of active genes results in strongly silenced fungal strains. These sites most probably allow the maximal dsRNA-trigger production (Kemppainen et al. 2009; Kemppainen and Pardo 2010).

Mycorrhization experiments of *Populus* with strongly *Lbnr*-silenced fungal strains revealed a systematic inhibition of symbiosis with nitrate as N source compared to the wild type. This inhibition of mycorrhization was reversed by an organic N source efficiently utilized by the fungus (i.e. L-asparagine). These observations strongly suggest that the plant is able to sense the nutritional status of a potential fungal symbiont avoiding the establishment of an unsatisfactory interaction. A control mechanism conducted by the plant would inhibit symbiosis when the metabolic profile of the fungal partner is not proper and mutual benefit from the symbiotic structure cannot be assured. These results are the first direct genetic proof showing that the alteration of expression of a fungal gene impairs mycorrhization and the first demonstration of the RNA silencing pathway in mycorrhizal fungi. They also highlight the great potential of RNAi technology for studying symbiotic interactions (Kemppainen et al. 2009).

However, the efficient use of RNA silencing requires a friendly silencing/transformation vector. Double-stranded hpRNA expression from stable integrated transgenes or from auto-replicative elements has been shown to be a widely efficient trigger in inducing RNA silencing in fungi. Vectors for plant silencing such as pHANNIBAL and pHELLSGATE (Wesley et al. 2001), and pSTARLING and pOpOFF (CSIRO) have been available for almost a decade and several silencing vectors for mammalian cells have been reported in the last few years (Wadhwa et al. 2004; Gou et al. 2007). RNA silencing vectors for fungi however did not exist before the launch of pSilent-1 (Nakayashiki et al. 2005) which has led to the successful use of RNA silencing as a genetic tool in filamentous ascomycetes. While an increasing number of silencing vectors for ascomycetes has been released during the last few years (Krajaejun et al. 2007; Nguyen et al. 2008; Shafran et al. 2008; Oliveira et al. 2008; Barton and Prade 2008) no silencing vectors adapted to basidiomycetes have been available.

The successful RNA silencing in *Laccaria* generated an urgent need for an easy-to-use RNA silencing/transformation vector compatible with AMT. Unfortunately, many commonly used ascomycete promoters are weakly recognized in filamentous basidiomycetes, especially when introduced predominantly as a single copy via AMT. This makes the use of ascomycete-adapted vectors difficult in basidiomycetes. Neither are these transformation vectors compatible with AMT. To fulfill this current gap of RNAi tools optimized for homobasidiomycetes we constructed the pSILBAγ silencing vector for efficient RNA silencing triggering in *L. bicolor* (Kemppainen and Pardo 2009). This cloning vector carries the *Agaricus bisporus gpdII*-promoter, two multiple cloning sites separated by a *L. bicolor* nitrate reductase intron and the *Aspergillus nidulans trpC* terminator. The pSILBAγ allows an easy oriented two-step PCR-cloning of hairpin sequences to be expressed in basidiomycetes. With one further cloning step into pHg, a pCAMBIA1300-based binary vector carrying a hygromycin B resistance cassette, the pHg/pSILBAγ plasmid can be used for AMT (Kemppainen and Pardo 2010). Besides, for RNAi triggering, the pHg/pSILBAγ can also be used for gene expression studies. Due to the widely recognized heterologous *A. bisporus gpdII*-promoter, both in the cloning and the transformants selection cassette, this vector system should also result as functional in other hygromycin B sensitive homobasidiomycete species.

We have shown that the dsRNA-triggered silencing pathway is functional in the model ectomycorrhizal fungus *L. bicolor* (Kemppainen et al. 2009; Kemppainen and Pardo 2010). This together with the availability of a easy-to-use basidiomycete-adapted silencing/AMT-vector, opens the possibility for efficient use of gene knock-down in genetic studies of *Laccaria*. More importantly, RNA silencing can result in dikaryotic fungal strains affected in their mycorrhization capacity. This sets the conditions for RNAi studies in ectomycorrhizal symbiosis, the research field that up today has been hindered by the lack of functional reverse genetic tools.

9.17 Conclusion

The symbiotic phase in the life-cycle of ectomycorrhizal basidiomycetes is the dikaryon. Thus, studies on symbiotic fungal gene function would require the inactivation of both gene copies in the dikaryotic mycelium. Due to the extremely low homologous recombination rate in this group of fungi, traditional gene knock-out experiments and particularly double gene interruption/replacement are rather difficult to achieve. However, an alternative approach, RNA silencing, has shed some light on the ectomycorrhizal research field.

In many organisms, including fungi, RNA silencing can be artificially triggered to target and degrade gene transcripts of interests, resulting in the so called gene knock-down. Most importantly, RNA silencing can act at the cytosolic level affecting mRNAs originating from several gene copies and different nuclei and it can thus offer an alternative and fast way for altering gene expression in dikaryotic, diploid, polyploid, and coenocytic organisms.

Laccaria bicolor, the first symbiotic fungus with its genome sequenced, has rapidly turned into a model fungus in ectomycorrhizal research. *Laccaria* possesses a complete set of genes known to be needed for RNA silencing in eukaryotic cells. We have demonstrated that RNA silencing is functional in *L. bicolor* and that it can be triggered via AMT. Moreover, targeted gene knock-down in dikaryotic mycelium can result in functional phenotypes altered in the symbiotic capacity confirming that RNA silencing is a powerful way to study symbiosis-regulated genes. These findings have now initiated the RNA silencing era in mycorrhizal research, a field that has been hindered by the lack of proper genetic tools.

References

Axtell MJ, Jan C, Rajagopalan R, Bartel DP (2006) A two-hit trigger for siRNA biogenesis in plants. Cell 127:565–577

Bailly J, Debaud J-C, Vegner M-C, Plassard C, Chalot M, Marmeisse R, Fraissinet-Tachet L (2007) How does a symbiotic fungus modulate expression of its host-plant nitrite reductase? New Phytol 175:155–165

Barnes SE, Alcocer MJC, Archer DB (2008) siRNA as a molecular tool for use in *Aspergillus niger*. Biotechnol Lett 30:885–890

Barton LM, Prade RA (2008) Inducible RNA interference of *brl*Aβ in *Aspergillus nidulans*. Eukaryot Cell 7:2004–2007

Bernstein E, Caudy AA, Hammond SH, Hannon GJ (2001) Role for a bidentate ribonuclease in the initiation step of RNA interference. Nature 409:363–366

Berry B, Deddouche S, Kirschner D, Imler JL, Antoniewski C (2009) Viral suppressors of RNA silencing hinder exogenous and endogenous small RNA pathways in *Drosophila*. PloS One 4: e5866

Bleys A, Van Houdt H, Depicker A (2006a) Down-regulation of endogenes mediated by a transitive silencing signal. RNA 12:1633–1639

Bleys A, Vermeersch L, Van Houdt H, Depicker A (2006b) The frequency and efficiency of endogene suppression by transitive silencing signals is influenced by the length of sequence homology. Plant Physiol 142:788–796
Bohse ML, Woods JP (2007) RNA interference-mediated silencing of the *YPS3* gene of *Hisoplasma capsulatum* reveals virulence defects. Infect Immun 75:2811–2817
Boutros M, Kiger A, Armknecht S, Kerr K, Hild M, Koch B, Haas S, Paro R, Perrimon N (2004) Genome-wide RNAi analysis of growth and viability in *Drosophila* cells. Science 303: 832–835
Bühler M (2009) RNA turnover and chromatin-dependent gene silencing. Chromosoma 118: 141–151
Cardoza RE, Vizcaino JA, Hermosa MR, Sousa S, Gonzalez FJ, Llobell A, Monte E, Gurierrez S (2006) Cloning and characterization of the *erg1* of *Trichoderma harzianum*: Effect of the *erg1* silencing on ergosterol biosynthesis and resistance to terbinafine. Fungal Genet Biol 43:164–178
Caribé dos Santos AC, Sena JA, Santos SC, Dias CV, Pirovani CP, Pungartnik C, Valle RR, Cascardo JC, Vincentz M (2009) dsRNA-induced gene silencing in *Moniliophthora perniciosa,* the causal agent of witches' broom disease of cacao. Fungal Genet Biol 46:825–836
Carmell MA, Hannon GJ (2004) RNase III enzymes and the initiation of gene silencing. Nat Struct Mol Biol 11:214–218
Carthew RW, Sontheimer EJ (2009) Origins and mechanisms of miRNAs and siRNAs. Cell 136:642–655
Castanotto D, Rossi JJ (2009) The promises and pitfalls of RNA interference-based therapeutics. Nature 457:426–433
Cecere G, Cogoni C (2009) Quelling targets the rDNA locus and fuctions in rDNA copy number control. BMC Microbiol 9:44
Chen J, Shi X, Padmanabhan R, Wang Q, Wu Z, Stevenson SC, Hild M, Garza D, Li H (2008) Identification of novel modulators of mitochondrial function by genome-wide RNAi sreen in *Drosophila melanogaster*. Genome Res 18:123–136
Choudhary S, Lee H-C, Maiti M, He Q, Cheng P, Liu Q, Liu Y (2007) A double-stranded-RNA response program important for RNA interference efficiency. Mol Cell Biol 27:3995–4005
Cogoni C, Macino G (1999) Homology-dependent gene silencing in plants and fungi: a number of variations on the same theme. Curr Opin Microbiol 2:657–662
Cogoni C, Irelan JT, Schumacher M, Schmidhauser TJ, Selker EU, Macino G (1996) Transgene silencing of the *al-1* gene in vegetative cells of *Neurospora* is mediated by a cytoplasmic effector and does not depend on DNA-DNA interactions or DNA methylation. EMBO J 15:3153–3163
Cooper KG, Woods JP (2009) Secreted dipeptidyl peptidase IV activity in the dimorphic fungal pathogen *Histoplasma capsulatum*. Infect Immun 77:2447–2454
Costa AM, Mills PR, Bailey AM, Foste GD, Challen MP (2008) Oligonucleotide sequences forming short self-complimentary hairpins can expedite the down-regulation of *Coprinopsis cinerea* genes. J Microbiol Methods 75:205–208
Cronin SJF, Nehme NT, Limmer S, Liegeois S, Pospisilik JA, Schramek D, Leibbrandt A, de Matos SR, Gruber S, Puc U, Ebersberger I, Zoranovic T, Neely GG, von Haeseler A, Ferrandon D, Penninger JM (2009) Genome-wide RNAi screen identifies genes involved in intestinal pathogenic bacterial infection. Science 325:340–343
Dalmay T, Hamilton A, Rudd S, Angell S, Baulcombe DC (2000) An RNA-dependent RNA polymerase gene in *Arabidopsis* is required for post-transcriptional gene silencing mediated by transgene but not by a virus. Cell 101:543–553
de Jong JF, Deelstra HJ, Wösten HAB, Lugones LG (2006) RNA-mediated gene silencing in monokaryons and dikaryons of *Schizophyllum commune*. Appl Environ Mircrobiol 72:1267–1269
Díaz-Pendón JA, Ding SW (2008) Direct and indirect roles of viral suppressors of RNA silencing in pathogenesis. Annu Rev Phytopathol 46:303–326
Djupedal I, Ekwall K (2009) Epigenetics: heterchromatin meets RNAi. Cell Res 19:282–295

Erental A, Harel A, Yarden O (2007) Type 2A phosphoprotein phosphatase is required for asexual development and pathogenesis of *Sclerotinia sclerotiorum*. Mol Plant Microbe Interact 20:944–954

Fire A, Xu S, Montgomery MK, Kostas SA, Driver SE, Mello CC (1998) Potent and specific genetic interference by double-stranded RNA in *Caenorhabditidis elegans*. Nature 391:806–811

Fitzgerald A, van Kan JAL, Plummer KM (2004) Simultaneous silencing of multiple genes in the apple scab fungus *Venturia inaequalis* by expression of RNA with chimeric inverted repeats. Fungal Genet Biol 41:963–971

Fulci B, Macino G (2007) Quelling: post-transcriptional gene silencing guided by small RNAs in *Neurospora crassa*. Curr Opin Microbiol 10:199–203

Girard A, Sachidanandam R, Hannon GJ, Carmell MA (2006) A germline-specific class of small RNAs binds mammalian Piwi proteins. Nature 442:199–202

Gitlin L, Andino R (2003) Nuclei acid-based immune system: the antiviral potential of mammalian RNA silencing. J Virol 77:7159–7165

Gobeil S, Zhu X, Doillon CJ, Green MR (2008) A genome-wide shRNA screen identifies GAS1 as a novel melanoma metastasis suppressor gene. Genes Dev 22:2932–2940

Golden DE, Gerbasi VR, Sontheimer EJ (2008) An inside job for siRNAs. Mol Cell 31:309–312

Goldoni M, Azzalin G, Macino G, Cogoni C (2004) Efficient gene siliencing by expression of double stranded RNA in *Neurospora crassa*. Fungal Genet Biol 41:1016–1024

Gong X, Fu Y, Jiang D, Li G, Yi X, Peng Y (2007) L-arginine is essential for conidiation in the filamentous fungus *Coniothyrium minitans*. Fungal Genet Biol 44:1368–1379

Gonzalez S, Pisano DG, Serrano M (2008) Mechanistic principles of chromatin remodeling guided by siRNAs and miRNAs. Cell Cycle 7:2601–2608

Gorlach JM, McDade HC, Perfect JR, Cox GM (2002) Antisense repression in *Cryptococcus neoformans* as a laboratory tool and potential antifungal strategy. Microbiology 128:213–219

Gou D, Weng T, Wang Y, Wang Z, Zhang H, Gao L, Chen Z, Wang P, Liu L (2007) A novel approach for the construction of multiple shRNA expression vectors. J Gene Med 9:751–763

Guang S, Bochner AF, Pavelec DM, Burkhart KB, Harding S, Lachowiec J, Kennedy S (2008) An argonaute transports siRNAs from the cytoplasm to the nucleus. Science 321:537–541

Guescini M, Pierloeni R, Palma F, Zeppa S, Vallorani L, Potenza L, Sacconi G, Giomaro G, Stocchi V (2003) Characterization of the *Tuber borchii* nitrate reductase gene and its role in ectomycorrhizae. Mol Genet Genomics 269:807–816

Guescini M, Zeppa S, Pierleoni R, Sisti D, Stocchi L, Stocchi V (2007) The expression profile of the *Tuber borchii* nitrite reductase suggests its positive contribution to the host plant nitrogen nutrition. Curr Genet 51:31–41

Guescini M, Stocchi L, Sisti D, Zeppa S, Polidori E, Ceccaroli P, Satarelli R, Stocchi V (2009) Characterization and mRNA expression profile of the *TbNre1* gene of the ectomycorrhizal fungus *Tuber borchii*. Curr Genet 55:59–68

Hamada W, Spanu PD (1998) Co-suppression of the hydrophobin gene *HCf-1* is correlated with antisense RNA biosynthesis in *Cladosprorum fulvum*. Mol Gen Genet 259:630–638

Hammond SM, Boettcher S, Caudy AA, Kobayashi R, Hannon GJ (2001) Argonaute2, a link between genetic and biochemical analyses of RNAi. Science 293:1146–1150

Hammond TM, Andrewski MD, Roossinck MJ, Keller NP (2008a) *Aspergillus* mycoviruses are targets and suppressors of RNA silencing. Eukaryot Cell 7:350–357

Hammond TM, Bok JW, Andrewski MD, Reyes-Domínguez Y, Scazzocchio C, Keller NP (2008b) RNA silencing gene truncation in the filamentous fungus *Aspergillus nidulans*. Eukaryot Cell 7:339–349

Hutvagner G, Simard MJ (2008) Argonaute proteins: key players in RNA silencing. Nat Rev Mol Cell Biol 9:22–32

Jackson AL, Burchard J, Schelter J, Chau BN, Cleary M, Lim L, Linsley PS (2006) Widespread siRNA "off-target" transcript silencing mediated by seed region sequence complementarity. RNA 12:1179–1187

Jagannath A, Wood MJA (2009) Localization of double-stranded small interfering RNA to cytoplasmic processing bodies Is Ago2 dependent and results in up-regulation of GW182 and Argonaute-2. Mol Biol Cell 20:521–529

Janus D, Hoff B, Hofmann E, Kück U (2007) An efficient fungal RNA-silencing system using the DsRed reporter gene. Appl Environ Microbiol 73:962–970

Jöchl C, Rederstorff M, Hertel J, Standler PF, Hofacker IL, Schrettl M, Haas H, Hüttenhofer A (2008) Small ncRNA transcriptote analisis from *Aspergillus fumigatus* suggests a novel mechanism for regulation of protein syntesis. Nucleic Acids Res 36:2677–2689

Jorgensen R (1990) Altered gene expression in plants due to trans interactions between homologous genes. Trends Biotechnol 8:340–344

Kadotani N, Nakayashiki H, Tosa Y, Mayama S (2003) RNA silencing in the phytophatogenic fungus *Magnaporthe oryzae*. Mol Plant Microbe Interact 16:769–776

Kalantidis K, Schumacher HT, Alexiadis T, Helm JM (2008) RNA silencing movement in plants. Biol Cell 100:13–26

Kaldorf M, Schmelzer E, Bothe H (1998) Expression of maize and fungal nitrate reductase genes in arbuscular mycorrhiza. Mol Plant Microbe Interact 11:439–448

Kamath RS, Fraser AG, Dong Y, Polin G, Durbin R, Gotta M, Kanapin A, Le Bot N, Moreno S, Sohrmann M, Welchman DP, Zipperlen P, Ahringer J (2003) Systemic functional analysis of the *Caenorhabditis elegans* genome using RNAi. Nature 421:231–237

Kemppainen M, Pardo AG (2010) pHg/pSILBAγ vector system for efficient gene silencing in homobasidiomycetes: optimization of ihpRNA-triggering in the mycorrhizal fungus *Laccaria bicolor*. Microb Biotechnol 3:178–200

Kemppainen M, Circosta A, Tagu D, Martin F, Pardo AG (2005) *Agrobacterium*-mediated transformation of the ectomycorrhizal symbiont *Laccaria bicolor* S238N. Mycorrhiza 16: 19–22

Kemppainen M, Duplessin S, Martin F, Pardo AG (2008) T-DNA insertion, plasmid rescue and integration analysis in the model mycorrhizal fungus *Laccaria bicolor*. Microb Biotechnol 1:259–269

Kemppainen M, Duplessin S, Martin F, Pardo AG (2009) RNA silencing in the model mycorrhizal fungus *Laccaria bicolor*: gene knock-down of nitrate reductase results in inhibition of symbiosis with *Populus*. Environ Microbiol 11:1878–1896

Khatri M, Rajam MV (2007) Targeting polyamines of *Aspergillus nidulans* by siRNA specific to fungal ornithine decarboxylase gene. Med Mycol 45:211–220

Kim VN (2005) MicroRNA biogenesis: coordinated cropping and dicing. Nat Rev Mol Cell Biol 6:376–385

Kim DH, Sætrom P, Snøve O, Rossi JJ (2008) MicroRNA-directed transcriptional gene silencing in mammalian cells. PNAS 105:16230–16235

Krajaejun T, Gauthier GM, Rappleye GA, Sullivan TD, Klein BS (2007) Development and application of a green fluorescent protein sentinel system for identification of RNA interference in *Balstomyces dermatitidis* illuminates the role of septin in morphogenesis and sporulation. Eukaryot Cell 6:1299–1309

Krishnan MN, Ng A, Sukumaran B, Gilfoy FD, Uchil PD, Sultana H, Brass AL, Adametz R, Tsui M, Qian F, Montgomery RR, Lev S, Mason PW, Koski RA, Elledge SJ, Xavier RJ, Agaisse H, Fikrig E (2008) RNA interference screen for human genes associated with West Nile virus infection. Nature 455:242–245

Laurie JD, Linning R, Bakkeren G (2008) Hallmarks of RNA silencing are found in the smut fungus *Ustilago hordei* but not in its close relative *Ustilago maydis*. Curr Genet 53:49–58

Laverman AM, Zoomer HR, van Verseveld HW, Verhoef HA (2000) Temporal and spatial variation of nitrogen transformations in a coniferous forest soil. Soil Biol Biochem 32: 1661–1670

Lewis BP, Burge CB, Bartel DP (2005) Conserved seed pairing, often flanked by adenosines, indicates that thousands of human genes are microRNA targets. Cell 120:15–20

Li WX, Li H, Lu R, Li F, Dus M, Atkinson P, Brydon EW, Johnson KL, García-Sastre A, Ball LA, Palese P, Ding SW (2004) Interferon antagonist proteins of influenza and vaccine virus are suppressors of RNA silencing. Proc Natl Acad Sci USA 101:1350–1355

Lippman Z, Martienssen R (2004) The role of RNA interference in heterochromatic silencing. Nature 431:364–370

Liu H, Cottrell T, Pierini LM, Goldman WE, Doering T (2002) RNA interference in the pathogenic fungus *Cryptococcus neoformans*. Genetics 160:463–470

Martin F, Aerts A, Ahren D, Brun A, Danchin EGJ, Duchaussoy F, Gibon J, Kohler A, Lindquist E, Pereda V, Salamov A, Shapiro HJ, Wuyts J, Blaudez D, Buée M, Brokstein P, Canbäck B, Cohen D, Courty PE, Coutinho PM, Delaruelle C, Detter JC, Deveau A, DiFazio S, Duplessis S, Fraissinet-Tachet L, Lucic E, Frey-Klett P, Fourrey C, Feussner I, Gay G, Grimwood J, Hoegger PJ, Jain P, Kilaru S, Labbé J, Lin YC, Legué V, Le Tacon F, Marmeisse R, Melayah D, Montanini B, Muratet M, Nehls U, Niculita-Hirzel H, Oudot-Le Secq MP, Peter M, Quesneville H, Rajashekar B, Reich M, Rouhier N, Schmutz J, Yin T, Chalot M, Henrissat B, Kües U, Lucas S, Van de Peer Y, Podila GK, Polle A, Pukkila PJ, Richardson PM, Rouzé P, Sanders IR, Stajich JE, Tunlid A, Tuskan G, Grigoriev IV (2008) The genome of *Laccaria bicolor* provides insights into mycorrhizal symbiosis. Nature 452:88–92

Matityahu A, Hadar Y, Dosoretz CG, Belinky PA (2008) Gene silencing by RNA interference in the white-rot fungus *Phanerochaete chrysosporium*. Appl Environ Microbiol 74:5359–5365

McDonald T, Brown D, Keller NP, Hammond TM (2005) RNA silencing of mycotoxin production in *Aspergillus* and *Fusarium* species. Mol Plant Microbe Interact 18:539–545

Mello CC, Conte D Jr (2004) Revealing the world of RNA interference. Nature 431:338–342

Miki D, Itoh R, Shimamoto K (2005) RNA silencing of single and multiple members in a gene family of rice. Plant Physiol 138:1903–1913

Moazed D (2009) Small RNAs in transcriptional gene silencing and genome defence. Nature 457:413–420

Mochizuki K, Gorovsky MA (2005) A dicer-like protein in *Tetrahymena* has distinct functions in genome rearrangement, chromosome segregation, and meiotic prophase. Genes Dev 19:77–89

Moissiard G, Voinnet O (2006) RNA silencing of host transcripts by cauliflower mosaic virus requires coordinated action of the four *Arabidopsis* Dicer-like proteins. Proc Natl Acad Sci USA 103:19593–19598

Moriwaki A, Ueno M, Arase S, Kihara J (2007) RNA-mediated gene silencing in the pythopathogenic fungus *Bipolaris oryzae*. FEMS Microbiol Lett 269:85–89

Mouyna I, Henry C, Doering TL, Latgé J-P (2004) Gene silencing by RNA interference in the human pathogenic fungus *Aspergillus fumigatus*. FEMS Microbiol Lett 237:317–324

Mummery-Widmer JL, Yamazaki M, Stoeger T, Novatchkova M, Bhalerao S, Chen D, Dietzl G, Dickson BJ, Knoblich JA (2009) Genome-wide analisis of Notch signallingin *Drosophila* by transgenic RNAi. Nature 458:987–992

Murata T, Kadotani N, Yamaguchi M, Tosa Y, Mayama S, Nakayashiki H (2007) siRNA-dependent and independent post-transcriptional cosuppression of the LTR-retrotransposon MAGGY in the phytopathogenic fungus *Magnaporthe oryzae*. Nucleic Acids Res 35:5987–5994

Nakayashiki H, Kadotani NSM (2006) Evolution and diversification of RNA silencing proteins in fungi. J Mol Evol 63:127–135

Nakayashiki H, Nguyen QB (2008) RNA interference: roles in fungal biology. Curr Opin Microbiol11:494–502

Nakayashiki H, Hanada S, Nguyen BQ, Kadotani N, Tosa Y, Mayama S (2005) RNA silencing as a tool for exploring gene function in ascomycete fungi. Fungal Genet Biol 42:275–283

Namekawa SH, Iwabata K, Sugawara H, Hamada FK, Koshiyama A, Chiku H, Kamada T, Sakaguchi K (2005) Knockdown of *LIM/15DMC1* in the mushroom *Coprinus cinereus* by double-stranded RNA-mediated gene silencing. Microbiology 151:3669–3678

Nguyen QB, Kadotani N, Kasahara YT, Mayama S, Nakayashiki H (2008) Systematic functional analysis of calcium-signalling proteins in the genome of the rice-blast fungus, *Magnaporthe oryzae*, using a high-throuhgput RNA-silencing system. J Microbiol 68:1348–1365

Nicolás FE, Torres-Martínez S, Ruiz-Vázquez RM (2003) Two classes of small antisense RNAs in fungal RNA silencing triggered by non-integrative transgenes. EMBO J 22:3983–3991

Nicolás FE, Calo S, Murcia-Flores L, Garre V, Ruiz-Vázquez RM, Torres Martínez S (2008) A RING-finger photocarotenogenic represor involved in asexual sporulation in *Mucor circinelloides*. FEMS Microbiol Lett 280:81–88

Nygren CMR, Eberhardt U, Karlsson M, Parret JL, Lindahl BD, Taylor AFS (2008) Growth on nitrate and occurrence of nitrate reductase-encoding genes in a phylogenetically diverse range of ectomycorrhizal fungi. New Phytol 180:875–889

Ohrt T, Mütze J, Staroske W, Weinmann L, Höck J, Crell K, Meister G, Schwille P (2008) Fluorescence correlation spectroscopy and fluorescence cross-correlation spectroscopy reveal the cytoplasmic origination of loaded nuclear RISC in vivo in human cells. Nucleic Acid Res 36:6439–6449

Oliveira JM, van der Veen D, de Graaff LH, Qui L (2008) Efficient cloning system for construction of gene silencing vectors in *Aspergillus niger*. Appl Microbiol Biotechnol 80:917–924

Orban TI, Izaurralde E (2005) Decay of mRNAs targeted by RISC requires XRN1, the Ski complex, and the exosome. RNA 11:459–469

Ossowski S, Schwab R, Weigel D (2008) Gene silencing in plants using artificial microRNAs and other small RNAs. Plant J 53:674–690

Pak J, Fire A (2007) Distinct populations of primary and secondary effectors during RNAi in *C. elegans*. Science 315:241–244

Panepinto J, Komperda K, Frases S, Park YD, Djordjevic JT, Casadevall A, Williamson PR (2009) Sec6-dependent sorting of fangal extracellular exosomes and laccase of *Cryptococcus neoformans*. Mol Microbiol 71:1165–1176

Petersen BO, Albrechtsen M (2005) Evidence implying only unprimed RdRP activity during transitive gene silencing in plants. Plant Mol Biol 58:575–583

Petersen CP, Bordeleau ME, Pelletier J, Sharp PA (2006) Short RNAs repress translation after initiation in mammalian cells. Mol Cell 21:533–542

Pfeffer S, Zavolan M, Grässer FA, Chien M, Russo JJ, Ju J, John B, Enright AJ, Marks D, Sanders C, Tuschl T (2004) Identification of virus-encoded microRNAs. Science 304:734–736

Raponi M, Arndt GM (2003) Double-stranded RNA-mediated gene silencing in fission yeast. Nucleic Acid Res 31:4481–4489

Rappleye CA, Engle JT, Coldman WE (2004) RNA interference in Hitoplasma capsulatum demonstrates a role for alpha-(1, 3)-glucan in virulence. Mol Microbiol 53:153–165

Romano N, Macino G (1992) Quelling: transient inactivation of gene expression in *Neurospora crassa* by transformation with homologous sequences. Mol Microbiol 6:3343–3353

Ruby JG, Jan C, Plaver C, Axtell MJ, Lee W, Nusbaum C, Ge H, Bartel DP (2006) Large-scale sequencing reveals 21U-RNAs and additional microRNAs and endogenous siRNAs in *C. elegans*. Cell 127:1193–1207

Ruiz M, Voinnet O, Baulcombe DC (1998) Initiation and maintenance of virus-induced gene silencing. Plant Cell 10:937–946

Schramke V, Allshire R (2003) Hairpin RNAs and retrotransposons LTRs effect RNAi and chromatin-based gene silencing. Science 301:1069–1107

Schümann J, Hertweck C (2007) Molecular basis of cytochalasin biosynthesis in fungi: gene cluster analysis and evidence for the involvement of a PKS-NRPS hybrid synthase by RNA silencing. J Am Chem Soc 129:9564–9565

Schuurs TA, Schaeffer EAM, Wessels JGH (1997) Homology-dependent silencing of the *SC3* gene in *Schizophyllum commune*. Genetics 147:589–596

Segers GC, van Wezel R, Zhang X, Hong Y, Nuss DL (2006) Hypovirus papain-like protease p29 suppresses RNA silencing in the natural fungal host and in a heterologous plant system. Eukaryot Cell 5:896–904

Segers CG, Zhang X, Deng F, Sun Q, Nuss DL (2007) Evidence that RNA silencing functions as an antiviral defense mechanism in fungi. Proc Natl Acad Sci USA 104:12902–12906

Shafran H, Miyara I, Eshed R, Prusky D, Sherman A (2008) Development of new tools for stydying gene function in fungi based on the Gateway system. Fungal Genet Biol 45:1147–1154

Sigova A, Rhind N, Zamore P (2004) A single argonaute protein mediates both transcriptional and posttranscriptional silencing in *Schizosaccharomyces pombe*. Genes Dev 18:2359–2367

Sijen T, Fleenor J, Simmer F, Thijssen KL, Parrish S, Timmons L, Plasterk RH, Fire A (2001) On the role of RNA amplification in dsRNA-triggered gene silencing. Cell 107:465–476

Sijen T, Steiner FA, Thijssen KL, Plasterk HA (2007) Secondary siRNAs result from unprimed RNA synthesis and from a distinct class. Science 315:244–247

Stark JM, Hart SC (1997) High rates of nitrification and nitrate turnover in undisturbed coniferous forests. Nature 385:61–64

Stark GR, Kerr IM, Williams BR, Silverman RH, Schreiber RD (1998) How cells respond to interferons. Annu Rev Biochem 67:227–264

Tanguay P, Bozza S, Breui C (2006) Assessing RNAi frequency and efficiensy in *Ophiostoma floccosum* and *O. piceae*. Fungal Genet Biol 43:802–804

Tomari Y, Zamora PD (2005) Perspective: machines for RNAi. Genes Dev 19:517–529

Travella S, Keller B (2009) Down-regulation of gene expression by RNA-induced gene silencing. Methods Mol Biol 478:185–199

Verdel A, Vavasseur A, Le Gorrec M, Touat-Todeschini L (2009) Common themes in siRNA-mediated epigenetic silencing pathways. Int J Dev Biol 53:245–257

Wadhwa R, Kaul SC, Miyagishi M, Taira K (2004) Vectors for RNA interferente. Curr Opin Mol Ther 6:367–372

Wälti MA, Villalba C, Buser RM, Grünler A, Aebi M, Künzler M (2006) Targeted gene silencing in the model mushroom *Coprinus cinerea* (*Coprinus cinereus*) by expression of homologous hairpin RNAs. Eukaryot Cell 5:732–744

Wassenegger M, Heimes S, Riedel L, Sanger HL (1994) RNA-directed *de novo* methylation of genomic sequences in plants. Cell 76:567–576

Weinberg MS, Villeneuve LM, Ehsani A, Amarzguioui M, Aagaard L, Chen Z-X, Riggs AD, Rossi JJ, Morris KV (2006) The antisense strand of small interfering RNAs directs histone methylation and transcriptional gene silencing in human cells. RNA 12:256–262

Weinmann L, Höck J, Ivacevic T, Ohrt T, Mütze J, Schwille P, Kremmer E, Benes V, Urlaub H, Meister G (2009) Importin 8 is a gene silencing factor that targets Argonaute proteins to distinct mRNAs. Cell 136:496–507

Wesley SV, Helliwell CA, Smith NA, Wang MB, Rouse DT, Liu Q, Gooding PS, Singh SP, Abbott D, Stoutjesdijk PA, Robinson SP, Gleave AP, Green AG, Waterhouse PM (2001) Construct design for efficient, effective and high-throughput gene silencing in plants. Plant J 27:581–590

Wu L, Fan J, Belasco JG (2006) MicroRNAs direct rapid deadenylation of mRNA. Proc Natl Acad Sci USA 103:4034–4039

Yu D, Fan B, MacFarlane SA, Chen Z (2003) Analisis of the involvement of an inducible *Arabidopsis* RNA-dependent RNA polymerase in antiviral defense. Mol Plant Microbe Interact 16:206–216

Zhang X, Segers GC, Sun Q, Deng F, Nuss DL (2008) Characterization of hypovirus-derived small RNAs generated in the chestnut blight fungus by an inducible DCL-2-dependent pathway. J Virol 82:2613–2619

Zhou H, Xu M, Huang Q, Gates AT, Zhang XD, Castle JC, Stec E, Ferrer M, Strulovici B, Hazuda DJ, Espeseth AS (2008) Genome-scale RNAi screen for host factors required for HIV replication. Cell Host Microbe 2:495–504

Part III
Functions and Interactions

Chapter 10
Ectomycoremediation: An Eco-Friendly Technique for the Remediation of Polluted Sites

Heike Bücking

10.1 Introduction

Estimates of the number of sites vary considerably, but approximately 450,000 to 1 million sites in the United States and 750,000 sites in Europe are suspected or known to be contaminated with a diverse group of pollutants due to previous anthropogenic impacts (De Sousa 2001; GAO 2004). Fuel hydrocarbons, polycyclic aromatic hydrocarbons (PAHs), polychlorinated biphenyls (PCBs), chlorinated aromatic hydrocarbons (CAHs), detergents, and pesticides are the most dominant organic pollutants in soils. The main inorganic pollutants are heavy metals, such as cadmium, lead and mercury, nitrates and phosphate, and radionuclides. The United States alone spends annually \$6–8 billion and global costs range between \$25 and 50 billion for efforts to remediate contaminated sites (Doty 2008). Costs for the restoration of all contaminated sites in the United States have been estimated to be approximately \$1.7 trillion (Kuiper et al. 2004). Conventional techniques for the remediation of polluted sites typically include: soil excavation, transport, washing and extraction, pumping and treating of contaminated water, and the addition of chemical reactants such as hydrogen peroxide or potassium permanganate, and incineration. However, these techniques are expensive and not always sufficient, very invasive, and can lead to the release of pollutants into the air or leaching into the ground water (Kuiper et al. 2004).

H. Bücking
Biology and Microbiology Department, South Dakota State University, Brookings, SD 57007, USA
e-mail: heike.bucking@sdstate.edu

M. Rai and A. Varma (eds.), *Diversity and Biotechnology of Ectomycorrhizae*, Soil Biology 25, DOI 10.1007/978-3-642-15196-5_10,

10.2 Phytoremediation: An Alternative Method for the Remediation of Contaminated Sites

Phytoremediation can represent a sustainable and cost-effective alternative to conventional remediation technologies, and includes all plant-induced biological, chemical, and physical processes that contribute to the remediation of contaminated sites (Cunningham et al. 1996). The following processes can be distinguished:

- *Phytostabilization.* Containment of pollutants by reducing the dislocation to other ecosystem compartments by, for example, mechanical stabilization of the soil, binding to the plant surface, or by reducing the leaching of pollutants into aquifers.
- *Phytoextraction.* Removal of pollutants from the soil by uptake from the environment and accumulation in the plant biomass, preferably in the above-ground plant tissue.
- *Phytotransformation.* Chemical transformation of pollutants through plant metabolism, resulting in inactivation, degradation (phytodegradation), or immobilization (phytostabilization) of pollutants.
- *Phytovolatilization.* Uptake of pollutants from the soil and release into the atmosphere, sometimes after transformation into a more volatile form.
- *Phytostimulation.* Increase in biodegradation through the stimulation of microbial communities in the rhizosphere. Since most of the processes that lead to the degradation of pollutants by plants occur in the rhizosphere and not in the whole plant per se, the term rhizoremediation is also often used (Meharg and Cairney 2000; Kuiper et al. 2004).

Compared to conventional techniques, phytoremediation has several advantages (1) it is much cheaper on a per volume basis (estimated costs of phytoremediation depend on site and pollutant but vary between 10 and 50% of the costs for conventional techniques; Sadowsky 1999); (2) less secondary waste is generated, and instead a useful product, such as wood, pulp, or bioenergy can be produced; (3) it retains the top soil and is noninvasive; and (4) it has a high public acceptance. Phytoremediation is suitable for large contaminated sites with low-to-moderate pollutant concentration particularly in top layers of the soil. However, polluted sites are often contaminated with a mixture of various problematic compounds, and the low nutrient availability and poor soil structure of, e.g., former industrial sites can prevent a successful establishment of plants (Meharg and Cairney 2000). Phytoremediation is generally slower than conventional practices, and it is not without environmental implications, since pollutants can get integrated into the wild-life food chains via the plant, and/or toxic intermediates can be formed by incomplete degradation pathways (Trapp and Karlson 2000).

10.3 Ectomycorrhizal Associations and Their Significance for Phytoremediation

Numerous studies have shown that plants can successfully be used for the remediation of soils contaminated with a variety of pollutants, such as salts, agrochemicals, nitroaromatics, chlorinated compounds, heavy metals, and hydrocarbons (Tsao 2003). Particularly, fast growing trees, such as *Populus* and *Salix*, are promising candidates for phytoremediation or dendroremediation (the remediation of polluted sites by trees, Komives and Gullner 2006) due to their deep root system, and their high biomass production and transpiration activity (Tlustoš et al. 2006). Laureysens et al. (2004), who examined the biomass production of 17 poplar clones on a waste disposal site, found an annual biomass production of up to 11.4 t ha^{-1}. For both plant species transpiration rates of approximately 100 l water per day have been reported (Wullschleger et al. 1998).

However, the majority of tree species that have been considered as potential candidates for phytoremediation live naturally in symbiosis with ectomycorrhizal (ECM) and/or arbuscular mycorrhizal (AM) fungi (Smith and Read 2008). For instance, *Populus* that has extensively been studied for its suitability in phytoremediation also due to its accessibility for genetic engineering (e.g., Rugh et al. 1998; Doty et al. 2007) forms ECM interactions with more than 60 different fungal species, and some of these fungi have been shown to increase the growth of poplar seedlings by more than 400% (Cripps 2003). Under natural conditions, more than 60–80% of the root systems of poplar and willow, respectively, are colonized with ECM fungi. Additionally, approximately 10% of the root system is colonized with AM fungi (van der Heijden and Vosatka 1999; Kaldorf et al. 2002; Becerra et al. 2009).

Mycorrhizal fungi play a key role in nutrient cycling and ecosystem functioning and have a profound effect on the composition of plant and microbial communities and thereby also on the capability of these communities to degrade anthropogenic pollutants. ECM fungi enhance the uptake of phosphate (P) and nitrogen (N), and can also contribute to the supply of the host with trace elements, such as copper (Cu) and zinc (Zn) (Smith and Read 2008). Furthermore, mycorrhizal plants have a higher resistance against abiotic (e.g., drought, heavy metals) and biotic (plant pathogens) stresses (Smith and Read 2008). In return, for these beneficial effects on nutrient uptake and stress resistance, the plant transfers between 10% and 20% of its photosynthetically fixed carbon to the fungus (Finlay and Söderström 1992). However, despite the widely acknowledged significance of ECM fungi for plant growth, fitness and community composition particularly in difficult environments, their contribution to phytoremediation processes has so far often been ignored.

The capability of tree species, such as *Populus* and *Salix*, to remediate contaminated soils has extensively been studied (for review see, e.g., Pulford and Watson 2003; Tlustoš et al. 2006), but the contribution of their ECM communities to the observed degradation processes has only rarely been examined (e.g., Sell et al. 2005). ECM fungi have been shown to degrade a variety of environmentally important organic pollutants, such as 2,4-dichlorophenol (Meharg et al. 1997b), 2,4,6-trinitrotoluene (Scheibner et al. 1997), PCB (Donnelly and Fletcher 1995), 4-fluorobiphenyl (Green et al. 1999), and some PAHs (Braun-Lüllemann et al. 1999) and the capability to degrade organic pollutants seems to be relatively common for this group of fungi. Meharg and Cairney (2000), for example, screened different ECM fungal species for their capability to degrade various classes of organic pollutants (e.g., PAHs, PCB) and found that 33 out of 42 species were able to degrade one or more classes of these contaminants. However, most of these studies were performed under pure culture conditions (e.g., Meharg et al. 1997b; Gramss et al. 1999) that do not necessarily reflect the capability of the fungus to degrade recalcitrant compounds in the symbiotic stage under field conditions (Meharg et al. 1997a; Dittmann et al. 2002). However, it is crucial to consider both, plant and fungus, as one unit, and to assess the capability of plants and fungi to degrade xenobiotics in the symbiotic stage to use both partners to their full potential for the remediation of polluted sites. This is important for the following reasons:

- Mycorrhizal fungi are ubiquitous in the soil and plants are under natural conditions normally associated with mycorrhizal fungi. However, the colonization with ECM fungi can have a synergistic (Meharg et al. 1997a; Sell et al. 2005), but also an antagonistic effect (Genney et al. 2004; Koivula et al. 2004; Joner et al. 2006; Gunderson et al. 2007) on the capability of the plant to remediate polluted sites.
- ECM roots represent the major portion of the nutrient-absorbing surface area in tree roots (Taylor and Peterson 2000), and thereby also of the root system that is potentially able to absorb and to extract pollutants from the soil. The fungal sheath of ECM roots represents for ions a significant apoplastic barrier (Bücking et al. 2002), and can be expected to restrict the movement of larger molecules, such as PAHs, into the root cortex. In the symbiotic stage, the nutrient uptake by the host is mainly controlled and regulated by the mycorrhizal fungus (Smith and Read 2008), and this can increase or reduce the bioavailability of the pollutant for the host (Sell et al. 2005; Gunderson et al. 2007).
- The interaction of a host plant with a mycorrhizal fungus has a much stronger impact on the whole plant physiology than previously suggested. Additionally, to the positive effects on nutrient uptake and biotic and abiotic stress resistance, the symbiosis affects, for example, the photosynthetic activity, transpiration, stomatal conductance, carbon allocation, and N metabolism of the host (Smith and Read 2008).
- The symbiosis with a host plant changes the physiology of the mycorrhizal fungus and could, for example, affect the expression of degrading enzymes (Meharg and Cairney 2000). In the symbiosis the fungus receives carbon from its host, and this could potentially improve the degradation rate by an increase in the fungal biomass, but could also reduce the degradation rate, because the more readily available substrates are generally preferred (Joner and Leyval 2003).
- Approximately 8,000 plant species and 7,000–10,000 fungal species form ECM associations (Taylor and Alexander 2005), but only a small fraction of potential

plant/fungal interactions has been tested so far. However, there is a considerable physiological diversity between individual plant and fungal species and also between different ECM plant/fungal associations (e.g., Sell et al. 2005), and the potential degradation rate of these systems is also dependent on the pollutant, its concentration, and the environmental conditions (e.g., soil characteristics, nutrient availability).

- So far, it has been difficult to translate remediation strategies that have been proven to be successful under laboratory or greenhouse conditions to the field and commercialization. Gerhardt et al. (2009) suggested that this is due to additional stress factors for the plant, or to inadequate methods to analyze remediation under field conditions. Discrepancies between the degradation rates under laboratory and under field conditions can be the result of many different factors, for example, soil characteristics that affect the bioavailability or weathering of soil pollutants. However, it is also possible that mycorrhizal communities play a much greater role, and that experiments with nonmycorrhizal plants or plants that are only colonized with one ECM partner do not reflect root systems under field conditions that are colonized on a very small spatial scale with different taxa and species of ECM fungi. Goodman and Trofymow (1998) identified, for example, 35 different ECM morphotypes in 3 l of soil from a mature spruce stand.
- ECM fungi influence both plant and microbial communities in the soil (Finlay 2008), and this will also have an effect on the capability of these communities to remediate polluted soils. Both symbiotic partners interact in the soil with a variety of microorganisms, and it has been suggested that the diverse microbial communities in the rhizosphere or mycorrhizosphere have a much greater impact on biodegradation of organic pollutants than the plant (Heinonsalo et al. 2000).

For the successful application of ECM plants for the phytoremediation of polluted sites, several interrelated factors need to be considered, including the properties, concentration, distribution, bioavailability, and toxicity of the pollutant; soil characteristics such as water and nutrient availability; the effect of the ECM system on the activity of microbial communities in rhizosphere and mycorrhizosphere; and the activity of both partners to degrade pollutants in the symbiotic stage.

10.3.1 The Contribution of Rhizosphere and Ectomycorrhizosphere to Bioremediation

The root rhizosphere with its stimulating effect on microbial activities has often been described as the main contributor to the phytoremediation of contaminated sites. Wang et al. (2008), for example, reported that in the rhizosphere three to four times more petroleum hydrocarbon (PHC) was degraded than in unplanted soil, and similar effects have also been observed for other pollutants (e.g., Gunderson et al.

2007). However, the mechanisms that cause a stimulation of degradation in the rhizosphere are complex and only poorly understood. Putative explanations include (1) the direct degradation by plant-derived enzymes; (2) the stimulation of microbial activity and changes in the microbial community composition due to root exudates; (3) the release of metabolic precursors, for example, phenolics, that induce the activity of enzymes that degrade pollutants; and (4) the establishment of environmental conditions that stimulate degradation (e.g., pH, O_2/CO_2 concentration, and osmotic potential) (Joner and Leyval 2003).

Plant roots release approximately 10–20% of their photosynthetically fixed carbon into the rhizosphere (Bais et al. 2006; Haichar et al. 2008), and the secreted organic compounds, such as sugars, sugar alcohols, amino acids, fatty acids, and proteins represent a significant carbon source and promote microbial proliferation and diversity in the rhizosphere. Root exudates have also been shown to shape the microbial community composition and its capability to degrade recalcitrant compounds. For instance, phenolic compounds in root exudates stimulate the microbial degradation of PCBs in the rhizosphere by serving as growth substrates for PCB-degrading bacteria and/or as inducer of the PCB-degradation pathway (Donnelly et al. 1994; Gilbert and Crowley 1997). Similar effects were also shown for the biodegradation of PAHs, such as naphthalene. The expression of naphthalene dioxygenase in *Pseudomonas fluorescens* was differentially regulated in response to different root exudate compounds, and the higher overall expression when root exudates were available resulted in increased naphthalene-degradation rates in the root rhizosphere (Kamath et al. 2004).

The plant rhizosphere has also been discussed as a potential way to introduce and to stabilize communities of genetically engineered microorganisms in the soil that could carry or transfer degradation genes to indigenous microbial communities ("bioaugmentation"; Sarand et al. 1998, 1999). Without the positive effect of the root rhizosphere, the number of introduced degrading microorganisms often decreases shortly after inoculation, due to competition with indigenous microbial communities for limited nutrient resources, antagonistic interactions, and other stresses (Gentry et al. 2004).

However, the mycorrhizosphere represents in most soils an even larger surface for the establishment of microbial communities than the rhizosphere. The extraradical mycelium acts as an extension of the root system and creates a large interface between hyphae and soil, the mycorrhizosphere that can be defined as "the zone of soil where the physical, chemical and microbiological processes are influenced by plant roots and their associated mycorrhizal fungi" (Giri et al. 2005). Rousseau et al. (1994) estimated, for example, that the extraradical mycelium of *Pisolithus tinctorius* represents 99% of the nutrient-absorbing surface length of pine roots. Mycorrhizal fungi and their extraradical mycelium represent a main part of the microbial biomass particularly in the top soil, and the fungal biomass in the soil can reach 700–900 kg ha^{-1} (Wallander et al. 2001) and accounts for at least 32% of the total microbial biomass (Högberg and Högberg 2002).

Ectomycorrhizal fungi are part of complex microbial communities in the soil, and the interactions between the diverse members of this community can be synergistic, competitive, or antagonistic depending on the species and the soil conditions. The presence of an ECM fungus leads to quantitative and qualitative changes in the microbial community composition (Heinonsalo et al. 2004; Assibetse et al. 2005) and can result in a shift towards catabolic microbial communities, and this could also result in changes in the capability of these communities to degrade pollutants in the soil. ECM roots and the extraradical mycelium are major sinks for photosynthetically fixed carbon in mycorrhizal root systems and it has been shown that the quantity and quality of root exudates is altered in response to ECM colonization. Rygiewicz and Anderson (1994) found that mycorrhizal plants transfer 23% more carbon from the shoot into the root and that more carbon is allocated into pools with a higher turnover rate.

Fungal hyphae and the extraradical mycelium release energy-rich compounds into the mycorrhizosphere and can act as conduit and provide carbon sources to heterotrophic bacterial and microbial communities and their metabolic activities in larger distance from the root (Heinonsalo et al. 2004; Finlay 2008). Olsson and Wallander (1998) reported that the microbial activity in the soil near to mycorrhizal roots and also in further distance from the root is higher than in noncolonized root systems. Heinonsalo et al. (2000) examined the bacterial density in different compartments in microcosms with a petroleum-contaminated soil and found the lowest number of bacteria in the bulk soil. The rhizosphere and mycorrhizosphere supported a higher number of bacterial cells and the bacterial density was 4–5 times higher in the mycorrhizosphere than in the rhizosphere. The mycorrhizosphere provides an ecological niche and nutritionally favorable conditions for diverse microbial communities, and transmission electron microscopy of ECM hyphae from PHC-contaminated soils revealed a microbial biofilm at the soil/fungal interface. Isolates from these communities were able to grow on *m*-toluate and *m*-xylene (two intermediates of BTEX degradation) as sole carbon source and contained plasmids with marker genes involved in the degradation of mono-aromatics (Sarand et al. 1998). Sarand et al. (1999) were able to establish in the mycorrhizosphere of Scots pine and the ECM fungus *Suillus bovinus*, communities of a *Pseudomonas fluorescens* isolate carrying the TOL plasmid pWW0 that encodes enzymes for the degradation of toluene, *p*- and *m*-xylenes, and *m*-toluate. Plant and fungus alone were not able to degrade *m*-toluate, but the bacteria with the catabolic plasmid were able to degrade this pollutant in the mycorrhizosphere. Free-living microbial communities may act in concert with ECM fungi and facilitate the degradation of recalcitrant soil pollutants (Sarand et al. 1999; Meharg and Cairney 2000). The partial degradation of persistent aromatic pollutants by ECM fungi could also provide intermediate products that can be further broken down by microbial communities in the mycorrhizosphere (Cairney and Meharg 2002).

The ECM colonization can lead not only to quantitative but also qualitative changes in the bacterial community composition and the competition for limited soil nutrients, or the biosynthesis of antibiotics by ECM fungi, can also result in a decrease of the microbial activity in the mycorrhizosphere (Olsson et al. 1996). Assibetse et al. (2005) examined the impact of the ECM fungus *Pisolithus albus* on the bacterial community composition in the mycorrhizosphere of *Acacia auriculiformis* and found that the number of actinomycetes was reduced in ECM root systems. Such antagonistic effects of ECM systems on microbial diversity in the mycorrhizosphere could potentially also reduce the capability of these communities to degrade recalcitrant compounds. Saprophytic actinomycetes are involved in the breakdown of complex biopolymers, such as lignin, hemicelluloses, pectin, and chitin and have also been shown to be able to degrade recalcitrant pollutants such as pesticides (De Schrijver and De Mot 1999), hydrocarbons (McCarthy and Williams 1992), and 1,4-dioxane (Mahendra and Alvarez-Cohen 2005).

10.3.2 Ectomycoremediation of Organic Xenobiotics

White rot fungi (WRF), such as *Phanerochaete chrysosporium* or *Trametes versicolor*, or other saprophytic fungi have often been studied for their potential to degrade recalcitrant organic pollutants in soils. These saprophytic basidiomycetes are characterized by their production of extracellular lignin- and humic acid-degrading enzymes, such as lignin peroxidase, manganese peroxidase, and laccase, that lack specificity for particular substrates and have been shown to attack a variety of other compounds, including xenobiotics, such as lindane, benzo(a)-pyrene, and dichlorodiphenyltrichloroethane (DDT) (Majcherczyk et al. 1993).

ECM fungi play a key role for nutrient cycling in boreal and temperate forest ecosystems and are also able to utilize complex organic molecules as a source for reduced carbon (Durall et al. 1994) or other nutrients such as nitrogen or phosphate (Perez-Moreno and Read 2000). Similar to WRF, ECM fungi also produce a variety of ligninolytic or cell wall-degrading enzymes (for review see Read and Perez-Moreno 2003) that could also enable them to metabolize recalcitrant organic pollutants with structural similarities to lignin, such as PAHs, PCBs, 2,4,6-trinitrotoluene (TNT), or DDT.

The genes encoding a lignin peroxidase and manganese peroxidase have already been identified in different orders of ECM fungi, and also a partial laccase sequence has been described (Chen et al. 2001; Ramesh et al. 2008). This seems to indicate that the evolutionary relationship between ECM fungi and other saprophytic fungi is much closer and that a clear distinction into the two different functional groups of symbiotic or saprophytic fungi is not as clear as previously been suggested (Chen et al. 2001). However, the activity of ligninolytic enzymes or other phenol oxidizing enzymes, such as tyrosinase, catechol oxidase, or ascorbate oxidase in ECM roots or mycelium has mainly been detected in nonsterile soils (e.g., Bending and Read 1997; Timonen and Sen 1998) and could also have been the result of the microbial activity of the rhizosphere or mycorrhizosphere.

The ligninolytic and facultative saprophytic lifestyle of ECM fungi is currently under debate (Baldrian 2009) and it is not clear whether these enzymes will also be expressed in the symbiotic stage, when the ECM fungus is supplied with carbon by its host. WRF or ericoid fungi are generally more effective in degrading lignin or phenolic compounds (Haselwandter et al. 1990; Bending and Read 1997), and facultative ECM fungi have been shown to have a higher activity than obligate ECM fungi (Haselwandter et al. 1990). Recent results of Cullings et al. (2008) indicate that the enzymatic activity of ECM fungi will be affected in the symbiotic stage. They found that the ECM fungus *Suillus granulatus* expressed D-glucosidase, laccase, manganese peroxidase, lignin peroxidase, and protease in symbiosis with *Pinus contorta*, and that the activity increased when the photosynthetic capacity of the host was reduced by partial defoliation. Similar effects of the carbon supply on the enzymatic activity have also been reported during tree bud break, when the carbon flux to the mycorrhizal fungus is reduced (Courty et al. 2007).

However, even if ECM fungi have in the symbiotic stage a lower activity of ligninolytic or phenol oxidizing enzymes than WRF or other wood and straw-degrading basidiomyetes, the activity of these enzymes may still enable ECM fungi to successfully degrade organic pollutants in the soil. The sustainability of ECM fungi in soils may still favor their use in bioremediation over WRF, which are generally nonedaphic living organisms and require exogenous carbon sources to facilitate remediation.

10.3.2.1 Petroleum Hydrocarbons

Petroleum hydrocarbons (PHCs) are complex mixtures of numerous alicyclic, aliphatic, and aromatic compounds and mainly contain paraffins, aromatics, naphthenics, asphaltenes, and heavy metals in different proportions. Although the toxicity of individual compounds is known, it is extremely difficult to assess the toxicity of PHCs, because the chemical composition varies considerably, and our knowledge about additive, synergistic, or antagonistic effects of the various compounds is limited (Robertson et al. 2007). The inconsistent composition also makes general statements about potential strategies to remediate contaminated soils extremely difficult and may have also been the reason why the described effects of ECM fungi on PHC degradation are inconsistent (see below). An incomplete biotransformation of PHCs can also lead to potentially toxic metabolites that further limit the capability of plants, fungi, and other microorganisms to degrade PHCs and its refined products (Robertson et al. 2007).

PHC contaminations in the soil often cause a decrease in the number of indigenous bacteria, followed later by an increase in the number of bacterial species that can use PHCs as carbon source. Fungi are generally more tolerant to PHCs than bacteria (Robertson et al. 2007), but ECM fungi differ in their sensitivity. In liquid cultures, the growth of some ECM fungi is reduced or completely inhibited by increasing crude oil concentrations, while the growth of other species is stimulated (Nicolotti and Egli 1998). Crude oil up to a concentration of 50 g kg^{-1} did not have an effect on the ECM colonization of *Populus* and only slightly reduced the ECM colonization of spruce after a longer exposure (Nicolotti and Egli 1998). Diesel oil also did not affect the ECM colonization of *Populus* (Gunderson et al. 2007). However, changes in the ECM community composition after exposure to PHCs have been observed (Nicolotti and Egli 1998).

By contrast, PHCs can significantly reduce the germination and growth of plants (Nicolotti and Egli 1998; Choi et al. 2005), but the colonization with ECM fungi can increase the resistance of plants against PHCs. Gunderson et al. (2007) suggested that the positive effect of ECM associations on plant resistance is mainly due to a better supply with P and N, because the nutrient uptake is often reduced in soils that are contaminated with hydrocarbons. Despite the positive effect of ECM fungi on plant growth, the ECM colonization can reduce the capability of plants to remediate soils that are contaminated with PHCs. The presence of nonmycorrhizal hybrid poplars led to a significant degradation of diesel oil, but the effect of mycorrhizal plants was significantly lower (Gunderson et al. 2007). Instead, three times more PHC was accumulated in mycorrhizal roots, and it has been suggested that PHCs could get absorbed by hydrophobins in the fungal sheath (Gunderson et al. 2007). Hydrophobins are fungal cell wall proteins that are involved in cell-to-cell-surface contact in fungal mycelia and can reduce the apoplastic permeability of the fungal sheath (Bücking et al. 2002), and thereby could also reduce the bioavailability of PHCs for the mycorrhizal host plant. Antagonistic effects of ECM pine and beech seedlings on the degradation of PHCs have also been described by Vavrek et al. (2001) and could have also been caused by the competition of ECM plants and microbial communities for limited soil nutrients (the bioavailability of phosphate, potassium, calcium, and magnesium in the used soil was low). By contrast, in forest humus ECM Scots pine seedlings were able to stimulate the degradation of PHCs. The higher degradation was correlated to increases in the microbial population and changes in the carbon source utilization patterns of microbial communities in the mycorrhizosphere (Heinonsalo et al. 2000). The mycorrhizosphere in PHC-contaminated soil has been shown to support morphologically diverse bacterial populations and microbial biofilms. Isolated bacteria of these biofilms were able to use *m*-toluate and *m*-xylene as sole carbon source and expressed marker genes that are involved in the degradation of mono-aromatics (Sarand et al. 1998).

10.3.2.2 Polycyclic Aromatic Hydrocarbons

Polycyclic aromatic hydrocarbons (PAHs) are ubiquitous pollutants that represent an important risk to the environment and human health (Kuiper et al. 2004; Koivula et al. 2004). PAHs consist of a diverse group of organic molecules with a wide range of chemical properties (number of aromatic rings, molecular weight, and structural configuration), water solubility, and volatility. Particularly, higher-ringed PAHs are recalcitrant to degradation and include toxic, mutagenic, teratogenic, and carcinogenic compounds. PAH contaminations of the soil are mainly the result of leakage from old storage tanks or natural oil reservoirs, oil spills, road surfaces, domestic waste, incomplete fossil fuel combustion, former gas plant facilities, and tanker accidents (Morgan and Watkinson 1989).

A wide variety of bacteria, fungi, and algae can degrade and utilize PAHs as a catabolic carbon source through ring fission. Gramss et al. (1999) tested the ability of 58 fungi, including 22 ECM fungi and other wood- and straw-degrading, terricolous, and mitosporic fungi to degrade the PAHs phenanthrene, anthracene, fluronthene, pyrene, and perylene in liquid cultures. The capability of ECM fungi to degrade PAHs was generally lower than for wood- and straw-degrading basidiomycetes, but the majority of ECM fungi was able to degrade at least four or all of the tested PAHs to a certain degree. *Hebeloma crustuliniforme, H. hienale*, and *Lactarius deliciosus* showed the highest removal rates of the examined ECM fungi, and their degradation rates were comparable to rates that have been observed for wood- and straw-degrading fungi (Gramss et al. 1999). This also confirms the results of Braun-Lüllemann et al. (1999), who tested the capability of 27 ECM strains from 16 different species to degrade benzo(*a*)pyrene, phenanthrene, pyrene, and chrysene and found that the majority of ECM fungal species were able to degrade these pollutants (70% of the fungal species degraded phenanthrene, 65% of the species pyrene). There were species- and strain-specific differences in the capability of different ECM fungi, but the degradation potential was independent of the condensation grade and the number of aromatic rings of the compound. The degradation, however, was dependent on the nutrient supply. A strain of *Suillus grevillei* metabolized more than 50% of the supplied benzo(*a*)pyrene, phenanthrene, and pyrene in a nitrogen-depleted medium, but showed no degradation in a nitrogen-supplied medium (Braun-Lüllemann et al. 1999).

The described activities, however, were measured under controlled conditions in pure culture experiments (Braun-Lüllemann et al. 1999; Gramss et al. 1999), but the degradation rates in the soil will vary considerably from those in liquid cultures due to the adsorption to the soil matrix and the low water solubility particularly of PAHs with a high number of fused rings (Kuiper et al. 2004). Koivula et al. (2004) found no positive effect of Scots pine and the ECM fungus *Paxillus involutus* on the mineralization of pyrene in microcosms with forest humus, and the degradation was rather reduced by the plant and the ECM fungus. The same fungal species was able to degrade pyrene under pure culture conditions (Gramss et al. 1999). A reduction in the degradation of PAHs by plants and their associated ECM communities has also been observed by Genney et al. (2004) and Joner et al. (2006). Genney et al. (2004) reported that ECM pine seedlings had no effect on the mineralization of naphthalene, but reduced the degradation of fluorene by 35% compared to non-planted microcosms. Instead, both ECM fungi metabolized fluorine to the dead-end products, 9-fluorenone and 9-hydroxyfluorene, which can themselves cause problems and could be responsible for the overall lower mineralization rate of fluorene in the microcosms with ECM plants. A reduced degradation of anthracene, anthraquinone, chrysene, and dibenz (*a*,*b*)anthracene was also found in the ectomycorrhizosphere of *Suillus bovinus* and may have been the result of a reduced microbial activity due to the competition for limited nutrients. The microbial degradation of PAHs has been shown to be dependent on the nutrient supply conditions (Joner et al. 2002). Also the capability of plants to extract PAHs from the soil can be affected by a mycorrhizal association. Binet et al. (2000) demonstrated that the colonization of ryegrass with the AM fungus *Glomus mosseae* reduced the transport of PAHs to the shoot, but increased the adsorption to the root system.

10.3.2.3 Nitro-aromatics

Former production and assembly sites for explosives and ammunition during World War II are still heavily contaminated with nitro-aromatics such as 2,4,6-trinitrotoluene (TNT). The contamination is often heterogeneously distributed in the soil (Scheibner et al. 1997; Koehler et al. 2002) and the removal of the recalcitrant xenobiotic TNT has been declared by environmental agencies to an urgent priority due to its toxicity and its mutagenic impacts. Several ECM fungi have been tested for their potential to degrade TNT, and the capability to reduce TNT to aminodinitrotoluenes seems to be a ubiquitous pathway among all ecological and taxonomic groups of fungi. Based on their observation that especially litter and wood-degrading fungi showed a high potential to mineralize TNT, Scheibner et al. (1997) suggested that ligninolytic enzymes (see above) could play a key role for the mineralization process. By contrast, Meharg et al. (1997b) assumed that ligninolytic enzymes are not involved in the initial stages of TNT degradation and that TNT could be reduced via the redox potential at the ECM fungal plasma membrane. The capability of plants to remediate TNT-contaminated soils depends on the plant species (Koehler et al. 2002; Schoenmuth and Pestemer 2004), but it has been shown that the capability of *Populus* can be increased by the colonization with an unidentified ECM fungus (presumably *Hebeloma* sp.) and that the mineralization rate of these ECM systems was higher than for WRF (Dobner 2003). When ^{14}C-labeled TNT was supplied to the root system, only a small proportion was transferred to the shoot, and the accumulation into the root biomass did not differ between mycorrhizal and nonmycorrhizal roots (Dobner 2003). This indicates that in these microcosm experiments, ECM fungi were able to facilitate the mineralization of TNT. However, a similar stimulation in TNT degradation by ECM plants could not be confirmed by in situ experiments at a former TNT production site, presumably because the TNT was heterogeneously distributed in the soil (Dobner 2003).

10.3.2.4 Chlorinated Aromatic Hydrocarbons

Chlorinated aromatic hydrocarbons (CAHs) are highly toxic and represent a major group of chemicals responsible for the pollution of soils. CAHs, such as chlorophenols, chlorobenzenes, chloronitrobenzenes, chloroaniline, and PCBs, are highly resistant to degradation and can bioaccumulate in human and animal tissue. Many CAHs have commonly been used as herbicides (e.g., atrazine, 2,4-dichlorophenoxyacetic acid or 2,4-D), or as dielectric or hydraulic fluids (PCBs) due to their chemical and thermal stability. PCBs are considered the most widespread pollutant on the planet, and were banned in most countries in 1979 (Mackova et al. 2006).

Plants, ECM fungi, ericoid mycorrhizal fungi, and saprophytic wood-degrading fungi have been shown to degrade a variety of different CAHs, such as atrazine or 2,4-D. The degradation rate depended on the compound, its concentration, the nutrient availability, and the plant and fungal species, but there was no difference between saprophytic or symbiotic fungi (Donnelly et al. 1993; Dittmann et al. 2002; Mackova et al. 2006). Generally, ECM basidiomycetes have a lower capability to degrade CAHs than WRF or ericoid mycorrhizal fungi (Donnelly et al. 1993; Dittmann et al. 2002), but the degradation rate of ECM fungi for certain compounds and concentration ranges is comparable to that of WRF. Donnelly and Fletcher (1995) screened 21 ECM fungi for their capability to degrade various PCB congeners and found that 14 ECM fungi were able to metabolize various PCBs and that the lower chlorinated congoners were more easily degradable than the higher chlorinated congoners. The ECM fungus *Radiigera atrogleba* degraded seven out of 19 PCB congeners, whereas *Phanerochaete chrysosporium*, a model organism for xenobiotic biodegradation studies, only metabolized three out of the 19 congeners (Donnelly and Fletcher 1995). The capability of ECM fungi to degrade CAHs seems to increase when the ECM fungus is in symbiosis with its host plant. Dittmann et al. (2002) found that *Suillus bovinus* was not able to degrade 3-chlorobenzoic acid (3-CBA) in pure cultures, but that mycorrhizal pines had a limited potential to degrade 3-CBA in microcosm experiments. Since the mycorrhizal plants were not continuously kept under sterile conditions, the reported increased degradation in the symbiotic stage could also be the result of microbial activity in the ectomycorrhizosphere. However, similar effects have also been reported by Meharg et al. (1997a), who found that the mineralization of 2,4-D by *Paxillus involutus* and *Suillus variegatus* was significantly higher in symbiosis than under pure culture conditions (these mycorrhizal systems were cultured under sterile conditions).

A large percentage of the removal of CAHs from the soil matrix by mycorrhizal systems is not the result of mineralization but also of incorporation into the fungal and plant biomass (Donnelly et al. 1993; Meharg et al. 1997a; Schnabel and White 2001; Huang et al. 2007). For instance, Meharg et al. (1997a) found that 40% of the supplied 2,4-D was incorporated into the fungal biomass and not mineralized. However, incorporation into the fungal biomass would still contribute to a phytostabilization of CAHs in the ectomycorrhizosphere and would thereby also make CAHs accessible for further microbial degradation. Dittmann et al. (2002) showed that 3-CBA is not only incorporated into the ECM fungal sheath, but also taken up and transferred via the stele to the needles. The transport to the above-ground plant part could contribute to CAH removal from the soil by phytoextraction.

10.3.3 Ectomycoremediation of Heavy Metals

The contamination of soils with the nonradioactive metals arsenic (As), cadmium (Cd), copper (Cu), mercury (Hg), lead (Pb), and zinc (Zn) and the radioactive metals strontium (Sr), caesium (Cs), and uranium (U) represents a major environmental and human health problem (Raskin et al. 1997).

Suitable plant species for the phytoremediation of these sites should be able to extract and to tolerate high heavy metal concentrations and should accumulate these metals mainly in their shoot (phytoextraction). However, plants often accumulate heavy metals in their root system to protect the shoot from toxic heavy metal concentrations (Bücking and Heyser 1994). This is the reason why in the past, studies on the phytoremediation of heavy metal-contaminated soils were mainly conducted with hyperaccumulating plants, such as *Thlaspi caerulescens*, which are able to tolerate and to concentrate high heavy metal concentrations in the shoot (Robinson et al. 1998). However, the small biomass development of these species significantly reduces their potential to extract significant amounts of heavy metals from contaminated soils. Fast growing trees, such as *Populus* and *Salix*, could potentially be used for the phytoremediation of these sites, because (1) both are known to naturally colonize areas with high metal soil concentrations such as active and inactive smelter sites (Cripps 2003), (2) they are genetically transformable (Doty 2008), and (3) the high biomass development can compensate for the moderate heavy metal concentrations in the shoot (Pulford and Watson 2003; Tlustoš et al. 2006).

Depending on the conditions, the utilization of mycorrhizal systems can assist the phytoremediation of heavy metal polluted soils by the positive effect of ECM fungi on plant tolerance, but could also limit the plants ability to extract heavy metals from the soil by reducing the uptake and the transfer into the shoot. The colonization of tree species with ECM fungi is often essential for the reforestation of old mine sites, because ECM fungi are able to ameliorate the toxicity of heavy metals (Marx 1975). On the other hand, ECM fungi have been shown to enhance the uptake of metals by plants particularly when the exogenous supply is low (Colpaert and Van Assche 1992), but can also reduce the uptake into the plant and the transport into the shoot when the external supply is high (Bücking and Heyser 1994; Krznaric et al. 2009). Several mechanisms have been described to be involved in the reduced heavy metal uptake of ECM plants (1) the larger cell wall surface that can bind heavy metals, (2) the filter function of the fungal sheath that restricts the apoplastic movement of heavy metals into the root cortex, (3) the extracellular precipitation of heavy metals, and (4) the intracellular chelation and compartmentation with, e.g., polyphosphates in the fungal vacuole (Hartley et al. 1997).

Whether ECM fungi are able to contribute to the phytoextraction of heavy metals from the soil despite their widely accepted effect on plant tolerance and uptake depends on the heavy metal, and the fungal and plant species involved. For instance, Sell et al. (2005) showed that by the ECM colonization with *Paxillus involutus*, the capability of *Populus canadensis* to extract Cd was increased, but that the same fungus had no effect on the Cd uptake by *Salix viminalis*. The effect of *P. involutus* on metal uptake and distribution in *Salix* varied depending on the fungal strain and its adaptation to heavy metals (Baum et al. 2006). However, the twofold increase in the capability of mycorrhizal *Populus* to extract Cd from the soil would substantially increase the potential of the plant to remove this metal from polluted soils (Sell et al. 2005).

Zimmer et al. (2009) showed that an ECM fungus can simultaneously increase the heavy metal tolerance of the plant and the accumulation of heavy metals in the plant biomass. An ECM colonization of *Salix viminalis* × *caprea* with *Hebeloma crustuliniforme* resulted in an increase in the plant biomass under metal stress, and increased the Cd and Zn accumulation in the stems. Interestingly, a dual inoculation of the root system with the bacterial strains *Micrococcus luteus* and *Sphingomonas* sp. 23L can further increase the effect of the ECM fungus on heavy metal accumulation (Zimmer et al. 2009). The increase by a factor of 4.7 and 3.4 for the Cd and Zn accumulation in the stem, respectively, suggests that a combination of ECM fungi and associated bacteria may represent a promising approach to increase the capability of plants to extract heavy metals from the soil. The bacteria could facilitate the ECM colonization (mycorrhiza helper bacteria) or could increase the bioavailability of heavy metals for the mycorrhizal plant.

Two thousand fungal strains belonging to 98 genera of fungi have been isolated from the Chernobyl Atomic Energy Station after its nuclear accident in 1986. Some isolates showed a growth promotion when exposed to ionizing radiation, and the hyphal growth was directed towards the radiation source (Zhdanova et al. 2004). By contrast, control isolates were inhibited or showed no response when exposed to ionizing radiation (Tugay et al. 2006). Particularly ECM basidiomycetes accumulate or hyperaccumulate radionuclides in their fruitbodies, and a directed growth would allow them to absorb radionuclides from radioactive hotspots in the soil. Fruitbodies are mainly formed when the fungus is in symbiosis with its host plant and could be collected from the site to extract radionuclides from the soil. However, the potential consumption of these fruitbodies by animals or humans also poses a substantial risk and could transfer these radionuclides into the food chain (Gray 1998).

10.4 Conclusions

So far, studies about the role of ECM fungi in phytoremediation have mainly focused on their effect on plant tolerance against environmental stresses and on recultivation of polluted sites to prevent soil erosion. The impact of ECM communities on the capability of plants to remediate contaminated sites has only rarely been studied, and our current understanding of the potential of ECM fungi to degrade recalcitrant xenobiotics and to facilitate ectomycoremediation is mainly based on the results of pure culture or microcosm experiments that do not necessarily reflect the potential of ECM communities to degrade pollutants. ECM fungi can play a key role in the application of woody plant species for the remediation of polluted sites, based on their ubiquitous nature in soils, their positive effect on abiotic stress resistance, their impact on microbial and plant communities, and their capability to degrade recalcitrant xenobiotics. ECM fungi can stimulate the ability of plants and of microbial communities in the mycorrhizosphere to degrade organic pollutants (Meharg et al. 1997a; Sell et al. 2005), but can also have an antagonistic effect on soil remediation (Genney et al. 2004; Joner et al. 2006). And while for certain pollutants, consistently higher degradation rates by ECM systems were found, the results for other contaminants were inconsistent and varied depending on the plant and fungal species, the nutrient availability in the soil, and its effect on the microbial community composition. A better understanding of the physiology, biochemistry, molecular genetics, and regulation of potential degradation pathways in ECM fungi and mycorrhizal plants and of microbial communities in the mycorrhizosphere is crucial for a successful application of these associations for the phytoremediation of contaminated sites. The potential of plants and their ECM communities for the remediation of polluted sites depends on several interrelated factors, such as properties, concentration, toxicity and bioavailability of the pollutant, soil characteristics, such as nutrient and water availability, and the interactions between ECM plants and indigenous microbial communities in rhizosphere and mycorrhizosphere and their capability to facilitate remediation. However, the multifaceted effects of ECM fungi on the capability of plants to remediate soils and their ubiquitous nature in soils should make them a prime target for further research to identify efficient and sustainable strategies for the remediation of contaminated sites by plants and their microbial communities.

References

Assibetse K, Gueye M, Thioulouse J, Duponnois R (2005) Soil bacterial diversity responses to root colonization by an ectomycorrhizal fungus are not root-growth dependent. Microb Ecol 50:350–359

Bais HP, Weir TL, Perry LG, Gilroy S, Vivanco JM (2006) The role of root exudates in rhizosphere interactions with plants and other organisms. Annu Rev Plant Biol 57:233–266

Baldrian P (2009) Ectomycorrhizal fungi and their enzymes in soils: is there enough evidence for their role as facultative soil saprotrophs? Oecologia 161:657–660

Baum C, Hrynkiewicz K, Leinweber P, Meissner R (2006) Heavy metal mobilization and uptake by mycorrhizal and non-mycorrhizal willows (*Salix* × *dasyclados*). J Plant Nutr Soil Sci 169:516–522

Becerra AG, Nouhra ER, Silva MP, McKay D (2009) Ectomycorrhizae, arbuscular mycorrhizae, and dark-septate fungi on *Salix humboldtiana* in two riparian populations from central Argentina. Mycoscience 50:343–352

Bending GD, Read DJ (1997) Lignin and soluble phenolic degradation by ectomycorrhizal and ericoid mycorrhizal fungi. Mycol Res 101:1348–1354

Binet P, Portal JM, Leyval C (2000) Fate of polycyclic aromatic hydrocarbons (PAH) in the rhizosphere and mycorrhizosphere of ryegrass. Plant Soil 277:207–213

Braun-Lüllemann A, Hüttermann A, Majchercyzk A (1999) Screening for ectomycorrhizal fungi for degradation of polycyclic aromatic hydrocarbons. Appl Microbiol Biotechnol 53:127–132

Bücking H, Heyser W (1994) The effect of ectomycorrhizal fungi on Zn uptake and distribution in seedlings of *Pinus sylvestris* L. Plant Soil 167:203–212

Bücking H, Kuhn AJ, Schröder WH, Heyser W (2002) The fungal sheath of ectomycorrhizal roots: an apoplastic barrier for the entry of calcium, magnesium and potassium into the root cortex. J Exp Bot 53:1659–1669

Cairney JWG, Meharg AA (2002) Interactions between ectomycorrhizal fungi and soil saprotrophs: implications for decomposition of organic matter in soils and degradation of organic pollutants in the rhizosphere. Can J Bot 80:803–809

Chen DM, Taylor AFS, Burke RM, Cairney JWG (2001) Identification of genes for lignin peroxidases and manganese peroxidases in ectomycorrhizal fungi. New Phytol 152:151–158
Choi W-J, Chang SX, Hao X (2005) White spruce response to co-composted hydrocarbon-contaminated drilling waste: effects of compost age and nitrogen fertilization. J Environ Qual 34:1319–1327
Colpaert JV, Van Assche JA (1992) Zinc toxicity in ectomycorrhizal *Pinus sylvestris*. Plant Soil 143:201–211
Courty P-E, Bréda N, Garbaye J (2007) Relation between oak tree phenology and the secretion of organic matter degrading enzymes by *Lactarius quietus* ectomycorrhizas before and during bud break. Soil Biol Biochem 39:1655–1663
Cripps CL (2003) Native mycorrhizal fungi with aspen on smelter-impacted sites in the Northern Rocky Mountains: occurrence and potential use in reclamation. In: Proceedings of the 2003 National meeting of the American Society of Mining and Reclamation and the 9th Billings Land Reclamation Symposium. American Society for Mining and Reclamation, Lexington, USA
Cullings K, Ishkhanova G, Henson J (2008) Defoliation effects on enzyme activities of the ectomycorrhizal fungus *Suillus granulatus* in a *Pinus contorta* (lodgepole pine) stand in Yellowstone National Park. Oecologia 158:77–83
Cunningham SD, Anderson TA, Schwab AP, Hsu FC (1996) Phytoremediation of soils contaminated with organic pollutants. Adv Agron 56:55–114
De Schrijver A, De Mot R (1999) Degradation of pesticides by actinomycetes. Crit Rev Microbiol 25:85–119
De Sousa C (2001) Contaminated sites: the Canadian situation in an international context. J Environ Manage 62:131–154
Dittmann J, Heyser W, Bücking H (2002) Biodegradation of aromatic compounds by white rot and ectomycorrhizal fungal species and the accumulation of chlorinated benzoic acid in ectomycorrhizal pine seedlings. Chemosphere 49:297–306
Dobner I (2003) Der Einsatz mykorrhizierter Gehölze in biologischen Sanierungsverfahren unter dem Aspekt TNT-belasteter Böden. Ph.D. thesis, University of Bremen, Germany. http://elib.suub.uni-bremen.de/publications/dissertations/E-Diss696.dobner.pdf. Accessed 30 Jan 2010
Donnelly PK, Fletcher JS (1995) PCB metabolism by ectomycorrhizal fungi. Bull Environ Contam Toxicol 54:507–513
Donnelly PK, Entry JA, Crawford DL (1993) Degradation of atrazine and 2, 4-dichlorophenoxyacetic acid by mycorrhizal fungi at three nitrogen concentrations in vitro. Appl Environ Microbiol 59:2642–2647
Donnelly PK, Hegde RS, Fletcher JS (1994) Growth of PCB-degrading bacteria on compounds from photosynthetic plants. Chemosphere 28:981–988
Doty SL (2008) Enhancing phytoremediation through the use of transgenics and endophytes. New Phytol 179:319–333
Doty SL, James CA, Moore AL, Vajzovic A, Singleton GL, Ma C, Khan Z, Xin G, Kang JW, Park JY, Meilan R, Strauss SH, Wilkerson J, Farin F, Strand SE (2007) Enhanced phytoremediation of volatile environmental pollutants with transgenic trees. PNAS 104:16816–16821
Durall DM, Todd AW, Trappe JM (1994) Decomposition of ^{14}C-labelled substrates by ectomycorrhizal fungi in association with Douglas fir. New Phytol 127:725–729
Finlay RD (2008) Ecological aspects of mycorrhizal symbiosis: with special emphasis on the functional diversity of interactions involving the extraradical mycelium. J Exp Bot 59:1115–1126
Finlay R, Söderström B (1992) Mycorrhiza and carbon flow to the soil. In: Allen MJ (ed) Mycorrhizal functioning. Chapman and Hall, New York, pp 134–160
GAO (2004) Brownfield redevelopment. United States Government Accountability Office GAO-05-94. http://www.gao.gov/new.items/d0594.pdf. Accessed 15 Dec 2009

Genney DR, Alexander IJ, Killham K, Meharg AA (2004) Degradation of the polycyclic aromatic hydrocarbon (PAH) fluorine is retarded in a Scots pine ectomycorrhizosphere. New Phytol 163:641–649

Gentry TJ, Rensing C, Pepper IL (2004) New approaches for bioaugmentation as a remediation technology. Crit Rev Environ Sci Technol 34:447–494

Gerhardt KE, Huang X-D, Glick BR, Greenberg BM (2009) Phytoremediation and rhizoremediation of organic soil contaminants: potential and challenges. Plant Sci 176:20–30

Gilbert ES, Crowley DE (1997) Plant compounds that induce polychlorinated biphenyl biodegradation by *Arthrobacter* sp. strain B1B. Appl Environ Microbiol 63:1933–1938

Giri B, Giang PH, Kumari R, Prasad R, Sachdev M, Garg AP, Oelmüller R, Varma A (2005) Mycorrhizosphere: strategies and functions. In: Buscot F, Varma A (eds) Soil biology, vol 3, Microorganisms in soils: roles in genesis and functions. Springer, Berlin, pp 213–252

Goodman DM, Trofymow JA (1998) Comparison of communities of ectomycorrhizal fungi in old-growth and mature stands of Douglas-fir at two sites on southern Vancouver Island. Can J For Res 28:574–581

Gramss G, Kirsche B, Voigt K-D, Günther T, Fritsche W (1999) Conversion rates of five polycyclic aromatic hydrocarbons in liquid cultures of fifty-eight fungi and the concomitant production of oxidative enzymes. Mycol Res 103:1009–1018

Gray SN (1998) Fungi as potential bioremediation agents in soil contaminated with heavy or radioactive metals. Biochem Soc Trans 26:666–670

Green NA, Meharg AA, Till C, Troke J, Nicholson JK (1999) Degradation of 4-fluorobiphenyl by mycorrhizal fungi as determined by ^{19}F nuclear magnetic resonance spectroscopy and ^{14}C radiolabelling analysis. Appl Environ Microbiol 65:4021–4027

Gunderson JJ, Knight JD, Van Rees KCJ (2007) Impact of ectomycorrhizal colonization of hybrid poplar on the remediation of diesel-contaminated soil. J Environ Qual 36:927–934

Haichar FeZ, Marol C, Berge O, Rangel-Castro JI, Prosser JI, Balesdent J, Heulin T, Achouak W (2008) Plant host habitat and root exudates shape soil bacterial community structure. ISME J 2:1221–1230

Hartley J, Cairney JWG, Meharg AA (1997) Do ectomycorrhizal fungi exhibit adaptive tolerance to potentially toxic metals in the environment? Plant Soil 189:303–319

Haselwandter K, Bobleter O, Read DJ (1990) Degradation of ^{14}C-labelled lignin and dehydropolymer of coniferyl alcohol by ericoid and ectomycorrhizal fungi. Arch Microbiol 153:352–354

Heinonsalo J, Jørgensen KS, Haahtela K, Sen R (2000) Effects of *Pinus sylvestris* root growth and mycorrhizosphere development on bacterial carbon source utilization and hydrocarbon oxidation in forest and petroleum-contaminated soils. Can J Microbiol 46:451–464

Heinonsalo J, Frey-Klett P, Pierrat J-C, Churin J-L, Vairelles D, Garbaye J (2004) Fate, tree growth effect and potential impact on soil microbial communities of mycorrhizal and bacterial inoculation in a forest plantation. Soil Biol Biochem 36:211–216

Högberg MN, Högberg P (2002) Extramatrical ectomycorrhizal mycelium contributes one-third of microbial biomass and produces, together with associated roots, half the dissolved organic carbon in a forest soil. New Phytol 154:791–795

Huang Y, Zhao X, Luan S (2007) Uptake and biodegradation of DDT by 4 ectomycorrhizal fungi. Sci Total Environ 385:235–241

Joner EJ, Leyval C (2003) Phytoremediation of organic pollutants using mycorrhizal plants: a new aspect of rhizosphere interactions. Agronomy 23:495–502

Joner EJ, Corgié SC, Amellal N, Leyval C (2002) Nutritional constraints to degradation of polycyclic aromatic hydrocarbons in a simulated rhizosphere. Soil Biol Biochem 34:859–864

Joner EJ, Leyval C, van Colpaert J (2006) Ectomycorrhizas impede phytoremediation of polycyclic aromatic hydrocarbons (PAHs) both within and beyond the rhizosphere. Environ Pollut 142:34–38

Kaldorf M, Fladung M, Muhs H-J, Buscot F (2002) Mycorrhizal colonization of transgenic aspen in a field trial. Planta 214:653–660

Kamath R, Schnoor JL, Alvarez PJJ (2004) Effect of root-derived substrates on the expression of nah-lux genes in *Pseudomonas fluorescens* HK44: implications for PAH biodegradation in the rhizosphere. Environ Sci Technol 38:1740–1745

Koehler H, Warrelmann J, Frische T, Behrend P, Walter U (2002) In-situ phytoremediation of TNT-contaminated soil. Acta Biotechnol 22:67–80

Koivula TT, Salkinoja-Salonen M, Peltola R, Romantschuk M (2004) Bioremediation and biodegradation. Pyrene degradation in forest humus microcosms with our without pine and its mycorrhizal fungus. J Environ Qual 33:45–53

Komives T, Gullner G (2006) Dendroremediation: the use of trees in cleaning up polluted soils. In: Machova M, Dowling D, Macek T (eds) Phytoremediation and rhizoremediation. Springer, Dordrecht, pp 23–31

Krznaric E, Verbruggen N, Wevers JHL, Carleer R, Vangronsveld J, Colpaert JV (2009) Cd-tolerant *Suillus luteus*: a fungal insurance for pines exposed to Cd. Environ Pollut 157:1581–1588

Kuiper I, Lagendijk EL, Bloemberg GV, Lugtenberg BJJ (2004) Rhizomediation: a beneficial plant-microbe interaction. MPMI 17:6–15

Laureysens I, Bogaert J, Blust R, Ceulemans R (2004) Biomass production of 17 poplar clones in a short-rotation coppice culture on a waste disposal site and its relation to soil characteristics. For Ecol Manage 187:295–309

Mackova M, Barriault D, Francova K, Sylvestre M, Möder M, Vrchotova B, Lovecka P, Najmanová J (2006) Phytoremediation of polychlorinated biphenyls. In: Mackova M, Dowling D, Macek T (eds) Phytoremediation and rhizoremediation. Springer, Dordrecht, pp 143–167

Mahendra S, Alvarez-Cohen L (2005) Pseudonocardia dioxanivorans sp. nov., a novel actinomycete that grows on 1, 4-dioxane. Int J Syst Evol Microbiol 55:593–598

Majcherczyk A, Zeddel A, Kelschebach M, Loske D, Hüttermann A (1993) Abbau von polyzyklischen aromatischen Kohlenwasserstoffen durch Weißfäulepilze. Bioengineering 2:27–31

Marx DH (1975) Mycorrhizae and establishment of trees on strip-mined land. Ohio J Sci 75:288–297

McCarthy AJ, Williams ST (1992) Actinomycetes as agents of biodegradation in the environment – a review. Gene 115:189–192

Meharg AA, Cairney JWG (2000) Ectomycorrhizas – extending the capabilities of rhizosphere remediation? Soil Biol Biochem 32:1475–1484

Meharg AA, Cairney JWG, Maguire N (1997a) Mineralization of 2, 4-dichlorophenol by ectomycorrhizal fungi in axenic culture and in symbiosis with pine. Chemosphere 34:2495–2504

Meharg AA, Dennis GR, Cairney JWG (1997b) Biotransformation of 2, 4, 6-trinitrotoluene (TNT) by ectomycorrhizal basidiomycetes. Chemosphere 35:513–521

Morgan P, Watkinson PJ (1989) Hydrocarbon degradation in soils and methods for soil treatment. Crit Rev Biotechnol 8:305–353

Nicolotti G, Egli S (1998) Soil contamination by crude oil: impact on the mycorhizosphere and on revegetation potential of forest trees. Environ Pollut 99:37–43

Olsson PA, Wallander H (1998) Interactions between ectomycorrhizal fungi and the bacterial community in soils amended with various primary minerals. FEMS Microbiol Ecol 27:195–205

Olsson PA, Chalot M, Bååth E, Finlay RD, Söderström B (1996) Ectomycorrhizal mycelia reduce bacterial activity in a sandy soil. FEMS Microbiol Ecol 21:77–86

Perez-Moreno J, Read DJ (2000) Mobilization and transfer of nutrients from litter to tree seedlings via the vegetative mycelium of ectomycorrhizal plants. New Phytol 145:301–309

Pulford ID, Watson C (2003) Phytoremediation of heavy metal-contaminated land by trees – a review. Environ Int 29:529–540

Ramesh G, Sweera R, Sudhakara Reddy M (2008) Enhancement of laccase in ectomycorrhizal fungus *Hebeloma cylindrosporum* in presence of different substrates. Adv Environ Biol 2:115–120

Raskin I, Smith RD, Salt DE (1997) Phytoremediation of metals: using plants to remove pollutants from the environment. Curr Opin Biotechnol 8:221–226

Read DJ, Perez-Moreno J (2003) Mycorrhizas and nutrient cycling in ecosystems – a journey towards relevance? New Phytol 157:475–492

Robertson SJ, McGill WB, Massicotte HB, Rutherford PM (2007) Petroleum hydrocarbon contamination in boreal forest soils: a mycorrhizal ecosystems perspective. Biol Rev 82:213–240

Robinson B, Leblanc M, Petit D, Brooks RR, Kirkman JH, Gregg PEH (1998) The potential of *Thlaspi caerulescens* for phytoremediation of contaminated soils. Plant Soil 203:47–56

Rousseau JVD, Sylvia DM, Fox AJ (1994) Contribution of ectomycorrhiza to the potential nutrient-absorbing surface of pine. New Phytol 128:639–644

Rugh CL, Senecoff JF, Meagher RB, Merkle SA (1998) Development of transgenic yellow poplar for mercury phytoremediation. Nat Biotechnol 16:925–928

Rygiewicz PT, Anderson CP (1994) Mycorrhizae alter quality and quantity of carbon allocated below ground. Nature 369:58–60

Sadowsky MJ (1999) Phytoremediation: past promises and future practices. In: Bell CR, Brylinsky M, Johnson-Green P (eds) Microbial biosystems: new frontiers. Proceedings of the 8th International Symposium on Microbial Ecology, Atlantic Canada Society for Microbial Ecology, Halifax

Sarand I, Timonen S, Nurmiaho-Lassila E-L, Koivula T, Haahtela K, Romantschuk M, Sen R (1998) Microbial biofilms and catabolic plasmid harbouring degradative fluorescent pseudomonads in Scots pine mycorrhizospheres developed on petroleum contaminated soil. FEMS Microbiol Ecol 27:115–126

Sarand I, Timonen S, Koivula T, Peltola R, Haahtela K, Sen R, Romantschuk M (1999) Tolerance and biodegradation of m-toluate by Scots pine, a mycorrhizal fungus and fluorescent pseudomonads individually and under associative conditions. J Appl Microbiol 86:817–826

Scheibner K, Hofrichter M, Herre A, Michels J, Fritsche W (1997) Screening for fungi intensively mineralizing 2, 4, 6-trinitrotoluene. Appl Microbiol Biotechnol 47:452–457

Schnabel WE, White DM (2001) The effect of mycorrhizal fungi on the fate of PCBs in two vegetated systems. Int J Phytoremed 3:203–220

Schoenmuth BW, Pestemer W (2004) Dendroremediation of trinitrotoluene (TNT). Part 2: fate of radio-labelled TNT in trees. Environ Sci Pollut Res 11:331–339

Sell J, Kayser A, Schulin R, Brunner I (2005) Contribution of ectomycorrhizal fungi to cadmium uptake of poplars and willows from a heavily polluted soil. Plant Soil 277:245–253

Smith SE, Read DJ (2008) Mycorrhizal symbiosis, 3rd edn. Academic, Amsterdam

Taylor AFS, Alexander I (2005) The ectomycorrhizal symbiosis: life in the real world. Mycologist 19:102–112

Taylor JH, Peterson CA (2000) Morphometric analysis of *Pinus banksiana* Lamb. root anatomy during a 3-month field study. Trees 14:239–247

Timonen S, Sen R (1998) Heterogeneity of fungal and plant enzyme expression in intact Scots pine-*Suillus bovinus* and –*Paxillus involutus* mycorrhizospheres developed in natural forest humus. New Phytol 138:355–366

Tlustoš P, Pavlíková D, Száková J, Fischerová Z, Balík J (2006) Exploitation of fast growing trees in metal remediation. In: Mackova M, Dowling D, Macek T (eds) Phytoremediation and rhizoremediation. Springer, Dordrecht, pp 83–102

Trapp S, Karlson U (2000) Phytoremediation organischer Schadstoffe. Z Umweltchem Ökotox 12:245–255

Tsao DT (2003) Overview of phytotechnologies. Adv Biochem Eng Biotechnol 78:1–50

Tugay T, Zhdanova NN, Zheltonozhsky V, Sadovnikov L, Dighton J (2006) The influence of ionizing radiation on spore germination and emergent hyphal growth response reactions in microfungi. Mycologia 98:521–527

van der Heijden EW, Vosatka M (1999) Mycorrhizal associations of *Salix repens* L. communities in succession of dune ecosystems. II. Mycorrhizal dynamics and interactions of ectomycorrhizal and arbuscular mycorrhizal fungi. Can J Bot 77:1833–1841

Vavrek MC, Colgan W III, Campbell WJ (2001) The role of plant-bacteria-fungal interaction in remediation of oak-hickory-pine systems. Louisiana Applied Oil Spill Research and Development Program, OSRADP Technical Report Series, Louisiana

Wallander H, Nilsson LO, Hagerberg D, Bååth E (2001) Estimation of the biomass and seasonal growth of external mycelium of ectomycorrhizal fungi in the field. New Phytol 151:753–760

Wang J, Zhang Z, Su Y, He W, He F, Song H (2008) Phytoremediation of petroleum polluted soils. Petrol Sci 5:167–171

Wullschleger S, Meinzer F, Vertessy RA (1998) A review of whole-plant water use studies in trees. Tree Physiol 18:499–512

Zhdanova NN, Tugay T, Dighton J, Zheltonozhsky V, McDermott P (2004) Ionizing radiation attracts soil fungi. Mycol Res 108:1089–1096

Zimmer D, Baum C, Leinweber P, Hrynkiewicz K, Meissner R (2009) Associated bacteria increase the phytoextraction of cadmium and zinc from a metal-contaminated soil by mycorrhizal willows. Int J Phytoremed 11:200–213

Chapter 11
Metal Elements and the Diversity and Function of Ectomycorrhizal Communities

Alexander Urban

11.1 Introduction

A few metal elements, Al, Fe, Ca, Na, Mg and K, are major components of the continental earth crust, while the majority of metals are present in traces only (Wedepohl 1995). Certain metal elements are required as essential nutrients (e.g. Na, K, Ca, Mg, Fe, Zn, Cu) while others have no known biological function (e.g. Al, Pb, Hg). All trace metals can be toxic, depending on concentration and speciation. Human activities have resulted in the release of trace metals into the environment far above natural levels. Metal and metalloid pollution is highly persistent and constitutes a major ecotoxicological concern, particularly in case of highly toxic elements such as As, Pb, Cd, Cu, Ni and Zn. The natural biogeochemical cycling of metals, that is the transformation and translocation of metal elements in the biosphere, has increasingly turned into a recycling of anthropogenic pollutants.

Soils are major sinks of metals emitted by human activities (Berthelsen et al. 1995). The largest quantities of metals are released in the northern hemisphere (Europe, North America, and, increasingly, Asia; Pacyna and Pacyna 2001). Large areas of metal-contaminated terrestrial environments are actually or potentially vegetated with ectomycorrhizal (ECM) forests. Metal-contaminated environments are dynamic systems, the speciation and compartmentation of metal elements depends on interactions of physico-chemical soil properties and biotic activities (Fomina et al. 2007). Recently, free-living and symbiotic fungi have been recognized as important agents in metal biogeochemistry, since they are involved in metal mobilization, transformation and immobilization (Gadd 2007). Ectomycorrhizal fungi (ECMF) are the dominant microbial component of ECM forest soils and key players in metal biogeochemical cycles in these environments, due to their important biomass and due to their manifold interactions with the organic and inorganic soil matrix. Increased interest in ECM–metal interactions was triggered

A. Urban
Department of Systematic and Evolutionary Botany, University of Vienna, Rennweg 14, 1030 Vienna, Austria
e-mail: alexander.urban@univie.ac.at

M. Rai and A. Varma (eds.), *Diversity and Biotechnology of Ectomycorrhizae*, Soil Biology 25, DOI 10.1007/978-3-642-15196-5_11,

by the potential applications of ECMF in phytoremediation of polluted sites. The evolution of metal tolerance in ECMF and potential host protection by ECMF and the role of ECM fungi in metal biogeochemical cycles (Gadd 2007) were covered by several reviews (Hartley et al. 1997; Leyval et al. 1997; Godbold et al. 1998; Jentschke and Godbold 2000; Meharg 2003; Bellion et al. 2006). Here I want to focus on studies that link various aspects of trace metal toxicity to population biology, community composition, biodiversity of ECMF and on the integration of ECMF into food webs.

11.2 Metal Element Cycling in ECM Forests

11.2.1 Biogeochemical Transformations of Metals in ECM Forest Soil

Transformations and transport of metals are part of natural biogeochemical cycles. Changes of metal speciation in biochemical cycles have important effects on solubility, bioavailability and toxicity. The soil pools of trace elements are distributed across different phases and are associated with different matrices: soil solution, living biomass, organic matter, organomineral complexes, clay minerals, and hydrous oxides of Al, Fe and Mn. Main processes that control the distribution, speciation and availability of metals are adsorption–desorption, precipitation–dissolution, complexation, redox reactions and volatilization (Gadd 2007; Huang 2008). Adsorption to minerals, organic matter and microbial or plant cell walls limits the transport in aqueous soil solution, reducing leaching to deeper soil horizons and water bodies. Cu, Pb and Cr are strongly retained by the forest floor, while Zn and Cd are regarded as more mobile. Fungi can bind metals to cell walls or extracellular polysaccharides. Precipitation of metals in the soil solution may occur under the influence of sulphide in reduced environments or by biotic oxalic acid exudation. Complexation reactions are highly diverse and they may involve specialized chelates such as siderophores or diverse organic acids (Gadd 2007; Huang 2008).

11.2.2 Weathering and Mineralization

Weathering of the parental rock matrix is an essential source of most plant nutrients, including biogenic metal elements (Landeweert et al. 2001). Weathering is important in replacing element losses that may occur naturally or anthropogenically through leaky nutrient cycles, leaching, removal of biomass, or as a result of forest fires and subsequent discharge of ashes.

The discovery of micrometer-scale tunnels inside mineral grains (feldspar and hornblende, which contain K, Ca and Mg) in the elution horizons of podzols led to the hypothesis that ECM fungi can actively penetrate and dissolve minerals and transport nutrients directly from minerals to ECM host plants (Jongmans et al. 1997; Landeweert et al. 2001). The direct link between rock and trees was thought to improve tree growth by protecting trees from nutrient deficiency and Al toxicity. It was demonstrated that certain ECM fungi are able to mobilize P or cationic nutrients from minerals such as apatite, biotite, goethite and muscovite. Fungal mobilization and translocation of K, Ca, Mg via ECM hyphae appeared to be driven by reduced bioavailability of these elements (van Schöll et al. 2008). Pot experiments with muscovite as only K source demonstrated an increase of weathering rates by tree seedlings by a factor 1.7, inoculation with *Paxillus involutus* increased weathering to a factor of 3.3, compared to abiotic weathering. Inoculation with *Suillus bovinus* or *Piloderma croceum* did not change weathering rates (van Schöll et al. 2006). Fungi target hyphal growth and weathering agents (organic acids, particularly oxalic acid, and siderophores) at minerals containing nutrient elements (P, K, Ca, Mg). Fungi can dissolve minerals and metal compounds by acidolysis, complexolysis, redoxolysis and metal accumulation. Combined acidolysis and complexolysis by organic acids serving as a source of both protons and ligands was found more effective than protonation alone (Gadd 2007). The process of heterotrophic leaching had been first studied from a biotechnological perspective (Burgstaller and Schinner 1993), before it was recognized as an important ecosystem process. Many minerals contain both nutrients and potentially toxic metal elements. Weathering, whether mediated by ECMF or not, is likely to increase the bioavailability of toxic metals. Al, the most abundant metal element in the earth crust, is toxic when present as free ion. Acidity increases Al availability but may reduce Al toxicity. Dicarboxylic and tricarboxylic acids such as oxalic acid and citric acid are involved in both weathering and detoxification. Complexation with citric acid (Tahara et al. 2005) and precipitation with oxalic acid are effective means of Al detoxification (Gadd 2007).

(Fomina et al. 2005) inferred metal tolerance of ECM fungi from growth rates in metal-spiked growth media and found that tolerant strains solubilized metal-containing minerals more effectively. Opposing results were obtained in an experiment with the ericoid mycorrhizal fungus *Oidiodendron maius*. Metal-tolerant strains from polluted soils did hardly respond to ZnO and $Zn_3(PO_4)_2$ by ligand excretion, while strains from unpolluted habitats increased ligand excretion and Zn solubilization (Martino et al. 2003). The lack of responsiveness to Zn minerals in metal-tolerant *Oidiodendron* is possibly a strategy to save organic carbon, since low-molecular weight organic acid (LMWOA) excretion can constitute a major carbon sink (van Hees et al. 2005). Gibson and Mitchell (2004) confirmed the Glucose dependency of Zn phosphate solubilization in *Hymenoscyphus ericae*.

11.2.3 Litter Decomposition

Litter is a major source of nutrients in forest ecosystems and litter decomposition is essential in nutrient cycling, particularly in ecosystems that are less herbivore-driven. Potentially toxic elements may accumulate in tree foliage and litter in several ways: through uptake from the soil solution via the xylem mass flow, through atmospheric deposition onto the tree canopies, or via translocation of elements from the soil solution to the decomposing litter. Atmospheric deposition is the main route for anthropogenically mobilized metal elements to enter forest ecosystems, with tree canopies being important scavengers of pollutants. The process of litter decomposition involving saprotrophic fungi, bacteria and soil arthropods results in changes of metal speciation, availability and concentration. ECM fungi potentially compete with saprotrophic fungi for nutrients released from decomposed litter and have been reported to partially inhibit litter decomposition (Gadgil and Gadgil 1971; Koide and Wu 2003). Lindahl et al. (2001) analysed the transfer of ^{32}P-labelled P in interactions between saprotrophic and ECM mycelia. The direction and quantity of P transfer was influenced by resource availability. Recent studies by Lindahl et al. (2007) demonstrated a spatial separation of litter decomposition and ECM nitrogen uptake. Saprotrophic fungi were essentially confined to more recently (<4 years) shed litter. Organic carbon was mineralized during decomposition, while N was retained, most likely within fungal mycelia. The dwindling source of organic C as expressed by a decreasing C:N ratio was hypothesized to favour the succession of ECM fungi.

Litter decomposition can be a long-lasting process, particularly in temperature- and nutrient-limited environments. Brun et al. (2008) measured element contents in conifer litter bags exposed on forest soils for as long as 8 years and recorded different types of evolution of mass-normalized concentrations in different groups of elements. Nutrient elements tended to decrease, along with the unessential or toxic elements Rb, Sc, Sr and Tl. The concentrations of most unessential and potentially toxic elements increased during decay. The concentrations of Cd, Hf, Hg, Ta and Zr initially increased, followed by a net decrease. The anthropogenic component of atmospheric deposition resulted in elevated concentrations of Pb, Cd and Hg.

11.2.4 Metal Element Cycling and Translocation by ECMF

ECMF are considered highly efficient recyclers of mineral nutrients, which minimize element losses through leaching. Very high affinities to certain nutrients can result in the uptake of similar non-essential elements, for example Rb and Cs that are accumulated by certain ECMF along with K. High bioconcentration factors (BCFs) for certain non-essential trace elements are a side effect of high affinity essential element capture. The efficiency of ECM element recycling can be illustrated by the observation that radioactive Cs has hardly been leached from forest soils more than 20 years after the fall out impact caused by the accident in the nuclear power plant in Chernobyl (Dighton et al. 2008). The ability of ECM fungi to capture cationic nutrient elements and to limit losses due to leaching is particularly important when the retention capacity of the soil matrix is low and when climatic conditions enhance the leaching potential.

Fungal translocation of metal elements might explain some surprising results from both weathering and litter decomposition studies. Budget studies in boreal podzols indicated an upward transport of Al and Fe from mineral soil to the organic soil layer, a phenomenon that cannot be explained by passive leaching. Smits and Hoffland (2009) tested the hypothesis that fungal mycelia in the E horizon connected to ECM roots in the O horizon might be responsible for the uplift of Al. A two-compartment in vitro experiment demonstrated that *Rhizopogon roseolus* and one isolate of *Paxillus involus* transported Al, while *Piloderma croceum*, *Hebeloma longicaudum* and *Suillus bovinus* did not. Ga, an element with similar properties as Al, was used as a tracer in a pot experiment with *Pinus sylvestris* seedlings planted into a reconstructed podzol. Ga uptake by *Pinus sylvestris* seedlings was largest when naturally present ECMF mycelia but not the tree roots had access to the compartment with Ga added. Comparisons between hyphal transfers of Ga and Al suggested that Ga is a good proxy for Al.

Scheid et al. (2009) compared accumulation and solubility of metals during the decomposition of leaf litter harvested from trees grown in non-polluted and polluted sites and exposed in non-polluted and polluted soils. Decomposition rates were not changed by elevated metal contents, suggesting that the function of decomposer communities was not affected. Leaves from trees grown on polluted soil had larger initial Zn, Cd and dissolved organic carbon concentrations. Total metal contents increased during decomposition in polluted soils. The solubility of metals decreased over time, indicating that litter acted as a temporal sink for metals from the soil. The sorbed metals were strongly bound in the litter 2 years after decomposition. The significant rise of metal contents in uncontaminated litter exposed on polluted soil documents the tight communication of soil and decomposing litter in interchanging elements. The agents of decomposition and metal translocation were not identified in this study, but it is well established that saprotrophic fungi have a primordial role in the earlier phases of litter decomposition, while ECMF take over in later phases.

11.2.5 Relevance of Fungal Soil Metal Transformations

Manifold experiments have demonstrated the potential of ECMF to dissolve nutrient-containing minerals by means of mechanical and chemical attack. Weathering by ECMF is fuelled by host tree organic carbon, weathering agents such as LMWOAs are important sinks of reduced carbon (van Hees et al. 2005). The relative proportion of ECMF weathering is still a matter of controversy. Earlier estimations that ECMF contributed to mineral weathering in a podzol via tunnelling alone by 50% (van Breemen et al. 2000) were probably far too high. Smits et al. (2005) quantified the contribution of ECM mineral tunnelling to total weathering by image analysis of soil thin sections and concluded that this process accounted for less than 0.5% of total feldspar weathering in the upper 2 cm of the mineral horizon of the investigated soil. Surface mineral weathering by ECM fungi was suggested to be quantitatively much more important than tunnelling (Smits et al. 2005). Pot experiments demonstrated that ECM weathering by exudation of LMWOAs can be upregulated under deficiency conditions, this regulatory potential might increase the relevance of ECMF weathering for forest health (van Schöll et al. 2008).

Anthropogenical trace metal contaminations are likely to be metabolized by ECM forest soils in similar ways like naturally occurring metal species, as long as the impact on the forest ecosystem does not result in disruption of ecosystem functions. Berthelsen et al. (1995) investigated a forest in southern Norway that had been submitted to long-term heavy metal emissions. They concluded that nearly all Cu in the upper humus layer might be contained in ECM structures, for Zn and Cd the respective proportions were between 30 and 40%. Only 2% of Pb was found associated with fungal biomass. Importantly, this study included morphotype-based identifications of the ECMF and linked the contribution of individual ECMF species to soil trace metal budgets. The considerable biomass and the larger surface: biomass ratio of ECMF mycelia compared to host root systems may explain the significant metal-binding capacity of ECMF. Most quantitative information about adsorption of trace metals to fungal mycelia was obtained in wastewater treatment studies. High-affinity adsorption of certain trace metals to the mycelia of fast growing saprotrophic fungi was found by Kapoor and Viraraghavan (1997). In ECMF, the adsorption potential is potentially limited by the availability of sorption sites and might be rapidly saturated in metalliferous soils (Godbold et al. 1998). Precipitation of trace metals on or near fungal hyphae, typically as oxalates, is not limited by hyphal surface area.

11.3 Metal Tolerance of ECM Associations

11.3.1 Metal Toxicity

Metal toxicity can interfere with the essential physiological and reproductive processes in fungi (Gadd 1993; Amir and Pineau 1998). Metal toxicity-based antimycotica illustrates the relevance of fungal metal sensitivity. Cu(II)sulphate, a fungicide still used in viticulture, is effective in controlling plant parasitic fungi at concentrations which are not toxic to the host plant.

Metal toxicity depends largely on speciation. Free metal ions, oxyanions and certain organic metal compounds such as methylated Hg are particularly toxic (Gadd 2007).

Mechanisms of metal toxicity comprise the elicitation of oxidative stress (even in non-redox-active metals such as Cd), depletion of antioxidant pools, competitive inhibition of the uptake of essential elements, denaturation of proteins, interference with functional groups of proteins by displacement of essential cationic cofactors, precipitation of P inducing P deficiency and membrane disruption. Cytoplasmatic metal homeostasis is an essential cellular function given the dual role of many trace metals as nutrients and potential toxicants.

11.3.2 Mechanisms of Metal Tolerance in ECMF

Metal tolerance may be defined as the genetically conditioned ability to grow and reproduce in environments with high concentrations of potentially toxic metals (Hartley et al. 1997). A metal-tolerant organism must be able to maintain metal homeostasis in the presence of high metal concentrations, controlling the concentrations of free metal ions in the cytosol. A multitude of mechanisms were proposed to contribute to metal tolerance in ECMF (Hartley et al. 1997; Leyval et al. 1997). Avoidance mechanisms reduce the exposure of ECMF cells to toxic metals and limit metal entry into the cell (Gadd 1993; Hartley et al. 1997). Avoidance mechanisms include the extracellular biochemical transformations of metals discussed above and the regulation of metal uptake: extracellular ligation of metal ions to di- and tricarboxylic acids and chelation with siderophores; extracellular metal immobilization by adsorption to cell walls, pigments and extracellular polysaccharides (fungal slimes); extracellular precipitation as oxalates; restriction of net metal uptake by reduced influx or increased efflux, through changes in the activity or specificity of metal transport channels (Bellion et al. 2006).

Cytoplasmatic and vacuolar sequestration of metals reduces the concentration of free ions in the cytosol. Mechanism of cellular sequestration and detoxification of metals comprise cytoplasmatic chelation by thiols (metallothioneins, glutathione and similar oligopeptides) and metal sequestration in the vacuole. Metal coordination within cells of ECMF was recently analysed by Fomina et al. (2007).

Increased production of cellular redox buffers such as glutathione and of the enzyme superoxide dismutase protects the cells from metal-induced oxidative damage, adding another line of defence against metal toxicity (Ott et al. 2002; Bellion et al. 2006).

The basic mechanisms of cytoplasmatic metal homeostasis used by fungi are shared with other eukaryotes. Studies in model organisms indicate that more cellular functions and molecules are involved in fungal metal tolerance. Kennedy et al. (2008) found many genes involved in Cd tolerance through a screen of knockout mutants of *Schizosaccharomyces pombe*. Their results suggested inter alia an involvement of coenzyme Q_{10} (ubiquinone) in Cd tolerance.

Most metal tolerance mechanisms imply metabolic costs: oxalate and exopolysaccharide production requires organic carbon and increased efflux by metal transport channels is energy dependent. Intracellular metal chelation with metallothioneins and glutathione requires considerable amounts of cysteine. Under acute metal stress, other cysteine-requiring cellular activities such as hydrophobin synthesis are down-regulated in support of intracellular thiol levels (Bellion et al. 2006).

Metal adsorption to cell walls does not require additional synthetic effort but might be of limited relevance due to rapid saturation of potential sorption sites in highly metalliferous soils (Jentschke and Godbold 2000). Down-regulation of the expression of metal influx channels would save metabolic energy and might be the most economic adaptation to constantly increased substrate metal concentrations. The evolution of higher specificity of metal influx channels might improve the discrimination of non-essential and essential elements, but evidence for such a process in ECMF is not yet available.

Hartley et al. (1997) and Meharg (2003) pointed out that ECM and ERM fungi share a long evolutionary history of exposition to toxic metal concentrations, since ECM forests with understoreys of ERM plants cover huge areas with highly acidic soils, and soil acidity increases the concentration of free ions of Al, Fe and Mn. Mechanisms of Al detoxification and of Fe and Mn homeostasis were supposed to confer cotolerance to other metal elements. More particularly, sites with geogenically elevated levels or toxic trace metals are potential hotspots for the evolution of tolerance of certain trace metals. Serpentine soils, the most widespread type of geogenically trace metal enriched substrate, are characterized by elevated levels of Mn, Ni and Cr (Urban et al. 2008; Moser et al. 2008). Other types of metalliferous outcrops are relatively rare and more restricted in surface.

The weathering of soil minerals by certain ECM fungi is considered important for the acquisition of essential cationic elements (van Schöll et al. 2008). By fungal attack on rock, potentially toxic metal ions can be liberated too, depending on the chemical composition of the rock material. Similar mechanisms are involved in ECM weathering and in the detoxification of metal ions by fungi, most importantly the exudation of LMWOAs. Citrate is the main ligand of Al^{3+} in podzolized forest soils (Landeweert et al. 2001) and oxalic acid is the main component of mycogenic precipitates of various metals (Gadd 2007). Extracellular ligation and precipitation, two essential mechanisms of avoidance of metal toxicity, may thus be an evolutionary by-product of the involvement of ECMF in mineral transformations driven by nutrient foraging.

Hartley et al. (1997) and Meharg (2003) hypothesized that cotolerance against several metals was likely to occur and would facilitate the evolution of metal tolerance. This argument may apply at least to very common mechanisms involving simple molecules, for example the up-regulation of oxalic acid exudation. On the other hand, it was frequently observed that increased metal tolerance in ECMF is metal specific, and that specificity in metal tolerance and local metal exposition can be correlated. Metal tolerance in *Suillus* spp. was found strain and metal specific and could be linked to the respective histories of metal exposure (Adriaensen et al. 2005; Krznaric et al. 2009). Recently, it could be demonstrated that metallothioneins of the ECMF *Paxillus involutus* and *Hebeloma cylidrosporum* and the regulation of metallothionein gene expression are metal specific (Bellion et al. 2007; Ramesh et al. 2009). More information on specific mechanisms of metal homeostasis is available for model organisms. A Cd regulated Cd efflux system based on a P1B-type ATPase (PCA1) was recently reported from *Saccharomyces cerevisiae* (Adle et al. 2006). Lin et al. (2008) discovered that a single amino acid change in the vacuolar Zn transporter ZRC1 changed the substrate specificity of the transporter from Zn to Fe.

11.3.3 Protection of Host Trees

Schramm (1966) investigated plant colonization of barren coal mine spoils in Pennsylvania, a highly acidic substrate with toxic metal concentrations. He observed that both planted and spontaneously established trees (*Pinus* spp., *Quercus borealis*, *Betula populifolia*) were consistently associated with ECMF. Fruit bodies of *Pisolithus tinctorius*, *Thelephora terrestris*, *T. caryophyllea*, *Astraeus hygrometricus* and *Inocybe* sp. appeared near young trees established on barren substrate. The majority of spontaneously established pine seedlings were mycorrhized, despite apparent limitation of ECM inoculum. The non-mycorrhized seedlings were reported as stunted and chlorotic, in contrast to the mycorrhized seedlings. Ever since it was observed many times that trees mycorrhized spontaneously or inoculated with appropriate ECMF species resist much better to extreme soil conditions, including metal toxicity. Trees were considered to tolerate metal contamination by means of phenotypic plasticity and ECM symbiosis. It was hypothesized that the presumably shorter life cycles of ECM fungi would open up more opportunity for genetic adaptation. Adaptive genetic change in ECM communities was considered essential for the understanding of the survival of ECM tree species challenged with metal toxicity (Wilkinson and Dickinson 1995). This hypothesis raised numerous questions, some of them being still under discussion.

11.3.4 Are ECMF More Resistant to Toxic Metals Than Their Host Trees?

Hartley et al. (1997) reviewed available data and concluded that a wide range of metal sensitivity can be found in both trees and ECMF, but upper tolerance limits appear to be far lower in the tree species. Growth-inhibiting Cd concentrations range from 0.3 to 30 μM for non-mycorrhized trees and 0.1 to 90 μM for ECM fungi in liquid media. In case of Pb exposition, the respective values were 0.5–230 μM for trees and 125–960 μM for ECMF in liquid media. Hartley et al. (1997) attributed the higher adaptability of ECMF as expressed by higher upper limits of metal resistance to the higher diversity of ECMF compared to ECM trees, and not to the presumably shorter fungal generation cycles (Wilkinson and Dickinson 1995).

11.3.5 Can Metal-Resistant ECMF Confer Resistance to Their Host Trees?

From a co-evolutionary point of view, it seems obvious that a metal-resistant ECMF which supplies mineral nutrients to its host and which depends on organic carbon obtained from the host would benefit more if it alleviates metal stress in its host tree, with other words, selection is likely to favour host protection by ECMF, especially in pioneer populations, where only one tree individual might be available as carbon source. However, there is some experimental evidence that the most resistant ECMF species is not necessarily the most protective one (Jones and Hutchinson 1986). Godbold et al. (1998) concluded that only in a small number of experiments, amelioration of metal toxicity could be demonstrated, and that this was the case for specific metals and certain fungi only. The statement that amelioration of metal toxicity is highly species, strain and metal specific is still valid. However, it has to be considered that many earlier experiments with negative results had used ECMF from unpolluted sites, while amelioration of metal toxicity under experimental conditions had been recorded most often when fungi from metalliferous were used. Later studies provided unequivocal evidence that certain ECMF protect their host trees highly efficiently against specific metals (Adriaensen et al. 2004, 2005, 2006; Krznaric et al. 2009). Given recent evidence that ECMF communities in highly metalliferous soils can be surprisingly diverse, the question arises, if all those metal-tolerant ECMF have a similarly beneficial effect on their host trees. (Colpaert and van Assche 1993) inferred from experimental results in a semi-hydroponic system that species with abundant mycelia have the most beneficial effect. Field observations suggest that species with abundant mycelia such as *Suillus* spp. can be highly metal tolerant and beneficial for their host trees (Colpaert 2008). However, if ECM trees are competent of rewarding the most beneficial ECMF through selective organic carbon allocation, the diversity of metal tolerant ECMF might be per se beneficial.

ECM-associated microbes are likely to be important, too. Coinoculation with the ECM-associated bacterium *Pseudomonas putida* improved the growth promoting effect of *Amanita rubescens* on *Pinus sylvestris* exposed to Cd (Kozdroj et al. 2007).

11.3.6 ECMF and Host Tree Nutrient Status and Metal Uptake

Many metal polluted sites are poor in essential nutrients, and toxic metals can interfere with the uptake of essential nutrients. It is not easy to disentangle the beneficial effects of improved access to nutrients and/or reduced metal uptake. If revegetation of a devastated area is the first goal, this distinction might seem secondary. If metal cycles and the potential contamination of food webs or applications such as phytoextraction are considered, the quantity of metal uptake is of high practical relevance. Experimental results on the transfer of metal elements to trees via ECM fungi are manifold, but in many cases the toxic element filter hypothesis (Turnau et al. 1996) might be applicable. ECM fungi can reduce levels of available metals in soil by precipitation and by binding to organic compounds (Huang 2008), they can control symplastic metal transfers and cell wall components with high-metal affinity are likely to reduce apoplastic transport to the fine root. A clear amelioration of Cd toxicity in *Picea abies* seedlings by *Paxillus involutus* was found (Godbold et al. 1998), but Cd uptake was not decreased. Shoot metal concentrations are not necessarily reduced due to ECM colonization, effects on total shoot metal contents can be very variable. Ahonen-Jonnarth and Finlay (2001) observed a positive growth response of Ni and Cd exposed *Pinus sylvestris* seedling upon inoculation with *Laccaria bicolor*. Shoot metal concentrations were not affected, resulting in enhanced total metal uptake. In accumulating tree species, evidence for an ECM filtering effect may be lacking too (Krpata et al. 2009). Again, the metal exposition histories of both the ECMF and the host tree are essential to interpret experimental results.

11.3.7 Can ECM Symbiosis Confer Resistance to Sensitive Host Tree Genotypes?

The role of host sensitivity in the success of ECM associations was rarely investigated. Brown and Wilkins (1985) found increased Zn tolerance due to ECM inoculation in both tolerant and non-tolerant *Betula*. The translocation of Zn to the shoots of *Betula* was reduced, but Zn accumulated in the ECM. The differences of tolerance of the trees as expressed in growth rate and the respective limitations of leaf Zn concentrations were largely maintained. Adaptive tolerance of metal toxicity was suggested to occur in populations of *Pinus ponderosa* (Wright 2007) and *P. balfouriana* (Oline et al. 2000) growing in serpentine soils, but the potential role of ECMF was not assessed. Kayama et al. (2006) observed significantly reduced ECM colonization in non-tolerant *Picea abies* planted into serpentine soil, while ECM colonization was not decreased in serpentine adapted *Picea glehnii*. Two alternative hypotheses are proposed here to be tested in future studies: (a) metal uptake exceeds a critical threshold despite ECM colonization due to the low tolerance of non-adapted trees; (b) the naive, non-adapted host tree fails to select the most beneficial, toxic metal filtering ECMF via selective carbon allocation. In both cases, the metal sensitive tree will fail to grow normally despite nutrients offered by metal tolerant ECMF, the consequent reduction of photosynthesis and carbon supply will reduce colonization intensity and growth of fungal mycelia and destabilize the symbiosis.

11.4 Population Genetics of Adaptive Metal Tolerance in ECMF

Metal-contaminated soils are an attractive model system to investigate environment-driven population genetic processes, and important new insights into the microevolution of adaptive metal tolerance were reported for a few ECM model species. In subpopulations of the ECM basidiomycete *Suillus luteus* growing in Zn polluted soils, considerable genetic diversity was found and no reduction of genetic diversity compared to control populations could be detected using AFLP (Muller et al. 2004) and microsatellite (Muller et al. 2007) population markers. In contrast to a priori expectations, there was no evidence for clustering of subpopulations from polluted vs. unpolluted sites, despite significant differences in metal tolerance. It was concluded that metal pollution had a limited effect on the genetic structure of *S. luteus* populations, and that extensive gene flow and a high frequency of sexual reproduction allowed rapid evolution of tolerance while maintaining high levels of genetic diversity (Muller et al. 2007).

Adaption to Ni toxicity in naturally metalliferous soils was demonstrated in the ECM ascomycete *Cenococcum geophilum* (Gonçalves et al. 2009). Mean in vitro 50% growth-inhibiting concentrations of Ni were about seven times higher in isolates from serpentine (23.4 μg/ml) than in control isolates (3.38 μg/ml). Furthermore, a marginally significant ($P = 0.06$) trend towards a negative correlation between Ni tolerance and growth rates in non-toxic conditions was found. This trade-off had been postulated earlier (Hartley et al. 1997) in order to explain why tolerance of metal toxicity fails to become a frequent trait in non-exposed populations. Moderate costs of metal tolerance are compatible with the observation of considerable variation in metal tolerance in non-exposed populations (Colpaert 2008). In contrast to the results of Gonçalves et al. (2009), Colpaert et al. (2005) found no reduction of growth rates at low Zn levels linked to reduced Zn uptake in Zn-tolerant strains of *Suillus* spp. However, in vitro experiments can at best partially reproduce selective forces in ECM symbioses in natural environments.

Results concerning the population genetics of serpentine colonizing *C. geophilum* are rather contradictory. Panaccione et al. (2001) detected genetic divergence between *C. geophilum* from serpentine and from control sites, while Gonçalves et al. (2007) found no differences linked to serpentine, but this result might be due to a limited sample size. Furthermore, Douhan et al. (2007) reconfirmed that *C. geophilum* s.l. is a species complex and recommended caution when conducting population genetic studies in *C. geophilum* due to the risk of comparing unrelated isolates.

The above-mentioned results suggest that in both anthropogenically and geogenically metal-contaminated soils, tolerance of very high levels of toxic metals can be acquired by adaptive evolution, with or without high rates of genetic exchange with non-exposed populations. It is not clear if adaptive evolution of metal tolerance is widespread among ECMF. Strains of *Paxillus involutus* collected from Zn polluted sites were as Zn sensitive as control strains (Colpaert 2008). In certain ECMF species, constitutive levels of metal tolerance seemingly suffice to survive and compete in contaminated sites, while in others the frequently observed variation of metal tolerance in non-exposed populations may be the base for rapid selection of highly metal-tolerant genotypes.

Adaptive evolution of metal tolerance might be compatible with different population genetic patterns, respectively, reproductive systems and life history traits. *Suillus luteus* is a panmictic, sexually reproducing ECMF, *C. geophilum* is a complex of cryptic species possibly lacking a sexual state. Population genetics might rather be conditioned by life cycle traits than by metal stress. The probability of evolving distinct, specialized genotypes and genetically isolated populations might rather be determined by the structure of reproductive systems than by the nature and intensity of environmental selection pressure. At present, there is no evidence for serpentine driven speciation in ECMF (Urban et al. 2008) and the genetic structure of metal-tolerant *Suillus luteus* is similar like in pseudo-metallophytes (Colpaert 2008).

11.5 Distribution of Metal Elements in ECMF

Most studies on the elemental composition of ECMF were based on sporophores, motivated by the interest in nutritional value and ecotoxicological concerns, since wild edible fungi are an important component of human diets in many parts of the world. Earlier studies on metal contents in macrofungi were reviewed by Kalač and Svoboda (2000). They summarized that BCFs in wild edible mushrooms were found high for Cd (50–300), which is probably the most problematic element in mushrooms, and Hg (30–500) but low for Pb (10^{-2}–10^{-1}). Melgar et al. (2009) confirmed that all fungal species investigated accumulated Hg (BCF $>$ 1). Highest values were found in ECM *Boletus pinophilus* and *B. aereus* and in saprotrophic *Agaricus macrosporus* and *Lepista nuda* (mean BCF between 300 and 450 in the hymenophore). Other ECM species had generally lower BCF values than saproptrophs.

Borovička and Řanda (2007) found Fe accumulation in *Hygrophoropsis aurantiaca* and Zn accumulation in ECM *Russula atropurpurea*. Generally, lower Se acumulation was found in checked ECMF compared to saprotrophs, but some of the highest Se concentrations were recorded in ECMF (*Boletus edulis*, *Boletus pinophilus*, *Amanita strobiliformis*, *Albatrellus pes-caprae*).

Metal hyperaccumulation by fungi was rarely reported. Stijve et al. (1990) measured As hyperaccumulation in *Sarcosphaera coronaria* (100–7,000 ppm) and Borovička et al. (2007) reported Ag hyperaccumulation in *Amanita strobiliformis* (mostly 200–700 mg/kg, highest value 1,253; mg/kg; BCF 800–2,500). Ag is microbicidal at low concentrations. Very high BCFs (about 1,000) were reported for the K analogue Rb in *Suillus grevillei* (Chudzynski and Falandysz 2008).

Bioavailability, nutritional value and toxicity of metals and metalloids in ECMF depend on speciation. Slejkovec et al. (1997) found some relation between As speciation and phylogenetic relationships in mushrooms. Mutanen (1986) reported low bioavailability of fungal Se. The metalloid Se is of interest due to frequent Se deficiency of human nutrition. Serafín Muñoz et al. (2006) suggested that a major part of Se in *Pleurotus ostreatus* is bound to the cell wall component chitin. Serafín Muñoz et al. (2007) demonstrated a speciation-dependent protective role of Se against oxidative damage induced by Cd and Ag in liquid cultures of *Pleurotus ostreatus*.

Krpata et al. (2009) compared Zn and Cd concentrations in fruit bodies of ECM fungi and in leaves of their host trees, metal-accumulating *Populus tremula*, in locations highly contaminated by Pb/Zn smelting. BCFs were based on total metal concentrations in the mor-type organic layer (BCF_{tot}) or on NH_4NO_3-extractable metal concentrations in mineral soil (BCF_{lab}). When plotted on log–log scale, a linear model described well the decrease of BCFs with increasing soil metal concentrations. A better correlation was found for BCF_{lab} than for BCF_{tot}. The observation of decreasing BCFs with increasing substrate metal concentrations is not uncommon in fungi (Gast et al. 1988) and plants (Langer et al. 2009). Differences between fungal genera were found in Zn-BCFs but not in Cd-BCFs. *Tricholoma scalpturatum* (762 ppm), *Scleroderma verrucosum* (598–777 ppm) and *Amanita vaginata* (403–571 ppm) had highest Zn concentrations, *Laccaria laccata* (12.3–93.3 ppm) and *Amanita vaginata* (10.1–48.5 ppm) had highest Cd values. Concentrations of Zn were in the range reported for fungal fruit bodies, Cd concentrations were highly elevated. Cd levels above 10 ppm are typically found in contaminated sites (Gast et al. 1988; Svoboda et al. 2006).

Accumulation and BCFs of Zn and Cd in the host trees were of the same order of magnitude as in the ECM fungi. Studies on metal element concentrations in sporophores demonstrated differences in metal affinities between various fungal species and strains and high metal specificity in fungal BCFs. Fungi growing in substrates with excessive metal concentrations usually have drastically reduced BCFs, probably a result of physiological control of metal uptake. The relative contributions of phenotypic plasticity and population genetic factors to the control of metal uptake rates as expressed by BCFs may vary according to the species concerned. Colpaert et al. (2005) found reduced Zn uptake in Zn-tolerant strains of *Suillus* spp. at low and high Zn concentrations and concluded that partial Zn exclusion contributed most to Zn tolerance. Zn tolerance as expressed by reduced Zn accumulation was specific, the concentrations of other micronutrients were not affected. Despite reduced Zn uptake in tolerant strains, the Zn concentrations in the mycelia of all treatments were very high, reaching up to 15.57 mg/g, a concentration representative of plant hyper-accumulators. Turnau et al. (2001) used micro-proton-induced X-ray emission (PIXE) true elemental maps to quantify metals in cryo-fixed *S. luteus* mycorrhizas collected from Zn wastes. They found similarly elevated Zn concentration in *Suillus* rhizomorphs, in average 12.83 mg/g. Colpaert and van Assche (1992) detected high concentrations of Zn in *Suillus* ECM grown in Zn-spiked substrate, while transfer to the host plant remained low. Representative Zn concentrations in *Suillus* sporophores are 30–150 mg/kg (Kalač and Svoboda 2000), a value about 100 times lower than in *Suillus* mycelia. Studies on metal concentrations in environmental samples of ECM mycelia are still scarce, despite growing awareness of their ecological significance (Finlay 2008). Wallander et al. (2003) investigated the elemental composition of ECM mycelia grown in contact with wood ash or apatite in forest soil. They measured high K accumulation by mycelium of *Suillus granulatus* and high concentrations of Ca, Ti, Mn and Pb in *Paxillus involutus* rhizomorphs. *Piloderma croceum* appears to accumulate and translocate Ca, an element that is scarce in podzols (Blum et al. 2002; Hagerberg et al. 2005).

11.6 Transfer of Trace Metals to Vertebrate Food Webs via ECMF

ECM fungi are a part of terrestrial food webs and of various pathways of human exposure to soil borne contaminants. Apart from direct use of fungi as food, the consumption of partly fungivorous game may be a major source of contaminant exposure in some populations.

The knowledge about heavy metal transfers via food webs to game is still fragmentary, despite concerns about elevated heavy metal concentrations in various species of game.

Pokorny et al. (2004) found a correlation between amounts of fungal spores and Hg levels in row deer (*Capreolus capreolus*) faeces collected in a moderately polluted area in Slovenia. Fungal spores were present in 89% of all faecal samples. Following fungal genera were reported: *Lycoperdon, Calvatia, Hypholoma, Coprinus, Russula, Elaphomyces, Xerocomus, Entoloma, Amanita, Cortinarius, Agaricus, Inocybe, Boletus, Macrolepiota, Suillus* and *Pluteus*. Analyses of heavy metal concentrations in fungal sporophores and in important food plants of row deer showed that Hg and As concentrations were in average by about two to three orders of magnitude higher in fungi, and Cd concentrations were elevated by about one order of magnitude. Pb was accumulated in saprotrophic puffballs (*Lycoperdon perlatum, Calvatia utriformis*) only, but these species accounted for a major part of the spores found in row deer faeces. These results explained well the late aestival and autumnal increase of Hg observed in row deer kidneys. Surprisingly, no parallel increase of As levels was found. This might indicate lower bioavailability of fungal As, or higher relevance of other As sources than those covered by this study.

Cs transfer from soil to roe deer was suggested by Kiefer et al. (1996), based on circumstantial evidence such as elevated Cs concentrations in *Xerocomus badius*, annual variations of roe deer Cs contamination and correlations with precipitation. Hohmann and Huckschlag (2005) investigated the pathways of ^{137}Cs contamination in wild boars in Rhineland-Palatinate, Germany. They found deer truffles (*Elaphomyces granulatus*) in significantly higher proportions (average weight proportion 15.3%) in highly contaminated stomach contents (345–1,749 Bq/kg) and in lower proportions (average weight proportion 1.6%) in less contaminated stomach contents ($\leq$20–199 Bq/kg, median $\leq$20 Bq/kg). ^{137}Cs activity concentrations in *Elaphomyces granulatus* were in average 6,030 Bq/kg (800–18,800 Bq/kg). These results explain why ^{137}Cs concentrations in wild boar meat remain high in many regions more than 20 years after the fallout of the Chernobyl nuclear accident. ECM fungi play a paramount role in the long-term recycling and retention of highly leachable ^{137}Cs in soil organic matter and as a gateway of this radioactive element to vertebrate food webs. Top predators such as *Lynx lynx* may accumulate extremely high concentrations of ^{137}Cs, up to 125,000 Bq/kg, depending on food choice (Åhman et al. 2004).

11.7 Diversity and Structure of ECM Communities Exposed to Metal Toxicity

Decreases in species diversity of ECMF are a common result of metal pollution, suggesting a selective elimination of more sensitive fungi (Gadd 1993; Godbold et al. 1998). Field studies on sporophore production along heavy-metal gradients in north and south Sweden found strong evidence for decreasing ECM diversity and productivity at elevated toxic metal concentrations. Most ECM species were found to decrease with increasing metal (Cu, Zn) concentration, for example *Cantharellus cibarius*, *Cortinarius* spp., *Gomphidius* spp., *Lactarius* spp. and *Russula* spp. Among the more tolerant taxa were *Albatrellus ovinus*, *Amanita* spp., *Cantharellus tubaeformis*, *Laccaria laccata*, *Leccinum* spp. (Rühling et al. 1984; Rühling and Söderström 1990). Colpaert (2008) reported that only four morphotypes of *Pinus sylvestris* ECM were detected in the most polluted area along the Zn pollution gradient in northern Limburg, Belgium. A dark ascomycete from the *Rhizoscyphus ericae* aggregate was a frequent mycorrhiza former in the most polluted plots only, confirming earlier observations by Vralstad et al. (2002).

Recent studies provided deeper insights into the below-ground diversity of ECMF in geogenically and anthropogenically metal enriched soils. Mleczko (2004) described 14 ECM morphotypes associated with *Pinus sylvestris* and *Betula pendula* growing on Zn wastes in southern Poland. Staudenrausch et al. (2005) analysed the ECM community of Birch regenerating on the wastes of a uranium mine in Thuringia, Germany. Twenty-three ECM morphotypes were distinguished, 14 of them were identified by ITS sequence analysis. Total ECM colonization and ECM diversity were lowest on a mine heap without organic layer, intermediate on mine wastes with organic layer and highest in a reference site nearby. All ECMF dominating on the mine heap were present in the reference site, indicating that fungi with high resistance against U toxicity and other stress factors associated with barren substrate had been selected from the local ECM community. Urban et al. (2008) identified 18 different ECMF from eight fungal orders associated with *Pinus sylvestris* in a highly metalliferous serpentine soil. Statistic estimators of species richness indicated the presence of about 30 (Jackknife estimator) to 40 (Michaelis–Menten estimator) ECMF species. Moser et al. (2008) differentiated 18 ECM morphotypes associated with *Quercus garryana*. Fifteen of these morphotypes were found in serpentine soil and 13 in non-serpentine soil. Multivariate analysis detected no differences between the communities. In contrast to the results of Urban et al. (2008), pezizalean hypogeous fungi were frequent and diverse. Brearley (2006) found higher percentage ECM colonization and higher ECM diversity in a pot experiment with *Dryobalanops lanceolata* (Dipterocarpaceae) grown in Ni- and Mg-rich ultra-mafic soil compared to seedlings grown in tropical ultisol. Relatively high diversity of ECMF is not limited to naturally metalliferous serpentine. Hrynkiewicz et al. (2007) detected 14 different ECM fungi, mainly *Tomentella* spp. and *Cortinariaceae* associated with *Salix caprea* at a heavy metal polluted former ore mining site near Freiburg in Germany. Krpata et al. (2008) differentiated as many as 54 ECMF associated with *Populus tremula* in a severely heavy metal-contaminated (Pb, Zn, Cd) site. The community was rich in Basidiomycota and dominated by *Cenococcum geophilum* and tomentelloid fungi. The study site had a very long pollution history of about 500 years, and the mining sites were not far away, hence there had been good opportunity for local microevolution of metal tolerance in many species.

The finding of phylogenetically diverse ECM communities in metal enriched soils indicates that the potential of metal tolerance is widespread among ECMF. In plants on the contrary, the potential to evolve increased metal tolerance appears to be limited to few phylogenetic groups, for example Brassicaceae or Poaceae. Plant communities on metalliferous soils such as serpentine are typically limited in diversity but rich in specialized metallophytes. ECMF communities might be highly diverse, but no metal-specific fungi have been found as yet. Patterns of ECM diversity do not seem to agree with patterns of vascular plant diversity in metalliferous soils.

Are there fungal species, genera or larger phylogenetic groups that are excluded by metal toxicity? The representation of major phylogenetic groups in ECM communities can be very different in similar edaphic conditions. Urban et al. (2008) found most major ECMF orders associated with pines and oaks on serpentine, except Pezizales and Gomphales. Moser et al. (2008) found several pezizalean ECMF associated with oaks on serpentine. Possibly, the potential of various species to colonize metalliferous soils is conditioned by the local evolutionary history rather than by genetic predispositions. At present, our knowledge about constitutive and adaptive mechanisms of tolerance is limited to a few well-studied model organisms. It is unknown whether different ECMF species use similar or diverse tolerance mechanisms, and if their population structures are similar. It appears that metal-tolerant ECMF genotypes are typically derived from local species pools (Staudenrausch et al. 2005; Colpaert 2008).

11.8 Biodiversity and Conservation

Geogenically metalliferous sites are potential evolutionary hotspots of metal tolerance traits, and they use to harbour unique organisms and communities. The essential genetic adaptations and innovations that confer metal tolerance are likely to have evolved originally in naturally metal enriched soils (Ernst 2000). The proximity of natural outcrops of ore and historic mining and smelting sites might have facilitated colonization of mine spoils by locally adapted metal-tolerant plants and fungi. Some historic mining and metal processing sites might merit protection, like naturally metalliferous soils, since they can host specialized communities of metal-tolerant plants and fungi.

The effects of widespread background contamination of soils with toxic metals on ECMF biodiversity, community composition and potential species losses are difficult to estimate. It can be inferred from the relatively rapid recovery of ECMF diversity in historically polluted sites that many species can adapt to high levels of toxic metals. If generic, constitutive mechanisms of metal tolerance suffice to resist moderate levels of metal toxicity, the challenge might not be fatal for most ECM fungal species. Are toxic trace metals involved in the decline of certain rare ECM species such as stipitate hydnoids and certain *Tricholoma* species? The enormous potential of ECMF from different phylogenetic groups to adapt to metalliferous soils might suggest that this risk is of minor concern, and that other types of pollution, particularly N emissions, are more important (Arnolds 1991). However, rare fungi with small population sizes might lack the genetic diversity necessary to adapt to metal toxicity. Given the high BCFs of certain fungal species challenged with elements to which they are not adapted, the accumulation of toxic metal concentrations cannot be excluded. The effects of experimental Ni pollution on an ECM community were assessed by Markkola et al. (2002), but the analysis of the ECM community was too coarse to allow conclusions about the effect of low pollution levels on rare fungi. Nevertheless, this study design hints at an important question: what is the fate of biodiversity, when pollution starts? Science usually enters the stage when public awareness and funding are available, for example when the results of pollution restrictions should be evaluated. Most studies on ECMF diversity have assessed effects of historic pollution events, the impact of ongoing pollution due to emerging industrialization in hitherto uncontaminated areas is rarely documented, but such studies would be required to know more about the impact of metal pollution on pristine ECMF communities. The study of sites with a very long or terminated pollution history rather informs about resilience of ECMF communities, long term impacts on nutrient cycling (Jamnická et al. 2007) and microevolutionary processes (Muller et al. 2007).

11.9 Conclusions

The effects of metal inputs to ecosystems far above natural levels are a priori unpredictable. The study of naturally or anthropogenically metal enriched sites, may assist in understanding the fundamental processes of the biogeochemical cycling of potentially toxic metals in ecosystems and in modelling the potential pathways of metal pollutants. The ECM association of trees and fungi is successful in recycling scarce essential metal elements and in colonizing soils with high levels of toxic metals. Uptake and translocation of metal elements by ECMF account for major deviations from simple models of soil metal budgets. Despite considerable conceptual and analytical progress concerning the fundamental biogeochemical and cellular processes, the role of ECM fungi in metal cycling and budgets has only started to be recognized by modellers (Rosling et al. 2009). Only a few studies attempt to link processes to budgets, and the inference of quantitative information about the role of ECMF in trace metal cycles is still largely based on microcosm experiments. New technologies for nano-scale element and isotope analysis may assist in demonstrating the role of ECMF in metal cycles in the field.

ECMF dispose of a variety of extracellular and cellular metal tolerance and homeostasis mechanisms. Some of these mechanisms can confer cotolerance of various metals, while others are metal specific. Metal-adapted ECMF generally provide better host protection than non-adapted strains.

Accumulation of certain essential and non-essential metals and metalloids (e.g. K, Rb, Cs, As, Se, Zn, Cd, Hg, Ag) is common in ECMF, while other elements tend to be excluded (e.g. Al, Pb). BCFs are species, strain and metal specific. A negative correlation between environmental metal concentrations and BCFs appears to be the rule. Metal hyperaccumulation is rare in fruit bodies of ECMF but appears to be rather common in ECM mycelia, which may accumulate a major part of the soil pools of certain trace metals. The role of ECMF as a nutrient source of rare elements (e.g. Se) and as an important gateway of toxic metals to vertebrate foodwebs in polluted areas merits further study.

ECM communities in metalliferous soils are surprisingly diverse, and the diversity of ECMF is likely to be important in alleviating metal toxicity in host trees. The potential to colonize metalliferous soils is widespread in various phylogenetic groups of ECMF. The microevolution of metal tolerance does not require populational differentiation. From a population genetic perspective, metal toxicity is just another environmental gradient exerting selection pressure. Reports on speciation in ECMF driven by metal tolerance are lacking thus far.

References

Adle DJ, Sinani D, Kim H, Lee J (2006) A Cadmium-transporting P1B-type ATPase in Yeast *Saccharomyces cerevisiae*. J Biol Chem 282:947–955

Adriaensen K, van der Lelie D, Van Laere A, Vangronsveld J, Colpaert JV (2004) A zinc-adapted fungus protects pines from zinc stress. New Phytol 161:549–555

Adriaensen K, Vralstad T, Noben JP, Vangronsveld J, Colpaert JV (2005) Copper-adapted *Suillus luteus*, a symbiotic solution for pines colonizing Cu mine spoils. Appl Environ Microbiol 71:7279

Adriaensen K, Vangronsveld J, Colpaert JV (2006) Zinc-tolerant *Suillus bovinus* improves growth of Zn-exposed Pinus sylvestris seedlings. Mycorrhiza 16:553–558

Åhman B, Wright SM, Howard BJ (2004) Radiocaesium in lynx in relation to ground deposition and diet. Radiat Environ Biophys 43:119–126

Ahonen-Jonnarth U, Finlay RD (2001) Effects of elevated nickel and cadmium concentrations on growth and nutrient uptake of mycorrhizal and non-mycorrhizal *Pinus sylvestris* seedlings. Plant Soil 236:129–138

Amir H, Pineau R (1998) Effects of metals on the germination and growth of fungal isolates from New Caledonian ultramafic soils. Soil Biol Biochem 30:2043–2054

Arnolds E (1991) Decline of ectomycorrhizal fungi in Europe. Agric Ecosyst Environ 35:209–244

Bellion M, Courbot M, Jacob C, Blaudez D, Chalot M (2006) Extracellular and cellular mechanisms sustaining metal tolerance in ectomycorrhizal fungi. FEMS Microbiol Lett 254:173–181

Bellion M, Courbot M, Jacob C, Guinet F, Blaudez D, Chalot M (2007) Metal induction of a *Paxillus involutus* metallothionein and its heterologous expression in *Hebeloma cylindrosporum*. New Phytol 174:151–158

Berthelsen BO, Olsen RA, Steinnes E (1995) Ectomycorrhizal heavy metal accumulation as a contributing factor to heavy metal levels in organic surface soils. Sci Total Environ 170:141–149

Blum JD, Klaue A, Nezat CA, Driscoll CT, Johnson CE, Siccama TG, Eagar C, Fahey TJ, Likens GE (2002) Mycorrhizal weathering of apatite as an important calcium source in base-poor forest ecosystems. Nature 417:729–731

Borovička J, Řanda Z (2007) Distribution of iron, cobalt, zinc and selenium in macrofungi. Mycol Prog 6:249–259

Borovička J, Řanda Z, Jelínek E, Kotrba P, Dunn CE (2007) Hyperaccumulation of silver by *Amanita strobiliformis* and related species of the section Lepidella. Mycol Res 111:1339–1344

Brearley FQ (2006) Differences in the growth and ectomycorrhizal community of *Dryobalanops lanceolata* (Dipterocarpaceae) seedlings grown in ultramafic and non-ultramafic soils. Soil Biol Biochem 38:3407–3410

Brown MT, Wilkins DA (1985) Zinc tolerance of mycorrhizal *Betula*. New Phytol 99:101–106

Brun CB, Åström ME, Peltola P, Johansson M (2008) Trends in major and trace elements in decomposing needle litters during a long-term experiment in Swedish forests. Plant Soil 306:199–210

Burgstaller W, Schinner F (1993) Leaching of metals with fungi. J Biotechnol 27:91–116

Chudzynski K, Falandysz J (2008) Multivariate analysis of elements content of Larch Bolete (*Suillus grevillei*) mushroom. Chemosphere 73:1230–1239

Colpaert JV (2008) Heavy metal pollution and genetic adaptations in ectomycorrhizal fungi. In: Simon V Avery, Malcolm Stratford, Pieter Van West (eds) Stress in Yeasts and Ilamentous Fungi (eds) British Mycological Symposia Series. Academic Press, Elsevier Ltd., London

Colpaert JV, van Assche JA (1992) Zinc toxicity in ectomycorrhizal Pinus sylvestris. Plant Soil 143:201–211

Colpaert JV, van Assche JA (1993) The effects of cadmium on ectomycorrhizal *Pinus sylvestris* L. New Phytol 123:325–333

Colpaert JV, Adriaensen K, Muller LAH, Lambaerts M, Faes C, Carleer R, Vangronsveld J (2005) Element profiles and growth in Zn-sensitive and Zn-resistant Suilloid fungi. Mycorrhiza 15:628–634

Dighton J, Tugay T, Zhdanova N (2008) Fungi and ionizing radiation from radionuclides. FEMS Microbiol Lett 281:109–120

Douhan GW, Huryn KL, Douhan LI (2007) Significant diversity and potential problems associated with inferring population structure within the *Cenococcum geophilum* species complex. Mycologia 99:812–819

Ernst WH (2000) Evolution of metal hyperaccumulation and phytoremediation hype. New Phytol 146:357–358

Finlay RD (2008) Ecological aspects of mycorrhizal symbiosis: with special emphasis on the functional diversity of interactions involving the extraradical mycelium. J Exp Bot 59:1115

Fomina MA, Alexander IJ, Colpaert JV, Gadd GM (2005) Solubilization of toxic metal minerals and metal tolerance of mycorrhizal fungi. Soil Biol Biochem 37:851–866

Fomina MA, Charnock J, Bowen AD, Gadd GM (2007) X-ray absorption spectroscopy (XAS) of toxic metal mineral transformations by fungi. Environ Microbiol 9:308–321

Gadd GM (1993) Interactions of fungi with toxic metals. New Phytol 124:25–60

Gadd GM (2007) Geomycology: biogeochemical transformations of rocks, minerals, metals and radionuclides by fungi, bioweathering and bioremediation. Mycol Res 111:3–49

Gadgil RL, Gadgil PD (1971) Mycorrhiza and litter decomposition. Nature 233:133

Gast CH, Jansen E, Bierling J, Haanstra L (1988) Heavy metals in mushrooms and their relationship with soil characteristics. Chemosphere 17:789–799

Gibson BR, Mitchell DT (2004) Nutritional influences on the solubilization of metal phosphate by ericoid mycorrhizal fungi. Mycol Res 108:947–954

Godbold DL, Jentschke G, Winter S, Marschner P (1998) Ectomycorrhizas and amelioration of metal stress in forest trees. Chemosphere 36:757–762

Gonçalves SC, Portugal A, Gonçalves MT, Vieira R, Martins-Loução MA, Freitas H (2007) Genetic diversity and differential in vitro responses to Ni in *Cenococcum geophilum* isolates from serpentine soils in Portugal. Mycorrhiza 17:677–686

Gonçalves SC, Martins-Loução MA, Freitas H (2009) Evidence of adaptive tolerance to nickel in isolates of *Cenococcum geophilum* from serpentine soils. Mycorrhiza 19:221–230

Hagerberg D, Pallon J, Wallander H (2005) The elemental content in the mycelium of the ectomycorrhizal fungus *Piloderma* sp. during the colonization of hardened wood ash. Mycorrhiza 15:387–392

Hartley J, Cairney JW, Meharg AA (1997) Do ectomycorrhizal fungi exhibit adaptive tolerance to potentially toxic metals in the environment? Plant Soil 189:303–319

Hohmann U, Huckschlag D (2005) Investigations on the radiocaesium contamination of wild boar (*Sus scrofa*) meat in Rhineland-Palatinate: a stomach content analysis. Eur J Wildl Res 51:263–270

Hrynkiewicz K, Haug I, Baum C (2007) Ectomycorrhizal community structure under willows at former ore mining sites. Eur J Soil Biol 44:37–44

Huang PM (2008) Impacts of physicochemical-biological interactions on metal and metalloid transformations in soils: an overview. In: Violante A, Huang PM, Gadd GM (eds) Biophysico-chemical processes of heavy metals and metalloids in soil environments. Wiley, Hoboken, NJ, pp 3–52

Jamnická G, Bučinová K, Havranová I, Urban A (2007) Current state of mineral nutrition and risk elements in a beech ecosystem situated near the aluminium smelter in Žiar nad Hronom, Central Slovakia. For Ecol Manage 248:26–35

Jentschke G, Godbold DL (2000) Metal toxicity and ectomycorrhizas. Physiol Plant 109:107–116

Jones MD, Hutchinson TC (1986) The Effect of Mycorrhizal Infection on the Response of *Betula papyrifera* to Nickel and Copper. New Phytol 102:429–442

Jongmans AG, van Breemen N, Lundstrom U, van Hees PAW, Finlay RD, Srinivasan M, Unestam T, Giesler R, Melkerud P, Olsson M (1997) Rock-eating fungi. Nature 389:682–683

Kalač P, Svoboda L (2000) A review of trace element concentrations in edible mushrooms. Food Chem 69:273–281

Kapoor A, Viraraghavan T (1997) Heavy metal biosorption sites in *Aspergillus niger*. Bioresour Technol 61:221–227

Kayama M, Choi D, Tobita H, Utsugi H, Kitao M, Maruyama Y, Nomura M, Koike T (2006) Comparison of growth characteristics and tolerance to serpentine soil of three ectomycorrhizal spruce seedlings in northern Japan. Trees 20:430–440

Kennedy PJ, Vashisht AA, Hoe K, Kim D, Park H, Hayles J, Russell P (2008) A genome-wide screen of genes involved in cadmium tolerance in *Schizosaccharomyces pombe*. Toxicol Sci 106:124–139

Kiefer P, Pröhl G, Müller H, Lindner G, Drissner J, Zibold G (1996) Factors affecting the transfer of radiocaesium from soil to roe deer in forest ecosystems of Southern Germany. Sci Total Environ 192:49–61

Koide RT, Wu T (2003) Ectomycorrhizas and retarded decomposition in a *Pinus resinosa* plantation. New Phytol 158:401–407

Kozdroj J, Piotrowska-Seget Z, Krupa P (2007) Mycorrhizal fungi and ectomycorrhiza associated bacteria isolated from an industrial desert soil protect pine seedlings against Cd (II) impact. Ecotoxicology 16:449–456

Krpata D, Peintner U, Langer I, Fitz WJ, Schweiger P (2008) Ectomycorrhizal communities associated with *Populus tremula* growing on a heavy metal contaminated site. Mycol Res 112:1069–1079

Krpata D, Fitz W, Peintner U, Langer I, Schweiger P (2009) Bioconcentration of zinc and cadmium in ectomycorrhizal fungi and associated aspen trees as affected by level of pollution. Environ Pollut 157:280–286

Krznaric E, Verbruggen N, Wevers JH, Carleer R, Vangronsveld J, Colpaert JV (2009) Cd-tolerant *Suillus luteus*: a fungal insurance for pines exposed to Cd. Environ Pollut 157:1581–1588

Landeweert R, Hoffland E, Finlay RD, Kuyper TW, van Breemen N (2001) Linking plants to rocks: ectomycorrhizal fungi mobilize nutrients from minerals. Trends Ecol Evol 16:248–254

Langer I, Krpata D, Fitz WJ, Wenzel WW, Schweiger PF (2009) Zinc accumulation potential and toxicity threshold determined for a metal-accumulating *Populus canescens* clone in a dose-response study. Environ Pollut 157:2871–2877

Leyval C, Turnau K, Haselwandter K (1997) Effect of heavy metal pollution on mycorrhizal colonization and function: physiological, ecological and applied aspects. Mycorrhiza 7:139–153

Lin H, Kumanovics A, Nelson JM, Warner DE, Ward DM, Kaplan J (2008) A single amino acid change in the yeast vacuolar metal transporters Zrc1 and Cot1 alters their substrate specificity. J Biol Chem 283:33865–33873

Lindahl BD, Stenlid J, Finlay R (2001) Effects of resource availability on mycelial interactions and 32P transfer between a saprotrophic and an ectomycorrhizal fungus in soil microcosms. FEMS Microbiol Ecol 38:43–52

Lindahl BD, Ihrmark K, Boberg J, Trumbore SE, Högberg P, Stenlid J, Finlay RD (2007) Spatial separation of litter decomposition and mycorrhizal nitrogen uptake in a boreal forest. New Phytol 173:611–620

Markkola AM, Ahonen-Jonnarth U, Roitto M, Strömmer R, Hyvärinen M (2002) Shift in ectomycorrhizal community composition in Scots pine (*Pinus sylvestris* L.) seedling roots as a response to nickel deposition and removal of lichen cover. Environ Pollut 120:797–803

Martino E, Perotto S, Parsons R, Gadd GM (2003) Solubilization of insoluble inorganic zinc compounds by ericoid mycorrhizal fungi derived from heavy metal polluted sites. Soil Biol Biochem 35:133–141

Meharg A (2003) The mechanistic basis of interactions between mycorrhizal associations and toxic metal cations. Mycol Res 107:1253–1265

Melgar MJ, Alonso J, Garcia MA (2009) Mercury in edible mushrooms and underlying soil: bioconcentration factors and toxicological risk. Sci Total Environ 407:5328–5334

Mleczko P (2004) Mycorrhizal and saprobic macrofungi of two zinc wastes in southern Poland. Acta Biol Crac Ser Bot 46:25–38

Moser AM, Frank JL, D'Allura JA, Southworth D (2008) Ectomycorrhizal communities of *Quercus garryana* are similar on serpentine and nonserpentine soils. Plant Soil 315:185–194

Muller LAH, Lambaerts M, Vangronsveld J, Colpaert JV (2004) AFLP-based assessment of the effects of environmental heavy metal pollution on the genetic structure of pioneer populations of *Suillus luteus*. New Phytol 164:297–303

Muller LAH, Vangronsveld J, Colpaert JV (2007) Genetic structure of *Suillus luteus* populations in heavy metal polluted and nonpolluted habitats. Mol Ecol 16:4728–4737

Mutanen M (1986) Bioavailability of selenium in mushrooms, *Boletus edulis,* to young women. Int J Vitam Nutr Res 56:297–301

Oline DK, Mitton JB, Grant MC (2000) Population and subspecific genetic differentiation in the Foxtail pine (*Pinus balfouriana*). Evolution 54:1813–1819

Ott T, Fritz E, Polle A, Schützendübel A (2002) Characterisation of antioxidative systems in the ectomycorrhiza-building basidiomycete *Paxillus involutus* (Bartsch) Fr. and its reaction to cadmium. FEMS Microbiol Ecol 42:359–366

Pacyna JM, Pacyna EG (2001) An assessment of global and regional emissions of trace metals to the atmosphere from anthropogenic sources worldwide. Environ Rev 9:269–298

Panaccione DG, Sheets NL, Miller SP, Cumming JR (2001) Diversity of *Cenococcum geophilum* isolates from serpentine and non-serpentine soils. Mycologia 93:645–652

Pokorny B, Al Sayegh-Petkovšek S, Ribarič-Lasnik C, Vrtačnik J, Doganoc DZ, Adamič M (2004) Fungi ingestion as an important factor influencing heavy metal intake in roe deer: evidence from faeces. Sci Total Environ 324:223–234

Ramesh G, Podila GK, Gay G, Marmeisse R, Reddy MS (2009) Different patterns of regulation for the copper and cadmium metallothioneins of the ectomycorrhizal fungus *Hebeloma cylindrosporum*. Appl Environ Microbiol 75:2266–2274

Rosling A, Roose T, Herrmann AM, Davidson FA, Finlay RD, Gadd GM (2009) Approaches to modelling mineral weathering by fungi. Fungal Biol Rev 23:138–144. Available at: http://linkinghub.elsevier.com/retrieve/pii/S1749461309000207.

Rühling Å, Söderström B (1990) Changes in fruitbody production of mycorrhizal and litter decomposing macromycetes in heavy metal polluted coniferous forests in North Sweden. Water Air Soil Pollut 49:375–387

Rühling Å, Bååth E, Nordgren A, Söderström B (1984) Fungi in metal-contaminated soil near the Gusum Brass Mill, Sweden. Ambio 13:34–36

Scheid S, Günthardt-Georg MS, Schulin R, Nowack B (2009) Accumulation and solubility of metals during leaf litter decomposition in non-polluted and polluted soil. Eur J Soil Sci 60:613–621

Schramm JR (1966) Plant Colonization Studies on Black Wastes from Anthracite Mining in Pennsylvania. American Philosophical Society. Available at: http://www.jstor.org/stable/1006024. Accessed 3 Dec 2009

Serafín Muñoz AH, Kubachka K, Wrobel K, Gutierrez Corona JF, Yathavakilla SKV, Caruso JA, Wrobel K (2006) Se-enriched mycelia of *Pleurotus ostreatus*: distribution of selenium in cell walls and cell membranes/cytosol. J Agric Food Chem 54:3440–3444

Serafín Muñoz AH, Wrobel K, Gutierrez Corona JF, Wrobel K (2007) The protective effect of selenium inorganic forms against cadmium and silver toxicity in mycelia of *Pleurotus ostreatus*. Mycol Res 111:626–632

Slejkovec Z, Byrne AR, Stijve T, Goessler W, Irgolic KJ (1997) Arsenic compounds in higher fungi. Appl Organomet Chem 11:673–682

Smits MM, Hoffland E (2009) Possible role of ectomycorrhizal fungi in cycling of aluminium in podzols. Soil Biol Biochem 41:491–497

Smits MM, Hoffland E, Jongmans AG, van Breemen N (2005) Contribution of mineral tunneling to total feldspar weathering. Geoderma 125:59–69

Staudenrausch S, Kaldorf M, Renker C, Luis P, Buscot F (2005) Diversity of the ectomycorrhiza community at a uranium mining heap. Biol Fertil Soils 41:439–446

Stijve T, Vellinga E, Herrmann A (1990) Arsenic accumulation in some higher fungi. Persoonia 14:161–166

Svoboda L, Havlícková B, Kalac P (2006) Contents of cadmium, mercury and lead in edible mushrooms growing in a historical silver-mining area. Food Chem 96:580–585

Tahara K, Norisada M, Tange T, Yagi H, Kojima K (2005) Ectomycorrhizal association enhances Al tolerance by inducing citrate secretion in *Pinus densiflora*. Soil Sci Plant Nutr 51:397–403

Turnau K, Kottke I, Dexheimer J (1996) Toxic element filtering in *Rhizopogon roseolus*/*Pinus sylvestris* mycorrhizas collected from calamine dumps. Mycol Res 100:16–22

Turnau K, Przybylowicz WJ, Mesjasz-Przybylowicz J (2001) Heavy metal distribution in *Suillus luteus* mycorrhizas – as revealed by micro-PIXE analysis. Nucl Instrum Methods Phys Res Sect B 181:649–658

Urban A, Puschenreiter M, Strauss J, Gorfer M (2008) Diversity and structure of ectomycorrhizal and co-associated fungal communities in a serpentine soil. Mycorrhiza 18:339–354

van Breemen N, Finlay R, Lundström U, Jongmans AG, Giesler R, Olsson M (2000) Mycorrhizal weathering: a true case of mineral plant nutrition? Biogeochemistry 49:53–67

van Hees PA, Jones DL, Finlay R, Godbold DL, Lundström US (2005) The carbon we do not see – the impact of low molecular weight compounds on carbon dynamics and respiration in forest soils: a review. Soil Biol Biochem 37:1–13

van Schöll LV, Smits MM, Hoffland E (2006) Ectomycorrhizal weathering of the soil minerals muscovite and hornblende. New Phytol 171:805–814

van Schöll LV, Kuyper T, Smits M, Landeweert R, Hoffland E, Breemen N (2008) Rock-eating mycorrhizas: their role in plant nutrition and biogeochemical cycles. Plant Soil 303:35–47

Vralstad T, Myhre E, Schumacher T (2002) Molecular diversity and phylogenetic affinities of symbiotic root-associated ascomycetes of the Helotiales in burnt and metal polluted habitats. New Phytol 155:131–148

Wallander H, Mahmood S, Hagerberg D, Johansson L, Pallon J (2003) Elemental composition of ectomycorrhizal mycelia identified by PCR-RFLP analysis and grown in contact with apatite or wood ash in forest soil. FEMS Microbiol Ecol 44:57–65

Wedepohl H (1995) The composition of the continental crust. Geochim Cosmochim Acta 59:1217–1232

Wilkinson DM, Dickinson NM (1995) Metal resistance in trees: the role of mycorrhizae. Oikos 72:298–300

Wright J (2007) Local adaptation to serpentine soils in Pinus ponderosa. Plant Soil 293:209–217

Chapter 12
A Conceptual Framework for Up-Scaling Ecological Processes and Application to Ectomycorrhizal Fungi

Virgil Iordache, Erika Kothe, Aurora Neagoe, and Felicia Gherghel

12.1 Introduction

Molecular biologists do not attempt to simply up-scale knowledge concerning macromolecules to the scale of organisms, but use intermediary theories, while ecologists make intensive efforts to up-scale knowledge concerning the functioning of tiny organisms to the large scale of ecosystems and landscapes. There is an implicit recognition of the fact that the first case (up-scaling from macromolecules to organisms) represents a nested hierarchy of systems, with each hierarchical level characterized by new, emergent, and in principle irreducible properties (although reduction attempts are heuristically valuable). In the second case (up-scaling from tiny organisms to ecosystems), the underlying assumption is that ecological processes are basically of the same type over a large range of scales. The problem is to find a meaningful way to aggregate the processes over different hierarchical levels. Hence the question: Is the concept "hierarchy of systems" consistent in biological sciences?

Mycorrhizal fungi, particularly the ectomycorrhizal fungi – (EMF), are well suited to study up-scaling because of their important role in controlling the functioning of forest ecosystems in view of climate change (see O'Neill et al. 1991). Recent reviews deal with methodological aspects of the problem (Pickles et al. 2009) and with mechanisms controlling structural changes of the communities over

V. Iordache (✉)
Department of Systems Ecology, University of Bucharest, Bucharest, Romania
e-mail: virgil.iordache@g.unibuc.ro

E. Kothe
Microbial Phytopathology, Institute of Microbiology, Friedrich Schiller University, Jena, Germany

A. Neagoe
Research Center for Ecological Services (CESEC), Faculty of Biology, University of Bucharest, Bucharest, Romania

F. Gherghel
UMR 1136 Interactions Arbres/Microorganismes, INRA 54280, Champenoux, France

M. Rai and A. Varma (eds.), *Diversity and Biotechnology of Ectomycorrhizae*, Soil Biology 25, DOI 10.1007/978-3-642-15196-5_12,

a wide range of scales (Wolfe et al. 2009). Wolfe et al. (2009) briefly discuss the functioning of mycorrhizal communities only to underline the need of research in this area.

In the present chapter, we focus on the functioning of EMF over scales, which, however, needs a clear concept about the structural aspects. The goal of the text is to provide an analytical framework for up-scaling processes involving EMF. Such a framework has not been proposed in literature but can be developed and then checked for consistency with the existing literature. As the processes involving EMF are ecological, one cannot look for the existence of an analytical framework specific only to EMFs. Consequently, we present a general up-scaling framework and then apply it to EMF.

We start with an overview of approaches to ecological up-scaling (Sect. 12.2), introduce a framework (Sect. 12.3), and within this framework identify the relevant scales for EMF functioning (Sect. 12.4.1) from a structural (Sect. 12.4.1.1) and functional (Sect. 12.4.1.2) point of view. The problems in interpretation of EMF responses to disturbances and their succession over scales (Sect. 12.4.2), of their management for ecosystem services (Sect. 12.4.3), and modeling (Sect. 12.4.4) provide a complementary perspective. Research directions are presented (Sect. 12.5) and conclusions end the text.

12.2 Addressing the Up-Scaling Problem

The idea to understand the relationship between patterns and processes at various scales within a hierarchy of nested systems is currently common to all ecological disciplines, including EMF ecology (Dahlberg 2001). The study of this relationship can start with the pattern (detection, description, and analyses of the underlying process) or with the process (description, simulation, and pattern generation) (Schroder and Seppelt 2006). The second approach is more interesting scientifically because it allows for mechanistic, theory-based research questions, a recognized need in the field of EMF ecology (Bruns and Kennedy 2009).

An objective hierarchy of systems has widely been assumed, but some researchers have arrived at the conclusion that "ecological hierarchy and associated scales do not exist per se; rather they represent an instrument constructed by the observer or modeler" (Lischke et al. 2006), and "hierarchy theory is but one way of viewing nature" (O'Neill 1996). Indeed, if one looks at the hierarchical approaches available in the literature, there is no general approach to the scale problem. The classic hierarchy of increasing scale systems accepted by most researchers is organism–population–community–ecosystem–landscape. The diversity of views comes when putting into practice this formal pattern. This can be illustrated by the study of Bailey (1987), looking for objective criteria for delineating large scale landscapes (ecoregions). The spatial scales of the landscape units are the plant (10^{-4}–10^{-1} m^2), the patch (10^0–10^4 m^2), the flowpath

(10^2–10^5 m^2), the landscape (integrative flow system 10^6–4×10^6 m^2), and the region (10^7–10^{10} m^2) (Reynolds and Wu 1999). According to these authors, the ecosystem concept fits in the patch level (patch ecosystems form a flowpath). At the ecosystem scale, the search for objective criteria is related to the delineation of boundaries between ecosystems (e.g., Fortin et al. 2000). Within ecosystems, food webs "are portrayed as static networks with highly aggregated trophic groups over broad scales of time and space" (Berg and Bengtsson 2007). Rillig (2004) uses three hierarchical levels when conceiving the effect of mycorrhizal fungi on ecosystem processes: the mycorrhizal fungi, the plants (community and composition), and the ecosystem. Within the soil compartment of terrestrial ecosystems, Miller and Lodge (1997) use a complex hierarchy of soil ecology systems, starting with detritusphere, aggregatusphere, and rhizosphere, continuing with drillosphere and porosphere, and ending with the soil as a whole. When studying the role of arbuscular mycorrhizal fungi (AMF) in controlling the effects of CO_2 on plants and ecosystems, Rillig and Allen (1999) use a hierarchy with two branches: host plant–plant population–plant community and host plant–functional group–ecosystem. Ettema and Wardle (2002) identify as relevant for soil ecology four scales: the fine scale (at root, organic particle and soil structure level), the plot scale, the field scale, and the large scale (gradients of texture and soil carbon, topography, and vegetation). No explicit dimensions in space and time are given for these scales, but they also associate specific ecological processes to each of these scales and discuss mechanisms of process' disturbance at each scale and point out qualitatively the connection between scales ("spatial distribution of soil organisms influence both plant growth and plant community structure"). Scaling is also related to the research on species–area relationships (Dengler 2009a), including those of microbes (Zhou et al. 2008); these relationships depend both on grain size and on the type of organism (Rahbek 2005). A review on the sampling and mathematical issues of this research area (Dengler 2009b) suggests a stratified approach at large scale (by environmental gradients, vegetation types, and landscape sectors) coupled with a nested one at small scale with five subscales (0.01, 0.1, 1, 10, and 100 m^2), thus at least six scales in all (five small and one large).

The limits of such approaches are:

- If the hierarchy does not exist per se, then there is no objective character of the entities at the hierarchical levels used, and there is no objective knowledge. This is a conclusion that is difficult to accept within a science.
- If the ecosystems and landscapes include both abiotic and biotic parts, it is not clear why the classical nested hierarchy includes only biological systems (organisms) as its elementary subsystems. Not including the abiotic factors at all levels leads to bizarre conclusions that the elementary units of the landscapes, such as the "area" proposed by Lepczyk et al. (2008) in a landscape ontology, refer only to biological systems such as populations and communities, and the abiotic factors appear only at larger scales (ecosystem and landscape).

- The aggregated trophic groups in ecosystems, including those of soil (Klironomos et al. 1999), are characterized by very different biomass turnover rates and species' dynamics in time; so it is not appropriate to treat the food web as characterized by an aggregated space-time scale.
- Stating that there is an "influence" of processes occurring at different scales is far from being enough because this influence depends also on the plastic phenotypic answer of organisms to soil heterogeneity (Maestre et al. 2003).
- An explicit statement of the time scale of the pattern is needed as well because the patterns can change in time in function of cyclic (e.g., seasonal) or long term driving forces (Wong and Asseng 2006). The evolutionary time scale should not be forgotten because evaluating the significance of biodiversity only at the time scale of ecosystems may lead to the unproductive (from conservation point of view) idea of functional redundancy not only in general but also in the case of soil (Beare et al. 1995).

Pickles et al. (2009) formulate in the end of their review a number of research questions, the first one of which is crucial for linking space-time scale with function: what constitutes a belowground community of EMF, and is it possible to determine the limits of a given "community"? In the next part, we will try to answer this question and to produce a theoretical framework surpassing the limits of existing approaches. Our method follows and generalizes Pahl-Vostl (1995), who came up with a singular approach for identifying ecosystems and clearly delineating communities. Besides the functional niche in differentiating the modules, she proposes the use of biomass turnover rate (inversely related to the length of the life cycle) and of the location in space-time. The application of these criteria would lead to a "trophic-dynamic module" (TDM). She defines a TDM as the group of biological populations having (1) similar rates of biomass cycling (inversely correlated with lifetime of the individuals), (2) the same location in space and time, and (3) similar roles of the species in the food web. Application of criterion one leads to dynamic classes of populations, and further, application of criterion two leads to dynamic modules, which by criterion three are split into TDMs. Pahl-Vostl's method of systems identification in this formulation still has two problems (a) it does not explicitly address the abiotic part of the ecosystem, so productivity can be described only at large ecosystem scales (or the problem is to exactly up-scale from smaller scale to ecosystem scale), and (b) some populations can have structural parts with very different turnover rates, or with different functional niches, and it is not clear how a population can, in this case, be included into a single TDM.

It is clear that a conceptual framework for integrating biotic and abiotic processes at all scales is needed. This is obvious now at least for theoreticians and modelers who are in a position to integrate the available information for fundamental or applied purposes. For instance, Seppelt et al. (2009) argue that modeling with reliable simulations of the human–environmental interactions necessitates linking modeling and simulation research much more strongly to science fields such as landscape ecology, community ecology, ecohydrology, etc. Such linking cannot be done in the absence of a formal cross-scales conceptual framework.

12.3 An Analytic Conceptual Framework for Integration Across Scales

A thorough conceptual analysis of the terms used in up-scaling research cannot have a place here for reasons of space, but it is useful to note that even some classic concepts such as ecosystem are under strong criticism in the current literature. O'Neill (2001) discusses logical and scientific problems associated with this concept in an article entitled "Is it time to bury the ecosystem concept?" He proposes that an ecological system is composed of a range of spatial scales from the local system to the potential dispersal range of all of the species within the local system. With the same critical attitude, Reynolds and Wu (1999) ask: "Do landscape structural and functional units exist?"

Looking for the ontological status of ecosystems and landscapes, in the last decade, we have approached the up-scaling problem in various contexts and produced step by step, new theoretical elements. Based on the analyses of biodiversity structure in large rivers, we introduced the concept of emergent TDM, characterizing each hierarchical ecosystem level (Vadineanu et al. 2001). Then, in the context of the natural capital management, we discriminated between the concepts of natural capital and of ecological systems (Iordache and Bodescu 2005). More recently, we advanced ideas about the integration across scales in the context of integrated modeling in the biogeochemistry of metals (Iordache et al. 2009a) and for assessing the effects of disturbance on ectomycorrhiza diversity (Iordache et al. 2009b). Here, we synthesize and further develop these ideas.

12.3.1 Developmental System as the Basic Unit of Ecological Functioning

In the most general sense, when we declare of an organism that "it functions," we mean that it produces biomass, that it reproduces, and that it has biological productivity. This process conceptually supports both the standard view of ecosystem functions (flow of energy, cycling of matter, self-regulation) and evolution (recall Darwin's 1859 "law of growth with reproduction" characterizing the organisms), but at different space and time (ST) scales. However, an organism as an isolated system cannot "function," because when the organism grows and reproduces, it uses natural resources and services. Darwin did not state this explicitly, but it is an implicit assumption of the fight for existence introduced in his explanatory argument for natural selection (the fight for existence cannot be directly deduced from an empirical "law of growth with reproduction" without scarcity of resources). Darwin avoided putting growth and reproduction in functional terms (i.e., to say "the organism uses natural resources and services *in order to* grow and reproduce") because he avoided making use of any teleological principle, being a convinced Newtonian. But in the current biological thinking, we are used to speaking about teleonomic behavior of the organisms, the behavior *as if* they would pursue the goals to grow and to reproduce. From such a line of argument (developed in more detail in Iordache 2009b), we first provide a definition for the basic unit of ecological functioning: a developmental system (DS) is a teleonomic entity within its environment producing natural resources and services. Two remarks are needed now (1) if we include the abiotic resources in the developmental system, it is not clear why we would not include the biotic resources too because the DS productivity depends also on biotic resources, and (2) if we include entities with positive value (resources or entities producing services – e.g., dispersal services) for the biomass production, we should also include entities with negative value because both of them influence productivity. In definition, a DS is then a teleonomic entity (TE) and all environmental entities with value (EV) for it (Iordache 2009a). To avoid confusion, we mention that EVs are not isomorphic with the multidimensional niche in the ecological sense because they refer not only to external entities but also to internal ones. Another argument for this lack of overlapping is provided in Sect. 12.4.1.1.4.

DS is a useful theoretical concept because it fits the formal structure of any biological organism with structural (internal, e.g., genes) EVs and external EVs (as "perceived" in its environment), fits the formal structure of human organizations (with organizational leader in the position of TE and organizational goal as goal function, the organization system as an extended body with internal EVs, and the organizational environment in place of external EVs), and fits the structure of management projects (as short-lived organizations).

The fact that it accommodates both natural and human systems provides a common theoretical basis for describing coupled natural-human systems (socio-ecological systems), the connection between biological productivity and organizational productivity, and for modeling such coupled systems.

12.3.2 Epistemic Status of the Developmental Systems

Here, we provide a way to "translate" the conceptual framework introduced in Sect. 12.3.1 into measurable phenomena and empirically based scientific knowledge. A DS can be modeled by a state space and the "law of growth" can be formulated in a such state-space. Usually, any system's "goal" is formulated as an attractor point or region in the state space of the properties of the systems or as a maximization or minimization of an index derived from the space parameters (e.g., ascendency in ecology). Here, we introduce the goal of "growth" through an underived variable (e.g., biomass) or set of variables (accounting for the life cycle of the organism) in the state space and the laws of growth through mathematical functions relating the numerical values of goal variable(s) and the numerical values of other parameters of the state space. We mention without demonstration that efficiency and effectiveness of the DS can be introduced from the form of these mathematical functions, and apparent and general fitness can be related to efficiency and effectiveness, respectively; this is a way to a common formalization of ecological and evolutionary theory. As far as the "reproduction" part of the Darwinian law is concerned, it cannot be formulated in the same organismic state space but actually involves a multiplication of the organismic state space within the integrating populational system (formally as subspaces). The full "law of growth with reproduction" presupposes the existence (and, from an epistemic point of view, the modeling) of a system with at least three hierarchical levels: parts of DS (TE and EVs), the DS per se, and a population of DS. In order to model the relationships between the DSs of a population or of a community, one needs to integrate the state spaces of each DS in a higher dimensional mathematical space. Part of this mathematical space should describe the "objective" environment between the DSs and include the mathematical functions showing how the common use of this environment by different TEs leads to a change in the value of its entities for other TEs. The integrated model would include two types of models: teleonomic (subjectivistic) of the DS and objectivistic (without goal parameters) submodels (Fig. 4 in Iordache et al. (2009a) illustrates this point). Both types of models need the concept of physical time, but the concept of physical space is requested only by the objectivistic models. The objectivistic model may be needed in two types of situations (a) the environment of the organisms is very heterogenous, and the organisms may deal with different values of the environmental parameters, (b) the organisms perceive the environment in different ways, the value of the environment for each organism being different as a result of their perceptual differences. If one assumes that such situations do not occur, then we arrive to a simplified state space having only one set of parameters for the environment and several goal parameters (reflecting the number of organisms in the intra- or interspecific group). In the case of natural selection modeling, this simplification leads to the replecement of the DS with the organism as the selection unit. However, it is currently accepted that the manner in which organisms perceive the environment is in itself a trait supporting the sorting of organisms by natural selection, and the role of environmental heterogeneity in the selection process is documented as well. For theoretical reasons (including compatibility with cultural and economic selection theories), it is more appropriate to consider the developmental system as the unit of selection, which boils down, in biology, to organisms as units of selection in a simplified model of the environmental pressure, as we have seen. Describing the formal structure of these models and deriving from them the Price equation (Price 1995) is a matter of ongoing research. Other aspects of the epistemic status of DSs as related to their scale will be tackled in Sect. 12.3.4.

12.3.3 *Scale and Productivity*

We limit this discussion to biological productivity. The timescale of DS productivity is short and limited to the life of the systems. One scale effect on biological productivity is given by the ST scale of the TE (organism here). Another scale effect is related to the fact that populations of DS of the same species have a larger ST scale and productivity (due to intraspecific relationships and processes, including reproduction). Still another scale effect is related to the communities of DS of different species, larger in ST scale and having larger productivity as a result of interspecific relations. One could think further about metapopulations and metacommunities of DS and so on. All these systems with larger ST scales are ecological systems but with various degrees of complexity. This view of ecological systems is partly convergent with O'Neill's (2001) view mentioned above, according to which an ecological system crosses a range of scales. We underline that we have not introduced here the concept of "population" and "community" in the classic sense (of biological individuals) but as populations and communities of developmental systems, including both organisms and their environment. Also, there is not a simple ST nesting hierarchy of these systems, and actually there is no hierarchy at all (see next point for the hierarchy as epistemically conditioned) because each DS is characterized by its own ST parameters and usually does not "stay" within the limits of an ecosystem in the standard sense. The more appropriate representation is that of a network of overlapped basic ecological systems (DSs) functioning and interacting through objective (nonteleonomic) systems at various ST scales.

12.3.4 Epistemic Status of Complex Ecological Systems

The fact that the ecological systems more complex than a DS have no definite boundaries in the Newtonian space is a problem for their description and management. Operationally, one can study or manage only structures well defined in a three-dimensional physical space, not in an *n* dimensional mathematical space. Even one DS may spread over many scales, if the EVs from their ET's developmental system spread over many scales, namely from scales smaller than ET's source scale (e.g., fungi for plants) to scales much larger than the source scale (large herbivorous mammals, or seed eating birds, for plants). The solution to this problem is a rough discretization, the modularization of the system (pointing out its "compartmentalized architecture" – Moore et al. 2007).

For a modularization in the interest of scientific description, we adapt Pahl-Vostl's (1995) trophic-dynamic concept (in line also with Godbold's 2005 and Satomura et al. 2006 views) in the following way: a functional dynamic modules (FDM) is a group of TEs or of parts of TEs having (1) similar rates of TE biomass cycling (inversely correlated with lifetime of the individuals), (2) the same location in space and time, and (3) similar functional niches, i.e., relations with TEs of the same or of different scales. We prefer the term "functional" to "trophic" because not just the trophic relations count for the differentiation of functional niches. Another reason is that we want this concept to be applicable to nonbiological TEs as well. With this concept, we tackle the TE part of the ecological systems. The abiotic parts (physical solids, liquids, and gasses) might be tackled by an apparent nested hierarchy approach at the scales resulting from the TE systems modularization (Iordache et al. 2011). Down-scaling and up-scaling in their case is a problem of physical modeling in a large sense.

Specific to the scientific modularization is that the ST scale is not constrained by the manageability of the delineated system. Some populations of ETs can be included in more than one FDM at the same time because of their internal structural diversity (Iordache et al. 2009b) For instance, populations of deciduous tree DSs have parts with very different rates of biomass cycling, like leaves and wood (criterion 1), as well as parts with different locations in space like below- vs. aboveground (criterion 2). Thus, the trees will belong to at least three FDMs: two aboveground and one belowground. The notions of "same order of magnitude," "same location in space and time," and "same role in food web" are to be defined by the researcher and can be applied more stringently or relaxed. In the most stringent application, they will lead to a model identical with the "reality" ("isomorphic" model). If relaxed too much, they will lead to a model too aggregated, having lost the key characteristics of the real system (simplistic model). Only at an appropriate intermediate level will they lead to a model simple enough for explanatory value but keeping the basic characteristic of the system ("homomorphic" model). The scale of the FDMs varies hugely, which implies that there is not one "true" scale for ecological processes. Rather, emergence of new structural (e.g., new FDMs) and functional (e.g., increase in overall biological productivity or changes in the rates of biogeochemical processes) properties should be defined and used to derive the mathematical function that links scale and emergence of new properties in different areas and in different periods of time ("emergence function").

As for the functional niche, we have to clarify what it can mean in a context where there are interactions with systems of many scales. Luck et al. (2003) introduced the concept of a service production unit (SPU) as a subsystem of or a full biological population directly contributing to the production of a resource or service perceived as such by humans. The concept of SPU can be generalized from the perspective of all other species. For instance, roots can be interpreted as an SPU for fungi, and sporcarps as SPUs for fungivorous mammals. This generalization allows for a precise delineation of what part of an EV organism provides value for a TE organism. For EVs with a scale larger than that of the source scale, we answer this question as follows: one has to either produce a model linking the EV with the source scale SPU providing direct services (this works for instance for fine roots of plants as SPUs for EM fungi) or consider the use of source scale TEs by the large scale entity as a biotic internal control parameter connected to a fungal SPU (this could work for consumption of fungal sporocarps by mammals, for instance). By internal control parameters, we mean those describing the influence from inside the DS model, but from scales different than the source TE scale; by external control parameters, we mean those describing the factors influencing from outside the DS model – e.g., large scale physical ones or human action. For entities with a smaller scale than the fungal source scale (e.g., bacteria, tiny invertebrates), one needs to produce a model linking the source scale individual with small scale organisms through the smaller scale SPU providing direct resources and services (e.g., organic exudates, hyphae) to these smaller scale organisms and then to up-scale the results of this model to the source scale to form another internal biotic control parameter. The particular interaction of the source scale individual with each small scale entity will not count, but the overall pattern resulting from the structural and functional characteristics of these small scale FDMs will. We can now interpret the functional niche of a TE as consisting of variables describing the source scale entities with value for the fungi, the source scale SPUs that are part of larger scale entities relevant for fungi, and the internal biotic control parameters. This concept of functional niche is an epistemic one, with no objective reality associated to it. It will not imply that the developmental system of fungi will not continue in reality to be spread across scales but that we have to modularize the scale continuum in order to obtain workable FDMs, which is a strategic part of the epistemic status of the DS. The DS of a TE (the DS's "world") will be modeled not only by the TE's FDMs but also by the entire homomorphic model reflecting also the EVs *direct* relevance for the TE, i.e., the epistemic status of a DS is related to the production of a structural homomorphic model and of the associated mathematical models (and of the hierarchically structured physical abiotic models). Of course, nobody produces such homomorphic models just for one DS, but for populations of DSs of different species. The homomorphic model for a single DS is a theoretical case with number of species one and number of individuals one. This theoretical case is important when one needs to move the discussion to natural selection, by providing a conceptual bridge between ecological and evolutionist theories. An important point is methodological: in an FDM of ecological use, the physical environment associated with it is considered homogenous and perceived identically not only for the individuals of the same species but also for all individuals of different species grouped in that FDM. This leads to the simplification of the state space used in modeling to make it workable for ecological purposes (where the evolution of the organisms is not taken into account) and especially for aggregation in view of up-scaling the ecological processes. Another point is that the space-time windows associated with the homomorphic models should be chosen, especially in the case of managerial modularization, such as to be compatible with the human practical possibilities of action.

The scientific or managerial modularization leads to a nested hierarchy of ecosystems but not a true one (Fig. 12.1). In this framework, it is not the case that "holons at level n form the entities at level $n + 1$" (Lischke et al. 2006) because besides the n level entities, there are also new larger scale entities forming the $n + 1$ level; each eco-level is characterized by structurally new types of FDMs that are not found at lower hierarchical levels (Fig. 12.2). Only the representational three dimensional physical spaces needed for scientific investigation or management are nested and not the productive systems analyzed within this three dimensional space. The standard homomorphic model of an ecological system (including compartments for primary producers, consumers, decomposers, etc.) will be then about interactions of different developmental ecosystems of different scales. What is not seen in the standard representation is the large number of smaller scale populations of TEs

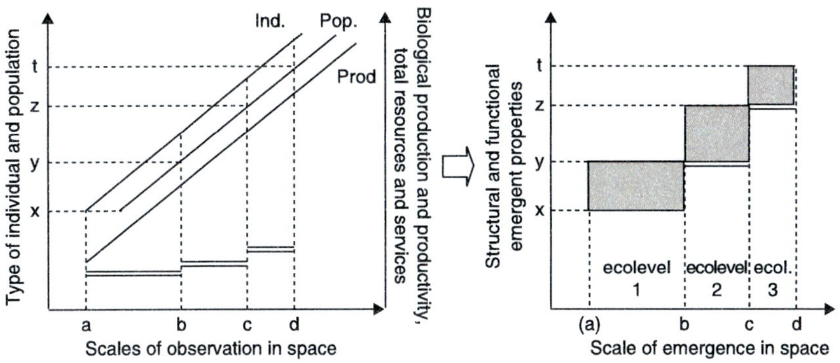

Fig. 12.1 The relationship between the scale of biological structural elements and processes (individuals, populations, *left graph* – left axes, production and productivity, *left graph* – right axes) and the hierarchical structure of ecosystems (*right graph*). At scales of observation from a to b (corresponding to ecological level 1), one can perceive all types of individuals (and their populations) from x to y but only some of the individual types from y to z (and not their populations). The FDMs including populations of type y to z are said to "emerge" at higher hierarchical ecological level 2. *Grey* areas on the *right graph* suggest the multidimensional spaces characterizing each ecological level in which the processes supporting the productivity of each level can be conceptualized. Note that the simplistic linear models (emergence functions) from the *left graph* can be cut in a different way, leading to alternative hierarchies. The real forms of the emergence functions are not linear and depend on the starting point of observation in space

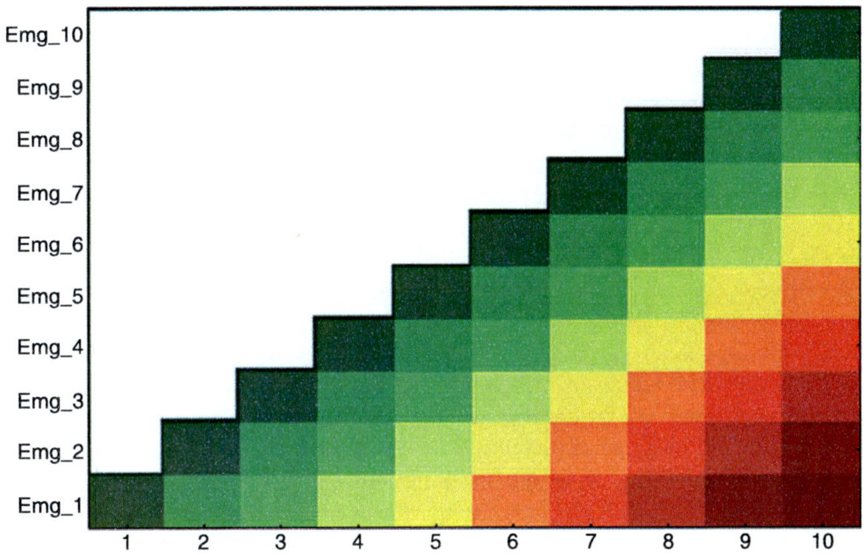

Fig. 12.2 Theoretical relationships between the number of eco-levels and the number of emergent FDMs (communities) of each type (Emg_1 to 10) within an eco-level (1 to 10). The color change from *green* to *red* indicates an increase. Biodiversity of the overall biocenoses (system of communities) is related to the number of FDM types and instances and to the species diversity inside each FDM

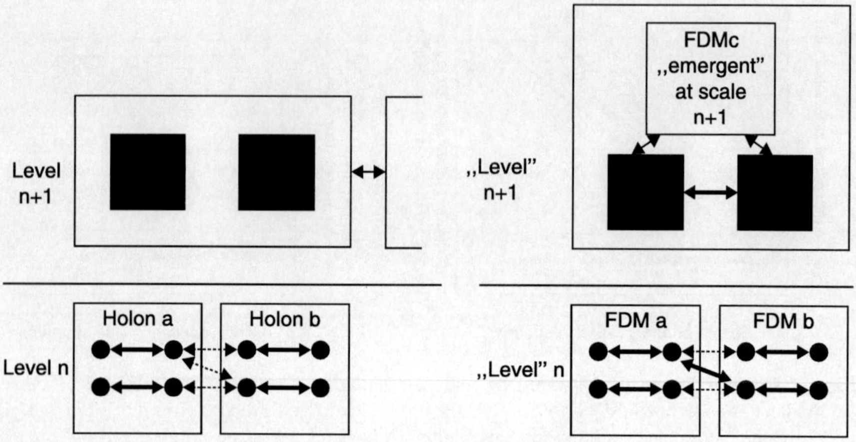

Fig. 12.3 Simplified representation of a true nested hierarchy (*left*, applicable, for instance, to molecules – level *n*, and cells level *n* + 1) and the apparent hierarchy of systems (*right*, resulted from the epistemic modularization of ecological systems). In the apparent hierarchy (abiotic physical parts not represented), the holon at *n* + 1 level includes subsystems that are not present at *n* level, which is not the case for true hierarchies. In the true hierarchy, the interactions within the subsystems of a holon are strong and between holons are weak, while in the apparent hierarchy, they may be very strong between holons (when, for instance, the FDMs include parts of the same population of DSs with different functional niches, or parts of the same DSs with different location in space and time, or different turnover rates)

compared to the large scale ones. Many small scale FDMs are coupled at the same time to a relatively larger FDM (Fig. 12.3).

Usually, the term ecosystem is used both for scientific and for management purposes. This leads to conceptual confusion and, as we have seen, even to proposals of rejecting the ecosystem concept with scientific arguments. Here, we separate the scientific use from the applied, managerial use, by the additional term natural capital. An ecosystem is natural capital when the modularization of the emergence function is done for management goals, i.e., of interest is the value of the ecosystem for humans. Promoting a separation between basic and applied studies is in line with other opinions from the literature. For instance, in a conceptual analysis Yarrow and Marin (2007) conclude that the concept of ecological boundary will find its primary utility within scientific circles, whereas the system-specific transition zone is quite useful in public discourse and socioeconomic decision-making. For a modularization in the interest of management, leading to a *theoretical* hierarchy of the natural capital, one has to use only scales appropriate for the coupling of natural DSs with human developmental systems (organizations, management projects) by natural resources and services relevant to humans. The theoretical natural capital hierarchy resulting from a managerial modularization of the emergence function is a hierarchy of the natural capital, with specific natural resources and services produced at each level (Iordache and Bodescu 2005) and with specific managerial organizations created for ("emergent at") each level. An emergence function for human developmental systems (organizations) can be constructed and modularized leading to a hierarchy of socioeconomic systems, with the importance difference being that this modularization is no longer at the latitude of the researcher but imposed by national and international institutional reality. Actually, the theoretical natural capital hierarchy may follow the socioeconomic current hierarchy for reasons of manageability of the natural modularized entities.

An analysis of the socioeconomic systems (the construction of the organizational emergence function just mentioned) reveals that the *real* structure of the natural capital is not the theoretical one, simply because the modularization of the natural emergence functions is not only a scientific process but also an institutional one associated with the functioning of a socioeconomic system that includes the scientists' organizations. What is interesting in real management is how particular organizations and projects deal with the ecological system, what they perceive as valuable, and how they interact when they want to maximize their separately perceived values (when they are in a conflict of interests, leading eventually to an environmental crisis). Exactly at this point, the key role of the theoretical concept of natural capital is revealed as an attractor for the structure of the real natural capital if one envisages the maximization of the privately and publicly intercepted natural resources and services. There is an institutional evolution of the real natural capital (Iordache 2004) and a cultural history of the natural capital as part of the overall cultural and institutional evolution. The theoretical concept of natural capital is useful also as a reference for conceiving the deterioration process – as structural change leading to the decrease of overall (whatever the intercepting organization is) natural resources and services produced by the natural capital – and for conceiving the restoration process (the inverse of deterioration).

Another important distinction to make is that between production and management of the natural resources and services. The production takes place at all scales of the ecological emergence function, but the management can occur only at the human-relevant scales. One problem is how to take into account the contribution of small scale organisms (like EMF) to the overall natural resources and services production at the management scale. This can be done through the SPU concept. From a managerial perspective, the TEs' populations of the same species (or fragments of such populations) found in an FDM constitute service productive units (SPUs). The SPU concept allows the identification of each species' contribution to the overall theoretical or real natural capital value, and on this basis, the design of targeted management measures. All services provided by SPUs are dependent on the production of biomass, and the material fluxes associated with it, but they can range from providing a source of carbon to a simple physical support (e.g., for smaller scale organisms). One needs long term field studies in order to appropriately assess the biomass production (Staddon et al. 2002), and hence the importance of long term ecological research. Besides this ecological knowledge, one will need for its application socioeconomic knowledge concerning the optimal institutional framework. Optimally, this framework has to allow for the integrated action of the organizations perceiving different values of the natural capital and the integrated implementation of management projects having different standpoints.

12.3.5 Up-Scaling the Ecological Processes

Up-scaling the biomass production function from original DSs of scale 1 to a larger target scale 2 is not a simple physical space-time (ST) procedure because it involves an increase in the complexity of the state space from the original to the target system. The scale issues are related to the position in space-time (ST) of the original TEs as grouped in FDMs of their internal and external EVs, of DSs of larger scale to which they are directly connected, and to the ST patterns of physical processes of larger scale controlling the value of abiotic parameters relevant for the original and large scale TEs.

Up-scaling the biomass production function from the FDMs of source scale to target scale may involve previous down-scaling of other processes. One of these situations occurs when the source scale DSs use as EV an organism of larger scale (for instance as a carbon source). In this case, one needs a model down-scaling the biomass production of that organism to the specific portion of biomass/carbon made available for source scale organisms, which is different from the model describing the overall biomass production of that large scale organism. Another situation is connected to the pattern (perceived at large scale) of distribution in space of the biomass of source scale FDMs, a pattern controlling the choice of mathematical tools for up-scaling (excellently reviewed by Lischke et al. 2006). This pattern may depend in time on external control parameters that can be modeled in space and time only at large, target scale. For instance, if one predicts with hydrogeomorphological models the dispersion of a toxic pollutant with a resolution of 50 $\times$ 50 m, this information should be down-scaled if the DSs controlled by this pollutant, whose up-scaling we are looking for, have a much smaller scale. The particular distribution at small scale can be influenced by the external control parameters, by changing the species location in space, changing the species location in time, or changing the turnover rate of the biomass, with overall consequences on the FDMs structure and associated productivity.

Within the presented analytical framework, the steps for up-scaling the biomass production (and related biogeochemical processes) from the original (source) DS systems could be based on *structural aspects* like (1) ST location and turnover rate of the source TEs (characterizing biomass turnover and position in ST of the TEs by one or several modules reflecting their life cycle and morphological properties), (2) entities providing resources and services to the original TEs (identifying the EVs relevant for each TE type by module and characterizing their ST position), (3) entities using resources and services of the original TEs (identifying the use of source ET modules as EVs by developmental systems centered on types of TE of larger scales and identifying the position of these large scale DSs in the system of target scale), (4) external control factors (identifying the large scale abiotic factors, described as external parameters, influencing the DS of the original scale identified above) and (5) the homomorphic model of the system of DSs and of its integrating system (identifying the overall ST scale of the system of FDMs characterizing the source DS, the relationships of these FDMs with larger scale ones and the external control parameters). In addition, *functional aspects* include (6) biomass production functions at source scale (characterization of the production functions at DS, populations of DSs, and FDMs of the source TEs at the original scale in the field, i.e., implicitly taking into account the influence of smaller scale DSs), (7) large scale DSs' influence on the production function (characterizing the mathematical functions describing the influence of large scale DSs on the productivity of the original DSs), (8) large scale abiotic factor's influence on the production function (characterizing the mathematical relationships underlying the influence of large scale abiotic parameters on DSs at the original scale, the space distribution of the abiotic parameters in the system of target scale with a resolution of original scale, and the mathematical functions predicting their space-time dynamic and the abiotic processes involved at the target scales) and (9) large scale abiotic factor's influence on large scale DSs connected to the source scale (the same as in point 6 done for the large scale FDMs mentioned at point 3 keeping into account only the abiotic processes influencing all DSs whatever the scale). The *integration* relies on (10) stratification (stratifying the system of target scale into strata having the scale of the system of FDMs characterizing the original DS as identified) , (11) production function by strata as controlled by large scale DSs and abiotic factors (assessment of the productivity or other associated processes of interest of original DSs by types of strata, as modeled by the system of FDMs developed within stratum under the control of larger scale DSs and abiotic parameters, providing the possibility to correct for results obtained in step 6 by results from steps 7, 8 and 9, and (12) model up-scaling (extrapolation of production function to the target area of interest by types of strata).

12.4 Application to Ectomycorrhizae

It follows from above that the up-scaling of the function of EMF cannot be done without modeling their function. We are still far from reaching this level of understanding EMF ecology. Below, we will concentrate on the structural and functional aspects, and less on the modeling side (touched on in Sect. 12.4.4) and in, particular, on the following questions:

- How many hierarchical levels do we need to construct and study from the natural emergence function in order to fully understand the role of EMF DSs in the ecological productivity of ecosystems and landscapes?
- What are the appropriate modularization scales for understanding the role of EMFs? ("which are the most relevant scales of analyses for these organisms?" Lilleskov et al. 2004)

- How many types of FDMs are needed for the structural and functional modeling of EMF communities?

These questions are directly relevant to the discussion of EMFs biodiversity. We have shown elsewhere that (Iordache et al. 2009b), for instance, it is not meaningful to interpret EMF patterns of alpha diversity at an ecosystem scale (10^4–10^6 m^2), but that this should be done at a much smaller scale (subtree level differentiated by soil layer). Here, we will not develop, however, this line of interpretation but only take the steps of the analytical framework introduced in the previous part.

12.4.1 Scales Relevant for EMF Developmental Systems

12.4.1.1 Structural Issues

The criteria for delineating FDMs are, as we have seen, the biomass turnover rate, the location in space-time, and the functional niche. The functional niche is discussed in terms of resources used by EMF and organisms using parts of EMF as service providing units.

Dimension and Turnover Rate of EMFs

The simple parameters characterizing an organism are difficult to determine in the case of fungi and even more difficult for EMF. The reasons for this are the complex structure and life cycle of these organisms and the underground location of most of their parts. An analysis of the EMF individual should in principle differentiate between fungal tissue (fruiting body – sporocarp, sclerotium, spore), the plant fungal interface (ectomycorrhizae (EM), also referred to as "root tips" in the literature and further below), and the soil fungal interface (extraradical mycelium or extraradical hyphae, which can differentiate rhizomorphs for long distance transport of nutrients and water) (Satomura et al. 2006). Anatomical details like the dimensions of hyphae, cell walls, etc., are provided in an excellent series of studies on ectomycorrhizae (e.g., Agerer and Weiss 1989 and references therein). The hyphae have diameters of 3–5 μm, smaller than many soil pores (Smith et al. 2010). The importance of each of these potential parts varies with the species. For instance, relatively few EMF are known to form sclerotia, but when present, they support resistance to disturbance and recolonization (Münzenberge et al. 2009). Rhizomorphs can be associated with root tips, with the extraradical hyphae network between trees, or with the sporocarps. Further complications (many plant fungal interfaces) occur in the case of fungi linking several plants. Ectomycorrhizae, rhizomorphs, hyphae, and sporocarps perform different functions, respond differently to environmental conditions, and their life span has consequences on soil biogeochemical processes (Treseder et al. 2005). Most microbial measurements to date, such as biomass or hyphal length, have been single time-point measurements (Allen et al. 2007).

The delineation of a fungal individual starting from its various parts can be done by direct physiological continuity or by genetic identity (Smith et al. 1992). Most of the reported sizes refer to the genets. This may be a false indicator when a physical fragmentation of an individual, especially at ERH level, occurs. In the case of plants, a distinction between genets and ramets (physiological individuals separated from the clone) is made. For fungi, the term "ramet" is not used since anastomoses of physically separated units may occur. Lamour et al. (2007) analyze, for instance, the network of an *Armillaria* species in two 25 m^2 plots of natural soil. They found a density of 4.3–6.1 rhyzomorphs per mm^2. At one site, the network consisted of 169 rhizomorphs as edges and 107 rhizomorphs as nodes. Only two critical rhizomorphal bridges would lead to the separation of significant physiologically independent fragments, so there was "low probability that amputation of a randomly chosen edge would separate the network into two disconnected components" (Lamour et al. 2007). So the dimension of a genet is a fairly good approximation for the dimension of the biological individual in the case of fungi.

Smith et al. (1992) reported a 15 ha large, 1,500 year old genet of a parasitic fungi with an estimated biomass of 10,000 kg. Similar investigations are missing for EMF. While ectomycorrhiza life span depends to some extent on the life span of the host tree, interconnections between trees are generally found and may be seen as indication of similar sizes and life times for EMF. Griffiths et al. (1996) reported that EMF mats may persist 2 years after their host trees have been cut down. The patch size of individual EMF mats studied by Agerer and Gőttlein (2003) were several dm^2. Within the mats, some species were positively correlated to $N\text{–}NH_4^+$ concentration, to total K, Na, Mg, Fe, Mn concentrations and pH, but other species distribution revealed no such correlation. The sizes of the genets of three EMF species studied by Redecker et al. (2001) were 1.5, 9.3, and 1.1 m^2. Bruns et al. (2002b) have mapped and genotyped the fruiting bodies of EMF in a forest in order to compare the pre- and postdisturbance distributions and identify the causes of community reestablishment (dispersal or regeneration from local forms of resistance), and the size of the genets ranged from 1 to 10^2 m^2 (estimates from their map). In reviews, the size of genets is estimated to vary from 10^{-1} to 10^2 m^2 (Lilleskov et al. 2004), 1 m^2 to more than 10^2 m^2 (Godbold 2005), and between 1 and 300 m^2 (Wolfe et al. 2009). Inside a genet, the root tip abundance itself may be patchy (Pickles et al. 2009).

The size of the genets is related to the physiological and reproduction strategies of the fungi (Table 12.1). The extent of extraradical mycelium for the same species

Table 12.1 Comparison of early- and late-stage ECM species characteristics (Iordache et al. 2009b). The relation to exploration types is hypothetical; the other relations are documented by the literature

Species/ characteristic	Reproduction	Genetic diversity	Requirement of C, N, P	Exploration type
Early	Primarily by spores	Higher	Small	Mainly medium and long distance
Late	Primarily by clonal expansion	Lower	Greater	Mainly contact and short distance

may depend on environmental variables (Scattolin et al. 2008), and consequently, the individuals of the same species may have different locations in space as a function of abiotic factors.

The information about the life time of a genet is very important for understanding the speciation process, but from a short and average timescale perspective, the turnover rate of the biomass of a genet's parts is of interest as well. The turnover rate may be estimated not only from the life span of the parts but also as a function of the decomposition of the associated organic matter.

Rygiewicz et al. (1997) found an average median lifetime for mycorrhizal tips of 139 days (lifespan + decomposition), ranging from 123 to 185 days for the eight treatments in the 18 month study period. These authors also review the older literature, some of which report a much longer life time of individual root tips (2–4 years). Treseder et al. (2004) screened the literature showing that ages extended from 1 to 6 years (including errors and sample variability), a range that overlaps with the 0.25–4 year life spans reported in other studies that have visually tracked ectomycorrhizae. The fine root turnover rate is dependent on external, large scale abiotic parameters such as CO_2 concentration (Tingey et al. 2000), with consequences on the root tip FDMs. But "ECM turnover need not precisely mirror root turnover. A number of ecological factors may influence fungal lifespan independently of roots, including life history of the fungal species, predation on the fungi, and shifts in the allocation of host plant C to the fungi" (Treseder et al. 2004).

The hyphal life span is reported to be 5–7 days (Godbold 2005), less than 1 week, although a subset can live for more than 1 month (Staddon et al. 2003). For rhizomorphs, this may be even longer. A discussion about the turnover rate of ERH is made in Godbold et al. (2006), mentioning that the turnover time may be longer than 30 days depending on the methodological details of the estimation. For comparison, turnover rates of leaves are about once per year and those of fine roots about three times per year (Godbold et al. 2006 and citations within). The turnover rate of organic matter derived from dead hyphae is much longer than the life of the hyphae because of recalcitrant substances like chitin. Due to the high turnover rate, "the mycorrhizal external mycelium was the dominant pathway, 62%, through which carbon entered the soil organic matter pool, exceeding the input via leaf litter and fine root turnover" (Godbold et al. 2006). Coutts and Nicoll (1990) report several months of life for rhizomorphs. In another study, rhizomorphs lived an average of 11 months in control plots, indicating that many individual rhizomorphs survive at least part of a nongrowing season (Treseder et al. 2005). Mushrooms (sporocarps) are short-lived, with an age of 1–2 weeks, and are formed by labile carbon derived from the ERH mycelium (Treseder et al. 2004). Based on the above information, we conclude that parts of EMF would fall, based on the biomass turnover criteria, into two dynamic classes: one with fast turnover rates (hyphae and sporocarps) and one with smaller turnover rates (ectomycorrhizae and rhizomorphs).

ST Location of EM Fungi

Locating an EMF individual in soil is a difficult task, taking into consideration its distribution between the interface with the plant, soil, and sporocarps (Pickles et al. 2009). There is an obvious difference between the location of aboveground sporocarps and belowground structures. For the belowground parts of the fungi, there is solid knowledge supporting the idea to differentiate between soil horizons, as documented below.

Vertical location A niche separation of EM species in coarse woody debris and mineral soil was reported by Tedersoo et al. (2003). The web of ERH not only colonized mineral soil, but was also abundant in litter and decaying wood (Buée et al. 2009). ERHs were differently distributed in logs, stumps, forest floors, and mineral soil (Goodman and Trofymov 1998). The vertical, gradual differentiation of EMF community structure with depth has been documented (Landeweert et al. 2003; Calvaruso et al. 2007; Courty et al. 2008). Dickie et al. (2002) also report the vertical niche differentiation of ERH in soil. In a detailed study, Gebhardt et al. (2009) analyzed only a 3 cm organic layer and cut it into 1 cm slices. Even at this resolution, they identified two organic sublayers with different EMF communities (only four species in common). The vertical patchiness of EMFs is related to the distribution of substrates (Genney et al. 2006) and soil horizon properties (Rosling et al. 2003; Baier et al. 2006). This vertical partitioning can be interpreted more generally as niche portioning based on soil chemistry: nitrogen (ammonium) content, base saturation, carbon age, and soil moisture (Peay et al. 2008). Specialization of ERH parts of EMF may also occur with respect to organic matter content, leaf litter type, and litter source (Rillig 2004). Vertical niche partitioning is thought to be one way by which the high species diversity of mycorrhizal fungi can be maintained at small spatial scale (Wolfe et al. 2009). The EMF role changes with depth, not only because some EMF prefer organic or mineral soil layers but also because the number of root tips and mycorrhized root tips (EV for EMFs) vary with depth (Scattolin et al. 2008).

From this information, we infer that the aboveground sporocarps should be included in a separate dynamic module, and the belowground root tips and ERHs should be split vertically at least into two dynamic modules as a function of the organic matter content of the soil layer.

Horizontal location The available information concerning the horizontal distribution of EMF refers to individuals (genets), populations, and communities. Additional information is provided indirectly by niche differentiation as a function of host species and by the possibility of connecting several hosts. Such information can be coupled with the hosts location in ST in order to locate the host-specific EMF.

At the individual level, a cluster of root tips colonized by the same species is likely to be colonized by the same genet (Godbold 2005). EMF with saprophytic abilities colonize (by their ERH) the litter layer and discrete patches of organic nutrients (Graham and Miller 2005). Genet size patterns may be different as a function of the site, and large genets may have smaller scale structures because of fragmentation and intense colonization of microsites (Lilleskov et al. 2004). Another problem in investigating the spatial structure of EMFs can also be the cryptic nature of some genera like *Cenococcum geophyllum* (Pickles et al. 2009). This genus was found to be distributed ubiquitously at a local tree plot scale, but patchily at microscale (within the 5 × 5 m, Matsuda et al. 2009), and with a large genetic diversity even within a single soil core (Douhan and Rizzo 2005). Clearly, genetic tools have to be applied for characterizing the ST location of such species (see Sect. 12.5, research directions). In a study of the genet distribution of sporocarps and ectomycorrhizae of *Suillus* species, Hirose et al. (2004) found in a 20 × 24 m plot four genets from sporcarps, which coincided with those identified for EMFs; the spatial distribution of EMFs of each genet were wider than those of sporocarps, the area occupied by each genet differed considerably within the plot.

Species are likely to differ in spatial colonization patterns because of different internal genet structure and rates of vegetative expansion (Lilleskov et al. 2004). These authors report that dominant EMF taxa showed patchiness at a scale of less than 3 m, with a range from 0 to more than 17 m. Minimal and maximal distances between cores for stand level EMF characterization are proposed to be 0.25 and 300 m (Lilleskov et al. 2004). Metapopulations of EMFs with epigeous fruiting bodies are genetically homogeneous over large distances (1 km), while those with hypogeous fruiting bodies tend to differentiate genetically at much smaller scales (Wolfe et al. 2009).

Griffiths et al. (1996) use two scales for the investigation of EMF: sampling within a stand of 2 × 10 m (to see the effects of forest floor attributes, understory vegetation, and other species of fungi) and a sampling of stands located in the forest (to see the effect of succession gradients). They found no correlation with the forest floor, but proximity between EMF mats, distance from the closest tree, and density of living trees in a stand had an impact. At scales of <4 m, there is a high community similarity, while at scales of <20 cm, the community is temporally dynamic, suggesting a high degree of species turnover probably due to root senescence (Wolfe et al. 2009). "Late stage fungi" can be found on roots closest to the trunk of the tree and "early stage fungi" on roots farthest from the base of the tree (Wolfe et al. 2009, see also Table 12.1), which was interpreted as a succession in ectomycorrhizal development. These patterns may be indicative of niche differentiation by host tree species and the influence of the neighboring tree of the same or of different species on the host tree community, as documented below.

The structure of an ectomycorrhizal community depends on the host trees and host range of the fungi (reviewed by Bruns et al. 2002a; Buée et al. 2009). Sympatric oak species had different EMF community structures (Morris et al. 2008), partly explainable by extractable phosphorus, but mainly attributable to the tree species. Host specificity is a niche dimension in itself (Peay et al. 2008). "Selection pressure for host specificity may not relate as much to interspecific interactions between trees in later stages of succession but rather to adaptations to marginal habitats (post disturbance) by the plant and its fungal symbionts" (Horton et al. 2005). Neighboring tree species identity shaped the EM community structure of the host, and the effects were specific to host–neighbor combination (Hubert and

Gehring 2008); tree species may serve as reservoirs of EM inoculation to one another. At the ecotone of a forest, there was an advancing front of EMFs by dispersal on barren soil, followed by trees, the invasiveness of a tree species being regulated by the spatial pattern of fungal inoculum in the soil (Thiet and Boerner 2007). Oak seedlings were less infected by EM fungi in a forest dominated by a different tree species than in oak forests, with consequences to the productivity (lower dry biomass, Lewis et al. 2008).

Another mechanism supporting the patterns is generated mainly by long range exploration fungi connecting trees and trees with mycoheterotrophic species. An implicit idea from Horton and Bruns (2001) is that this networking by fungi can be made not only by long distance exploration types but also by short distance types when the roots of the trees spatially overlap. Orchid EMF can simultaneously form ectomycorrhiza with forest trees (mycoheterotrophy, Bidartondo et al. 2004). The EMF can interconnect roots of the same or different species (Simard and Durall 2004). The same idea is also supported by Horton and Bruns (2001): dissimilar plants are associated with many of the same EMF on a small enough spatial scale to share those fungi. The observed structure at the tree and intertree levels may be different also because the observation of low abundance fungi as hyphae or tips is not possible. For instance, Koide et al. (2005) observed in a community some species as root tip, but not as hyphae, and vice versa. As long as we are interested in the function of EMF at an ecosystem level on short and average term, these methodological limits are acceptable.

Based on the above elements (which in the case of intertree location also include elements related to the functional niche), one has to separate the fungi species into potentially six dynamic modules in the two dimensional space:

- At tree level (about 4 m around the tree) and at plot level (group of trees and mycoheterotrophic plants occupying a surface of 400–900 m^2)
- Three types of functionally differentiated plots (a) with trees of the same species, (b) with trees of different species, and (c) intertree-mycoheterotrophic plant level
- Tree level dynamic modules in function of their position inside the forest: those at the ecotones between different vegetation patches should be placed in separate strata

Location in time Some EMF populations build up mostly in winter and others in summer periods (Courty et al. 2008; Buée et al. 2009 and the references within). Koide et al. (2007) found an even more complex temporal partioning of niches: three groups of EMF separated in time over 13 months with respect to their hyphae in the bulk soil (ERH) and two groups of EMF separated at the same time with respect to their parts associated to roots. EMF sporocarps may have a specific location in time. For instance, Nara (2008) reports on the seasonality of sporocarp formation in a volcanic desert on Mount Fuji, Japan.

Related to the location in time, some authors use the notion of "ectomycorrhiza turn-over," referring not to the biomass turnover but to the community change over time. Izzo et al. (2005), for instance, conclude that "annual occurrence of the dominant ectomycorrhizal species was constant at larger spatial scales but varied more across years at a fine spatial scale. Turnover of ectomycorrhizal species between years was observed frequently at scales <20 cm." Such information cannot be used for delineating the location in time of the communities unless some periodicity of community structure at the multi-year level is observed (and then the homomorphic model of the study system should be constructed with a multi-year characteristic timescale).

Based on the existing information, it seems that in some cases, several dynamic modules separated in time could be differentiated for each FDM separated in space. However, it is too early to generalize this. The decision in a real situation should be made after at least 2 years of monitoring the structure of an EMF community.

Resources for Ectomycorrhizal Fungi and Their Space-Time Location

In general terms, fungi are considered to be organisms that strongly influence the microscopic and the macroscopic world (Peay et al. 2008) and a good model for experimenting with the coupling between processes of different scales and connecting soil microbes with animal populations via the direct effects on plants (Smith et al. 2010). The EVs with positive value for EMF are soil abiotic mineral and organic parts, litter, plants, and bacteria.

Many basidiomycetes EMF have retained some of their saprophytic abilities, and thus have the potential to access organic sources of nitrogen and phosphorus and to degrade the lignocellulose fraction of dead plant material (Graham and Miller 2005). A key service provided to fungi by plants is the carbon transfer to the fungus in the fine roots (Dell 2002), with the extra carbon accumulating at the edge of the hyphal mat. Fungi also redistribute water from moist layers to upper dry layers ("hydraulic lift"), which is beneficial to soil microorganisms and increases the availability of nutrients to plants (Liste and White 2008). In addition, bacteria, archaea, phages, saprophytic fungi, and soil fauna may interact with EMF (Buée et al. 2009).

Biological factors influencing the structure of soil microbial communities (Buée et al. 2009) are: plant developmental stage, plant species, and plant cultivar (genetic diversity). Mycorrhizae affect the functional diversity of rhizosphere bacteria, fungi, and other microbes (Buée et al. 2009 and the references within). In turn, EMF can be supported by rhizosphere bacteria, a phenomenon that lead to the concept of "mycorrhization helper bacteria," reviewed in detail by Tarkka and Frey-Klett (2008). The distribution of "mycorrhization helper bacteria" followed the vertical EMF stratification both at tree level and between trees (Calvaruso et al. 2007).

A service provided to EMFs by other fungivorous organisms is spore dispersal. A variety of organisms have been shown to move viable spores of mycorrhizal fungi at scales from cms to kms (Wolfe et al. 2009). This is an important aspect because the scale of selection operating on EMF with direct consequences on genetic divergence (and in time speciation) is related to the maximal dispersal capability of the fungi. Direct evidence of EMF spore dispersal by mobile animals was produced and reviewed by many authors (e.g., Johnson 1996; Carrey and Harrington 2001). In an excellent study, Lilleskov and Bruns (2005) found that EMF spore densities were high in the guts of arthropod fungivores (mites, springtails, millipedes, beetles, fly larvae) but present also in arthropod and vertebrate predators (centipedes and salamanders). A low percent of the spores had intact nuclei in predators, but most of the spores in the fungivores had intact nuclei and seemed viable.

Organisms Using Resources and Services Provided by EMF

Organisms using EMF can be interpreted in many cases as EVs with negative value. In the particular case of trees and fungivorous animals, there is a reciprocal use (++ interspecific relationships).

The key organisms benefiting from EMF are trees. EMF assist forest trees in exploiting the soil, in uptaking nutrients by solubilizing soil minerals with organic acids (Buée et al. 2009) and in mobilizing organic forms of nutrients by enzymatic activities (Courty et al. 2005). The variation in EMF perceived by the host plant may be of a discrete (presence – absence of EMF) rather than continuous nature (variation in identity or abundance of EMF) (Karst et al. 2008). The ability of EMF to capture and transport nutrients is believed to be strongly related to the exploration ability and function of ERH. The relevant contribution to nutrient uptake is estimated by the proportion of root tips colonized (Graham and Miller 2005). However, root colonization may not be a good predictor for nutrients uptake (Graham 2008). Instead of this, measurement of lower level mechanisms for nutrient uptake being needed is preferred; but it is difficult to scale up the information thus obtained to the field. Secondary services are also provided. EMF species differ in the ability to capture nutrients, uptake water, protect against pathogens, and increase tolerance to heavy metals (Godbold 2005), unfavorable pH, or salinity (Dell 2002).

EMF can transfer carbon and nutrients between host plants or to mycoheterotrophic plants (Bidartondo et al. 2004; Buée et al. 2009; Leake and Cameron 2010) and can facilitate interplant transfer of carbon, nitrogen, phosphorus, and water, eventually following source-sink gradients between plants (Simard and Durall 2004). Mycoheterotrophic and mixotrophic plants are dependent on the transfer of carbon by EMF networks, which in turn depend on their host photosynthetic plants (review in Selosse et al. 2006). Hyphal connections can also maintain physiological continuity between ramets of plants (Hutchings and Bradbury 1986) or between a tree and seedlings (Simard et al. 1997).

EMF provide an energy supply to the detrital food web as a result of the large hyphae turnover, benefiting saprophytic microbes and other soil organisms (Dell 2002). The EMF mycelium constitutes the largest part of the biomass of most EMF species (Godbold 2005). The hyphal mantle mycelium and extraradical hyphae can have a biomass of 500–700 kg/ha (Godbold 2005). The EMF mycelium supports the activity of free living decomposers (Buée et al. 2009). EMF also provide food by mycophagy of the sporocarps (Dell 2002; details in Sect. 12.4.1.1.3 under the service of spore dispersal).

EMF provide indirect services to many organisms related to the structure of the soil (increasing the formation of soil aggregates) and to the cycling of nutrients (replenishment of the available nutrients pool, Dell 2002). Rillig and Mammey (2006) review how mycorrhizal fungi can influence soil aggregation at various scales. Many services provided by EMF SPUs might occur by this mechanism, but their understanding and quantification is still a matter of further research. EMF can also indirectly affect plants. In a review, Koricheva et al. (2009) analyze indirect effects of mycorrhizal fungi on insect herbivores. They describe significant effects on all aspects of insect herbivores performance, including consumption, growth rate, mass, fecundity, survival, and density. The scale of this influence is dependent on space and time distribution of the insects.

It is clearly documented that soil invertebrates disrupt the carbon flow through mycorrhizal networks by feeding on hyphae (e.g., collembola, – Hiol Hiol et al. 2004, orbatid mites – Schneider et al. 2005). But, taking into consideration the high turnover rate of ERH and its very complex network (Lamour et al. 2007), the negative influence of this consumption on the connections between plants suggested by several authors is questionable (Moore et al. 1985; Johnson et al. 2005). However, predation at the rhizosphere level may have an aboveground effect on plant primary production if such interspecific interactions have an important influence on the overall pattern occurring at the root FDMs level (Moore et al. 2003). Earthworms (Szlavecz et al. 2009) may also disrupt the mycelia through soil mixing and burrowing, and changes in nutrient availability by altering litter quality and quantity occur, resulting in a shift in the composition of the fungal community. Humans also use EMF, but we do not discuss this here because the mechanisms supporting the use of EMF are not biological alone (see Sect. 12.4.3). In light of the literature reviewed here, the functional niches occupied by the EMF seem to be no more diverse than those already identified in Sect. 12.4.1.1.2 based on the species of the host plants and the number of hosts. Before looking at the structure of the resulting homomorphic model (Sect. 12.4.1.1.6), we will briefly mention the external control factors influencing the EMF system.

External Control Factors

External factors should not be confused with the parameters describing them. The parameters resulting from the effect of the factors can be measured at all scales (e.g., concentrations of toxic metals), including within the EMF developmental system, but the action of the factor (pollution with metals by dry deposition) is from outside the system. From a scale larger than their DSs, EM fungi are affected by fire, increase of CO_2 in air, warming, drought, nitrogen deposition, deposition of toxic substances, and management practices. There can be direct effects or/ indirect effects of one factor (e.g., by plants). The effect of coupled external control factors may be different than the separately considered effects. More information on the references supporting this statement and about the effects of these factors is provided in Sect. 12.4.2 under the heading of disturbance mechanisms, because the external control factors are investigated in fungal research only in the context of their effects.

The Homomorphic Model for Up-Scaling

The homomorphic model of the community of fungi DSs is presented in Fig. 12.4. Some of these FDMs could be split into several ones separated in time with seasonal periodicity (see Sect. 12.4.1.1.2).

The extraradical FDMs, located between trees, and the sporocarps' FDMs do not include only the EM fungi but also other fungi if present in the same ST location, The root FDMs, as well, may not only be limited to the EM fungi present in the rhizosphere but could also include other fungi (eventually mycorrhizal) along the mutualism–parasitism continuum (Schulz and Boyle 2005), if present in the same

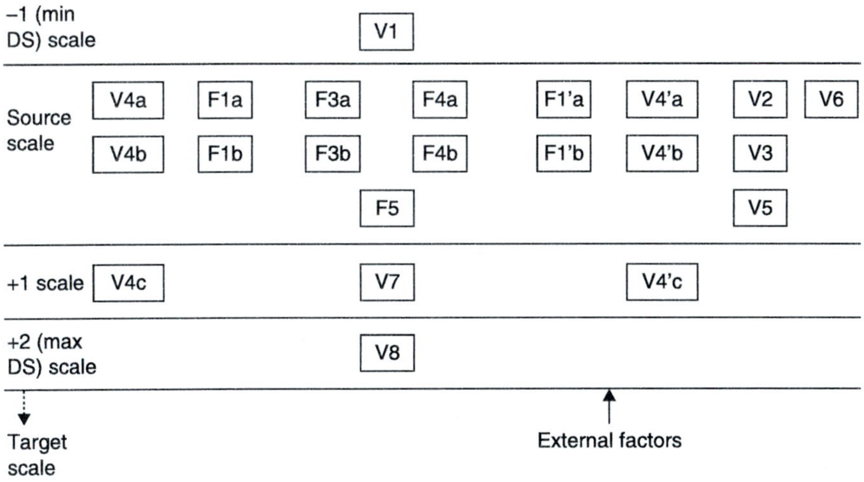

Fig. 12.4 Maximal homomorphic model of a community of EM fungi DSs (relationships not represented for reason of visibility; a connectivity matrix can be easily constructed using the information presented in the text). The scale refers to scale in space, not in time. The physical part is not represented. The part of the model within the source scale (within a stratum) is the homomorphic model for up-scaling under the constraints from FDMs of different scales and external factors. *Legend*: the ET is noted with F (from fungi), the EVs with V (from value); F1a,b = fungi parts (root tips in the case of EMF) in two vertical layers at the roots of tree 1, F1′a,b = fungi parts in two vertical layers at the roots of tree 1′, F3a,b = hyphae in the extra-radical mycelium in two vertical layers, F4a,b = rhizomorphs in the extraradical mycelium in two vertical layers, F5 = sporocarps, V1 = bacteria, V2 = mineral P and N, V3 = organic P and N, V4 = trees (a,b = roots by layer, c = aboveground parts), V5 = mycoheterotrophic plants, V6 = soil microinvertebrates, V7 = fungivorous invertebrates, V8 = fungivorous mammals

ST location. To the extent that interspecific interactions within the fungi species of a FDM are important, they should be taken into consideration when assessing the roles of an individual EMF species, or the EMF community as a whole by each FDM.

We need four (apparent) hierarchical levels for understanding the ecological functioning of EMF communities, and one more level eventually for management purposes, if the target scale (of management) is larger than the maximal DS scale (given by the size of fungivorous mammal populations). One more scale should be added if one looks also for speciation processes (Sect. 12.4.1.2.2). There is a convergence with the opinion of other authors about the number of needed scales but not about exactly what those scales are. Wolfe et al. (2009), for instance, "believe four scales are most relevant for the discussion of spatial pattern and process of mycorrhizal population and communities (1) across landscapes, (2) within plant communities/ ecosystems, (3) within an individual host root system, and (4) within an individual mycelium."

12.4.1.2 Functional Issues

Stratification

According to the review of the literature, the basic unit for stratification of the target system in the case of EM fungi is a 400–900 m^2 tree plot. However, as we have shown in Sect. 12.4.1.1.2, several classes of tree plots should be differentiated as a function of the position in the center or at the ecotone of the target system and as a function of the structure of the vegetation. Other criteria might include management practices. As the differences between the soil horizons of the EMF community can be modified by liming, in addition to the tree host species (Rineau and Garbaye 2009), it means that one should stratify the target system both as a function of the vegetation structure and of (at least in part) the management measures applied. In general, the principle for stratification should be that one stratum type corresponds only to one type of homomorphic model of the EMF community, as reflected both by the fungi per se and by their symbiotic hosts.

Scale-Specific Mechanisms of EMF Productivity

We have touched on these mechanisms in the previous part because researchers rarely look only at structural issues without approaching the functional aspects as well. Here, we attempt to systematize the information from the scales and systems synthesized in Fig. 12.4. Most of the knowledge concerning the productivity mechanisms of EMF is in a verbal, nonmathematical form.

Organism and population level Function at the organism level is supported by metabolic processes and ecophysiological processes (in the classic sense). The scarcity of literature concerning the function of an individual fungi in an ecological context is probably due not only to experimental difficulties but also to a lack of concentrated effort on a model species. Fungi seem to not yet be in the focus of systems biologists. We have not insisted on the exploration of this type of literature. In a singular study, Ritter et al. (1989) analyzed in detail cytological aspects of the stages of ectomycorrhizal vitality. Smith et al. (2010) characterized plant-inducible phosphorus transporters in fungi and mycorrhiza-inducible transporters in plants. Lewis et al. (1994) studied the effect of CO_2 on an individual plant–fungi system; the pattern of results with the same control factor was found to show interspecific variation between EMF species. A new approach for the study of in vitro transport of amino acids in hyphae is proposed by Watkinson et al. (2005). Bonneville et al. (2009) report a first direct observation of the effect of a soil fungus on the surface of a mineral. A recent review (Rosling 2009) summarizes the current knowledge on fungal weathering as affected by experimental setup and conditions (pure or symbiotic growth, nitrogen source, the means of detecting weathering activity and the species examined).

The extra productivity (compared to the individual level) emerging at the population level is in principle related to the intraspecific mechanisms. We have not found literature about such interactions, and this is probably due to the methodological difficulties in delineating such individuals.

Tree and plot level We refer here to the fungal FDMs at tree and intertree levels, within each stratum. Each such FDM should correspond, in our opinion, to a community in the classical sense (it is the system within which alfa diversity should be measured). The extra productivity at this level is due to interspecific interactions; however, the literature does not always make clear what kind of diversity is discussed with respect to its effects.

A review of the EMF community ecology was done by Dahlberg (2001). Data aggregation at different scales may lead to different control variables for the community composition; soil parameters at subplot level, but plant community at plot level (Fitzsimons et al. 2008). We interpret this as being within FDM diversity at subplot level and between FDM diversity at plot level. Some EM species were found significantly more frequently as mycelia than as root tips, while others were less dominant as mycelia than as root tips (Kjoller 2006). We can interpret this as being a consequence of the exploration type in the FDMs structure. Kranabetter et al. (2009) found significant changes in an EMF community along a 25 km productivity gradient (to nitrogen rich sites). EMF species abundance in relation to site productivity included parabolic, negative, linear, and exponential curves. A functional diversity of EMFs was observed, with the specialization of EMF communities contributing to the successful soil exploitation by a single tree host species. The phenotypic plasticity of the tree species was much enhanced as a result of the interaction with the EMF. The diversity described by Kranabetter et al. (2009) seems to be of at least gamma type (aggregates of communities).

The detailed heterogeneity of abiotic parameters at the FDM level might influence the functioning of the community. A detailed review of the space-time heterogeneity of key abiotic parameters at rhizosphere level was done by Hinsinger et al. (2009). They point out that the integration of such models into root growth and root architecture models for up-scaling of rhizosphere processes is still a matter of future research, but no knowledge of fungal distribution at this fine scale (to be related to the abiotic parameters) is referred to.

The interaction of EMF with lower scale EVs increases their productivity. The coinoculation of EMF with bacteria increased mineral weathering, plant uptake of soluble forms of K and Mg, and tree biomass, as compared to simple EMF inoculation, while the bacteria alone had no effect (Koele et al. 2009). The relationship between two plants at the same time may increase the productivity too. Intertree EM linkages can reduce plant competition for resources, promote forest recovery, and influence the pattern of plant succession (Amaranthus and Perry 1994). Seedlings were hydraulically linked by the mycorrhizal network to large trees in a study by Warren et al. (2008).

Some generalizations accepted for AMF could also hold for EMF. The species composition and richness of AMF was found to be an important contributor to plant species composition and productivity, with an effect mainly on subdominant plant species, and with different plant species benefiting from different AMF taxa (van der Heijden et al. 1998). The external driving factors affect both AMF and EMF. For instance, Godbold et al. (1997) found changes in the relative proportion of EMF to AMF as a result of elevated CO_2 concentrations. Consequently, the roles of EMF and AMF in the functioning of the vegetation communities, especially in the up-scaling context, should probably be approached in an integrated manner. It is clear that the FDMs should be seen as based on fungi species, not just separately on EM, AM, or saprophytic fungi, especially in the case of the FDMs associated with the external mycelium.

Ecosystem level We refer here to systems of FDMs from the plot scale up to the "ecosystem" scale, to the control of EMF productivity by feedback from the host plants, and to indirect effects by changing the soil structure. The scale approached here is that between the stratum and the target scale envisaged in the up-scaling procedure. The intermediate scales to be tackled will depend on the organisms retained as relevant from the point of view of direct and indirect interactions with the EMF.

In an excellent review of mycorrhizal–plant–insect interactions, Gehring and Bennett (2009) summarize the proven effects of mycorrhizal plants as follows: negative on root herbivores, positive on pollinators, negative, neutral, or positive on herbivores, and positive or negative on herbivore enemies. Each such group is characterized by specific scales, and the decision whether or not to include these interactions in the overall homomorphic model for up-scaling should be done as a function of the feedback effects of these groups (as influenced by EMFs) on the primary productivity of the ecosystem and on the dynamic of carbon in the forest floor.

Other complex effects come from the action of external driving factors on several types of organisms of different scales. In many systems exposed to elevated CO_2, mycorrhizal fungi sequester increased amounts of carbon in living, dead, and residual hyphal biomass in the soil (Treseder and Allen 2000). When this is coupled with nitrogen deposition, an increased turnover rate of hyphal biomass can occur. The two processes are associated with a shift in the EMF community's composition as a result of physiological interspecific differences.

The facilitative effects of mycorrhizal fungal networks depend on the seedling species identity, mycorrhizal identity, plant species combination, and study system, but seedlings associated with EMF benefited in the majority of reported cases (van der Heijden and Horton 2009). Southworth et al. (2005) applied the network theory to mycorrhizal networks from a phytocentric and fungicentric perspective and concluded that all individual plants are more or less equal in linking fungi, but from a mycocentric perspective the network is scale free (Barabasi and Albert 1999; Barabasi and Bonabeau 2003), meaning that certain species of fungi act as hubs with frequent connections to the other elements of the network.

Productivity may be increased also by changing over time the structure of the soil. While AMF produce large quantities of glycoproteins in soil (1.45 $\times$ 106 g/ha, 3.2% of the total soil carbon in the 0–10 cm soil layer, Lovelock et al. 2004), EMF are characterized mainly by the production of extracellular enzymes responsible for the mobilization of organic carbon and associated nutrients (Maijala et al. 1991; Courty et al. 2005). Both mycorrhizal fungi types also produce effectors responsible for colonization (Martin and Nehls 2009). The influence of AMF's ERH on soil structure might be of even greater importance to the carbon stock than the influence of hyphal standing crops (Miller and Kling 2000), and this statement can hold as well for EMF. This is because, in both cases, there are feedback responses between the soil and canopy mediated by fungi and supported by mechanisms linked to nutrient acquisition and the allocation of the tree's assimilated carbon by mycorrhiza or indirectly by litter fall.

EMF may facilitate the dispersal of plants and thus increase the productivity of the target system. In an interesting article, Thiet and Boerner (2007) studied the EMF's role at the ecotones of a forest. The invasion of a pine species on an unforested area was dependent on the previous dispersals of EMF either as spores by animals or wind, or as hyphae from the ecotonal trees. The latter mechanism proved in this case to be the most important for pine seedlings, underlying the need for a stratified analysis of EMF's role in the central and peripheral parts of the forest.

At the ecosystem scale is supported also the stability of the fungal community, by dispersal between plot scale areas. With regard to the stability issue (and assuming that diversity is directly related to functional stability), in a study of tree islands ranging from <10 to >10,000 m^2 (Peay et al. 2007), the species area slope was similar to slopes for macroorganisms, suggesting that microbes are not ubiquitous even in suitable habitats (if small); the trade-off between dispersal and competition played an important role in structuring EMF assemblages.

Speciation of EMF We refer here to systems larger than ecosystems, at the scale of which speciation of EMF occur. This scale has not been discussed or included in Fig. 12.4 because the up-scaling is usually only needed for management over a short time, without taking into account speciation. Underlining of the importance of evolutionary processes is needed, both for diversity conservation and for the interpretation of the apparent redundancy of species within the FDMs. The large number of rare EMF species (as indicated by non saturated rarefaction curves) indicates a functional redundancy in the relatively short term, but this apparent redundancy in fact supports the stability of the ecological system in the average term and the potential for EMF's evolution in the long term.

High treeless ridgelines are effective barriers to EMF gene flow even at distances less than 65 km, whereas populations (according to Amend et al. 2010, but probably metapopulations) located within watersheds are structured at greater distances (125 km). So the scale at which speciation takes place is much larger than the ecosystem scale and is the maximal one indicated by Wolfe et al. (2009) to be included in the study of EMF. It is documented that the structure of the landscape influences the evolutionary outcome. For instance, Read and Perez-Moreno (2003) point out that selection has favored EMF systems with well-developed saprophytic capabilities in those ecosystems characterized by retention of nitrogen and phosphorus as organic complexes in the soil.

To sum up the information presented in Sect. 12.4.1.2.2, different productivity mechanisms involving EMF are distributed across many scales. The information is scarce at individual level, apparently absent at population level, richer at community and ecosystem levels, and scarce again at large landscape levels. The production takes place at all levels, but with maximal intensity and stability in time at the large landscape scale.

12.5 Disturbance and Succession of Ectomycorrhizal Systems

The change in the relative importance of fungi and bacteria in forest soils with succession seems to remain uninvestigated, but in many secondary succession grassland chronosequence studies, the soil microbial community tends to shift towards a less bacterial and more fungal-dominated food web (Maharning et al. 2009). Hypothetically, this may occur in forests too. A discussion of fungal succession is much more complicated, however, because it envisages processes at very different scales. Iordache et al. (2009b) presented and critiqued the early and late stage species approach for the succession of EMFs and then introduced an ecosystem approach to fungal succession in an improved framework based on Pahl-Vostl's TDM concept. We refer the reader to Iordache et al. (2009b) for aspects relating to succession. In this present text, we develop the analytical framework sketched there and apply it to the more general up-scaling problem of EMF functioning. Another subject discussed in the paper is EMF disturbance due to heavy metals, for which a data processing and interpretation framework was proposed. This picture is complemented here by a short review of disturbance factors relevant to EMF.

Cudlin et al. (2007) review the effects of acidic deposition, nitrogen deposition, increased ozone levels, elevated CO_2, and drought on fine roots and EMF. EMF colonization was not a suitable parameter for assessing the effects of these driving factors, but fine root length and biomass could be useful. This does not mean that the FDMs of EMF have not been affected because they are not delineated in space at the fine root level but at the plant root/rhizosphere level. The disturbance factors usually act at a large scale, but the mycorrhizal response is at an FDM scale and depends on the species composition of the EM community and the relationships of ectomycorrhizal FDMs with other types of FDMs (the structure of the FDM network at stratum scale).

Acid rain reduced the number, length, and biomass of lateral tree roots and the percent and number of EMF (Esher et al. 1992). Some EMF confer drought tolerance to their host (via influencing the plant's osmoregulation), while others confer drought avoidance (by hyphal transport via EMHs, Mudge et al. 1987). The experimental warming of root-associated fungal communities in an arctic region increased the density of different genotypes but did not affect the biodiversity within the time frame of the experiment (Fujimura et al. 2008). Burning either decreases or increases the colonization of EMF. The increase is attributed to the reduction of substances that inhibit germination (Cairney and Bastias 2007), and the decrease occurs mostly in the organic layer, not in the mineral soil. Peay et al. (2009) provide a comparison of disturbance factors by scale. With regard to fire, they conclude that spore heat resistance plays an important role in the disturbance-mediated assemblage shift of EMF. Fire disturbance favors competitively inferior species, keeping diversity of EMF at landscape scale. Nilsson and Wallander (2003) report a negative influence of nitrogen fertilization on the external mycelium of EMF, not directly by the soil nitrogen concentrations but rather by the nitrogen status of the trees.

The effect of enhanced CO_2 concentrations mediated by EM communities takes place through the modification of carbon inputs from plant to soil, with consequences on the biomass, infectivity, and species composition of the symbionts (Diaz 1996). Godbold and Berntson (1997) reported changes in EMF community structure as a result of elevated CO_2. A review by Staddon et al. (2002) about temperature and CO_2 effects on EM fungi concluded that they should involve the study at the individual-plant level, multiple species level, and community level. An interesting finding about the effect of CO_2 is that of Pritchard et al. (2008): "CO_2 enrichment increased mycorrhizal root tip production in deep soil, but did not influence it in shallow soil;" also "the rhizomorph turnover was accelerated in shallow soil, but effects on survivorship in deep soil varied according to diameter." These FDM-specific effects open the way to a line of research on how external control parameters especially influence some of the FDMs from the structure of the homomorphic model. For now, the experiments for assessing the effect of CO_2 have been performed mostly at small plot level (10^{-2} m^2) with fewer at "field" level (10^2 m^2) and especially with monoculture (Staddon et al. 2002).

Miller and Lodge (1997) review the fungal response to disturbances in agriculture and forestry. By disturbance they mean the physical and chemical phenomena that disrupt communities and ecosystems. Fungi are concluded to be control points in management practices (tillage and crop rotation, nutrient additions, air pollution, site preparation, woody debris, opening the canopy, and moisture fluctuations). Jones et al. (2003) review in detail the dynamics of ectomycorrhizal communities after clear-cut logging, identifying the amount and type of inoculum, and the changes in the soil abiotic and biotic environment as the major groups of factors controlling the succession without discussion of diversity indexes as aggregated indicators of ecosystem state. Studies of another forest management practice, gap opening, show a significant reduction in EMF diversity indices and a change in EMF and fine root dynamics compared to closed stands (Grebenc et al. 2009) but with some fungi preferring the new conditions.

Colpaert (2008) reviews the effects of metal on fungi and their adaptation. He concludes that there is true tolerance of EMF to metals. Investigations of EMF species at the community level have revealed wide inter- and intraspecific variations in sensitivity to metals (Hartley et al. 1997). The EMF community in a soil contaminated with metals was rich (not with just a few specialist fungi as expected) but did not vary with the soil horizon, season, or plot location in the forest (Krpata et al. 2008). This suggests an interesting hypothesis that the FDM structure of EMF communities is simplified as a result of toxic pressure compared to uncontaminated areas. Different metals controlled in a specific way the EMF community, the results of a multimetal stress being complex at forest levels and reflecting the distribution of metals at tree scale (Gherghel 2009). Alleviation of metal toxicity in plants through EMF has been demonstrated (review by Jentschke and Godbold 2000), but the mechanism remains unclear. Possibilities include immobilization in fungi, exudation of metal chelating substances in the soil, or nutritional and hormonal effects in plants mediated by fungi. EMF and associated bacteria protect pine seedlings against bioavailable forms of Cd, but there are differences in the level of protection provided by different fungi species (Kozdrój et al. 2007). Bioaccumulation factors for Zn and Cd in fruiting bodies of EMF decreased with increasing soil concentration, showing that in such a case fungi did not act by accumulation as an effective barrier against metal uptake by the symbiotic tree (Krpata et al. 2009).

12.5.1 Role of EMF in the Functioning of the Natural Capital

Through some of its EV, the fungal DS reaches the smaller scale directly relevant for the management of the natural capital (hectares to square kms, here by convention the ecosystem scale). Griffiths et al. (1996) propose an even larger system for analyzing the role of EM SPUs, namely the watershed. This is similar to the scale of EMF speciation. Fungi and their ecosystem services might be in jeopardy if habitat (tree patch) size is a strong determinant of fungal richness, as it seems to be (Peay et al. 2007, 2008).

As an example of the explicit connection between basic and applied ecology in our target domain, Dighton (2003) analyzes the role of fungi in the production of ecosystem services, including mycorrhizal fungi and their relationships with arthropods, as well as the consequences of human action on the relevant mechanisms. Recently, Jackson et al. (2008) also reviewed the effects of root processes (including the dependence on mycorrhization for nutrient acquisition) on ecosystem services. The value of EMF to people is related to local consumption and trade of sporocarps, to their use in medicine, biomonitoring and bioremediation, to their esthetic value, and to the services provided by EM SPUs to other biological compartments supporting the ecosystem function – implicitly the production of other biological resources and services directly relevant to the people (Dell 2002). Hall et al. (2003) reviews the literature about edible ectomycorrhizal mushrooms.

The EMF SPUs are perceived as useful mainly through their support of the resources and services associated with forest ecosystems. Management practices that create intense disturbance and loss of organic matter decrease the ability of plants to form EMF linkages, but management practices that retain living trees and shrubs and the input of organic matter facilitate EMF linkages (Amaranthus and Perry 1994). Helpful forestry management practices targeted to EMF can be taken at various scales: small ones (retaining refuge plants, mature trees) and larger ones (retaining old-growth forest, avoiding high intensity broadcast burn, and retaining the edge to area ratio of harvested areas within certain limits) (Wiensczyk et al. 2002).

Heneghan et al. (2008) review the use of soil ecological knowledge for restoration. From their perspective, when the goal is loosely defined (no specific recovery trajectory envisaged), less precise knowledge can be useful (for instance, the concept of "soil quality" is acceptable in this context for management issues), but when a complex outcome is desired, exact knowledge is needed (we would say, down to the SPU's level). Another point in their analysis (and of others, e.g., Neagoe et al. 2009) is that a restoration project should be seen as a scientific experiment and valued as such by the scientist.

In prescribed burning, unburned patches act as an inoculation source. The return to the preburn state of the EMF community takes place in about 15 years (Cairney and Bastias 2007). The role of fungi in ecosystem restoration after fire (Claridge et al. 2009) is related to the stabilization of the soil in the absence of plants (by some species of fungi), to nutrient acquisition from minerals, and to mycorrhizal function once the plants start to recover (other species of fungi). It can be said that the fungi are strongly involved in the secondary succession and restoration management after such a disturbance is in fact an attempt to control the secondary succession (Neagoe et al. 2009; Iordache et al. 2009b). Change in soil structure is an important process during succession and, as we have noticed in Sect. 12.4.1.1.4, the EM SPUs provide important services by influencing the soil structure.

The use of mycorrhizal fungi can also be targeted to restoration goals related to the control of toxic substances, metals, or hydrocarbons (Robertson et al. 2007; Ghergel 2009). Filamentous fungi and their enzymatic system were found to be a "potent tool to decrease the levels of contaminants in soils by degradation and stabilization" (review by Mougin et al. 2009). Measures taken in the frame of restoration projects, such as liming, strongly affect the EMF community (in this case by pH increase, Kjoller and Clemmensen 2009). The decline of the services valued by humans is, of course, related not to mycorrhizal fungi alone. For instance, Gilliam (2006) identifies six types of mechanisms supporting the deterioration of the herbaceous layer by nitrogen deposition (Fig. 4 in Gilliam 2006): interspecific competition, herbivory, mechanisms related to mycorrhizal infection, pathogenic fungal infection, species invasion, and exotic earthworm activity.

We conclude that understanding the effects of EMF-mediated deterioration at a large scale needs integration in the up-scaling model of all other mechanisms observable at intermediate scales related to SPUs of EMF. The integration of other ecological mechanisms, depending on the service and resources identified, will be needed as well for a full understanding and effective specific goal-directed management of that natural capital.

12.5.2 *Mathematical Modeling*

On the mathematical modeling side, Johnson et al. (2006) provide an excellent review of seven types of mycorrhizal models (that include mycorrhizal parameters in their structure), varying in their scale of resolution and dynamics and discuss approaches for integrating these models with each other and with general models of terrestrial ecosystems. They use the classical hierarchy (individual, populations, communities, and ecosystems). At individual and population levels, there are biomass allocation (functional equilibrium) models, economic biological market models, and integrative agent-based models. At the community scale, there are community feedback models, and co-evolutionary mosaic models. At the ecosystem scale, there are food web models and pedogenesis models. In the concluding part, Johnson et al. (2006) mention that "mycorrhizal effects on resource availability and biomass allocation patterns have not been included in these models, partly because of insufficient information but also because of scaling differences [...] mycorrhizal effects on soil properties, disease resistance, and trophic cascades are not emphasized to the same extent in current models," and "encourage future efforts to develop methods for measuring mycorrhizal structure and function at relevant spatial and organizational scales."

If one looks into each category of models for relevance of the mathematical formalism to EMF, the picture is not encouraging. Functional equilibrium models are exemplified by a conceptual model dedicated to AM fungi, and the primary source cited does not include mathematical relationships between the mycorrhizal biomass and the carbon and nutrient resources. Economic models and food-web models referred also to AM fungi. Agent-based models and pedogenesis models referred both to AM and EMF; community feedback models and co-evolutionary mosaic models referred to mycorrhizal fungi in general.

It seems that the state of mathematical models needed for up-scaling the EMF ecological processes from the stratum scale to the target scale is not yet appropriate, and that developing such models is an important research direction.

12.6 Research Directions

The proposed framework for conceptualizing the DS's functioning across scales is convergent with recent proposals for coupling traditionally small scale targeted ecophysiology to the functioning of ecosystems under the umbrella of a "macrophysiology" (Gaston et al. 2009).

For the up-scaling of EM fungi functions, we must (adapted from Lilleskov et al. 2004) identify individuals in tips, soil (as hyphae), and sporocarps, discern patch change over time, identify endogenous factors (intraspecific such as clonal expansion or high spore rain and interspecific interactions structured by a FDM approach), identify exogenous factors (patterns of resource availability, disturbance history, and current external driving factors), then on the resulting dynamic architecture add the functional information inside each FDM and between FDMs, and finally aggregate it from the original to the target scale. Doing this is still precluded by many gaps, both at structural and at functional levels. A structured analytical approach to the problem (like the one proposed here) might accelerate the knowledge development in the area.

Coupling observations at the molecular scale (plant–fungus gene expression, TE level, and internal EVs), interfacial scale (TE-external EV level, to quantify ion uptake by plants), experiments at pot scales (individual fungi–plant or FDM level), lysimeter scales (FDM level), and plot scales (system of FDMs level) with long term field ecosystem studies is crucial for obtaining the knowledge needed for integrating (up-scaling) lower level processes into the management (remediation, restoration, control of secondary succession) of the natural capital at ecosystem and landscape scales (Neagoe et al. 2005; Graham and Miller 2005; Meixner et al. 2006; Neagoe et al. 2009). Hinsinger et al. (2009) underline the need for the integration of ST models of abiotic parameters in the rhizosphere into root growth and root architecture models for up-scaling of rhizosphere processes. Graham (2008) states that "experimental design should either integrate multiple mechanisms of the landscape scale and include such measures as mycorrhizal influences on net primary production, evapotranspiration and nutrient cycling, or integrate measures of [...] fungal diversity into assessment of ecosystem function." We believe that an FDM approach allows for the structured investigation of functional diversity and complementarities between EMF species within FDMs (for example, some fungi may be effective in scavenging organic nitrogen, and others more effective in phosphorus uptake and transport – Buée et al. 2009), and the functional complementarities between different types of FDMs. Pool-flux classical ecosystem type research can be associated with a FDM based homomorphic model, and theoretical research on the characteristics of the fungal networks are also compatible with this framework.

A major limitation to scaling the mycorrhizal symbioses to higher organizational levels is knowledge about the fungal biomass in the characteristic FDMs. The ERH biomass can be assessed by total hyphal length (Graham and Miller 2005), using biochemical markers (chitin, ergosterol, or a specific fatty acid) or by competitive PCR (Godbold 2005). The background biomass of saprophytic fungi should also be determined. Satomura et al. (2006) present direct methods to quantify the fungal content in EM fine roots. One needs long term field studies in order to appropriately assess the biomass production (Staddon et al. 2002).

The available methods for investigating the distribution of functional parameters across scales summarized by Wolfe et al. (2009) include rotated cores (for nutrient uptake and decomposition), molecular approaches for expression of functional genes, and the use of natural gradients. To these, one can add extracellular enzymes profiles (e.g., microplate assays developed by Courty et al. 2005) and the assay developed by Rineau et al. (2008) for comparing iron chelation, free iron uptake, and oxalate production of freshly sampled EMs . Graham (2008) mentions additional methods for AMF, including in situ root observation windows, mesh dividers, bags and isotopic tracers, and signature fatty acids, the third of which can be applied to EMFs as well (Buée et al. 2009). A tool-box for mycorrhizal research at the ecosystem scale is also provided by Rillig (2004).

Scaling-up from sequence data to a whole plant and its functions requires a genomic-based approach and a systems approach to study the information flow (Graham and Miller 2005). There is also a need to understand the genetic basis of tolerance to metals in ECM symbiosis (Hartley et al. 1997). An important review of the molecular tools in EM ecology is made by Horton and Bruns (2001). An overview of the molecular techniques available for the analysis of fungal communities was done by Peay et al. (2008). The use of genomics for EMF ecological insights is essential (Martin and Nehls 2009). Tools allowing for the production of extensive data sets needed for models in order to couple the characterization of EMF with indicators of their functional rates have also started to be available. Vargas and Allen (2008), for instance, use CO_2 microsensors for characterizing respiration in an EM root system. The use of new sensor technologies is of great promise for the generation of both small scale intensive data sets (with structural and function significance – Allen et al. 2007; Hasselquist et al. 2009) and large scale ones (indicators of ecosystem functions or external control factors – Porter et al. 2005, 2009; Rundel et al. 2009).

Specific research directions/questions are:

- Describing the formal structure of the DS population's models and deriving from them the Price equation, if possible. Formulating a decoupling theory for the apparent ecological hierarchical levels based on the scale specificity of the types of abiotic and biological processes (Iordache et al. 2011).
- Exploring the potential of new theoretical tools (Barabasi and Albert 1999; Barabasi and Bonabeau 2003) for the conceptualization of fungal networks (and of the emergence function relevant for EMF). A theoretical network approach to the EMF – plant systems (and more generally MF-plant systems) at stratum level and interstratum level is proposed by Southworth et al. (2005). A cost benefit approach to the individual members of a mycorrhizal network is suggested van der by Heijden and Horton (2009).
- Study of taxa area relationships for EMF taxa based on functional genes (gene area relationships, Zhou et al. 2008).
- Linking proteomics and ecological processes with a focus on soil enzymes as mediators of decomposition, dissolved organic carbon production, and nitrogen and phosphorus mineralization (Allison et al. 2007).
- To what extent the driver/passenger hypotheses formulated for AMF (Hart et al. 2001; fungi drive the plant community or are just a by-product of changes in the plant community) could be relevant also to EMF?

- How the external control factors influence especially some of the FDMs from the structure of the homomorphic model (fire for the top FDMs and CO_2 for the deep FDMs).
- Is the gene flow affected in landscapes fragmented by contamination with metals? Is the FDM structure of EMF communities simplified as a result of toxic substances pressure?
- Is there an optimal form of natural capital's modularization maximizing the value produced by SPUs associated to EMF?

12.7 Conclusions

We have developed an analytical framework for up-scaling ecological processes and applied it to EMF. One has to construct four "hierarchical" levels in order to understand the ecological role of EMF in the ecological productivity of ecosystems, and one more if interested in evolutionary processes (gene flow, speciation). The modularization scales for understanding the role of EMF are those specific to bacteria, to fungi (FDM occupying surface of tenths of m^2, and tree plot of 400–900 m^2), to epigeous fungivorous invertebrates, and to fungivorous mammals, and, for speciation, to small catchments of several hundreds of km^2. The analysis showed that the source scale for up-scaling has to be a plot of 400–900 m^2. This plot has an associated homomorphic model with a maximum number of nine FDM for the structural and functional modeling of EMF communities. Only one modeling step is needed for up-scaling from the source scale (plot) to the ecosystem scale, but the model's construction involves the previous construction of several up-scaling and down-scaling models in order to quantify the effects on smaller and larger scale organisms on fungi.

Acknowledgments We warmly thank Professor Mahendra Rai for providing the opportunity to prepare this text, and a native English speaker for checking the language. The research leading to the results presented here was supported by German program DAAD (research visit awarded to the first author), the Romanian agencies UEFISCSU (project MECOTER code 1006 nr. 291/2007), CNMP (projects FITORISC 31012/2007, and PECOTOX 31043/2007), and the European program FP7 (project UMBRELLA nr. 226870 Grant Agreement 090528).

References

Agerer R, Gőttlein A (2003) Correlations between projection area of ectomycorrhizae and H_2O extractable nutrients in organic soil layers. Mycol Prog 2:45–52

Agerer R, Weiss M (1989) Studies on ectomycorrhizae. XX. Mycorrhizae formed by Thelephora terrestris on Norway spruce. Mycologia 81:444–453

Allen MF, Vargas R, Graham E, Swenson W, Hamilton M, Taggart M, Harmon TC, Rat'ko A, Rundel P, Fulkerson B, Estrin D (2007) Soil sensor technology: life within a pixel. Bioscience 57:859–867

Allison SD, Garner TB, Holland K, Weintraub M, Sinsabaugh RL (2007) Soil enzymes. Linking proteomics and ecological processes. In: Hurst CJ et al (eds) Manual of environmental microbiology, 3rd edn. American Society for Microbiology, Washington D.C., pp 704–711

Amaranthus MP, Perry DA (1994) The functioning of ectomycorrhizal fungi in the field: linkanges in space and time. Plant Soil 159:133–140

Amend A, Garbelotto M, Fang Z, Keeley S (2010) Isolation by landscape in populations of a prized edible mushroom *Tricholoma matsutake*. Conserv Genet 11(3):795–802

Baier R, Ingenhagg J, Blaschke H, Gőttlein A, Agerer R (2006) Vertical distribution of an ectomycorrhizal community in upper soil horizons of a young Norway spruce (*Picea abies* [L.] Karst.) stand of the Bavarian Limestone Alps. Mycorrhiza 16:197–206

Bailey RG (1987) Suggested hierarchy of criteria for multiscale ecosystem mapping. Landsc Urban Plan 14:313–319

Barabasi AL, Albert R (1999) Emergence of scaling in random network. Science 286:509–512

Barabasi AL, Bonabeau E (2003) Scale-free networks. Sci Am 288:50–59

Beare MH, Coleman DC, Crossley DA Jr, Henddrix PF, Odum EP (1995) A hierarchical approach to evaluating the significance of soil biodiversity to biogeochemical cycling. Plant Soil 170:5–22

Berg MP, Bengtsson J (2007) Temporal and spatial variability in soil food web structure. Oikos 116:1789–1804

Bidartondo MI, Burghardt B, Gerbauer G, Bruns RD, Read DJ (2004) Changing partners in the dark: isotopic and molecular evidence of ectomycorrhizal liaisons between forest orchids and trees. Proc R Soc Lond B 271:1799–1806

Bonneville S, Smits MM, Brown A, Harringston J, Leake JR, Brydson R, Benning LG (2009) Plant-driven fungal weathering: early stages of mineral alteration at the nanometer scale. Geology 37:615–618

Bruns TD, Kennedy PG (2009) Individuals, populations, communities and function: the growing field of ectomycorrhizal ecology. New Phytol 182:12–14

Bruns TD, Bidartondo MI, Taylor L (2002a) Host specificity in ectomycorrhizal communities: what do the exceptions tell us? Integr Comp Biol 42:352–359

Bruns TD, Tan J, Bidartondo M, Szaro T, Redecker D (2002b) Survival of Suillus pungens and Amanita francheti ectomycorrhizal genets was rare or absent after a stand-replacing wildfire. New Phytol 155:517–523

Buée M, De Boer W, Martin F, van Overbeek L, Jurkevitch E (2009) The rhizosphere zoo: an overview of plant-associated communities of microorganisms, including phages, bacteria, archaea, and fungi, and of some of their structuring factors. Plant Soil 321:189–212

Cairney JWG, Bastias BA (2007) Influences of fire on forest soil fungal communities. Can J For Res 37:207–215

Calvaruso C, Turpault MP, Leclerc E, Frey-Klett P (2007) Impact of ectomycorrhizosphere on the fnctional diversity od soil bacterial and fungal communities from a forest stand in relation to nutrient mobilization processes. Microb Ecol 54:567–577

Carrey AB, Harrington CA (2001) Small mammal in young forests: implications for management and sustainability. For Ecol Manag 154:289–309

Claridge AW, Trappe JM, Hansen K (2009) Do fungi have a role as soil stabilizers and remediators after forest fire? For Ecol Manag 257:1063–1069

Colpaert JV (2008) Heavy metal pollution and genetic adaptations in ectomycorrhizal fungi. In: Avery S et al (eds) Stress in yeasts and filamentous fungi. Elsevier, London, pp 157–173

Courty PE, Pritsch K, Schloter M, Hartmann A, Garbaye J (2005) Activity profiling of ectomychorrhiza communities in two forest soil using multiple enzymatic tests. New Phytol 167:309–319

Courty PE, Franc A, Pierrat JC, Garbaye J (2008) Temporal changes in the ectomycorrhizal community in two soil horizons of a temperate oak forest. Appl Environ Microbiol 74: 5792–5801

Coutts MP, Nicoll BC (1990) Growth and survival of shoots, roots and mycorrhizal mycelium in clonal Sitka spruce during the first growing season after planting. Can J For Res 20:861–868

Cudlin P, Kieliszewska-Rojucka B, Rudawska M, Grebenc T, Alberton O, Lehto T, Bakker MR, Borja I, Konopka B, Leski T, Kraigher H, Kuyper TW (2007) Fine roots and ectomycorrhizas as indicators of environmental change. Plant Biosyst 141:406–425

Dahlberg A (2001) Community ecology of ectomycorrhizal fungi: an advancing interdisciplinary field. New Phytol 150:555–562

Darwin C (1859) Origin of species. John Murray, London

Dell B (2002) Role of mycorrhizal fungi in ecosystems. CMU J 1:47–60

Dengler J (2009a) A flexible multi-scale approach for standardized of plant species richness patterns. Ecol Indic 9:1169–1178

Dengler J (2009b) Which function describes the species-area relationship best? A review and empirical evaluation. J Biogeogr 36:728–744

Diaz S (1996) Effects of elevated [CO_2] at the community level mediated by root symbionts. Plant Soil 187:309–320

Dickie IA, Bing Xu, Koide RT (2002) Vertical niche differentiation of ectomycorrhizal hyphae in soil as shown by T-RFLP analysis. New Phytol 156:527–535

Dighton J (2003) Fungi in ecosystem processes. Marcel Dekker, New York

Douhan GW, Rizzo DM (2005) Phylogenetic divergence in a local population of the ectomycorrhizal fungus *Cenococcum geophilum*. New Phytol 166:263–271

Esher RJ, Marx DH, Ursic J, Baker RL, Brown LR, Coleman DC (1992) Simulated acid rain effects on fine roots, ectomycorrhizae, microorganisms, and invertebrates in pine forests of the Southern United States. Water Air Soil Pollut 61:269–278

Ettema CH, Wardle DA (2002) Spatial soil ecology. Trends Ecol Evol 17(4):177–183

Fitzsimons MS, Miller RM, Jastrow JD (2008) Scale-dependent niche axes of arbuscular mycorrhizal fungi. Oecologia 158:117–127

Fortin MJ, Olsol RJ, Ferson S, Iverson L, Hunsaker C, Edwards G, Levine D, Butera K, Klemas V (2000) Issues related to the detection of boundaries. Landsc Ecol 15:453–466

Fujimura KE, Egger KN, Henry GH (2008) The effect of experimental warming on the root-associated fungal community of *Salix arctica*. ISME J 2:105–114

Gaston KJ, Chown SL, Calosi P, Bernardo J, Bilton DT, Clarke A, Trullas SC, Ghalambo CK, Konarzewsk M, Peck LS, Porter WP, Pörtner HO, Rezende EL, Schulte PM, Spicer JI, Stillman JH, Terblanche JS, van Kleunen M (2009) Macrophysiology: a conceptual reunification. Am Nat 174:595–612

Gebhardt S, Wőllecke J, Műnzenberger B, Hűttl RF (2009) Microscale spatial distribution patterns of red oak (Quercus rubra L.) ectomycorrhizae. Mycol Prog 8:245–257

Gehring C, Bennett A (2009) Mycorrhizal fungal-plant-insect interactions: the importance of a community approach. Environ Entomol 38(1):93–102

Genney DR, Anderson IC, Alexander IJ (2006) Fine-scale distribution of pine ectomycorrhizas and their extramatrical mycelium. New Phytol 170:381–390

Gherghel F (2009) Identification and characterization of Quercus robur ectomycorrhiza in relation to heavy metal contamination. Dissertation, Friedrich Schiller University, Jena, Germany

Gilliam FS (2006) Response of the herbaceous layer of forest ecosystems to excess nitrogen deposition. J Ecol 94:1176–1191

Godbold DL (2005) Ectomycorrhizal community structure: linking biodiversity to function. Prog Bot 66:374–391

Godbold DL, Berntson GM (1997) Elevated atmospheric CO_2 concentration changes ectomycorrhizal morphotype assamblages in *Betula papyrifera*. Tree Physiol 17:347–350

Godbold DL, Berntson GM, Bazzaz FA (1997) Growth and mycorrhizal colonization of three North American tree species under elevated atmospheric CO_2. New Phytol 137:433–440

Godbold DL, Hoosbeek MR, Lukac M, Cotrufo MF, Janssens IA, Ceulemans R, Andrea P, Velthorst EJ, Scarascia-Mugnozza G, De Angelis P, Miglietta F, Peressotti A (2006) Mycorrhizal hyphal turnover as a dominant process for carbon input into soil organic matter. Plant Soil 281:15–24

Goodman DM, Trofymov VA (1998) Distribution of ectomycorrhizas in micro-habitats in mature and old-growth stands of Douglas-fir on southeastern Vancouver Island. Soil Biol Biochem 30:2127–2138

Graham JH (2008) Scaling-up evaluation of field functioning of arbuscular mycorrhizal fungi. New Phytol 180:1–2

Graham JH, Miller RM (2005) Mycorrhizas: gene to function. Plant Soil 274:79–100

Grebenc T, Christensen M, Vilhar U, Cater M, Martin MP, Simoncic P, Kraingher H (2009) Response of ectomycorrhizal community structure to gap opening in natural and managed temperate beech-dominated forests. Can J For Res 39:1375–1386

Griffiths RP, Bradshaw GA, Marks B, Lienkaemper GW (1996) Spatial distribution of ectomycorrhizal mats in coniferous forests of the Pacific Northwest, USA. Plant Soil 180:147–158

Hall IR, Yun W, Amicucci A (2003) Cultivation of edible ectomycorrhizal mushrooms. Trends Biotechnol 21:433–438

Hart MM, Reader RJ, Klironomos JN (2001) Life-history strategies of arbuscular mycorrhizal fungi in relation to their successional dynamics. Mycologia 93:1186–1194

Hartley J, Cairney JWG, Meharg AA (1997) Do ectomycorrhizal fungi exhibit adaptive tolerance to potentially toxic metals in the environment? Plant Soil 189:303–319

Hasselquist NJ, Vargas R, Allen MF (2009) Using soil sensing technology to examine interactions and controls between ectomycorrhizal growth and environmental factors on soil CO_2 dynamics. Plant Soil. doi:10.1007/s11104-009-0183-y

Heneghan L, Miller SP, Baer S, Callaham MA Jr, Montgomery J, Pavao-Zuckerman M, Rhoades CC, Richardson S (2008) Integrating soil ecological knowledge into restoration management. Restor Ecol 16:608–617

Hinsinger P, Bengough AG, Vetterlein D, Young IM (2009) Rhizosphere: biophysics, biogeochemistry and ecological relevance. Plant Soil 321:117–152

Hiol Hiol F, Dixon RK, Curl EA (2004) The feeding preference of mycophagous collembola varies with the ectomycorrhizal symbiont. Mycorrhiza 5:99–103

Hirose D, Kikuchi J, Kanzaki N, Kazuyoshi F (2004) Genet distribution of sporocarps and ectomycorrhizas of *Suillus pictus* in a Japanese white plantation. New Phytol 164:527–541

Horton TR, Bruns TD (2001) The molecular revolution in ectomycorrhizal ecology: peeking into the black-box. Mol Ecol 10:1855–1871

Horton TR, Molina R, Hood K (2005) Douglas-fir ectomycorrhizae in 40- and 400-year-old stands: mycobiont availability to late successional western hemlock. Mycorrhiza 15:393–403

Hubert NA, Gehring CA (2008) Neighboring trees affect ectomycorrhizal fungal community composition in a woodland-forest ecotone. Mycorrhiza 18:363–374

Hutchings MJ, Bradbury IK (1986) Ecological perspectives on clonal perennial herbs. Bioscience 36:178–182

Iordache V (2004) The constitution from the point of view of human ecology (in Romanian), Sfera Politicii 106:29–32. http://www.sferapoliticii.ro/sfera/pdf/Sfera_106.pdf

Iordache V (2009a) The general concept of evolution. In Staicu L (ed) Rationality and evolution – philosophical exploration of complexity. Bucharest University Press, Bucharest, pp 83–155 (in Romanian)

Iordache V (2009b) Darwin's law of growth with reproduction in the Origin of Species: law of nature or hidden teleologic principle? Communication at the Conference "Darwin, the evolution of species and the evolutionary thinking", 20–22 Nov 2009. University of Bucharest, Romania

Iordache V, Bodescu F (2005) Emergent properties of the Lower Danube River System: consequences for the integrated monitoring system. Arch Hydrobiol Suppl Large Rivers 158: 95–128

Iordache V, Gherghel F, Kothe E (2009a) (2009b) Assessing the effect of disturbances on diversity of ectomycorrhiza. Int J Environ Res Public Health 6:414–432

Iordache V, Ion S, Pohoata A (2009b) Integrated modeling of metals biogeochemistry: potential and limits. Chem Erde-Geochem 69:125–169

Iordache V, Lacatusu R, Scradeanu D, Onete M, Purice D, Cobzaru I (2011) Scale-specific mechanisms of metals distribution and mobility in contaminated areas. In: Kothe E, Varma A (eds) Bio-geo-interactions in contaminated soils, Springer, in preparation

Izzo A, Agbowo J, Bruns TD (2005) Detection of plot-level changes in ectomycorrhizal communities across years in an old-growth mixed-conifer forest. New Phytol 166:619–630

Jackson LE, Burger M, Cavagnaro TR (2008) Roots, nitrogen transformations, and ecosystem services. Ann Rev Plant Biol 59:341–363

Jentschke G, Godbold DL (2000) Metal toxicity and ectomycorrhizas. Physiol Plant 109:107–116

Johnson CN (1996) Interactions between mammals and ectomycorrhizal fungi. Trends Ecol Evol 11:503–507

Johnson D, Krsek M, Wellington EMH, Stott AW, Cole L, Bardgett RD, Read DJ, DJ JR, Leake JR (2005) Soil invertebrates disrupt carbon flow through fungal networks. Science 309:1047

Johnson CN, Hoeksema JD, Bever JD, Chaudhary VB, Gehring C, Klironomos J, Koide R, Miller RM, Moore J, Moutoglis P, Schwartz M, Simard S, Swenson W, Umbanhowar J, Wilson G, Zabinski C (2006) From lilliput to brobdingnag: extending models of mycorrhizal function across scales. Bioscience 56:889–900

Jones MD, Durall DM, Cairney JWG (2003) Ectomycorrhizal fungal communities in young forest stands regenerating after clearcut logging. New Phytol 157:399–422

Karst J, Marczak L, Jones MD, Turkington R (2008) The mutual-parasitism continuum in ectomycorrhizas: a quantitative assessment using meta-analysis. Ecology 89:1032–1042

Kjoller R (2006) Disproportionate abundance between ectomycorrhizal root tips and their associated mycelia. Microbiol Ecol 58:214–224

Kjoller R, Clemmensen KE (2009) Belowground ectomycorrhizal fungal community respond to liming in three southern Swedish coniferous forest stands. For Ecol Manag 257:2217–2225

Klironomos JN, Rillig MC, Allen MF (1999) Designing belowground field experiments with the help of semi-variance and power analyses. Appl Soil Ecol 12:227–238

Koele N, Turpault MP, Hildebrand EE, Uroz S, Frey-Klett P (2009) Interactions between mycorrhizal fungi and mycorrhizosphere bacteria during mineral weathering: budget analysis and bacterial quantification. Soil Biol Biochem 41:1935–1942

Koide RT, Xu B, Sharda J (2005) Contrasting below-ground views of an ectomycorrhizal fungal community. New Phytol 166:251–262

Koide RT, Shumway DL, Xu B, Sharda J (2007) On temporal partitioning of a community of ectomicorrhizal fungi. New Phytol 174:420–429

Koricheva J, Gange AC, Jones T (2009) Effects of mycorrhizal fungi on insect herbivores: a meta-analysis. Ecology 90:2088–2097

Kozdrój K, Piotrowska-Seget Z, Krupa P (2007) Mycorrhizal fungi and ectomycorrhiza associated bacteria isolated from an industrial desert soil protect pine seedlings against Cd (II) impact. Ecotoxicology 16:449–456

Kranabetter JM, Durall DM, MacKenzie WH (2009) Diversity and species distribution of ectomycorrhizal fungi along productivity gradients of a southern boreal forest. Mycorrhiza 19:99–111

Krpata D, Peintner U, Langer I, Fitz WJ, Schweiger P (2008) Ectomycorrhizal communities associated with *Populus tremula* growing on a heavy metal contaminated site. Mycol Res 112:1069–1079

Krpata D, Fitz WJ, Peintner U, Langer I, Schweiger P (2009) Bioconcentration of zinc and cadmium in ectomycorrhizal fungi and associated aspen trees as affected by level of pollution. Environ Pollut 157:280–286

Lamour A, Termorshuizen AJ, Volker D, Michael JJ (2007) Network formation by rhizomorphs of *Amillaria lutea* in natural soil: their description and ecological significance. FEMS Microbiol Ecol 62:222–232

Landeweert R, Leeflang P, Kuyper TW, Hoffland E, Rosling A, Wernars K, Smit E (2003) Molecular identification of ectomycorrhizal mycelium in soil horizons. Appl Environ Microbiol 69:327–333

Leake JR, Cameron DD (2010) Physiological ecology of mycoheterotrophy. New Phytol 185:601–605

Lepczyk CA, Lortie CJ, Anderson LJ (2008) An ontology for landscapes. Ecol Complex 5: 272–279

Lewis JD, Thomas RB, Strain BR (1994) Effect of elevated CO_2 on mycorrhizal colonization of loblolly pine (*Pinus taeda*L.) seedlings. Plant Soil 165:81–88

Lewis JD, Licitra J, Tuininga AR, Sirulnik A, Turner GD, Johnson J (2008) Oak seedling growth and ectomycorrhizal colonization are less in eastern hemlock stands infested with hemlock woolly adelgid than in adjacent oak stands. Tree Physiol 28:629–636

Lilleskov EA, Bruns TD (2005) Spore dispersal of a resupinate ectomycorrhizal fungal, *Tomentella sublilacina*, via soil food webs. Mycologia 97:762–769

Lilleskov EA, Bruns TD, Horton TR, Taylor DL, Grogan P (2004) Detection of forest stand-level spatial structure in ectomicorrhizal fungal communities. FEMS Microbiol Ecol 49:319–332

Lischke L, Löffler TJ, Thornton PE, Zimmermann NE (2006) Model up-scaling in landscape research. In: Kienast F et al (eds) A changing world. Challenges for landscape research. Springer, Dordrecht, pp 259–282

Liste HH, White JC (2008) Plant hydraulic lift of soil water-implications for crop production and land restoration. Plant Soil 313:1–17

Lovelock CE, Wright SF, Clark DA, Ruess RW (2004) Soil stocks of glomalin produced by arbusclar mycorrhizal fungi across a tropical rain forest landscape. J Ecol 92:278–287

Luck GW, Daily GC, Ehrlich PR (2003) Population diversity and ecosystem services. Trends Ecol Evol 18:331–336

Maestre FT, Cortina J, Bautista S, Bellot J, Vallejo R (2003) Small-scale environmental heterogeneity and spatiotemporal dynamics of seedling establishment in a semiarid degreded ecosystem. Ecosystems 6:630–643

Maharning AR, Mills AAS, Adl SM (2009) Soil community changes during secondary succession to naturalized grasslands. Appl Soil Ecol 41:137–147

Maijala P, Fagerstedt KV, Raudaskoski M (1991) Detection of extracellular and proteolytic activity in ectomycorrhizal fungus *Heterobasidion annosum* (Fr.) Bref. New Phytol 117:643–648

Martin F, Nehls U (2009) Harnessing ectomycorrhizal genomics for ecological insight. Curr Opin Plant Biol 12:508–515

Matsuda Y, Hayakawa N, Ito S (2009) Local and microscale distributions of *Cenococcum geophilum* in soil of coastal pine forests. Fungal Ecol 2:31–35

Meixner H, Schnauder I, Bölscher J, Iordache V (2006) Hydraulic, sedimentological and ecological problems of multifunctional riparian forest management – RipFor Guidelines for End-Users. In: Berliner Geographische Abhandlungen, vol 66. ISBN3-88009-067-x. http://www.cesec.ro/pdf/RipForGL2006_b1.pdf

Miller RM, Kling M (2000) The importance of integration and scale in the arbuscular mycorrhizal symbiosis. Plant Soil 226:295–309

Miller RM, Lodge DJ (1997) Fungal responses to disturbance: agriculture and forestry. In: Wicklow DT, Söderström B (eds) Environmental and microbial relationships. Springer, Berlin, pp 65–84

Moore JC, St John TV, Coleman DC (1985) Ingestion of vesicular arbuscular mycorrhizal hyphae by soil microarthropods. Ecology 66:1979–1981

Moore JC, McCann K, Setala H, De Ruiter PC (2003) Top-down is bottom-up: does predation in the rhizosphere regulate aboveground dynamics? Ecology 84:846–857

Moore JC, McCann K, de Ruiter PC (2007) Soil rhizosphere food webs, their stability, and implications for soil processes in ecosystems. In: Cardon ZG, White JL (eds) Rhizosphere. An ecological perspective. Academic, Burlington, MA, pp 101–126

Morris AH, Smith ME, Rizzo DM, Rejmánek M, Bledsoe CS (2008) Contrasting ectomycorrhizal fungal community on the roots of co-occurring oaks (*Quercus spp.*) in a California woodland. New Phytol 178:167–176

Mougin C, Boukcim H, Jolivalt C (2009) Soil bioremediation strategies based on the use of fungal enzymes. In: Singh A et al (eds) Advances in applied bioremediation. Springer, Heidelberg, pp 123–149

Mudge KW, Diebolt KS, Whitlow TH (1987) Ectomycorrhizal effect on host plant response to drought stress. J Environ Hortic 5:183–187
Münzenberge B, Bubner B, Wollecke J, Sieber TN, Bauer R, Fladung M, Hüttl RF (2009) The ectomycorrhizal morphotype *Pinirizha sclerotia* is formed by *Acephalia macrosclerotium* sp. nov., a close relative of *Phialocephala fortinii*. Mycorrhiza 19:481–492
Nara K (2008) Community developmental patterns and ecological functions of ectomycorrhizal fungi: implications from primary succession. In: Varma A (ed) Mycorrhiza: state of the art, genetics and molecular biology, eco-function, biotechnology, eco-physiology, structure and systematics. Springer, Heidelberg, pp 581–600
Neagoe A, Ebena G, Carlsson E (2005) The effect of soil amendments on plant performance in an area affected by acid mine drainage. Chem Erde-Geochem 65:115–129
Neagoe A, Merten D, Iordache V, Buechel G (2009) The effect of bioremediation methods involving different degrees of soil disturbance on the export of metals by leaching and by plant uptake. Chem Erde-Geochem 69(S2):57–73
Nilsson LO, Wallander H (2003) Production of external mycelium by ectomycorrhizal fungi in a Norway spruce forest was reduces in response to nitrogen fertilization. New Phytol 158: 409–416
O'Neill RV (1996) Recent developments in ecological theory: hierarchy and scale. In: Scott JM et al (eds) Gap analysis: a landscape approach to biodiversity planning. American Society for Photogrammetry Remote Sensing, Bethesda, MD, pp 7–14
O'Neill RV (2001) Is it time to burry the ecosystem concept? Ecology 82:3275–3284
O'Neill EG, O'Neill RV, Norby RJ (1991) Hierarchy theory as a guide to mycorrhizal research on large-scale problems. Environ Pollut 73:271–284
Pahl-Vostl C (1995) The dynamic nature of ecosystems. Wiley, New York
Peay KG, Bruns TD, Kennedy PG, Bergemann SE, Garbelotto M (2007) A strong species-area relationship for eukaryotic soil microbes: island size matters for ectomycorrhzal fungi. Ecol Lett 10:470–480
Peay KG, Kennedy PG, Bruns TD (2008) Fungal community ecology: a hybrid beast with a molecular master. Bioscience 58:799–810
Peay KG, Garbelotto M, Bruns TD (2009) Spore heat resistence plays an important role in disturbance-mediated assemblage shift of ectomycorrhizal fungi colonizing *Pinus muricata* seedlings. J Ecol 97:537–547
Pickles BJ, Genney D, Anderson IC, Alexander IJ (2009) Spatial ecology of ectomycorrhizas: analytical strategies. In: Azcón-Aguilar C et al (eds) Mycorrhizas – functional processes and ecological impact. Springer, Berlin, pp 155–165
Porter J, Arzberger P, Braun HW, Bryant P, Gage S, Hansen T, Hanson P, Lin CC, Lin FP, Kratz T, Michener W, Shapiro S, Williams T (2005) Wireless sensor networks for ecology. Bioscience 55:561–572
Porter JH, Nagy E, Kratz TT, Hanson P, Scott L, Collins SI, Arzberger P (2009) New eyes on the world: advanced sensors for ecology. Bioscience 59:385–397
Price GR (1995) The nature of selection. J Theor Biol 175:389–396
Pritchard SG, Strand AE, McCormack ML, Davis MA, Oren R (2008) Mycorrhizal and rhizomorph dynamics in a loblolly pine forest during 5 years of free-air-CO_2-enrichment. Glob Chang Biol 14:1252–1264
Rahbek C (2005) The role of spatial scale and the perception of large-scale species-richness patterns. Ecol Lett 8:224–239
Read DJ, Perez-Moreno J (2003) Mycorrhizas and nutrient cycling in ecosystems – a journey towards relevance? New Phytol 157:475–492
Redecker D, Szaro TM, Rand B, Bruns TD (2001) Small genets of *Lactarius xanthogalactus*, *Russula cremoricolor* and *Amanita francheti* in late-stage ectomycorrhizal successions. Mol Ecol 10:1025–1034
Reynolds JF, Wu J (1999) Do landscape structural and functional units exist? In: Tenhunen HD, Kabat P (eds) Integrating hydrology, ecosystem dynamics, and biogeochemistry in complex landscapes. Wiley, West Sussex, pp 273–296

Rillig MC (2004) Arbuscular mycorrhizae and terrestrial ecosystem processes. Ecol Lett 7: 740–754

Rillig MC, Allen MF (1999) What is the role of arbuscular mycorrhizal fungi in plant-to-ecosystem responses to elevated atmospheric CO_2? Mycorrhiza 9:1–8

Rillig MC, Mammey DL (2006) Mycorrhizas and soil structure. New Phytol 171:41–53

Rineau F, Garbaye J (2009) Effects of liming on ectomycorrhizal community structure in relation to soil horizons and tree hosts. Fungal Ecol 2:103–109

Rineau F, Courty PE, Uroz S, Buée M, Garbaye J (2008) Simple microplate assays to measure iron mobilization and oxalate secretion by ectomycorrhizal tree roots. Soil Biol Biochem 40: 2460–2463

Ritter T, Weber G, Haug I, Kotke I, Oberwinkler F (1989) Interrelationship between vitality of ectomycorrhizae and occurrence of microfungi. Ann Sci For 46(Suppl):745s–749s

Robertson SJ, McGill WB, Massicotte HB, Rutherford PM (2007) Petroleum hydrocarbon contamination in boreal forest soils: a mycorrhizal ecosystems perspective. Biol Rev 82:213–240

Rosling A (2009) Trees, mycorrhiza and minerals-field relevance of in vitro experiments. Geobiomicrobiol J 26:389–401

Rosling A, Landeweert R, Lindahl BD, Larsson KH, Kuyper TW, Taylor AFS, Finlay RD (2003) Vertical distribution of ectomycorrhizal fungal taxa in a podzol soil profile. New Phytol 159: 775–783

Rundel PW, Graham EA, Allen MF, Fisher JC, Harmon TC (2009) Environmental sensor networks in ecological research. New Phytol 182:589–607

Rygiewicz PT, Johnson MG, Ganio LM, Tingey DT, Storm MJ (1997) Lifetime and temporal occurrence of ectomycorrhizae on ponderosa pine (*Pinus ponderosa* Laws.) seedlings grown under varied atmospheric CO_2 and nitrogen levels. Plant Soil 189:275–287

Satomura T, Hashimoto Y, Kinoshita A, Horikoshi T (2006) Methods to study the role of ectomycorrhizal fungi in forest carbon cycling 1: introduction to the direct methods to quantify the fungal content in ectomicorrhizal fine roots. Root Res 15:119–124

Scattolin L, Montecchio L, Mosca E, Agerer R (2008) Vertical distribution of the ectomycorrhizal community in the top soil of Norway spruce stands. Eur J For Res 127:347–357

Schneider K, Renker C, Maraun M (2005) Oribatid mite (Acari, Oribatida) feeding on ectomycorrhizal fungi. Mycorrhiza 16:67–72

Schroder B, Seppelt R (2006) Analysis of pattern–process interactions based on landscape models – Overview, general concepts, and methodological issues. Ecol Model 199:505–516

Schulz B, Boyle C (2005) The endophytic continuum. Mycol Res 109:661–686

Selosse MA, Richard F, Xand H, Simard SW (2006) Mycorrhizal networks: des liaisons dangereuses? Trends Ecol Evol 21:621–628

Seppelt R, Müller F, Schröder B, Martin Volk M (2009) Challenges of simulating complex environmental systems at the landscape scale: a controversial dialogue between two cups of espresso. Ecol Model 220:3481–3489

Simard SW, Durall DM (2004) Mycorrhizal networks: a review of their extent, function, and importance. Can J Bot 82:1140–1165

Simard SW, Durall DM, Jones MD (1997) Carbon allocation and carbon transfer between *Betula papyrifera* and *Pseudotsuga menziesii* seedlings using a ^{13}C pulse-labeling method. Plant Soil 191:41–45

Smith ML, Bruhn JN, Anderson JB (1992) The fungus *Armillaria bulbosa* is among the largest and oldest living organisms. Nature 356:428–431

Smith SE, Facelli E, Pope S, Smith FA (2010) Plant performance in stressful environments: interpreting new and established knowledge of the roles of arbuscular mycorrhizas. Plant Soil 326:3–20

Southworth D, He XH, Swenson W, Bledsoe CS, Horwath WR (2005) Application of network theory to potential mycorrhizal networks. Mycorrhiza 15:589–595

Staddon PL, Heinemeyer A, Fitter AH (2002) Mycorrhizas and global environmental change: research at different scales. Plant Soil 244:253–261

Staddon PL, Ramsey CB, Ostle N, Ineson P, Fitter AH (2003) Rapid turnover of hyphae of mycorrhizal fungi determined by AMS microanalysis of ^{14}C. Science 300:1138–1140

Szlavecz K, McCormick MK, Xia L, Whigham D (2009) Direct and indirect effects of earthworms on mycorrhizal fungi. 94th ESA annual meeting, PS 20-177. http://eco.confex.com/eco/2009/techprogram/P19960.HTM

Tarkka MT, Frey-Klett P (2008) Mycorrhiza helper bacteria. In: Varma A (ed) Mycorrhiza – state of the art, genetics and molecular biology, eco-function, biotechnology, eco-physiology, structure and systematics. Springer, Heidelberg, pp 113–134

Tedersoo L, Kőljalg U, Hallenberg N, Larsson KH (2003) Fine scale distribution of ectomycorrhizal fungi and roots across substrate layers including coarse woody debris in a mixed forest. New Phytol 159:153–165

Thiet RK, Boerner REJ (2007) Spatial patterns of ectomycorrhizal fungal inoculum in arbuscular mycorrhizal barrens communities: implications for controlling invasion by *Pinus virginiana*. Mycorrhiza 17:507–517

Tingey DT, Philips DL, Johnson MG (2000) Elevated CO_2 and conifer roots: effects of growth, life span, and turnover. New Phytol 147:87–103

Treseder KK, Allen MF (2000) Mycorrhizal fungi have a potential role in soil carbon storage under elevated CO_2 and nitrogen deposition. New Phytol 147:189–200

Treseder KK, Masiello CA, Lansing JL, Allen MF (2004) Species-specific measurements of ectomycorrhizal turnover under N-fertilization: combining isotopic and genetic approaches. Oecologia 138:419–425

Treseder KK, Allen MF, Ruess RW, Pregitzer KS, Hendrick RL (2005) Lifespans of fungal rhizomorphs under nitrogen fertilization in a pinyon-juniper woodland. Plant Soil 270:249–255

Vadineanu A, Cristofor S, Iordache V (2001) Lower Danube River System biodiversity changes. In: Gopal B et al (eds) Biodiversity in wetlands: assessment. Function and conservation. Backhuys, Leiden, pp 29–63

van der Heijden MGA, Horton TR (2009) Socialism in soil? The importance of mycorrhizal fungal networks for facilitation in natural ecosystems. J Ecol 97:1139–1150

van der Heijden MGA, Klironomos JN, Ursic M, Moutoglis P, Streitwolf-Engel R, Boller T, Wiemken A, Sanders IR (1998) Mycorrhizal fungal diversity determines plant biodiversity, ecosystem variability and productivity. Nature 396:69–72

Vargas R, Allen MF (2008) Dynamics of fine root, fungal rhizomorphs, and soil respiration in a mixed temperature forest: integrating sensors and observation. Vadose Zone J 7:1055–1064

Warren JM, Brooks JR, Meinzer FC, Eberhart JL (2008) Hydraulic redistribution of water from *Pinus ponderosa* trees to seedlings: evidence for an ectomycorrhizal pathway. New Phytol 178:382–394

Watkinson SC, Boddy L, Burton K, Darrah PR, Eastwood D, Fricker MD, Tlalka M (2005) New approaches to investigating the function of mycelial networks. Mycologist 19:11–17

Wiensczyk Am, Gamiet S, Durall DM, Jones MD, Simard SW (2002) Ectomycorrhizae and forestry in British Columbia: a summary of current research and conservation strategies. BC J Ecosyst Manag 2. http://www.forrex.org/JEM/ISS2/vol2_no1_art6.pdf

Wolfe BE, Parrent JL, Koch AM, Sikes BA, Gardes M, Klironomos JN (2009) Spatial heterogeneitiy of mycorrhizal populations and communities: scales and mechanisms. In: Azcón-Aguilar C et al (eds) Mycorrhizas – functional processes and ecological impact. Springer, Heidelberg, pp 167–186

Wong MTF, Asseng S (2006) Determining the causes of spatial and temporal variability of wheat yields at sub-field scale using a new method of upscaling a crop model. Plant Soil 283:203–215

Yarrow MM, Marin VH (2007) Toward conceptual cohesiveness: a historical analyses of the theory and utility of ecological boundaries and transition zones. Ecosystems 10:462–476

Zhou JZ, Kang S, Schadt CW, Garten CT (2008) Spatial scaling of functional gene diversity across various microbial taxa. Proc Natl Acad Sci USA 105:7768–7773

Chapter 13
Mycobioindication of Stress in Forest Ecosystems

Hojka Kraigher and Samar Al Sayegh Petkovšek

13.1 Introduction

Indication of the existence of an environmental stress situation is a prerequisite for its amelioration or adaptation to it. Environmental monitoring based on chemical and physical parameters can give a measure of a chemical in the environment, or of temperature in a certain period of time and (micro)location. However, such measurements do not provide any information on the effects of these factors on the living organisms, and on any synergistic or antagonistic effects on conversion products or pathways. The state or reaction of a living organism to physical and chemical environment, its survival, and early diagnosis of any disease can only be determined by using bioindicators and a biomonitoring approach.

Soil is one of the principal regulatory compartments in forest ecosystems in which below-ground processes are regulated and mediated to a large extent through the mycelia of mycorrhizal fungi. Early stress indication through changes in mycorrhizal structure and function, therefore, also contributes to an understanding of complex interactions in the mycelial wood-wide web and the functioning of an entire ecosystem. Since the functional compatibility and stress tolerance of ectomycorrhizal (ECM) types are species specific, information on the ECM community can not only be applied as an indication of stress and disturbance of forest soils, but it also adds to the understanding of processes in forest ecosystems.

We present a brief overview of the impacts of stress factors on mycorrhiza and of mycobioindication approaches developed for monitoring climate change and air pollution effects on forest ecosystems in Slovenia in the last two decades.

H. Kraigher (✉)
Slovenian Forestry Institute, Večna pot 2, 1000 Ljubljana, Slovenia
e-mail: hojka.kraigher@gozdis.si

S. Al Sayegh Petkovšek
ERICo Velenje, Ecological Research & Industrial Cooperation, Koroška 58, 3320 Velenje, Slovenia
e-mail: samar.petkovsek@erico.si

M. Rai and A. Varma (eds.), *Diversity and Biotechnology of Ectomycorrhizae*, Soil Biology 25, DOI 10.1007/978-3-642-15196-5_13,

13.2 Impact of Stress Factors on Mycorrhiza

Human-induced stress factors can affect ectomycorrhiza directly through reduced growth of fine roots and decreased uptake of nutrients, and/or indirectly through foliar damage to the host plant and, consequently, a changed C allocation below-ground (Kieliszewska-Rokicka et al. 1997; Brunner 2001; Leake 2001; Andersen 2003; Kernaghan 2005; Kraigher et al. 2007; Cudlin et al. 2007). When natural stress factors (e.g., water shortage, elevated and low temperature (T), and pathogens) interact, they can supplement the effects of pollutants or induce antagonistic effects. The impact of anthropogenic stress factors is most pronounced in effects on biodiversity of below-ground ECM communities and, consequently, on the sustainability, productivity, and vitality of forests (Kraigher et al. 1995, 2007; Kraigher 1997, 1999, 2001; Read 1998; Simončič et al. 1998; Kovacs et al. 2000; Erland and Taylor 2002; Cudlin et al. 2007).

13.2.1 Acid Deposition

In spite of the fact that slow acidification of soil is a natural process in most coniferous forests (Brunner 2001; Erland and Taylor 2002), the percentage of forests in Europe with soil pH below 4.2 increased toward the end of the millennium, due to the deposition of high atmospheric inputs of acidifying pollutants [mostly sulfur (S) and nitrogen (N) compounds] (Arnolds 1991). Moreover, the release of heavy metals, especially Al ions, into the soil solution as a consequence of acidification was determined (Brunner 2001). In general, acidification did not reduce the degree of mycorrhization of tree fine roots, which is close to 100% in most of the forests studied, but the total number of ECM root tips per soil volume and the number of morphotypes were reduced (Cudlin et al. 2007). A moderate reduction in pH does not drastically affect the diverse types of ectomycorrhizae, since ECM communities are mostly adapted to acid soil conditions (Read 1991). A pronounced acidification effect on ectomycorrhizal fungi (EMF) is species specific and can result in an increased abundance of selected types of ectomycorrhiza (e.g., *Russula ochroleuca* in Norway spruce plots subjected to acid irrigation), so it can trigger a shift in the community structure (Kraigher 1991, 1994, 1997; Erland and Taylor 2002).

13.2.2 Nitrogen Deposition and Fertilization

Nitrogen deposition, which has increased across a large number of temperate conifer and deciduous ecosystems, affects tree growth and nutrient cycles (Vitousek et al. 1997). Increased soil N concentrations are often correlated with changes in the number of ECM community attributes, such as decreased sporocarp production, lower community diversity, and shifts in the relative abundance of ECM community members.

The first documented biological effect of nitrogen deposition on the diversity of EMF was mainly limited to observations made in sporocarp surveys of EMF (Arnolds 1991). In nitrogen-supplemented plots, a rapid and substantial decrease in the diversity and production of ECM sporocarps of most species was observed. This issue was recently thoroughly reviewed by Lilleskov (2005). The response to fertilization seems to vary among species, with certain taxa declining in abundance and diversity (*Cortinarius*, *Hydnum*, *Russula*, and *Suillus*) and some continuing to fruit at higher N deposition levels (*Lactarius rufus*, *Lactarius theiogalus*, *Paxillus involutus*, and *Laccaria* species).

In addition to sporocarp surveys, research based on below-ground ECM observations has provided new information about richness or diversity in response to N inputs and revealed the lack of direct correspondence between the two approaches. Contrasting results between different studies have also appeared. Several reports have indicated a decrease in fractional colonization by EMF, or total numbers of ECM roots (Tetreault et al. 1978; Alexander and Fairley 1983; Rudawska 1986; Taylor and Alexander 1989; Haug et al. 1992) or have shown a decline in the abundance and diversity of EMF, or a shift in ECM community composition (Kraigher et al. 1996; Taylor et al. 2000; Peter et al. 2001; Lilleskov et al. 2002; Parrent et al. 2006). However, the decrease in ECM colonization has sometimes been fairly short-lived and disappeared a few years after treatment (Menge et al. 1977). The reduction in ECM root tips in soil cores in spruce and beech forests along transects through Europe, representing gradients in increasing mineral N deposition, was analyzed by Taylor et al. (2000). A reduction in biodiversity of ECM types in polluted areas was shown in spruce forests (Kraigher 1997), whereas in beech forests, no negative effects on biodiversity were determined (Kraigher et al. 2007). Within spruce forests, the decline in diversity toward the south of the transect was accompanied by increases in single species dominance. *Tylospora* sp. and *L. rufus* dominated the ECM community of the most polluted site, forming 75% of mycorrhizal tips. A reverse trend was determined for species from genera *Cortinarius*, *Tricholoma*, and *Russula*, these species appearing to be particularly sensitive to nitrogen enrichment (Taylor et al. 2000).

13.2.3 Metal Deposition

Atmospheric pollution leads to soil acidification and elevated concentrations of trace elements in the soil. Elevated concentrations affect soil microorganisms and microbial processes, reduce the development of mycorrhiza (Schneider et al. 1989; Perrin and Estivalet 1990), and decrease the abundance and biodiversity of sporocarps (Arnolds 1991). However, some fungal species tolerate increased metal levels in soil (e.g., *Amanita*, *Albatrellus ovinus*, and *Leccinum scabrum*) (Erland

and Taylor 2002). Mycorrhizal fungi can accumulate metals in their sporocarps (Kalač and Svoboda 2000; Falandysz et al. 2001, 2003; Al Sayegh Petkovšek et al. 2002; Johanson et al. 2004; Cocchi et al. 2006; Al Sayegh Petkovšek 2008). Moreover, they have the capacity to accumulate metals in the external mycelium (mostly Cu and to a lesser extent Zn and Cd) (Berthelsen et al. 1995). EMF establish metal tolerance by binding metals to electronegative sites on the cell walls of hyphae, or by binding to phosphates and sulfhydryl compounds within the cells (Brunner 2001).

13.2.4 Ozone

Tropospheric ozone (O_3) is a secondary atmospheric pollutant, generated from oxides of nitrogen and volatile organic compounds reacting in the presence of sunlight. It has been recognized as an increasing and damaging agent to plants (Karnosky et al. 2005). O_3 triggers physiologic changes in leaves, affecting carbon source strength, i.e., the amount of carbon available for allocation to sink tissues. Decreased carbon assimilation, increased metabolic costs for repair mechanisms, and decreased phloem loading all lead to decreased carbon allocation belowground, thus affecting roots, root symbionts, rhizodeposition, litter quality and quantity, and, consequently, the whole soil food web (Andersen 2003). Carbon source–sink relationships or the functional balance of roots and shoots has been reported as primary factors in continuous adjustments between root and shoot growth (Tingey et al. 1976), possibly acting through root to shoot signaling, including the hormonal regulation of root proliferation. In this context, the effects of ozone fumigation on the cytokinins (CK) of beech trees (Winwood et al. 2007) may be related to mycorrhiza-associated changes in cytokinin concentrations in the host plants (Kraigher et al. 1991, 1993, 2008) and ozone-induced changes in fine root growth and ECM community structure (Grebenc and Kraigher 2007a).

Sensitivity to ozone has been reported to differ between species (for review, see Andersen and Rygiewicz 1995; Bortier et al. 1999, 2000; Andersen 2003); between different clones and populations (Ballach et al. 1992; Coleman et al. 1996; Vanhatalo et al. 2003; Ottossen et al. 2003), experimental growth conditions, such as duration of the fumigation, light regime, irrigation, mineral nutrition, and the combination of various stress factors (McLaughlin and Downing 1995; Roth and Fahey 1998; Topa et al. 2004; Löw et al. 2006; Železnik et al. 2007); and between age-related physiologic differences within the same species (Matyssek et al. 2007). Ozone-caused reduced root growth has been found to alter the functioning of rhizosphere organisms and make them more susceptible to drought or nutrient deficiency. A carryover effect has also been established (Andersen and Rygiewicz 1991). The length of fumigation should be considered with a view to the life span of the fumigated plant (Kraigher et al. 2008).

In adult beech trees, a significant increase in the number of vital ECM root tips, of nonturgescent short roots, and of types of ectomycorrhiza has been observed.

Species richness appears to be affected only by O_3 in a normally humid year and not under conditions of drought stress (Grebenc and Kraigher 2007a). The same treatment also demonstrated an impact on individual ECM species abundance. An increased abundance of *Cenococcum geophilum*, *Russula fellea*, and *Russula illota* suggested their tolerance to the changed physiology of beech trees under ozone, while *Xerocomus* sp., *Russula cyanoxantha*, *R. ochroleuca*, and a *Lactarius* sp. decreased in abundance under increased ozone treatment or were only found with control trees (Grebenc and Kraigher 2007b). In terms of the nutrient status of roots, $\delta^{15}N$ measurements indicated a reduction in total nitrogen in the fine roots of ozone-treated adult trees (Haberer et al. 2007), while Mg concentration in seedlings was reduced under the same conditions (Železnik et al. 2007). In contrast to the results from adult trees, a significant reduction in number of vital ECM types and number of ECM root tips was observed in beech seedlings (ibid.).

Through its indirect effects, ozone imposes general stress on the below-ground community, often occurring in combination with other stresses, so there is no clear differentiation of ozone-sensitive and ozone-tolerant ECM species, such as have been proposed in acidic and nitrogen deposition studies (e.g., Kraigher et al. 1996). Functional groups of ECM types (exploration types as proposed by Agerer 2001) are perhaps more or less sensitive to changes in carbon allocation below-ground, and the physiology, and thus bioindication value, of each of the ECM types or groups, might be studied further.

13.2.5 Elevated CO_2

Ectomycorrhizal roots respond positively to elevated CO_2 (Alberton et al. 2005). An increase in root tip abundance under elevated CO_2 is consistent with reported changes in the root:shoot ratio, especially under conditions of severe nutrient limitation (Poorter and Nagel 2000; Ågren and Franklin 2003). It is important to note, however, that such an increased carbon allocation to ECM roots does not automatically translate into an increased uptake of limiting nutrients and hence to increased forest productivity. ECM forests can already be "saturated" with ECM (O'Neill 1994) and any further increase in ECM roots and mycelia would only increase nutrient limitation of the system. This phenomenon has been described as the progressive nitrogen limitation (PNL) hypothesis (Luo et al. 2004; Johnson 2006; Hu et al. 2006).

13.2.6 Drought

In a meta-analysis of the effects of several stress factors (Cudlin et al. 2007), the clearest effect found was a decrease in the fine root biomass during drought. A relative reallocation of growth to below-ground organs at the expense of above-ground ones during mild drought has often been found, and even absolute root growth may increase during mild drought (Becker et al. 1987). However, when water stress continues, the usual response is a reduction in root growth (Joslin et al. 2000). In contrast to fine root biomass, the ECM fractional colonization did not show a reduction but a slight (insignificant) increase. This may be due to a negative effect of drought on the total number of root tips. This kind of effect was shown in Norway spruce by Feil et al. (1988).

Cenococcum geophilum (*C. graniforme*) is often mentioned as a particularly drought-tolerant fungus. However, with some exceptions (Worley and Hacskaylo 1959), there are only a few reports based on exactly defined experimental conditions. Pigott (1982) showed that *Cenococcum* survived better than other mycorrhizal fungi during a drought episode. It has also been proposed that the mycelium of *Cenococcum* is more resistant to decomposition than that of some other species (Meyer 1987). Furthermore, the concepts of taxonomy of this fungus have varied with time, and morphological identification may have allowed possible confusion with other fungi. *Cenococcum* may also be tolerant to other stresses, such as increased ozone (Grebenc and Kraigher 2007a, b) or salinity (Saleh-Rastin 1976, cit. in Cudlin et al. 2007). While mycorrhiza formation in other species was strongly delayed in drought-treated plants, *Thelephora terrestris* formed abundant mycorrhizal systems, irrespective of watering treatment (Lehto 1992). Differences in community structure in drought conditions can, therefore, also be expected to occur in forests.

13.3 Mycobioindication in Forest Ecosystems

Forest sustainability, productivity, and vitality depend on the relationship with soil resources, the interface between soil nutrient pools and tree roots, as uptake organs to sustain above-ground growth (Bakker 1999). In temperate forests in Europe, the fine roots of stand-forming forest tree species are predominantly covered with a fungal sheath of a range of EMF, which take over the function of uptake and translocation of water and nutrients. As such, mycorrhizae are the main spatial and temporal linkage between the various constituents of a forest ecosystem (Dighton and Boddy 1988; Kraigher 1996; Dighton 2003).

The forest decline observed since the early 1980s in both Europe and North America has increased the interest of environmental research in bioindication methods (Tausz et al. 1996). Bioindication of environmental stresses provides a suitable means of evaluating the impact of these variables on biological systems (Arndt et al. 1987). Reactions that indicate a state of stress make the employment of sensitive plant, animal, fungal, or other species as bioindicators of environmental stress, or the use of living organisms or their parts as biomonitors possible.

Bioindicators are organisms or communities of organisms that react to environmental conditions by changing their vital functions and/or their chemical composition, thus making it possible to draw conclusions about the state of their environment (Arndt et al. 1987). Plants and fungi are recognized as suitable bioindicators with respect to forest ecosystems. Bioindicators (Fig. 13.1) can be differentiated into sensitive (response) and accumulative indicators, and are subdivided into three different groups (Arndt et al. 1987) (1) pointer organisms or ecological indicators provide evidence about the state of entire ecosystems; (2) tester organisms are used in standardized laboratory procedures; and (3) monitoring organisms are used to monitor the quality and quantity of harmful substances in the environment and to detect their effects. These monitoring organisms either already exist in the ecosystem (passive monitoring) or are introduced into the ecosystem in a standardized form (active monitoring).

Bioindicators can be obtained from any level of biological organization, from cell, tissue, organ, organism, and community levels (Fig. 13.2), and several

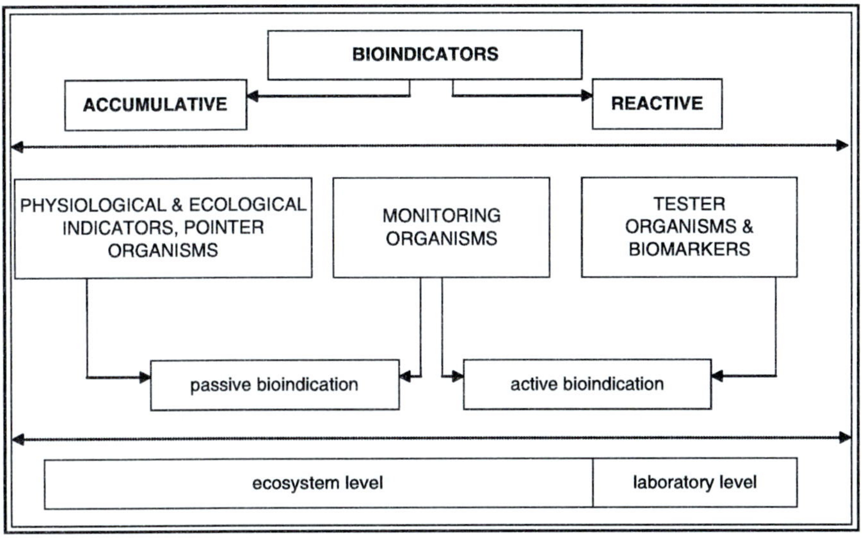

Fig. 13.1 Schematic presentation of bioindicators (modified after Arndt et al. 1987)

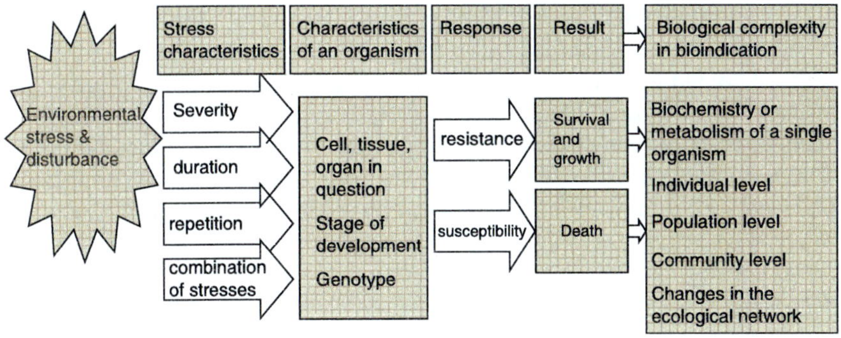

Fig. 13.2 How the response of an organism can be applied in bioindication of stress and disturbance at different levels of organization (modified from Buchanan et al. 2000 and Martínez-Crego et al. 2010)

summarizing indices that attempt to integrate relevant ecological information into an overall expression of biotic integrity are being developed (Martínez-Crego et al. 2010). Environmental stress differs in terms of its severity, duration, or number of exposures and a combination of stresses may occur with synergistic or antagonistic effects. The response of the organism on which the stress is imposed depends on the organ or tissue in question, stage of development, and genotype, which all influence the resistance or susceptibility to stress, and result in survival, modifications in growth, or death of the organism in question. The relevance of the bioindicator depends on its relevance to ecological integrity, broad-scale applicability, early detection capacity, feasibility of implementation, interpretability against reference conditions, and linking ecosystem degradation to its causative stress agents (Martínez-Crego et al. 2010).

Static and deterministic approaches have been developed as part of phytoindication, through, e.g., phytocoenological analyses, and "load capacity" estimations have been applied (Ellenberg 1972, cit. in Fränzle 2006). They depend on the disposition, susceptibility, and regeneration in predefined estimation scales. A combination of dynamic (reflecting the time evolution of perturbation, sensitivity, and adaptation) and statistical (defining probability distributions in order to calculate expected values) approaches are being developed (Fränzle 2006). Complex indicator approaches seem to reflect well the dynamics of ecosystem development and include fluctuations of energy, entropy production of ecosystems, fluctuations of selected micronutrients, duration of biogeochemical cycles, biomarkers (stress proteins and phytoalexines), changes in biodiversity of selected assemblages, dynamics of selected populations, changes in the competitive behavior of functionally important species, and modifications of the food web structure.

Decreases in species diversity and in the abundance of sporocarps of macromycetes in Europe were reported in the second half of the twentieth century (Arnolds 1988, 1991; Jaenike 1991) and the method of mycobioindication of forest site pollution was soon suggested by Fellner (1989) and Fellner and Peškova (1995), based on the impoverishment of ECM mycobiocenosis. Arnolds (1991) suggested an adaptation of Fellner's qualification of impoverishment, based on the various sensitivities of different ECM species. However, the occurrence of fruit bodies depends on a range of climatic factors in different years, while ECM types are present in the soils at any time of the year. Furthermore, a range of ECM fungi, such as the group commonly called Fungi Imperfecti (predominantly belonging to ascomycetes, but without a known anamorph), do not produce any sporocarps. As a result, more detailed studies of the ECM potential of forest sites by determination of ECM types were initiated in the 1990s, with a view to applying the results to bioindication (Fellner and Peškova 1995; Kraigher 1994; Kraigher et al. 1995, 1996; Al Sayegh Petkovšek 1997, 2004, 2005; Al Sayegh Petkovšek and Kraigher 2003; Taylor et al. 2000; Erland and Taylor 2002; Taylor and Alexander 2005).

In Slovenia, a series of studies was carried out by applying types of ECM as tools for the bioindication of stress in forest sites (Kraigher et al. 1995, 1996, 2007; Grebenc and Kraigher 2007a; Al Sayegh Petkovšek 2008). The application of ECM studies in bioindication was based on (1) the diversity of ECM types on Norway spruce (*Picea abies* (L) Karst) or beech (*Fagus sylvatica* L.) in forest stands (*in situ* ecological indicators); (2) the determination and quantification of selected pollution-sensitive or insensitive ECM types (*in situ* passive monitors); (3) the development of ectomycorrhiza on spruce seedlings, planted in studied sites (active monitors); (4) ECM or root growth parameters on tree seedlings, tested in an experimental setup (*ex situ* testers of substrate pollution); and (5) the bioaccumulation of metals in fungal sporocarps (*in situ* accumulative bioindicators).

13.3.1 Ecological Indicators and Passive Monitors of Stress in a Forest Ecosystem

13.3.1.1 Responsive *In Situ* Mycoindicators

Diversity indexes indicate the dynamics of an ecosystem, i.e., its potential to react to a changing environment (Atlas and Bartha 1981). The species richness index (*d*) links the number of species and their importance in the total community. The Shannon–Weaver index (*H*) also provides the relative abundance of each species, which indicates whether there are dominant populations in the sample. Both indexes are in general lower in populations in stressed environments. A reduction in the number of species in a community of mycorrhizal fungi can negatively influence the capacity of populations of mycorrhizal fungi and tree seedlings to form a functional symbiosis.

Pollution and other anthropogenic stresses have been found to diminish biodiversity indices of ECM types on Norway spruce (Kraigher 1999; Taylor et al. 2000; Peter et al. 2008), oak (Kovacs et al. 2000) and several other tree species (reviewed by Erland and Taylor 2002). A survey performed in three Norway spruce stands with different degrees of forest decline due to air pollution revealed that the number of living trees and their defoliation status may directly impact on the ECM species composition by affecting the amount of carbon delivered to the symbiotic fungal partners (Peter et al. 2008). The Shannon–Weaver index of adult trees was significantly lower in the heavily damaged site (only 3% of leafing trees); however, adult tress and seedlings were fully mycorrhizal. The most abundant species in all sites was *Tylospora fibrillosa*, especially in the most damaged site (it made up over 60% of root tips). It was emphasized that atheloids and thelephoroids (*Thelephora terrestris*, *Tylospora fibrillosa*, and *Thelephora asterophora*) and *C. geophilum* might play a crucial role in stressed forest ecosystems (*ibid.*).

Trends of diminishing ECM biodiversity were also clear in our studies in spruce stands. However, in European beech, no trend was detected (Table 13.1). Early studies (Kraigher et al. 1996) of ECM types on Norway spruce showed that several ECM types disappeared, while others proliferated in polluted sites. Additionally, some ECM types have been found to be restricted or mainly to occur either in polluted or in unpolluted sites, even if the identity of the fungus was not determined.

Table 13.1 Comparison of Shannon–Weaver indices (H) in soil samples sampled near different tree species from differenty polluted areas

Species	Unpolluted areas	Polluted areas and/or presence of stressors	References
Fagus sylvatica	2.8 (Gribskov, Denmark)	2.3 (Aubure, NE France)	Taylor et al. (2000)[a]
		3.2 (Collelongo, Italy)	
		3.9 (Schacht, Germany)	
	1.6–1.8 (Val di Sella, Italy; Idrija, Slovenia; Nizbor, Czech Republic		Pučko et al. (2004)[b]
		0.0–1.1 (Snežna jama, Slovenia)	Grebenc (2005)[c]
		0.6–1.6 (Rajhenavski Rog, Slovenia)	Grebenc and Kraigher (2007a, b)[c]
		0.9–1.3 (Krazberg, Germany)	Grebenc et al. (2009)[c]
	1.6–2.6 (Preža, Slovenia)	1.3–2.0 (Zavodnje, Slovenia)	Al Sayegh Petkovšek (2008)[d]
	1.9–2.8 (Moravške gredice, Slovenia)	1.6–2.3 (Dobovec, Slovenia)	
Picea abies	2.2 (Pokljuka, Slovenia)	2.3 (Zavodnje, Slovenia)	Kraigher (1999)
	2.2 (Mislinjski graben, Slovenia)		Kraigher et al. (2000)[e]
	3.1 (Pokljuka, Slovenia)		Vilhar et al. (2004)[f]
	3.5 (Ahden, N Sweden)	2.6 (Waldstein, Germany)	Taylor et al. (2000)[a]
		3.3 (Aubure, NE France and Klosterhede, Denmark)	
		1.1–2.0 (Mumlovska hora, Alzbentinka, Modry dul, Czech republic)	Peter et al. (2008)[g]
		0.7–1.2 (Hudobrežnikov vrh, Slovenia)*	Al Sayegh Petkovšek (2008)[d]
		1.0–1.3 (Hudobrežnikov vrh, Slovenia)**	
		0.2–3.0 (Veliki Vrh, Slovenia)*	
		0.2–0.7 (Veliki Vrh, Slovenia)**	
Quercus spp.		1.2–1.3 (Austria)*	Kovacs et al. (2000)[h]
		1.3–1.5 (Austria)**	
		1.1–1.4 (Spain)	de Roman and de Miguel (2005)[i]

[a]Study was performed along north–south transects in Europe regarding anthropogenic N-enrichment
[b]Identification of types of ectomycorrhizae on 7 year old seedlings in a beech provenance trial
[c]Canopy gaps with no natural regeneration (Snežna jama) and with natural regeneration (Rajhenavski Rog); 2× ambient ozone-treated plot (Krazberg)
[d]Research plot exposed to emissions of thermal power plants (air pollution); soil samples sampled near vital trees (*) and declining trees (**)
[e]Research plots exposed to emissions from thermal power plants (air pollution)
[f]Regeneration center in an autochothon Norway spruce forest on the Pokljuka plateau
[g]Damaged spruce stand in the Czech Republic due to air pollution
[h]Comparison of two oak stands (*Quercus petrarea* Liebl. *Ouercus robur* L.) regarding vital (*) and declining tress (**)
[i]Post-fire ectomycorrhizal community in a *Quercus ilex* L. stand over a 3-years period

This was the case with *Piceirhiza terraphila* and *Piceirhiza inflata* in polluted sites (Kraigher et al. 1996), and *Piceirhiza oleiferans* (Waller et al. 1993) in unpolluted plots. Similar to pollution-sensitive or -insensitive ECM species in spruce forests, we have suggested *Hydnum rufescens* (sensitive), *P. involutus* (insensitive), and *Elaphomyces* sp. in beech forests (insensitive).

However, since stress in natural conditions is complex, other factors, such as drought in the upper soil horizons (possibly influencing the high abundance of *C. geophilum* in unpolluted site), cannot be ruled out. Jany et al. (2003) determined that *C. geophilum* maintains the physiological integrity of beech roots facing drought stress. In an assessment of air pollution in the emission area of the thermal power plant Šoštanj (Al Sayegh Petkovšek 2008), C. *geophilum* was also determined as the dominant ECM type in soil cores from forest research plots exposed to air pollution and drought stress (Table 13.2). In addition, a statistically significant difference in the number of sclerotia of *C. geophilum* in ozone-fumigated

Table 13.2 List of types of ectomycorrhiza determined on European beech and Norway spruce, their bioindication applicability as a stress indicator and corresponding references

ECM type	*Fagus sylvatica*	*Picea abies*	Type of stress[a]	References
Cenococcum geophilum Fr.	+	+	2, 3	Cudlin et al. (2007), Kraigher et al. (2007), Grebenc and Kraigher (2007a, b), Al Sayegh Petkovšek (2008)
Dermocybe cinammomea (L.) Wünsche + *Picea abies* (L.) Karst	+	+	1, 2	Al Sayegh Petkovšek (2008)
Elaphomyces granulatus Fr. + *Picea abies* (L.) Karst.	+	+	1	Al Sayegh Petkovšek (2008)
Elaphomyces muricatus Fr. + *Fagus sylvatica* L.	+		1	Al Sayegh Petkovšek (2008)
Elaphomyces sp. 1	+	+	1	Al Sayegh Petkovšek (2008)
Elaphomyces sp. 2	+	+	1	Al Sayegh Petkovšek (2008)
Fagirhiza oleifera	+	+	1, 2	Al Sayegh Petkovšek (2008)
Fagirhiza spinulosa		+	1, 2	Al Sayegh Petkovšek (2008)
Paxillus involutus		+	1	Kraigher (1999), Kraigher et al. (2007)
Piceirhiza inflata		+	1	Kraigher (1999), Kraigher et al. (2007)
Piceirhiza terraphila		+	1	Kraigher (1999), Kraigher et al. (2007)
Russula fellea (Fr.: Fr.) Fr. + *Fagus sylvatica* L.	+		3	Grebenc and Kraigher (2007a, b)
Russula illota Romagn. + *Fagus sylvatica* L.	+		3	Grebenc and Kraigher (2007a, b)
Russula sp. 4		+	1	Kraigher (1999), Kraigher et al. (2007)
Xerocomus badius (Fr.Fr.) Gilb. + *Picea abies* (L.) Karst.	+	+	1	Kraigher (1999), Kraigher et al. (2007)

[a]Different stressors are marked with numbers (1 – polluted air, 2 – drought, and 3 – ozone)

containers when compared to that in nonfumigated controls (description available at: http://www.casiroz.de), especially at a depth of 2–8 cm, was found (Fig. 13.3).

The average number of short root tips in 274 ml of soil cores ranged from (400) 1,100 to 2,000 (5,600), indicating that the number of active short root tips in the upper 20 cm of forest soils should be between 3×10^5 and 5×10^6 per square meter. These numbers suggest that even types of ECM occurring in small percentages may form symbiosis on several hundred thousands of roots (i.e., *Lactarius lignyotus* on ca 230,000 root tips; Kraigher 1997). Such high numbers of root tips also indicate that it is impossible to analyze more than a small fraction of the total ECM fungal community on root tips. The high heterogeneity within forest soils also interacts with the heterogenous distribution of ECM types in time and space; therefore, representative sampling is difficult (Kraigher and Agerer 2000; Erland and Taylor 2002). However, stress-sensitive Norway spruce stands, with a very low ground vegetation diversity, showed clear shifts in the diversity indices of the very diverse below-ground ECM community (Kraigher 1999). On the contrary, in ground vegetation rich beech sites, the same approach failed as a bioindication method. The lack of changes in the ECM biodiversity indices in beech may reflect the well-adaptable buffering capacity of the forest soils; the presence of favorable humus forms, especially in older beech stands; the sustainability of natural beech forests; their adaptation to N deposits; or the fact that the diversity both above- and below-ground was too high to allow the detection of any changes (Al Sayegh Petkovšek 2004, 2008). Our standardized sampling protocols were designed for Norway spruce stands (Kraigher and Agerer 2000).

13.3.1.2 Accumulative In Situ Passive Monitors

Sporocarps of selected ECM fungi can be used as accumulative bioindicators (passive monitors) of forest site pollution, since they accumulate high concentrations

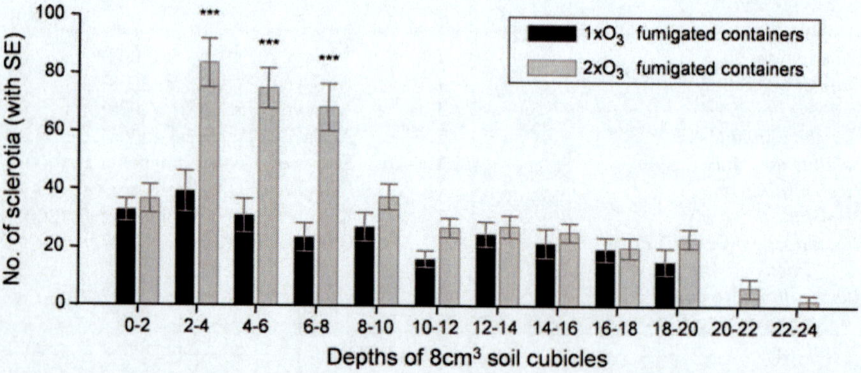

Fig. 13.3 Occurrence of sclerotia of *Cenococcum geophilum* in ozone-fumigated containers (*gray bars*) compared to the control (*black bars*) in the Free Air Ozone Fumigation System in Kranzberg Forest (for description, see http://www.casiroz.de)

of heavy metals. First of all, they are useful for distinguishing polluted and unpolluted areas (Kalač and Svoboda 2000; Rudawska and Leski 2005a, b); moreover, a survey performed in variously polluted areas in Slovenia suggests that several ECM species are employable as bioindicators of soil polluted with heavy metals (e.g., *Boletus edulis* and *Laccaria amethystina*) (Al Sayegh Petkovšek and Pokorny 2006; Al Sayegh Petkovšek 2008). In addition, the bioconcentration factor (BCF), calculated as the quotient between metals in the sporocarp and metal content in soil can be used as a tool for assessing forest site pollution. BCFs ranged from 50 to 300 (Cd), 30 to 500 (Hg), and 0.1 to 0.2 (Pb) and differ in relation to the metal levels in soil and ECM fungal species (Kalač and Svoboda 2000; Rudawska and Leski 2005a). It appears that the bioindicative value of BCF is fairly problematic, since BCF is not a constant value for a particular species. However, according to Rudawska and Leski (2005a), BCF factors may be useful for comparing fungal species from different sites with those from soil with uniform properties.

13.3.1.3 Conclusions on Indicators and Passive Monitors

Conclusions from the application of biodiversity indices and passive monitors in forest stands can be summarized as follows:

1. ECM species richness (*d*) for Norway spruce showed higher values in unpolluted sites than in NO_x and SO_x polluted ones.
2. ECM species richness for European beech was similar in polluted and unpolluted sites, so it cannot be applied as an ecological indicator of soil pollution.
3. Similar *H* values indicate whether there was no dominant ECM species in polluted vs. unpolluted sites.
4. Differences in diversity indices in polluted and unpolluted Norway spruce sites may indicate a potentially fast response toward degradation when a stress factor is imposed, while beech forest sites may be considered to be more resilient toward any stresses.
5. Biodiversity indices as ecological indicators of the impact of pollution on forest soils can be applied in various (but not all) forest ecosystems, show a reasonable sensitivity to stress, and can provide some insight into changes in below-ground forest ecosystem processes. However, they are species specific and should be complemented with other bioindication approaches, cannot be used as a single indicator of stress, and should be upgraded with functional diversity approaches.
6. Specific persistent structures, such as the sclerotia of *C. geophilum*, which are presumed to help this fungus to survive in unfavorable conditions, can be significantly multiplied in soils exposed to ozone fumigation and/or drought. They have potential as an applicable bioindicator after the system has been analyzed under various conditions of stress in comparison with controls – an active monitoring approach may develop.

7. The application of pollution-sensitive and -insensitive ECM species can reveal the ecology of a certain ECM species and is, therefore, important for understanding the processes in a forest ecosystem.

13.3.2 Active Monitors of Stress in Forest Ecosystems

13.3.2.1 Active monitors: *In Situ* Exposed Nonmycorrhizal Seedlings

Nonmycorrhizal Norway spruce seedlings were exposed and planted in variously polluted forest sites as active monitors of stress (Kraigher et al. 2007). In beech dominated sites, the mycorrhization percentage was significantly lower in the polluted plot (here identified as Zavodnje B) in two different years when testing was applied (Fig. 13.3). The results indicated that spruce seedlings and their mycorrhization can be applied as active monitors of pollution when the seedlings are planted in polluted and unpolluted stands in broadleaf forests. On the contrary, the mycorrhization percentage of spruce seedlings planted in a polluted spruce dominated site (Zavodnje A) was significantly higher than that of the unpolluted spruce stand (Pokljuka). The opposite result, i.e., a higher mycorrhization in Pokljuka, was discussed as being due to the presence of the same macrobiont in the plot for several centuries. Our results might, therefore, indicate less favorable growing conditions resulting from the high altitude of the plot (Pokljuka, 1,250 m/a. s.l.) and the relatively short vegetation period when compared to all other plots (Kraigher et al. 2000). In the within-treatment comparison (between 6 and 12 months of exposure), the average mycorrhization percentage was higher for all plots after longer exposure. The dry weight of needles was in close linear correlation with mycorrhization in all treatments. Seedlings with low mycorrhization also had less needle biomass and *vice versa*. As with the mycorrhization %, the calculated ratio was significantly higher in unpolluted broadleaf stands but not in spruce stands. The results indicate that spruce seedlings might serve as a good active monitoring indicator of pollution if planted in beech stands in which no specific ECM indicator types have been determined.

13.3.2.2 *Ex Situ* Testers: Mycorrhizal Inoculum Potential of Differently Polluted Soil Substrates

Mycorrhization inoculum potential is a method developed by Kropaček et al. (1989). It is based on naturally present mycorrhizal inoculum, existing in sieved soil substrate taken from a forest site and applied in a pot experiment with seedlings of a chosen forest tree species in a growth chamber under controlled environmental conditions.

Two differently polluted forest sites in the emission area of Thermal Power Plant Šoštanj were submitted to studies of mycorrhizal inoculum potential in different years, before and after the installation of cleaning blocks in the Thermal Power Plant. Emissions were reduced from 80.516 t SO_2 in 1994 to 44.253 t in 2000, and the tests were done in 1992–1993 and in 2002–2003. The percentage of mycorrhizal short roots of potted seedlings from Zavodnje (polluted substrate) was significantly lower ($p < 0.05$) in comparison with that of Pohorje (unpolluted), suggesting that the mycorrhizal potential of the more polluted area was lower (Fig. 13.4). Other growth parameters of the seedlings were also different: the fresh weights of roots, stems, and needles were higher in Pohorje (Al Sayegh Petkovšek and Kraigher 2003). Pollution influenced the mycorrhizal potential of forest soils and this negative impact was still present more than 5 years after the reduction of the emissions. We conclude that further analyses of mycorrhizal potential are recommended in order to monitor the changes in ECM composition continuously after the reduction of the emissions from Thermal Power Plant Šoštanj. These might also include testing the application of mycorrhizal fungi in bioremediation processes, as reviewed by Gadd (2005), for remediation of the thermal power plant wastes.

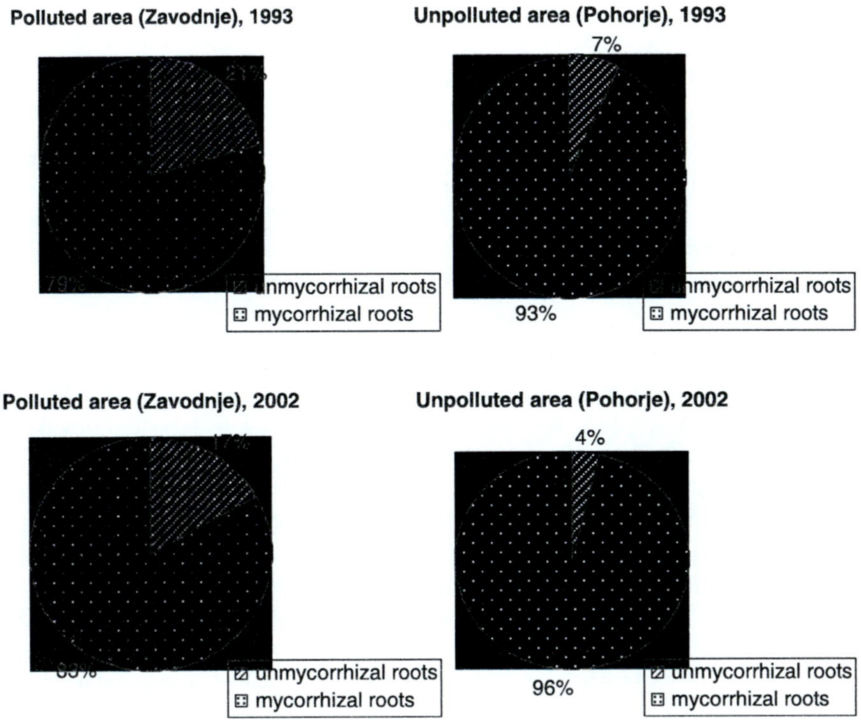

Fig. 13.4 Mycorrhization (%) of Norway spruce seedlings ($n = 25$) from the mycorrhizal inoculum potential of forest soil from 1993 to 2002. The differences among sites were significant (Fisher LSD, $p < 0{,}001$), differences among years showed trends to higher mycorrhization (Fisher LSD, $p = 0.077$)

13.3.2.3 Conclusions on Active Indicators in Mycoindication

Our conclusions from mycorrhizal inoculum potential tests can be summarized as follows:

1. Mycorrhizal potential as a bioassay for soil pollution (N and S deposit) is an acceptable mycoindication method.
2. The differences between the two sites were statistically significant, although the pollution effects were not highly destructive with respect to mycorrhizal soil inoculum potential.
3. Even though emissions from the thermal power plant were reduced in the last decade, the revitalization of soil substrates is a slow process, it might, therefore, take several decades before the mycorrhizal potential of the polluted and the unpolluted sites reaches the same level.
4. The types of ECM are of primary importance for the functioning of a forest ecosystem, although the simplified situation in pot studies of mycorrhizal potential supports the primary importance of the percentage of mycorrhizal infection.

13.4 Final remarks

Bioindicators have been proposed as "environmental fever thermometers" (Fränzle 2006), since they should respond to early stages of either exposure or effects without disclosing cause–effect relationships. As a follow-up action, preventive or corrective measures need to be initiated. Their qualities, demands, and limitations can be presented as summarized by Fränzle (2006), with respect to mycoindication approaches:

- Environmental observation techniques face a large number of preselected physical or chemical stress factors, with multiple synergistic and antagonistic effects on conversion products and pathways. Biomonitoring techniques provide a timely, widely applicable, low-cost, feasible way of observing deleterious effects. They involve passive and active approaches on different scales, from microcoenoses to ecosystems. Their main task is a general determination of physiological effects rather than measurement of concentrations of stress factors. The lack of specificity in early recognition perspective might be considered to be an advantage inducive of a subsequent systematic search for quantification of causal interrelationships. The disadvantage is the highly variable susceptibility of the multitude of species exposed to stress factors, resulting in the need for fuzzy logic approaches.
- The field of biomarkers has evolved rapidly in recent decades; a biomarker may be considered to be a biological response to chemical(s) that might provide its quantification and measure of toxicity. A parallel exposure to ambient stress factors and a control set would be ideal to estimate the ecological significance of

a slight increase or decrease in a measured parameter. In a mycobioindication, this would imply analysis of a natural mycorrhizal community structure with special emphasis on resistant and sensitive species, exposure of nonmycorrhizal seedlings in sites with a gradually increasing stress, and establishment of a controlled experimental system using the mycorrhizal inoculum potential of a systematically increased level of stress.

- The interpretation of the results of studies in complex ecosystems such as forests is problematic. Their functional redundancy might interfere with clear indication of stress, since the loss of functional capacity by one organism will immediately be compensated by increased activity of another. An intersystemic comparison of biotic reactions must also take into account natural successions, and spatial and temporal variability of ecosystems, whose communities are organized by competition and collaboration, and predation and disturbances. All these systems are therefore dynamic; there is no single stability, susceptibility, or vulnerability measure for a community or a whole ecosystem. Active and passive biomonitoring techniques demand representative sampling over temporal and spatial scales, and their interpretation through geostatistical requirements and kriging, both at the single species and the community level.
- However, each bioindication method contributes to an understanding of ecosystem functioning, community interactions, and single species physiology and ecology. A combination of approaches provides a timely and feasible interpretation of stresses, and their application leads to step-wise revelations of processes in highly complex below-ground processes.

Acknowledgments The study was part of the research program P4-0107 financed by the Slovenian Research Agency, summarizing the results of a number of national and international projects, with numerous collaborators contributing to obtain the data. Among these, we would like to acknowledge the collaboration of Dr. Tine Grebenc, Barbara Štupar, Melita Hrenko, and Jana Janša from the Slovenian Forestry Institute; Dr. Boštjan Pokorny from ERICo Ecological Research Institute; the coordinator of the EU project CASIROZ Prof. Dr. Rainer Matyssek; and his team. We would also like to acknowledge the early suggestions on mycorrhizal inoculum potential from Dr. Pavel Cudlin, instructions on identification of ectomycorrhizae from Prof. Dr. Reinhard Agerer, on the development of bioindication methods from Prof. Dr. Franc Batič and Prof. Dr. Dieter Grill, and Martin Creegen for final language corrections.

References

Agerer R (2001) Exploration types of ectomycorrhizae. A proposal to classify ectomycorrhizal mycelial systems according to their patterns of differentiation and putative ecological importance. Mycorrhiza 11:107–114

Ågren GI, Franklin O (2003) Root: shoot ratios, optimization and nitrogen productivity. Ann Bot 92:795–800

Al Sayegh Petkovšek S (1997) Mycorrhizal potential of two differently polluted forest sites in the emission region of the Thermal Power Plant Šoštanj. Zb GL 52:323–350

Al Sayegh Petkovšek S (2004) Biodiversity of types of ectomycorrhizae in fagus stands in differently polluted forest research plots. Zb GL 75:5–19

Al Sayegh Petkovšek S (2005) Belowground ectomycorrhizal fungal communities at fagus stands in differently polluted forest research plots. Zb GL 76:5–38

Al Sayegh Petkovšek S (2008) Fungi as sensitive and accumulative bioindicators of forest site pollution in the Šalek Valley. Doctoral dissertation, University of Ljubljana

Al Sayegh Petkovšek S, Kraigher H (2003) Mycorrhizal potential of two forest research plots with respect to reduction of the emissions from the Thermal Power Plant Šoštanj. Acta Biol Slov 46:9–16

Al Sayegh Petkovšek S, Pokorny B (2006) Fungi as sensitive and accumulative bioindicators of forest site pollutions in the Šalek Valley. Zb GL 81:61–71

Al Sayegh Petkovšek S, Pokorny B, Ribarič Lasnik C, Vrtačnik J (2002) Cd, Pb, Hg and As in fruiting bodies of higher fungi from the forest landscape of the Šalek Valley. Zb GL 67:5–46

Alberton O, Kuyper TW, Gorissen A (2005) Taking mycocentrism seriously: mycorrhizal fungal and plant responses to elevated CO2. New Phytol 167:859–868

Alexander IJ, Fairley RI (1983) Effects of N fertilisation on populations of fine roots and mycorrhizas in spruce humus. Plant Soil 71:49–53

Andersen CP (2003) Source–sink balance and carbon allocation below ground in plants exposed to ozone (Tansley review). New Phytol 157:213–228

Andersen CP, Rygiewicz PT (1991) Stress interactions and mycorrhizal plant response: understanding carbon allocation priorities. Environ Pollut 73:217–244

Andersen CP, Rygiewicz PT (1995) Effects of ozone on temporal allocation of carbon in mycorrhizal Pinus ponderosa seedlings. New Phytol 131:471–480

Arndt U, Nobel W, Schweizer B (1987) Bioindikatoren – Möglichkeiten, Grenzen und neue Erkentnisse. Ulmer, Stuttgart

Arnolds E (1988) The changing macromycete flora of the Netherlands. Trans Br Mycol Soc 90:391–406

Arnolds E (1991) Decline of ectomycorrhizal fungi in Europe. Agric Ecosyst Environ 35:209–244

Atlas R, Bartha R (1981) Introduction to microbiology. Addison-Wesley, Reading

Bakker MR (1999) Fine root parameters as indicators of sustainability of forest ecosystems. For Ecol Manage 122:7–16

Ballach HJ, Oppenheimer S, Mooi J (1992) Reactions of cloned poplars to air pollution: premature leaf loss and investigations of the nitrogen metabolism. Z Naturforsch 47c:109–119

Becker CA, Mroz CD, Fuller LG (1987) The effects of plant moisture stress on red pine (Pinus resinosa) seedling growth and establishment. Can J For Res 17:813–820

Berthelsen BO, Olsen RA, Steinnes E (1995) Ectomycorrhizal heavy metal accumulation as a contributing factor to heavy metal levels in organic surface soil. Sci Total Environ 170:141–149

Bortier K, Ceulemans R, De Temmerman L (1999) Effects of tropospheric ozone on woody plants. In: Agrawal SB, Agrawal M (eds) Environmental pollution and plant response. Lewis, Boca Raton, pp 153–182

Bortier K, De Temmerman L, Ceulemans R (2000) Effects of ozone exposure in open-top chambers on poplar (Populus nigra) and beech (Fagus sylvatica): a comparison. Environ Pollut 109:509–516

Brunner I (2001) Ectomycorrhizas: their role in forest ecosystems under the impact of acidifying pollutants. Perspect Plant Ecol Evol Syst 4(1):13–27

Buchanan B, Gruissens W, Jones R (2000) Biochemistry and molecular biology of plants. American Society of Plant Physiologists, Rockville, MD

Cocchi L, Vescovi L, Petrini LE, Petrinini O (2006) Heavy metals in edible mushrooms in Italy. Food Chem 98:277–284

Coleman MD, Dickson RE, Isebrands JG, Karnosky DF (1996) Root growth and physiology of potted and field-grown trembling aspen exposed to tropospheric ozone. Tree Physiol 16:145–152

Cudlin P, Kieliszewska-Rokicka B, Rudawska M, Grebenc T, Alberton O, Lehto T, Bakker MR, Börja I, Konopka B, Leski T, Kraigher H, Kuyper TW (2007) Fine roots and ectomycorrhizas as indicators of environmental change. Plant Biosyst 141(3):406–525 de Roman M, de Miguel AM (2005) Post-fire, seasonal and annual dynamics of the ectomycorrhizal community in a *Quercus ilex* L. over 3-year period. Mycorrhizae 15:471–482

Dighton J (2003) Fungi in Ecosystem Processes. Dekker, New York

Dighton J, Boddy L (1988) Role of fungi in nitrogen, phosphorus and sulphur cycling in temperate forest ecosystems. In: Boddy L, Marchant R, Read DJ (eds) Nitrogen, phosphorus and sulphur utilization by fungi. Symposium of the BMS, Cambridge University Press, Cambridge, pp 269–299

Erland S, Taylor AFS (2002) Diversity of ectomycorrhizal fungal communities in relation to the abiotic environment. In: van der Heijden M, Sanders T (eds) The ecology of ectomycorrhizas. Ecological studies Series, Vol 157. Chapter 7, Springer, pp 163–193

Falandysz J, Szymczyk K, Ichihashi H, Bielawski L, Gucia M, Frankowska A, Yamasaki S (2001) ICP/MS and ICP/AES elemental analyses (38 elements) of edible wild mushrooms growing in Poland. Food Addit Contam 18:503–513

Falandysz J, Kawano M, Swieczkowski A, Brzostowski A, Dadej M (2003) Total mercury in wild-growing higher mushrooms and underlying soil from Wdzydze Landscape Park, Northern Poland. Food Chem 81:21–26

Feil W, Kottke I, Oberwinkler F (1988) The effect of drought on mycorrhizal production and very fine root system development of Norway spruce under natural and experimental conditions. Plant Soil 108:221–231

Fellner R (1989) Mycorrhiza-forming fungi as bioindicators of air pollution. Agric Ecosyst Environ 28:115–120

Fellner R, Peškova V (1995) Effects of industrial pollutants on ectomycorrhizal relationships in temperate forest. Can J Bot 73(Suppl 1):1310–1315

Fränzle O (2006) Complex bioindication and environmental stress assessment. Ecol Indic 6:114–136

Gadd GM (2005) Microorganisms in toxic metal-polluted soils. In: Buscot F, Varma A (eds) Microorganisms in soils: roles in genesis and functions, vol 3, Soil biology. Springer, Berlin, pp 325–356

Grebenc T (2005) Tipi ektomikorize na bukvi (Fagus sylvatica L.) v naravnem in gospodarskem gozdu: doktorska disertacija. Univerza v Ljubljani, Biotehniška fakulteta, Interdisciplinaren podiplomski študij biotehnologije. Ljubljana

Grebenc T, Kraigher H (2007a) Types of ectomycorrhiza of mature beech and spruce at ozone-fumigated and control forest plots. Environ Monit Assess 128:47–59

Grebenc T, Kraigher H (2007b) Changes in the community of ectomycorrhizal fungi and increased fine root number under adult beech trees chronically fumigated with double ambient ozone concentration. Plant Biol (Stuttg) 9(2):279–287

Grebenc T, Christensen M, Vilhar U, Čater M, Martin MP, Simončič P, Kraigher H (2009) Response of ectomycorrhizal community structure to gap opening in natural and managed temperate beech-dominated forests. Can J For Res 39(7):1375–1386

Haberer K, Grebenc T, Alexou M, Gessler A, Kraigher H, Rennenberg H (2007) Effects of long-term free-air ozone fumigation on δ15N and total N in Fagus sylvatica and associated mycorrhizal fungi. Plant Biol 9:242–252

Haug I, Pritsch K, Oberwinkler F (1992) Der Einfluss der Düngung auf Feinwurzeln und Mykorrhizen im Kulturversuch und im Freiland. KfK-PEF 97:1–159

Hu S, Tu C, Chen X, Gruver JB (2006) Progressive N limitation of plant response to elevated CO2: a microbiological perspective. Plant Soil 289:47–58

Jaenike J (1991) Mass extinction of European fungi. Tree 6:174–175

Jany JL, Martin F, Garbaye J (2003) Respiration activity of ectomycorrhizas from Cenoccocum geophilum and Lactarius spec.: in relation to soil water potential in five beech forests. Plant Soil 225:487–494

Johanson KJ, Nikolova I, Taylor AFS (2004) Uptake of elements by fungi in the Forsmark area. Technical Report TR-04-26, Swedish Nuclear Fuel and Waste Management

Johnson DW (2006) Progressive limitation in forests: review and implications for long-term responses to elevated CO2. Ecology 84:4–75

Joslin JD, Wolfe MH, Hanson PJ (2000) Effects of altered water regimes on forest root systems. New Phytol 147:117–129
Kalač P, Svoboda L (2000) A review of trace element concentrations in edible mushrooms. Food Chem 69:273–281
Karnosky DF, Pregitzer KS, Zak DR, Kubiske M, Hendrey GR, Weinstein D, Nosal M, Percy KE (2005) Scaling ozone responses of forest trees to the ecosystem level in a changing climate. Plant Cell Environ 28:965–981
Kernaghan G (2005) Mycorrhizal diversity: cause and effect? Pedobiologia 49:511–520
Kieliszewska-Rokicka B, Rudawska M, Leski T (1997) Ectomycorrhizae of young and mature scots pine trees in industrial regions in Poland. Environ Pollut 98(3):315–324
Kovacs G, Pausch M, Urban A (2000) Diversity of ectomycorhizal morphotypes and oak decline. Phyton (Horn, Austria) 40(4):109–116
Kraigher H (1991) Mineral nutrients in mycorrhizal Norway spruce trees on Pohorje. MSc Thesis, University of Ljubljana, Ljubljana
Kraigher H (1994) Cytokinins and types of ectomycorrhiza in Norway spruce (Picea abies (L.) Karst) seedlings as indicators of forest site pollution. PhD Thesis, University of Ljubljana, Ljubljana
Kraigher H (1996) Tipi ektomikorize – taksonomija, pomen in aplikacije (Types of ectomycorrhizae – their taxonomy, role and application). ZbGL 49:33–66
Kraigher H (1997) Mycobioindication of pollution of two forest sites. In: Robek R (ed) Proučevanje propadanja gozdov v Sloveniji v obdobju 1985–1995. ZbGL 52:279–322
Kraigher H (1999) Diversity of types of ectomycorrhizae on Norway spruce in Slovenia. Phyton 39(3):199–202
Kraigher H (2001) Mycorrhizal fungi. In: Hlad B, Skoberne P (eds) Overview of biodiversity in Slovenia, 1st edn. Ministry for environment and spatial planning, Environmental Agency od RS, Ljubljana
Kraigher H, Agerer R (2000). Identification and characterisation of types of ectomycorrhizae. V: Methods in root–soil interactions research: protocols compiled for the Technical Workshop of the COST E6 Action Eurosilva. In: Martin MP (ed) Forest Tree Physiology Research held in Gozd Martuljek, Slovenia. Slovenian Forestry Institute, Ljubljana, 8 Sep 1999, pp 29–34
Kraigher H, Grayling A, Wang TL, Hanke DE (1991) Cytokinin production by two ectomycorrhizal fungi in liquid culture. Phytochemistry 30(7):2249–2254
Kraigher H, Strnad M, Hanke DE, Batič F (1993) Cytokiningehalte von Fichtennadeln (Picea abies [L.] Karst) nach Inokulation mit zwei Stämmen des Mykorrhizapilzes Thelephora terrestris (Ehrh.) Fr. Forstwiss Cbl 112:107–111
Kraigher H, Agerer R, Javornik B (1995) Ectomycorrhiza of Lactarius lignyotus on Norway spruce, characterised by anatomical and molecular tools. Mycorrhiza 5:175–180
Kraigher H, Batič F, Agerer R (1996) Types of ectomycorrhizae and mycobioindication of forest site pollution. Phyton 36(3):115–120
Kraigher H, Simončič P, Smolej I, Diaci J (2000) Conclusions, applications and perspectives of research. In: Kraigher H, Smolej I (eds) The rhizosphere: studies of forest soils and the rhizosphere and their influences on chosen physiological parameters of forest trees in selected forest ecosystems, forest types and developmental phases of the forest. Strokovna in znanstvena dela, vol 118. Gozdarski inštitut Slovenije, Ljubljana, pp 246–262
Kraigher H, Al Sayegh-Petkovšek S, Grebenc T, Simončič P (2007) Types of ectomycorrhiza as pollution stress indicators: case studies in Slovenia. Environ Monit Assess 128:31–45
Kraigher H, Grebenc T, Hanke DE (2008) Ozone stress and ectomycorrhizal root-shoot signaling. In: Varma A (ed) Mycorrhiza: state of the art, genetics and molecular biology, eco-function, biotechnology, eco-physiology, structure and systematics, 3rd edn. Springer, Berlin, pp 337–357
Kropaček K, Kristinova M, Chemelikova E, Cudlin P (1989) The mycorrhizal inoculation potential of forest soils exposed to different pollutions stress. Agric Ecosyst Environ 28:217–277

Leake JR (2001) Is diversity of ECM important for ecosystem function? New Phytol 152:1–3
Lehto T (1992) Mycorrhizas and drought resistance of Picea sitchensis. II. In conditions of adequate nutrition. New Phytol 122:669–673
Lilleskov EA (2005) How do composition, structure, and function of mycorrhizal fungal communities respond to nitrogen deposition and ozone exposure? In: Dighton J, Oudemans P, White J (eds) The fungal community: its organization and role in the ecosystem, 3rd edn. Dekker, New York, pp 769–801
Lilleskov EA, Fahey TJ, Horton TR, Lovett GM (2002) Belowground ectomycorrhizal fungal community change over a nitrogen deposition gradient in Alaska. Ecology 83:104–115
Löw M, Herbinger K, Nunn AJ, Häberle K-H, Leuchner M, Heerdt C, Werner H, Wipfler P, Pretzsch H, Tausz M, Matyssek R (2006) Extraordinary drought of 2003 overrules ozone impact on adult beech trees (Fagus sylvatica). Trees Struct Funct 20:539–548
Luo Y, Su B, Currie WS, Dukes JS, Finzi A, Hartwig U, Hungate B, McMurtrie RS, Oren R, Parton WJ, Pataki DE, Shaw R, Zak DR, Field CB (2004) Progressive nitrogen limitation of ecosystem responses to rising atmospheric carbon dioxide. Bioscience 54:731–739
Martínez-Crego B, Alcoverro T, Romero J (2010) Biotic indices for assessing the status of coastal waters: a review of strengths and weaknesses. J Environ Monit 12:1013–1028
Matyssek R, Bahnweg G, Ceulemans R, Fabian W, Grill D, Hanke DE, Kraigher H, Oßwald W, Rennenberg H, Sandermann H, Tausz M, Wieser G (2007) Synopsis of the CASIROZ Case Study: carbon sink strength of Fagus sylvatica L. in a changing environment – experimental risk assessment of mitigation by chronic ozone impact. Plant Biol 9:163–180
McLaughlin SB, Downing DJ (1995) Interactive effects of ambient ozone measured on mature forest trees. Nature 374:252–257
Menge JA, Grand LF, Haines LW (1977) The effect of fertilization on growth and mycorrhizae numbers in 11-year old loblolly pine plantations. For Sci 23:37–44
Meyer FH (1987) Der Verzweigungsindex, ein Indikator für Schäden am Feinwurzelsystem. Forstwiss Cbl 106:84–92
O'Neill EG (1994) Responses of soil biota to elevated atmospheric carbon dioxide. Plant Soil 165:55–65
Ottossen S, Wallin G, Skärby L, Karlsson P-E, Medin E-L, Räntfors M, Pleijel H, Selldén G (2003) Four years of ozone exposure at high or low phosphorus reduced biomass in Norway spruce. Trees 17:299–307
Parrent JL, Morris WF, Vilgalys R (2006) CO_2 enrichment and nutrient availability alter ectomycorrhizal fungal communities. Ecology 87(9):2278–2287
Perrin R, Estivalet D (1990) Mycorrhizal association and forest decline (yellowing of spruce). Agric Ecosyst Environ 28:381–387
Peter M, Ayer F, Egli S (2001) Nitrogen addition in a Norway spruce stand altered macromycete sporocarp production and below-ground ectomycorrhizal species composition. New Phytol 149:311–325
Peter M, Ayer F, Cudlin P, Egli S (2008) Belowground ectomycorrhizal communities in three Norway spruce stands with different degrees of decline in the Czech Republic. Mycorrhizae 18:157–169
Pigott CD (1982) Survival of mycorrhiza formed by Cenococcum geophilum Fr. in dry soils. New Phytol 92:513–517
Poorter H, Nagel O (2000) The role of biomass allocation in the growth response of plants to different levels of light, CO2, nutrients and water: a quantitative review. Aust J Plant Physiol 27:595–607
Pučko M, Grebenc T, Božič G, Brus R, Kraigher H (2004) Identifikacija tipov ektomikorize na sadikah v bukovem provenienčnem poskusu. Zb GL 75:87–104
Read DJ (1991) Mycorrhizas in ecosystems. Experientia 47(4):376–390
Read DJ (1998) Plants on the web. Nature 396:22–23
Roth DR, Fahey TJ (1998) The effects of acid precipitation and ozone on the ectomycorrhizae of red spruce saplings. Water Air Soil Pollut 103:263–276

Rudawska M (1986) Sugar metabolism of ectomycorrhizal Scots pine seedlings as influenced by different nitrogen forms and levels. In: Gianinazzi-Pearson V, Gianinazzi S (eds) Physiological and genetical aspects of Mycorrhizae. Proceedings of the 1st European Symposium on Mycorrhizae. INRA, Paris, pp:389–394

Rudawska M, Leski T (2005a) Trace elements in fruiting bodies of ectomycorrhizal fungi growing in Scots pine (Pinus sylvestris L.) stands in Poland. Sci Total Environ 339:103–115

Rudawska M, Leski T (2005b) Macro- and microelement contents in fruiting bodies of wild mushrooms from the Notecka forest in west-central Poland. Food Chem 92:499–506

Schneider BU, Meyer J, Schulze ED, Zech W (1989) Root and mycorrhizal development in healthy and declining Norway spruce stands. Forest decline air pollution, vol 77. Berlin, Springer, pp 370–391

Simončič P, Smolej I, Rupel M, Urbančič M, Kalan P, Kraigher H (1998) Kroženje hranil in pestrost ektomikorize v smrekovem gozdu na Pokljuki. In: Zbornik referatov "Gorski gozd". XIX. Gozdarski študijski dnevi, Logarska dolina, pp 207–221

Tausz M, Batič F, Grill D (1996) Bioindication at forest sites – concepts, practice and outlook. Phyton 36:7–14

Taylor AFS, Alexander IJ (1989) Demography and population dynamics of ectomycorrhizas of Sitka spruce fertilized with N. Agric Ecosyst Environ 28:493–496

Taylor AFS, Alexander I (2005) The ectomycorrhizal symbiosis: life in the real world. Mycologist 19:102–112

Taylor AFS, Martin F, Read DJ (2000) Fungal diversity in ectomycorrhizal communities of Norway spruce (Picea abies [L.] Karst.) and beech (Fagus sylvatca L.) along North-South transects in Europe. In: Schulze ED (ed) Carbon and nitrogen cycling in European forest ecosystems, vol 142, Ecological studies series. Springer, Heidelberg, pp 344–365

Tetreault JP, Bernier B, Fortin JA (1978) Nitrogen fertilization and mycorrhizae of balsam fir seedlings in natural stands. Naturaliste Can 105:461–466

Tingey DT, Wilhour RG, Standley C (1976) The effect of chronic ozone exposure on the metabolite content of ponderosa pine seedling. For Sci 22:234–241

Topa MA, McDermitt DJ, Yun SC, King PS (2004) Do elevated ozone and variable light alter carbon transport to roots in sugar maple? New Phytol 162(1):173–186

Vanhatalo M, Bäck J, Huttunen S (2003) Differential impacts of long-term (CO2) and O3 exposure on growth of northern conifer and deciduous tree species. Oecologia 17:211–220

Vilhar U, Smolej I, Trošt T, Kutnar L, Kraigher H (2004) Pestrost tipov ektomikorize v smrekovem sestoju na Pokljuki. Zb GL 75:71–85

Vitousek P, Aber J, Howarth R, Likens G, Matson P, Schindler D, Schlesinger W, Tilman D (1997) Human alteration of the global nitrogen cycle: sources and consequences. Ecol Appl 7:737–750

Waller K, Agerer R, Brand F, Taylor AFS, Wanner G (1993) Piceirhiza oleiferans, eine neue Ektomykorrhizen-Art an Picea abies. Sendtnera 1:11–22

Winwood J, Pate AE, Price AJ, Hanke DE (2007) Effects of long-term, free-air ozone fumigation on the cytokinin content of mature Beech trees. Plant Biol 9:265–278

Worley JF, Hacskaylo E (1959) The effect of the available soil moisture on the mycorrhizal association of Virginia pine. For Sci 5:267–268

Železnik P, Hrenko M, Then Ch, Koch N, Grebenc T, Levanič T, Kraigher H (2007) CASIROZ: root parameters and types of ectomycorrhiza of young beech plants exposed to different ozone and light regimes. Plant Biol 9:298–308

Chapter 14
Effects of Pesticides on the Growth of Ectomycorrhizal Fungi and Ectomycorrhiza Formation

Miguel Marin

14.1 Introduction

In all reforestation programs, the quality of the plant to be used is of vital importance. This quality is determined by the origin of plant material used and the management during the production in nursery phase. Often, the criteria for determining the quality of a plant had been limited to assessing the status and size of the shoot. In the recent years, the nurseries were also given sufficient attention to the quality of the root system, in recognition of their tremendous importance in the uptake of water and nutrients. However, the quality of the seedlings is not always satisfactory and the mortality of pine seedlings after planting in the forest site can be very high (Heiskanen and Rikala 2003). A term "transplant shock" has been used to describe a suite of symptoms in seedlings, including yellowing, short needles, and reduced growth, arising upon relocation from nursery to forest soil (Reitveld 1989). The formation of ectomycorrhizae is the natural state of almost all of the trees in forests, and therefore part of their root systems and a factor to take into account when assessing their quality.

Seedlings are nowadays produced in large commercial tree nurseries mainly as container seedlings. The proportion of bare-rooted seedlings has decreased over the last 20 years. In Finland, container seedlings represent 98% of Scots pine, 86% of Norway spruce, and 89% of silver birch production (Peltola 2001). The production takes place in tree nurseries where large quantities of seedlings are produced in controlled conditions using artificial growth substrate, irrigation, fertilization, pest control, and sometimes day–night shift treatments to promote seedling growth (Landis 1989). Normally in the production of seedlings, the seedlings spend between 1 and 3 years of cultivation in the nursery until they are transplanted to the field. Plant diseases may be the biggest problem in nurseries (Landis 1989) and their prevention in the large monoculture seedling beds has traditionally been

M. Marin
Department of Physiology, Genetics and Microbiology, University of Alicante, Carretera San Vicente del Raspeig s/n, E-03690, San Vicente del Raspeig (Alicante), Spain
e-mail: mmarinz@ua.es

M. Rai and A. Varma (eds.), *Diversity and Biotechnology of Ectomycorrhizae*, Soil Biology 25, DOI 10.1007/978-3-642-15196-5_14,

achieved using fungicides, insecticides, and other chemical products. Maintenance of nursery hygiene is also of major importance. The opportunistic plant control is done by herbicides. Under nursery conditions, ectomycorrhizal fungi (EMF) live in an artificial environment and are affected by pest control, watering, and fertilization (Marin 2009). Generally, all these factors are unfavorable to most EMF, but certain types are commonly found in the nurseries, e.g., *Thelephora* spp. and ectendomycorrhizal species (Laiho 1965; Landis 1989). Unfortunately, most of the so-called nursery fungi are not represented as dominant members in forest mycorrhizal communities, which suggest a poor competitive capacity after transplanting.

In 1984, Trappe et al. wrote a paper which includes everything related to the interaction between the pesticide and the mycorrhizae. This article discusses the effect of pesticides by type of mycorrhiza (ectomycorrhiza, endomycorrhiza, and ectendomicorriza), with in vitro and in vivo conditions, as the major groups of pesticides (general pesticides, fungicides, herbicides and insecticides). In this paper, the authors reviewed over 150 articles on every possible combination, of which over 60 are related with EMF. The present work covers articles published from 1984 to present. About 30 articles that have relation only with EMF/ectomycorrhizae and pesticides have been reviewed. It shows that the questions are still unresolved, apart from studying the effects of new pesticides on ectomycorrhization. It must be said that in recent years the study on the effect of pesticides has focused on the endomycorrhizae for their application in agriculture.

14.2 Forest Tree Mycorrhization

The first experimental work in the field of physiology of ectomycorrhizae carried out in the first half of the twentieth century, have already shown that mycorrhizal plants grow better than non-mycorrhizal and also contained the highest amounts of major nutrients per unit mass (Hatch 1937). Since then, Hatch was devoted to research of mycorrhizae and many of his works haven tested in different scenarios to explain the role of ectomycorrhizal symbiosis in the absorption of water and soil major nutrient (phosphorus and nitrogen) (Tarkka et al. 2005), the role of these fungi in the vitamin uptake of substances or plant growth regulators (Barker and Tagu 2000), relationships with other microorganisms in the rhizosphere (Perry et al. 1987), their role in biological control of certain root pathogens (Whipps 2004), and its role in the whole forest ecosystem, where EMF are involved in the renovation of biomass and nutrient cycling and in competitive relations or cooperation established between the plants for water and nutrient uptake (Van der Heijden and Sanders 2002).

14.2.1 Controlled Mycorrhization

Controlled mycorrhization has been used in some countries with the aim of introducing growth-promoting EMF into nurseries, and modifying seedling production procedures respectively, in order to achieve improved growth responses and rapid post-transplantation adaptation in the reforestation site (Marin 2009). In North America, the research and practical applications have been established for decades (Castellano and Molina 1989), while in Europe the procedure has been in commercial use since the last decade of last century (Le Tacon et al. 1997).

The potential advantages of controlled mycorrhization in the nursery are not only the positive growth response of the seedlings, but also improvement of nutrient stress under deficiency conditions; EMF can increase nutrient uptake (Jones et al. 1991) and under toxic conditions as with heavy metals, mycorrhizal fungi pull metals out of the soil and sequester them (Jones and Hutchinson 1986). EMF stimulates root production and considerably increases the volume of soil the plant can explore, and enhancement of rooting, and reduction of transplant shock (Teste et al. 2004). The increased root volume is able to take up more water; the mycorrhizal fungi also enhance the host's osmotic adjusting capabilities, allowing some plants to continue to extract water from soils as they become drier (Garbaye 2000). Mycorrhizae reduce directly and/or indirectly, antagonizing disease organisms, increase the number of biocontrol agents around the roots, occupy potential infection sites on the root, and increase host plant vigor to the extent that it can survive disease (Datnoff et al. 1995). Also, mycorrhizal fungi increase beneficial interactions with other microbes (Garbaye 1994); it also increases phosphate uptake to the plant by phosphate solubilizing bacteria.

Seedling performance in the field, after ectomycorrhizal inoculation, has been variable in European field studies. In the best cases, 6 years after out planting, an increase in tree volume of 90% has been observed, while in other conditions inoculated seedlings performed worse in the field compared to control ones (Le Tacon et al. 1992). Variable environmental growth conditions and differences in local mycoflora obviously explain part of the variation.

It is important that the planting of seedlings that have been inoculated with specific mycorrhizal strains, typically of exotic origin, into nature should not cause any negative impacts on the local, indigenous microflora. Ecological knowledge on the interactions of existing community and introduced strains should therefore be improved to prevent the risk of adverse effects. Understanding the survival and competitive abilities of selected strains under field conditions is of high priority in the risk management of controlled mycorrhization. As already mentioned, and we shall see, the use of pesticides in the nursery may play an important role in the mycorrhization of seedlings with desired EMF, creating added value and better marketing.

14.3 Pesticides

Since the 1960s, industrial chemistry has increased number of sophisticated chemicals for agricultural pest control. At that time, the future of these chemicals among chemists, agronomists, and resource managers seem to have been overly optimistic about the promise of agricultural chemicals as simple solutions to complex pest problems during the early part of the new era in agriculture. It is now apparent that pesticides can have unexpected effects on nontarget organisms and can thereby influence crop productivity as deeply or even more so than do the pests they are intended to control (Trappe et al. 1984). Wilde (1958) was one of the first to emphasize the deleterious effects of pesticides on the productivity of nursery soils. Today, persistence of many pesticides in the soil, their interactions within the soil, and the resulting effects on the growth and development of seedlings is of great concern.

14.3.1 Classification of Pesticides

Pesticide, or biocide, is a common name for compounds, which are used for protecting cultivated plants or for elimination of parasitic plants, fungi, insects, or other pests. Plant growth regulators (i.e. phytohormones) are also classified as pesticides. Pesticides are usually grouped according to their mode of action or the purpose of their use (e.g., fungicides are used against fungal disease, herbicides against weeds, and insecticides against insect pests). For instance, protectant is a fungicide that will shield healthy tissue from fungal invasion, whereas an eradicant will kill fungi that have already invaded the plant. Systematic pesticide spreads within the plant, thus compound applied onto foliage (e.g., leaf-action herbicide) will be transported to stem and root, or vice versa (e.g., soil applied herbicide). Pesticide can be specific when it affects selectively, only one species or group, or wide-spectral when it can affect a large number of species or groups of organisms.

Formulated pesticide products contain one or more active ingredients (abbr. a.i.) and some other inactive ingredients like solvents, diluents, carrier substances, non-forming, or acidity control chemicals, binding agents or pigments, which increase the effectiveness of the active ingredient. Many of these inactive ingredients may also be used as pesticide (Environmental Protection Agency of USA). Active ingredient can be naturally occurring compounds, like pyrethrin and antibiotics, or synthetic inorganic or organic compounds, or mixture of compounds or ligands. Some pesticides will be converted to active form after application by soil microbes, e.g., sesone to 2,4-D herbicide (Metcalf 1971). Inorganic pesticides have typically simple chemical structure like fungicides, e.g., copper oxychloride (Cu(OH)Cl) or so-called Bordeaux mixture ($[Cu(OH)_2]\cdot CaSO_4$). Majority of modern pesticides are, or have been various kinds of organic compounds (MetCalf 1971). More than 1,100 common names for pesticides have been assigned by International Organization of Standardization (ISO).

14.3.2 Effects of Pesticides on Soil Microorganisms

The repeated and long-term use of pesticides has raised the concern of accumulation of pesticides in soils and the side effects of pesticides on soil microorganisms and other biotypes (Smith et al. 2000). Simultaneous use of several pesticides may lead to synergetic or potential, or antagonistic effects of pesticides on various organisms. Pesticides may also have influence on the competitive position of soil microbes in the community by destroying or preventing the growth of some microbe groups, and thus enhancing the growth of other microbes (Schüepp and Bodmer 1991).

Most of the information about side effects of the pesticide has been obtained from the studies in agricultural fields, and only few studies have been performed also in forest nurseries and sapling stands, some of which are presented below. In general, tests with soils have shown that fungicides can have deleterious effects not only on fungi, but also on bacteria, and also herbicides and insecticides can affect bacteria and fungi. Commonly, this depression of microbial growth and activity has been reversible after some duration (Tu 1978; Ingham and Coleman 1984; Colinas et al. 1994a).

Tu (1993) studied the effects of two fungicides, captanol and chlorothalonil, and eight herbicides, including linuron, on microbial and enzymatic activities in soils. The pesticide application rate was 10 ppm in soil. Fungicides initially diminished both bacterial and fungal populations, but the recovery was rapid. This caused increase in oxygen consumption. Both fungicides suppressed invertase and amylase activities for 1 day, and captanol dehydrogenase activity for 4 days, but all of these were recovered equal to that of control. Some herbicides had effect on activities of bacteria and fungi for a week, but populations returned to levels similar to controls. After several herbicide treatments there was a slight depression in nitrification. In experiments with 11 insecticides, some pesticides diminished the populations of bacteria and fungi, but they were recovered after couple of weeks (Tu 1991). Bacteria seemed to be more sensitive to various insecticides. On the contrary, cypermethrin, chlordane, and chlorpyrifos stimulated fungal growth. Chakravarty and Chatarpaul (1990a) found herbicides glyphosate and hexazinone to have a significant short-time (2 months) decrease in soil microbial activity, which was recovered in 6 months.

Chen and Edwards (2001) studied the effects of three broad-spectrum fungicides, benomyl, captan, and chlorothalonil, on soil microbial activities and biomass. Benomyl and chlorothalonil had relatively transient effects, while captan had a greater and long-lasting overall influence on soil microbial activities. In a test with forest soil (Colinas et al. 1994b), fungicide captan reduced 90% of active fungal length compared to control, sampled during 5 days of incubation, although total hyphal length was not affected. Besides that, captan reduced 50% of the number of active bacteria and 30% of bacteria-feeding nematodes.

14.3.3 Pesticide Persistence in the Soil

Many pesticides are quite persistent in the environment and this is one explains, why they can be detected in surface waters (Burgoa and Wauchope 1995), groundwater

(Dörfler et al. 1997), in soils and sediments (Rostad 1997), and in precipitation or air samples (Dubus et al. 2000) for months or even years, after the last treatment. Besides its functional toxicity, a pesticide can be toxic or has other impacts to humans (Betarbet et al. 2000), animals (Elbetieha et al. 2001), or other organisms (DeLorenzo et al. 2002). Agricultural and forest pesticides are applied in the fields where they can also affect organisms other than the protected plants (McLaughlin and Mineau 1995).

Persistence of a pesticide is derived from its bioaccumulation that is related to degrability (half life time) and physicochemical properties (Gramatica and Di Guarno 2002). Climate and weather are also important factors, which affect the persistence of the pesticide and its degradation rates (Russell 1995). Soil structure (density and porosity), composition (organic C content), chemistry (pH) and microbiology, as well as the uptake of the pesticide by plants and soil animals also have a remarkable influence on pesticide persistence in the environment (Bergström and Stenström 1998).

Some persistent pesticides may accumulate in the soil from repeated application. Volatilization, leaching from the soil, and degradation by soil microorganisms may account for loss of a major part of some pesticides that disappear rapidly. The longer a pesticide persists in the soil, the greater the probability that several processes will become involved in its inactivation and disappearance. Some organic pesticides that persist for several months create a long period for the recolonization of soil microorganisms. Pesticides, such methyl bromide, may disappear within 2–4 days. The herbicide Eptam (s-ethyl dipropylthiocarbamate) disappears in 3–10 weeks, while others such as the fungicide benzene hexachloride and the insecticide chlordane may remain for a year or more. Many herbicides that persist from one season to the next can injure sensitive plants. The triazine herbicides (atrazine and simazine) applied as pre-emergent herbicides for selective weed control in corn, sometimes persist and injure sensitive crops the next year. Additionally, they can be spread as sprays, or volatilised and transported in air, or run off, or leached out from the application area. Occasionally, amounts of pesticides in surface waters (Gerecke et al. 2002) or in ground water (Gaus 2000) have been close to or exceed the levels known to have toxic impact on the aquatic flora or fauna (Kreuger 1998), or has been set for limits for drinking water.

Modern pesticides are usually more economic to use than the old ones, because the dosage of specific pesticides is much smaller, even one-hundredth of that, which was in old compounds (Chrispeels and Sadava 1994). However, this does not necessarily reduce environmental and public hazards of pesticides. In general, most modern pesticides are more than ten times as toxic to organisms than those used in 1950 (Pimentel et al. 1998). Increasing concern of the harmful side effects to non-target organisms, caused by more and more complex and environmentally stable synthetic compounds used as pesticides, have put pressure on manufacturers to develop pesticides, which are more specific, less toxic, and biologically degradable in nature.

It has been estimated that the concentration of pesticides after one application would leach into a soil top-layer of 5 cm depth in mineral and in organic soil. In most cases, the pesticide concentrations exceed 1 ppm (Flykt et al. 2008). Pesticide concentration may be even higher on the surface of plants roots, because the pesticides are generally sprayed on seeds, or seedlings (fungicides and insecticides), or on weeds (herbicides); seedlings may even be dipped into pesticide solution (insecticide). Pesticide solution or residues can leach into the container or soil either during the application or afterwards due to irrigation or rainwater.

As discussed above, the depths to which a pesticide can leach into the soil is dependent on the organic matter content of soil and the physico-chemical properties of the pesticide. In peat-pots, pesticides can be bound to the peat medium and only a small fraction may be detectable in leaching water. For example, in a trial with peat-pots seedlings, less than 1% of applied chlorothalonil, but almost 30% of applied propiconazole leached through the peat medium (Juntunen and Kitunen 2003). Soil microbial activity can release soil-bound pesticides back to undergo environmental interactions (Levanon et al. 1994), and thus, the mycorrhiza and mycelium, which are mainly present in this organic top-layer of the soil (Smith and Read 1997) can be exposed to pesticides. On pines treated with triadimefon in foliar sprays, Marx observed that this chemical was translocated downward from the needles and also was leached by water into the root zone, so a considerable amount of triadimefon is sprayed on the soil surface during its application on small pine seedlings.

However, Sidhu and Chakravarty (1990) performed a comparison of the results of mycorrhization of *Pinus contorta* var. *latifolia* and *Picea glauca* with *Suillus tomentosus* in the laboratory, greenhouse, and field conditions with herbicides hexazinone, glyphosate, and tryclopyr. The results showed that seedling growth and ectomycorrhizal formation was affected significantly in the laboratory and greenhouse, but not in field except at higher rates. The lower levels of the herbicides available to the seedlings in field at the time of planting probably accounted for the lower sensitivity of pine and spruce seedlings in the field.

14.4 Pesticides and EMF

As discussed at the beginning of this work, Trappe et al. (1984) conducted a first review study of the effect of pesticides on mycorrhizal fungi. This actual work covers articles published from 1984 to present, the last 25 years, and reviews about 30 articles that have a relation only between EMF and pesticides. Most of these works were performed between 1980s and 1990s, but even today new studies are conducted with the emergence of new pesticides (Laatikainen and Heinonen-Tanski 2002) and the high diversity of EMF and pesticides (Hutchison 1990). Tables 14.1–14.6 are a summary of articles concerning studies on the effect of pesticides on the ectomycorrhizal mycelium and the formation of ectomycorrhizae in forest plants. Research has been conducted both in vitro and in vivo concerning fungicides, herbicides, and insecticides (and other pesticides).

Table 14.1 Fungicide effects on ectomycorrhizal fungi in vitro culture

Fungicide	Concentrations (μg/ml)	Number of species	Effects on growth[a]	References
Benomyl	10–10,000	1	0	Chakravarty et al. (1990)
	10	96	−, 0	Hutchison (1990)
	2–3	7	−, 0	Niini and Raudaskoski (1993)
	10	105	−, 0	Dickinson and Hutchison (1997)
	1–10	15	−, 0, +	Laatikainen and Heinonen-Tanski (2002)
	1–1,000	2	0	Díaz et al. (2003)
Captan	1–1,000	2	0	Díaz et al. (2003)
Carbendazim	10–100	2	−	Zambonelli and Iotti (2001)
Chlorothalonil[b]	1–10	15	−	Laatikainen and Heinonen-Tanski (2002)
Copper oxychloride	1–10	15	−, 0	Laatikainen and Heinonen-Tanski (2002)
Cyproconazole[b]	0.1–10	7	−	Desprez-Loustau et al. (1992)
Flusilazole	0.1–10	7	−, 0	Desprez-Loustau et al. (1992)
Flutriafol	0.1–10	7	−, 0	Desprez-Loustau et al. (1992)
Hymexazol	1–1,000	2	−	Díaz et al. (2003)
Iprodione	1–1,000	2	−	Díaz et al. (2003)
Mancozeb	5–15	2	−	Reddy and Natarajan (1995)
	10–100	2	−, 0	Zambonelli and Iotti (2001)
Maneb	1–10	15	−, 0	Laatikainen and Heinonen-Tanski (2002)
Oxine benzoate[b]	10–10,000	1	0	Chakravarty et al. (1990)
Oxycarboxin	10–100	2	−, 0	Zambonelli and Iotti (2001)
Propamocarb hydroc.	1–1,000	2	0	Díaz et al. (2003)
Propiconazole	10–100	2	−	Zambonelli and Iotti (2001)
	0.1–10	15	−, +	Laatikainen and Heinonen-Tanski (2002)
	0.0015	1	−	Lekounougou et al. (2008)
Tebuconazole	0.1–10	7	−	Desprez-Loustau et al. (1992)
Tolclofos-methyl	10–100	2	−, 0	Zambonelli and Iotti (2001)
Thiram	1–1,000	2	0	Díaz et al. (2003)
Triadimenol	0.1–10	2	−	Marx et al. (1986)
	0.1–10	7	−	Desprez-Loustau et al. (1992)
Triadimefon[b]	0.1–10	2	−	Marx et al. (1986)
	0.1–10	7	−, 0	Desprez-Loustau et al. (1992)

[a]Fungicide effects to low concentrations (<10 μg/ml). 0: no effect; +: growth increased; −: growth decreased

[b]Chlorothalonil: not listed; Cyproconazole: expires 30/06/11; Oxine benzoate: not defended their inclusion; Triadimefon: not defended their inclusion (Directive 91/414/CEE, October 2009). The rest of pesticides are accepted

There have been many studies on the in vitro effects of pesticides on the EMF; in vitro tests for mycelial growth can be used when estimating pesticide effects on mycorrhiza, but for example, Unestam et al. (1989) did not recommend using them when deciding pesticide usefulness in forest nurseries or plantations. In some

Table 14.2 Herbicide effects on ectomycorrhizal fungi in vitro culture

Herbicide	Concentrations (μg/ml)	Number of species	Effects on growth[a]	References
2,4-D	1–10,000	3	–, 0	Estok et al. (1989)
	200–800	2	–, 0	Donnelly et al. (1993)
Atrazine[b]	200–800	2	0	Donnelly et al. (1993)
Chlorthiamid[b]	1	15	0	Laatikainen and Heinonen-Tanski (2002)
Glyphosate	0.1–1,000	5	–, 0	Chakravarty and Sidhu (1987)
	0.1–1,000	5	–, 0	Chakravarty and Chatarpaul (1990a)
	1–10,000	3	–, 0	Estok et al. (1989)
	1	15	–, 0, +	Laatikainen and Heinonen-Tanski (2002)
	1–1,000	2	0	Díaz et al. (2003)
Hexazinone	0.1–1,000	5	–, 0	Chakravarty and Sidhu (1987)
	0.1–1,000	5	–, 0	Chakravarty and Chatarpaul (1989)
	1–10,000	3	–, 0, +	Estok et al. (1989)
	1	15	0, +	Laatikainen and Heinonen-Tanski (2002)
	1–1,000	2	0	Díaz et al. (2003)
Linuron	1	15	–, 0	Laatikainen and Heinonen-Tanski (2002)
Simazine[b]	1–1,000	2	0	Díaz et al. (2003)
Terbuthylazine[b]	1	15	–, 0, +	Laatikainen and Heinonen-Tanski (2002)
Triclopyr	0.1–1,000	5	–	Chakravarty and Sidhu (1987)
	1–10,000	3	–, +	Estok et al. (1989)

[a]Herbicide effects to low concentrations (<10 μg/ml). 0: no effect; +: growth increased; –: growth decreased

[b]Chlorthiamid: not listed; Atrazine: not defended their inclusion; Simazine: excluded; Terbuthylazine: expires 30/06/11 (Directive 91/414/CEE, October 2009). The rest of pesticides are accepted

Table 14.3 Insecticides effects on ectomycorrhizal fungi in vitro culture

Insecticide	Concentrations (μg/ml)	Number of species	Effects on growth[a]	References
Cypermethrin	1–10	15	–, 0	Laatikainen and Heinonen-Tanski (2002)
Fenvalerate	0.25–1.5	2	–	Reddy and Natarajan (1994)
Nocodazole[b]	2	7	–	Niini and Raudaskoski (1993)

[a]Insecticide effects to low concentrations (<10 μg/ml). 0: no effect; +: growth increased; –: growth decreased

[b]Nocodazole: not listed (Directive 91/414/CEE, October 2009). The rest of pesticides are accepted

studies, pesticide toxicity to EMF in vitro has been confirmed with experiment in vivo (Marx et al. 1986; Chakravarty and Sidhu 1987; Chakravarty and Chatarpaul 1988). However, the influence of pesticide on the growth of EMF can be different when pure culture mycelium in vitro is tested compared to tests by using inoculated

Table 14.4 Fungicide effects on ectomycorrhizal formation

Fungicide	Application	Host species[a]	Effects on mycorrhizae[b]	References
Benomyl	Commercial labels	Pire	0, +	Chakravarty et al. (1990)
	1–100 ppm	Pist	+	De la Bastide and Kendrick (1990)
	500 ppm/pot	Thpl	–	Cade-Menun and Berch (1997)
	20–150 kg/ha	Piel	0	Pedersen and Sylvia (1997)
	Not known	Pisi	–	O'Neill and Mitchell (2000)
	Commercial labels	Piab	–	Flykt et al. (2008)
Captan	25 ppm dry soil	Psme	0, +	Colinas et al. (1994b)
	Not known	Pisi	+	O'Neill and Mitchell (2000)
Chlorothalonil	2,000 g ai/ha	Pisy	0	Laatikainen et al. (2000)
	Not known	Pisy	0	Aleksandrowicz-Trzcinska (2007)
	Commercial labels	Piab	–	Flykt et al. (2008)
Copper oxychloride	588 ppm	Pisy	0	Manninen et al. (1998)
Ferbam	Unspecific	Piel, Pita	0	Marx et al. (1986)
Fenhexamid	Commercial labels	Piab	–	Flykt et al. (2008)
Iprodione	1 g/Kg	Quro	–	Garbaye et al. (1992)
	Commercial labels	Piab	–	Flykt et al. (2008)
Mancozeb	800–2,400 ppm	Pipa	–	Reddy and Natarajan (1995)
Oxine benzoate	Commercial labels	Pire	–	Chakravarty et al. (1990)
Oxycarboxin[c]	100 mg ai/l sprayed	Qupu	–, 0	Zambonelli and Iotti (2001)
Propiconazole	250 ppm	Pisy	–	Manninen et al. (1998)
	125 g ai/ha	Pisy	0	Laatikainen et al. (2000)
	125–375 ppm	Psme	–	Teste et al. (2006)
	Commercial labels	Piab	–	Flykt et al. (2008)
	25 mg ai/m^2	Bepe		Nerg et al. (2008)
Pyrimethanil	Commercial labels	Piab	–	Flykt et al. (2008)
spi + teb + tri[d]	Not known	Pisy	0	Aleksandrowicz-Trzcinska (2007)
Tolylfluanid	Commercial labels	Piab	–	Flykt et al. (2008)
Thiophanate-methyl	350–1,050 ppm	Psme	–	Teste et al. (2006)
	Commercial labels	Piab	–	Flykt et al. (2008)
Triadimefon	0.56 Kg ai/ha	Piel, Pita	–	Marx et al. (1986)
	6, 12 and 24 oz/ac	Pita	–, 0	Kelley (1987)
	Not known	Pipo	–	Page-Dumroese et al. (1996)
	Commercial labels	Piab	–	Flykt et al. (2008)

[a]*Betua pendula: Bepe*; *Picea abies*: Piab; *Picea sitchensis*: Pisi; *Pinus elliotti*: Piel; *Pinus patula*: Pipa; *Pinus ponderosa*: Pipo; *Pinus resinosa*: Pire; *Pinus strobus*: Pist; *Pinus sylvestris*: Pisy; *Pinus tadea*: Pita; *Pseudotsuga menziesii*: Psme; *Quercus pubescens*: Qucu; *Quercus robur*: Quro; *Thuja plicata*: Thpl
[b]0: no effect; +: mycorrhiza formation increased; –: mycorrhiza formation decreased
[c]Oxycarboxin: not defended their inclusion (Directive 91/414/CEE, October 2009). The rest of pesticides are accepted
[d]Spiroxamine + tebuconazole + triadimenol

seedlings. Marx and Rowan (1981) studied the influence of four fungicides on two EMF, *Pisolithus tinctorius* and *Thelephora terrestris*, by infesting the nursery soil with these fungi and growing *Pinus tadea* seedlings in this soil. Ectomycorrhizal development of *P. tinctorius* was depressed by benomyl and captan, but enhanced by benodanil, whereas ectomycorrhizal development of *T. terrestris* was the greatest in plots with benomyl and captan applications. In the laboratory test, benomyl had slight inhibitory and captan no effect on the mycelial growth of both *P. tinctorius* and *T. terrestris*.

Table 14.5 Herbicides effects on ectomycorrhizal formation

Herbicide	Application	Host species[a]	Effects on mycorrhizae[b]	References
Clopyralid	Commercial labels	Piab	–	Flykt et al. (2008)
Glyphosate	0.1–100 ppm	Pico, Pigl	–, 0	Sidhu and Chakravarty (1990)
Hexazinone	1.0–4.0 kg ai/ha drench	Pico, Pigl	–, 0	Sidhu and Chakravarty (1990)
Imazapyr	1.1–2.2 kg ai/ha	Abco, Pipo, Psme	0	Busse et al. (2004)
Phosphinothricin[c]	Not known	Pira	0	Gonzalez-Moro et al. (2000)
Sulfometuron methyl[c]	0.14–0.28 kg ai/ha	Abco, Pipo, Psme	0	Busse et al. (2004)
Triazine[c]	Commercial labels	Piab	–	Flykt et al. (2008)
Triclopyr	0.1–100 ppm	Pico, Pigl	–, 0	Sidhu and Chakravarty (1990)
	4.5–9.0 kg ai/ha	Abco, Pipo, Psme	0	Busse et al. (2004)

[a]*Abies concolor*: Abco; *Picea abies*: Piab; *Picea glauca*: Pigl; *Pinus contorta*: Pico; *Pinus ponderosa*: Pipo; *Pinus radiata*: Pira; *Pseudotsuga menziesii*: Psme
[b]0: no effect; +: mycorrhiza formation increased; –: mycorrhiza formation decreased
[c]Phosphinothricin: not listed; Sulfometuron methyl: not listed; Triazine: not listed (Directive 91/414/CEE, October 2009). The rest of pesticides are accepted

Table 14.6 Insecticide and others chemicals (O) effects on ectomycorrhizal formation

Chemical	Application	Host species[a]	Effects on mycorrhizae[b]	Reference
Chloropicrin (O)[c]	0.66–3.3 ml/kg stump	Psme	0	Massicote et al. (1998)
Deltamethrin	Commercial labels	Piab	0	Flykt et al. (2008)
Dimethoate	200 ppm dry soil	Psme	–, 0	Colinas et al. (1994b)
	Commercial labels	Piab	0	Flykt et al. (2008)
Fenvalerate	Commercial labels	Pipa	–	Reddy and Natarajan (1994)
Fumagillin (O)[c]	10 ppm dry soil	Psme	0, +	Colinas et al. (1994b)
Oxydemeton-methyl[c]	Commercial labels	Piab	0	Flykt et al. (2008)
Quinoclamine (O)	Commercial labels	Piab	0	Flykt et al. (2008)
Permethrin	Commercial labels	Piab	0	Flykt et al. (2008)
Vapam (O)	Not known	Pipo	–	Page-Dumroese et al. (1996)

[a]*Picea abies*: Piab; *Pinus patula*: Pipa; *Pinus ponderosa*: Pipo; *Pseudotsuga menziesii*: Psme
[b]0: no effect; +: mycorrhiza formation increased; –: mycorrhiza formation decreased
[c]Chloropicrin: expires 30/06/11; Fumagillin: not listed; Oxydemeton-methyl: excluded (Directive 91/414/CEE, October 2009). The rest of pesticides are accepted

Pesticides may influence the growth of mycorrhizal plants by affecting the nutrient uptake and allocation by suppressing the mycorrhizal symbiosis (Schweiger and Jakobsen 1998). The detrimental effect on ectomycorrhiza can be seen as reduced growth in seedlings (Reddy and Natarajan 1995).

14.4.1 Fungicides

Generally, the fungicides proved to be toxic to EMF, presumably due to their general mode of actions, and such fungicides can inhibit fungal cell division (benomyl), impair ergosterol biosynthesis (propiconazole), inactivate fungal cell thiols (chlorothalonil), cause protein damage (copper oxychloride), or bind to cell copper compounds (maneb). Thus, chlorothalonil is viewed as an effective fungicide against a broad spectrum of plant pathogens. Propiconazole, which is a systemic fungicide, belonging to the group called sterol synthesis inhibitors also has a broad range of activity.

Fungicides can be useful in studies of controlled mycorrhization with different inoculated EMF because not all fungi are equally sensitive to fungicides: zygomycetes are particularly susceptible to most of benzimidazole fungicides (Trappe et al. 1984). Benomyl is generally known to inhibit the growth of ascomycetes fungi like *Cenoccocum geophilum*, *Tuber* spp., and ectendomycorrhizae, and ascomycetes tend to be more sensitive in low concentrations than basidiomycetes (Niini and Raudaskoski 1993). Similarly, in axenic culture, some EMF respond differently to fungicides than others (Hutchison 1990); the differential sensitivity of EMF to fungicides can be a valuable tool in controlling mycorrhizal formation in the nursery and in the field (Marx and Rowan 1981; Trappe et al. 1984). This genetical distance from the other EMF could account for its different response from all other identified fungi when tested with the pesticides. It has been observed that in field conditions, benomyl increases the development of ectomycorrhizae that are nursery-grown (Trappe et al. 1984; Chakravarty et al. 1990; De la Bastide and Kendrick 1990) and associated to decrease the rhizosphere competition. However, O'Neill and Mitchell (2000) observed that benomyl has deleterious effect on root length and mycorrhizal colonization of *Picea sitchensis*.

Edible EMF is cultivated by establishing plantations with inoculated plants (Hall and Wang 1998). A major problem during the cultivation process is the replacement of the inoculated fungus by more aggressive competing fungi. Selective fungicides might therefore be used to control competitive EMF during the production of edible EMF, *Tuber* spp. infected plants, and later in plantations of mycorrhizated trees. For example, *Tuber melanosporum* can be replaced by *Scleroderma* spp. (Hall and Wang 1998), *T. brumale* (Chevalier et al. 1982; Etayo and De Miguel 1998), or *Hebeloma* spp. (Granetti and Angelini 1992). Similar problems have been detected during the cultivation of *T. magnatum* and *T. uncinatum* (Capecchi et al. 1999).

There is every reason to suspect that the same problems will occur during the cultivation of other EMF such as *Tuber borchii*, which was cultivated by Zambonelli et al. (2000). In greenhouse experiments, *Hebeloma sinapizans* actively competed with *T. borchii* not only during the initial phases of ectomycorrhizal formation, but also when *H. sinapizans* was introduced after infection had been established (Zambonelli and Govi 1991). Zambonelli and Iotti (2001) demonstrated that fungicides such as oxycarboxin, which have relatively little effect on *T. borchii*, can be used to selectively suppress ectomycorrhizal development of contaminant basidiomycetes such as *H. sinapizans*. This fungicide is commonly used to control plant pathogenic basidiomycetes. These results support and extend Unestam's findings that oxycarboxin also suppresses the development of ectomycorrhizal basidiomycetes (Unestam et al. 1989). This opens up the possibility of using selective fungicides, such as oxycarboxin, during the commercial production of plants infected with *T. borchii* and other species of *Tuber*. Although it is difficult to predict the field competition of greenhouse contaminant fungi, many authors (see Govi et al. 1997) agree that their presence should be avoided in *Tuber* infected plants.

In a field experiment, propiconazole and copper oxychloride reduced ectomycorrhizal growth of Scots pine seedlings growing in sand-filled-pots (Manninen et al. 1998). In the field experiment with Loblolly and Slash pine seedlings, the ectomycorrhizal development with *P. tinctorius* and naturally occurring fungi were both suppressed by triadimefon treatments, which confirmed the results from in vitro tests. Reddy and Natarajan (1995) examined the effect of fungicide mancozeb on *Laccaria laccata* and *Thelephora terrestris* both in vitro and inoculated with *Pinus patula* seedlings. Fungicide inhibited the growth of mycelia of both fungi, as well as, mycorrhization of seedlings roots was clearly reduced. Colonization with *L. laccata* was totally inhibited and colonization with *T. terrestris* reduced more than 50% at the recommended fungicide dose. In a greenhouse experiment with container-grown *Pinus palustris* seedlings, benomyl stimulated the ectomycorrhizal development with tested fungi, *P. tinctorius* and *T. terrestris* (Pawuk et al. 1980). Theodorou and Skinner (1976) found seed dressing with fungicides captan, zineb, and thiram to inhibit mycorrhizal development on *Pinus radiata* seedlings with various inoculated mycorrhizal fungi, while natural mycorrhization in soil were not affected. However, studies in Spain with *Pinus halepensis* (Aleppo pine) and the use of fungicides (benomyl, captan, and thiram) in the nursery, necessary for its cultivation , given its sensitivity to fungal pathogens at this stage of development does not adversely affect the process of controlled mycorrhization (Carrillo 2000).

14.4.2 Herbicides

Herbicides such as glyphosate and hexazinone have been tested with various EMF in pure culture tests, e.g., *Hebeloma crustuliforme*, *L. laccata*, and *S. tomentosus*, and they inhibited all of the tested fungi at concentrations above 10 ppm (Chakravarty and Sidhu 1987), and in a second experiment they had inhibitory effects on *C. geophilum*, *Hebeloma longicaudum* and *P. tintorius* at concentrations below 100 ppm (Estok et al. 1989), though such high concentrations may not be particularly relevant to the situation in forest nurseries.

The growth stimulation of some EMF strains was caused mainly by herbicides glyphosate, terbuthylazine, and hexazinone (*Suillus* species), but in a few rare cases also by the insecticides, cypermethrin, and all of the fungicides, except chlorothalonil. The growth stimulation of EMF might indicate that these fungi are able to degrade pesticides, but this stimulation noted in the laboratory may be less likely to occur in field conditions if there is a lack of nutrients such as potassium or phosphorus in the soil (DaSilva et al. 1977). In some cases pesticide molecule is not mineralized, but can become incorporated into the fungal tissue (Donnelly et al. 1993). Ectomycorrhizae having the potential to degrade and mineralize pesticides and other persistent organic pollutants could also be used in bioremediation (Meharg et al. 1997a). For example, *Amanita species*, *P. involutus*, and *Suillus* species are known to degrade several organic pollutants (Meharg and Cairney 2000), and *C. geophilim* can immobilize, for example, hexazinone (Donnelly and Fletcher 1994). The growth of EMF in symbiosis with *Pinus sylvestris* has stimulated even greater pollutant mineralization than in pure cultures (Meharg et al. 1997b). The ability to degrade some aromatic herbicides appears to be dependent on the specific EMF and the herbicide. When the EMF is growing with the host plant, it has an ample supply of carbohydrates provided by the host plant for its growth. It is know that extracellular enzymatic activity of the fungus dramatically increases when the EMF is growing with the host plant vs. growing in pure culture (Donnelly et al. 1993).

Estok et al. (1989) have studied the effects of herbicides 2,4-D, glyphosate, hexazinone, and triclopyr on the growth of *C. geophilum*, *Pisolithus tinctorius* and *Hebeloma longicaudum*. *C. geophilum* was the least sensitive to tested herbicides but the growth of other two tested fungi was slightly reduced already at the concentration of 1 ppm. Generally, the low concentrations of herbicides could promote the growth, whereas concentrations of 10–1,000 ppm strongly inhibited the growth of all five EMF. The order of decreasing sensitivity of fungi to tested herbicides was *Suillus tomentosus*, *Thelephora americana*, *Hebeloma crustuliniforme*, *Laccaria laccata*, and finally, *Thelephora terrestris*. Later, hexazinone at the recommended field rates of applications (4–18 ppm) was found to reduce the mycorrhizal development of inoculated *L. laccata* with *Pinus resinosa*, as well as inhibit naturally occurring mycorrhizal fungi in the field (Chakravarty and Chatarpaul 1988).

During the other experiment, Chakravarty and Chatarpaul (1990a) examined in vitro the effects of glyphosate and hexazinone on the growth of five EMF, *Cenococcum graniforme*, *Hebeloma crustuliniforme*, *Laccaria laccata*, *Suillus tormentosus*, and *Paxillus involutus*, and all were significantly reduced at the concentrations of 50 ppm. Glyphosate was also tested in the field experiment with *Pinus resinosa* seedlings inoculated with *Paxillus involutus* (Chakravarty and Chatarpaul 1990b). Neither seedling growth nor ectomycorrhizal development was affected by glyphosate treatment. In the field experiment, twofold higher recommended dose of copper oxychloride and propiconazole reduced ectomycorrhizal development and growth of Scot pine (*Pinus sylvestris* L.) seedlings (Manninen et al. 1998), and propiconazole, as in the case of some fungicides, selectively killed ascomycete symbionts, while basidiomycete symbionts were less affected.

14.4.3 Insecticides and Other Pesticides

There were not very many studies made with insecticides and EMF. The cases studied show the inhibitory effect of insecticides on mycelial growth at low concentrations, contradicting what is observed by Trappe et al (1984). This fact is consistent with the observation mentioned before, in that new pesticides can affect EMF despite the use of small doses of the product.

As in the case of fungicides, insecticides and other pesticides listed in Table 14.6, when applied in forest plant nursery on mycorrhized plants, their effect on mycorrhizal colonization is not significant or very low.

14.5 Pesticides and *Lactarius* spp.

Here, a special case of in vitro effect of some fungicides and herbicides on various species of *Lactarius* and strains of *Lactarius deliciosus* is shown. The basidiomycete *L. deliciosus* is an ectomycorrhizal fungus principally on *Pinus* spp. roots in the Mediterranean forests (Sánchez et al. 1994). This species is socioeconomically important in Spain and others countries of the world, because it is a popular edible wild mushroom that gives a new value to our forest ecosystems (Singer 1986). *L. deliciosus* has been used in mycorrhization of pine seedlings in nursery (Guerin-Laguette et al. 2000).

Tables 14.7 and 14.8 shows the effect of some pesticides on mycelial growth on species of the genus *Lactarius* at different concentrations. At the same time, the fungistatic effect of each of the pesticide on the fungus was observed by transferring the piece of agar with mycelium (not grown on culture medium with pesticide) to a new culture medium without pesticide, and noting the presence or absence of mycelial growth, thus showing that the fungus was not dead. The effects of each pesticide on the growth of different *Lactarius* spp. in axenic culture were similar, but differences between the species studied and even among isolates of *L. deliciosus* were also observed.

14.5.1 Fungicides

Benomyl, a common systemic fungicide in agricultural practices (Torstensson and Wessén 1984) has been extensively studied in axenic culture on several EMF (see Trappe et al. 1984, and Table 14.1). *Lactarius* isolates had a strong inhibition of growth to low concentrations of fungicide (Table 14.7). Hutchison (1990) found between the Lactarii, a similar sensitive response to benomyl. In addition, this author found that species of *Lactarius* of section Dapetes (Singer 1986) where carpophore exudes bright-colored latex when cut, exhibited tolerance to benomyl. Certainly, more resistance is found in some isolates of *L. deliciosus* and the isolate

Table 14.7 Fungicide effects on mycelial diameter growth (*d*) of *Lactarius* isolates on fungicide-amended BAF for 6 weeks

	Ld CBS 334.65	Ld DAOM 197163	Ld UAMH 5547	Ls CBS 409.75	Lh Baar	Lc CECT 20239
BEN (μg/ml)	*d* (mm)					
0	28.5	22.0	19.8	15.3	21.7	17.3
1	38.2	18.0	23.0**	15.8	23.8	10.3
10	(33.3)[a]	18.3	ng	6.2*	(100)	(100)
50	ng	(100)	ng	(100)	(66.7)	(100)
100	ng	(100)	ng	(100)	(33.3)	(100)
500	ng	(100)	ng	(100)	ng	ng
1,000	ng	ng	ng	ng	ng	ng
FAL (μg/ml)	*d* (mm)					
0	28.5	22.0	19.8	15.3	21.7	17.3
1	38.7	22.7	15.0	25.3**	15.3	11.0
10	41.3**	18.2	24.7	17.0	10.0*	1.5*
50	21.8	18.2	13.5	17.5	13.0*	2.5*
100	30.8	17.8	(66.7)	15.7	8.8*	(100)
500	(33.3)	6.7*	ng	7.7*	(100)	(100)
1,000	ng	(66.7)	ng	(100)	ng	ng
PRO (μg/ml)	*d* (mm)					
0	28.5	22.0	19.8	15.3	21.7	17.3
1	35.0	23.5	20.0	17.8	17.3	11.3
10	30.5	23.8	24.2	17.0	33.2**	15.2
50	37.8	27.3	26.5	19.2	33.5**	11.5
100	33.0	32.3**	29.0	19.0	20.8	10.0
500	21.2*	28.8	23.3	15.2	25.5	6.8*
1,000	ng	27.8	23.5	13.5	16.0	3.2*
THI (μg/ml)	*d* (mm)					
0	28.5	22.0	19.8	15.3	21.7	17.3
1	40.0**	23.3	18.8	18.3**	23.8	10.5
10	ng	10.3*	ng	7.2*	ng	ng
50	ng	(100)	ng	(66.7)	ng	ng
100	ng	(33.3)	ng	(33.3)	ng	ng
500	ng	ng	ng	ng	ng	ng
1,000	ng	ng	ng	ng	ng	ng

Significant differences compared to control (0 μg/ml fungicide-medium) are indicated with "*" for low values and "**" for high values ($P < 0.05$ Tukey or KW)

CBS Centraalbureau voor Schimmelcultures, Baarn, The Netherlands; *CECT* Spanish Collection of Type Cultures, València, Spain; *DAOM* Canadian Collection of Fungus Cultures, Ottawa, Canada; *UAMH* University of Alberta Microfungus Collection, Alberta, Canada; *Baar* provide by J. Baar, originating from a nitrogen-enriched Scots pine

ng not growing

BEN benomyl; *FAL* fosetyl-aluminum; *PRO* procymidone; *THI* thiram

[a]Percentage of samples that recovered growth after reharvesting in new media BAF without biocides

of *L. sanguifluus*, but this characteristic is not general in the section Dapetes of Lactarii (Hutchison 1990).

Fosetyl-Al is a systemic fungicide active against oomycetes. Jalai-Hare and Kendrick (1987) found that fosetyl-Al stimulated the growth of *Glomus intraradices* in allium roots. This fungicide had only fungitoxic effect on *Lactarius* spp. at

Table 14.8 Herbicide effects on mycelial diameter growth (*d*) of *Lactarius* isolates on herbicide-amended BAF for 6 weeks

	Ld CBS 334.65	Ld DAOM 197163	Ld UAMH 5547	Ls CBS 409.75	Lh Baar	Lc CECT 20239
GLU (μg/ml)	*d* (mm)					
0	28.5	22.0	19.8	15.3	21.7	17.3
1	35.8**	16.0*	12.8*	18.0**	25.8	5.8*
10	ng	ng	7.8*	(100)	6.8*	3.0*
50	ng	ng	(100)[a]	(100)	(66.7)	(100)
100	ng	ng	ng	ng	ng	(100)
500	ng	ng	ng	ng	ng	ng
1,000	ng	ng	ng	ng	ng	ng
GLY (μg/ml)	*d* (mm)					
0	28.5	22.0	19.8	15.3	21.7	17.3
1	39.3	18.5	16.3*	18.2	24.3	16.7
10	38.2	25.5	17.0*	5.8*	8.8*	9.2
50	40.2	20.5	13.0*	4.5*	9.3*	10.5
100	41.0	20.8	10.5*	5.3*	7.3*	9.0
500	15.0	4.5*	(100)	ng	(100)	5.5*
1,000	ng	ng	(100)	ng	(100)	2.0*
MCPA (μg/ml)	*d* (mm)					
0	28.5	22.0	19.8	15.3	21.7	17.3
1	37.7	16.8*	19.3	16.2	17.0	11.2
10	36.3	17.3	18.7	17.3	11.3*	10.3
50	35.0	20.5	10.5*	16.5	4.3*	8.3*
100	27.2	13.5*	10.5*	10.5	ng	8.0*
500	ng	ng	ng	ng	ng	ng
1,000	ng	ng	ng	ng	ng	ng
OXY (μg/ml)	*d* (mm)					
0	28.5	22.0	19.8	15.3	21.7	17.3
1	28.2	12.8*	19.3	20.7**	23.3	9.0
10	24.8	16.2*	(100)	8.2*	(100)	5.0*
50	4.5*	ng	(100)	(66.7)	(100)	ng
100	(100)	ng	ng	ng	ng	ng
500	ng	ng	ng	ng	ng	ng
1,000	ng	ng	ng	ng	ng	ng

Significant differences compared to control (0 μg/ml herbicide-medium) are indicated with "*" for low values and "**" for high values ($P < 0.05$ Tukey or KW)

CBS Centraalbureau voor Schimmelcultures, Baarn, The Netherlands; *CECT* Spanish Collection of Type Cultures, València, Spain; *DAOM* Canadian Collection of Fungus Cultures, Ottawa, Canada; *UAMH* University of Alberta Microfungus Collection, Alberta, Canada; *Baar* provide by J. Baar, originating from a nitrogen-enriched Scots pine

GLU glufosinate; *GLY* glyfosate; *MCPA* MCPA; *OXY* oxyfluorfen

ng not growing

[a]Percentage of samples that recovered growth after reharvested in new media BAF without biocides

high concentrations of the product. The fungicide effects of fosetyl-Al have not been shown previously in EMF.

Procymidone is an example of pesticide that can be utilized as energy sources for some fungi (Altman 1969). Four of the six strains tested had not accepted at concentrations of 1,000 ppm. Procymidone is a systemic fungicide used on lupins, grapes, stone fruit, strawberries, and some vegetables. It is widely used in horticulture, either as a seed dressing, pre-harvest spray, or post-harvest dip. As well, procymidone has not been shown previously in direct fungitoxicity on EMF.

Thiram is a preventive fungicide of large spectrum, and similarly in the studied strains of *Lactarius*. A high fungitoxicity is observed to low concentrations in EMF in axenic culture (Trappe et al. 1984; Díaz et al. 2003). In addition, negative effects have been found in mycorrhiza formation (Trappe et al. 1984).

14.5.2 Herbicides

Glufosinate (Table 14.8) is a non-selective herbicide that inhibits the glutamine synthetase activity, produce accumulation of ammonia, and inhibit formation of glutamine amino acid (Montanini et al. 2003). *Lactarius* isolates are affected strongly by this herbicide. However, glufosinate has low residual effect after reharvesting in control media, except over some *L. deliciosus*. Direct fungitoxicity on EMF has not been shown previously.

Glyphosate is an herbicide that is very commonly used in forest and agricultural activity (Sidhu and Chakravarty 1990). Table 14.2 shows that the effect in axenic culture at low concentrations of this herbicide on mycelial growth of EMF was different according to several studies. Díaz et al (2003) observed in *L. deliciosus* not effect, perhaps, stimulating mycelial growth. In this experience, inhibition or reduction of mycelial growth in medium or high concentrations was found.

MCPA is a phenoxyacetic herbicide, as 2,4-D, aromatic herbicides. MCPA is similar to 2,4-D, except in-group chloride on position 2. Donnelly et al. (1993) found that 2,4-D was degraded by incorporation of herbicide carbon into tissue. Donnelly and Fletcher (1995) found that the degradation capacity of these herbicides depends of the number and the position of chloride groups in the aromatic molecule. Dasilva et al. (1977) found strong stimulation of growth to low concentrations of MCPA on two *Boletus* spp., while with 2,4-D, the stimulation was lower. Similarly, the growth of *Lactarius* isolates was also stimulated. In principle, it would be prudent to replace 2,4-D by MCPA in agriculture and forestry practices when the plant is to be mycorrhized with EMF.

Oxyfluorfen produced the rupture of cellular membrane in presence of light as all the diphenyl ethers (Choi et al. 1999). This group of herbicides produced in EMF a strong inhibition of growth in axenic conditions (Kelley and South 1980) like we have observed in *Lactarius* spp.

14.6 Conclusions

In recent years, some interest among scientists have been lost in studying the effects of pesticides on the EMF, and consequently on the formation of ectomycorrhizae in forest plants. Recent studies have developed in the production of mycorrhized forest plant, especially with container seedlings. However, most of the forest productions still need the use of pesticide. However, at present, the use of pesticides is being reduced in general agricultural practices and particularly in forest productions, but this will be still difficult to occur in all situations and nurseries. The study of Trappe et al. (1984) showed the irrationality of the research on the effect of pesticides on mycorrhiza, and that these would increase as new substances are used as pesticides. This situation continued to occur over recent years, and the variety and diversity of studies makes it difficult to rationalize the observations found. Based on the variety of forest plants, soils, and climates under pesticides are used, it is difficult to propose one type of pesticide for each of the multiple combinations EMF/plant.

There are a number of parameters that are observed among all studies, and more important as expected is that fungicides have a greater effect on EMF than other pesticides. Knowledge on the capabilities of EMF to tolerance of pesticides might be useful in deciding which pesticides would be less harmful to mycorrhizal fungi when used in forest nurseries and in afforested fields. Pesticides may affect directly the seedling roots, and thus pure culture tests alone are not recommended when estimating the effects of pesticides on mycorrhizae and deciding the usefulness of different pesticides in the nursery or new plantations.

It should be noted that production of container seedlings is more similar to horticultural production than to agricultural production. Seedlings are started in greenhouses, and depending on seedling type, they are raised in greenhouses from 2 to 6 months with peat as the growing media. When container seedlings are sprayed with pesticides, most of the pesticides are sprayed on very densely growing seedlings. In production of container seedlings the pesticides leaching to the ground include both pesticides leaching from the growth medium and pesticides applied directly to aisles and other empty space around the container blocks. For production of bareroot seedlings, the situation is similar to that on agricultural fields. On bareroot fields, seedlings grow at much lower density than in containers, and there are empty areas between seedling rows and beds. Much of the pesticide suspension is applied directly on the ground than in container production.

A better integration of systems of mycorrhized plant production and use of pesticides can be carried out by a better development of Integrated Pest Management (IPM), a system that combines cultural, biological, and chemical technologies to reduce insect, fungal, and weed populations. Finally, pesticides cause many complex reactions of all organisms in the forest plant production, and generalization should be made with caution. General biocides, fungicides, herbicides, and insecticides cause complex reactions to be studied in each case on a small scale.

References

Aleksandrowicz-Trzcinska M (2007) The effect of fungicides used in the protection of natural regenerations of pine against Lophodermium needle cast on mycorrhizae and seedling growth. Sylwan 151:27–34

Altman J (1969) Predisposition of sugarbeets to *Rhizoctonia* damping-off with herbicides. Phytophathology 59:1015

Barker SJ, Tagu D (2000) The role of auxins and cytokinines in mycorrhizal symbioses. J Plant Growth Regul 19:144–154

Bergström L, Stenström J (1998) Environmental fate of chemicals in soil. Ambio 27:16–23

Betarbet R, Sherer TB, MacKenzie G, Garcia-Osuna M, Panov AV, Greenamyre JT (2000) Chronic systematic pesticide exposure reproduces features of Parkinson's disease. Nat Neurosci 3:1301–1306

Burgoa B, Wauchope RD (1995) Pesticides in run-off and surface waters. In: Roberts TR, Kearney PC (eds) Environmental behaviour of agrochemicals. Wiley, Chichester, pp 131–184

Busse MD, Fiddler GO, Ratcliff AW (2004) Ectomycorrhizal formation in herbicide-treated soils of differing clay and organic matter content. Water Air Soil Pollut 152:23–34

Cade-Menun BJ, Berch SM (1997) Response of mycorrhizal western red cedar to organic phosphorus sources and benomyl. Can J Bot 75:1226–1235

Capecchi M, Vecchiati P, Zambonelli A (1999) Dinamica della micorrizazione in due tartufaie sperimentali di *Tuber magnatum* Pico in provincia di Modena. Micol Ital 17:9–18

Carrillo C (2000) Producción de inóculo de hongos ectomicorrícicos y micorrización controlada de *Pinus halepensis* Miller en vivero. (Production of ectomycorrhizal fungal inoculum and controlled mycorrhization of *Pinus halepensis* Miller in nursery). Ph.D. thesis, University of Murcia, Spain, 241 pp

Castellano MA, Molina R (1989) Mycorrhizae. In: Landis TD, Tinus RW, McDonald SE, Barnett JP (eds) The container tree nursery manual, vol 5, Agricultural handbook 674. Department of Agriculture, Forest Service, Washington, DC, pp 101–167

Chakravarty P, Chatarpaul L (1988) The effects of Velpar L. (hexazinone) on seedling growth and ectomycorrhizal symbiosis of *Pinus resinosa*. Can J For Res 18:917–921

Chakravarty P, Chatarpaul L (1990a) Non-target effect of herbicides: I. Effect of glyphosate and hexazinone on soil microbial activity. Microbial population, and *in vitro* growth of ectomycorrhizal fungi. Pestic Sci 28:233–241

Chakravarty P, Chatarpaul L (1990b) Non-target effect of herbicides: II. The influence of glyphosate on ectomycorrhizal symbiosis of red pine (*Pinus resinosa*) under greenhouse and field conditions. Pestic Sci 28:243–247

Chakravarty P, Sidhu SS (1987) Effect of glyphosate, hexazinone and triclopyr on *in vitro* growth of five species of ectomycorrhizal fungi. Eur J For Pathol 17:204–210

Chakravarty P, Peterson RL, Ellis BE (1990) Integrated control of *Fusarium* damping-off in red pine seedlings with the ectomycorrhizal fungus *Paxillus involutus* and fungicides. Can J For Res 20:1283–1288

Chen SK, Edwards CA (2001) A microcosm approach to assess the effects of fungicides on soil ecological processed and plant growth: comparisons of two soil types. Soil Biol Biochem 33:1981–1991

Chevalier G, Giraud M, Bardet MC (1982) Interactions entre les mycorhizes de Tuber melanosporum et celles d'autres champignons ectomycorhiziens en sols favorables à la truffe. In: INRA (ed) Les Mycorhizes: biologie et utilisation, vol 13. Les Colloques de l'INRA, Paris, pp 313–321

Choi JS, Hee HJ, Hwang IT, Pyon JY, Cho KY (1999) Differential susceptibilities of wheat and barley to diphenyl ether herbicide oxyfluorfen. Pestic Biochem Phys 65:62–67

Chrispeels MJ, Sadava DE (1994) Plants, genes and agriculture. Jones and Bartlett, Boston, p 478

Colinas C, Ingham E, Molina R (1994a) Population responses of target and non-target forest soil organisms to selective biocides. Soil Biol Biochem 26:41–47

Colinas C, Molina R, Trappe J, Perry D (1994b) Ectomycorrhizas and rhizosphere microorganisms of seedlings of *Pseudotsuga menziesii* (Mirb.) Franco planted on a degraded site and inoculated with forest soils pretreated with selective biocides. New Phytol 127:529–537

DaSilva EJ, Henriksson LE, Urdis M (1977) Growth responses of mycorrhizal *Boletus* and *Rhizopogon* species to pesticides. Trans Br Mycol Soc 68:434–437

Datnoff LE, Nemec S, Pernezny K (1995) Biological control of fusarium crown and root rot of tomato in Florida using *Trichoderma harzianum* and *Glomus intraradices*. Biol Control 5:427–431

De la Bastide PY, Kendrick B (1990) The *in vitro* effects of benomyl on disease tolerance, ectomycorrhiza formation, and growth of white pine (*Pinus strobus*) seedlings. Can J Bot 68:444–448

DeLorenzo ME, Taylor LA, Lund SA, Pennington PL, Strozier ED, Fulton MH (2002) Toxicity and bioconcentration potential of the agricultural pesticide endosulfan in phytoplankton zooplakton. Arch Environ Contam Toxicol 42:173–181

Desprez-Loustau M-L, Dupuis F, Viguié A (1992) Evaluation of single annual applications of sterol-inhibiting fungicides for control of pine twisting rust. Plant Dis 76:376–382

Díaz G, Carrillo C, Honrubia M (2003) Differential responses of ectomycorrhizal fungi to pesticides in pure culture. Cryptogam Mycol 24:199–211

Dickinson TA, Hutchison LJ (1997) Numerical taxonomic methods cultural characters, and the systematics of ectomycorrhizal agarics, boletes and gastermycetes. Mycol Res 101:477–492

Donnelly PK, Fletcher JS (1995) PCB metabolism by ectomycorrhizal fungi. Bull Environ Contam Toxicol 54:507–513

Donnelly PK, Entry JA, Crawford DL (1993) Degradation of atrazine and 2, 4-dochlorophenoxyacetic acid by mycorrhizal fungi at three nitrogen concentrations in vitro. Appl Environ Microbiol 59:2642–2647

Donnelly PK, Fletcher JS (1994) Potential use of mycorrhizal fungi as bioremediation agents. In: Anderson TA, Coats JR (eds), Bioremediation Through Rhizosphere Technology, American Chemical Society, Washington, pp 93–99

Dörfler U, Feicht EA, Scheunert I (1997) S-triazine residues in groundwater. Chemosphere 35:99–106

Dubus IG, Hollis JM, Brown CD (2000) Pesticides in rainfall in Europe. Environ Pollut 110:331–344

Elbetieha A, Da'as SI, Khamas W, Darmani H (2001) Evaluation of the toxic potencials of cypermethrin pesticide on some reproductive and fertility parameters in the male rats. Arch Environ Contam Toxicol 41:522–528

Estok D, Freedman B, Boyle D (1989) Effects of the herbicides 2, 4-D, glyphosate, hexazinone, and triclopyr on the growth of three species of ectomycorrhizal fungi. Bull Environ Contam Toxicol 42:835–839

Etayo ML, De Miguel AM (1998) Estudio de las ectomycorrizas en una trufera cultivada situada en Olóriz (Navarra). Publicaciones de Biologia de la Universidad de Navarra. Serie Botanica 11:55–114

Flykt E, Timonen S, Pennanen T (2008) Variation of ectomycorrhizal colonisation in Norway spruce seedlings in Finnish forest nurseries. Silva Fenn 42:571–585

Garbaye J (1994) Helper bacteria – a new dimension to the mycorrhizal symbiosis. New Phytol 128:197–210

Garbaye J (2000) The role of ectomycorrhizal symbiosis in the resistance of forests to water stress. Outlook Agric 29:63–69

Garbaye J, Churin J-L, Duponnois R (1992) Effects of substrate sterilization, fungicide treatment and mycorrhization helper bacteria on ectomycorrhizal formation of pedunculate oak (*Quercus robur*) inoculated wtih *Laccaria laccata* in two peat bare-root nurseries. Biol Fertil Soils 13:55–57

Gaus I (2000) Effects of water extraction in a vulnerable phreatic aquifer: consequences for groundwater contamination by pesticides, Sint-Jansteen area, The Netherlands. Hydrogeol J 8:218–229

Gerecke AC, Schärer M, Singer HP, Müller SR, Schwarzenbach RP, Sägesser M, Ochsenbein U, Popow G (2002) Sources of pesticides in surface waters in Switzerland: pesticide load through waste water treatment plants – current situation and reduction potential. Chemosphere 48:307–315

Gonzalez-Moro MB, Iribeei N, Duñabeitia MK, Loureiro-Beldarrain I, González-Murua C (2000) Effect of phosphinothricin herbicide on nitrogen metabolism in *Pinus radiata* and *Laccaria bicolor*. Phyton (Special Issue Root-soil interactions) 40:71–77

Govi G, Bencivenga M, Granetti B, Pacioni G, Palenzona M, Tocci A, Zambonelli A (1997) Metodo basato sulla caratterizzazione morfologica delle micorrize. In: Toscana R (ed) Il Tartufo. Florence, Centro Stampa Giunta Regionale, pp 49–62

Gramatica P, Di Guarno A (2002) Screening of pesticides for environmental partitioning tendency. Chemosphere 47:947–956

Granetti B, Angelini P (1992) Competizione tra alcuni funghi ectomicorrizici e *T. melanosporum* in una tartufaia coltivata. Micol Vegetazione Mediterr 7:173–188

Guerin-Laguette A, Plassard C, Mousain D (2000) Effects of experimental conditions on mycorrhizal relationships between *Pinus sylvestris* and *Lactarius deliciosus* and unprecedented fruit-body formation of the Saffron milk cap under controlled soilless conditions. Can J Microbiol 46:790–799

Hall IR, Wang Y (1998) Methods for cultivating edible ectomycorrhizal mushrooms. In: Varma A (ed) Mycorrhiza manual. Springer, Berlin, pp 99–114

Hatch AB (1937) The physiological basis of mycotrophy in the genus *Pinus*. Black Rock For Bull 6:168

Heiskanen J, Rikala R (2003) Effect of peat-based container media on establishment of Scots pine, Norway spruce and silver birch seedlings. Tree Planter's Notes 50:28–33

Hutchison LJ (1990) Studies on the systematics of ectomycorrhizal fungi in axenic culture. IV. The effect of some selected fungitoxic compounds upon linear growth. Can J Bot 68:2172–2178

Ingham ER, Coleman DC (1984) Effects of streptomycin, cycloheximide, fungizone, captan, carbofuran, cygon, and PCNB on soil microorganisms. Microb Ecol 10:345–358

Jalai-Hare SH, Kendrick WB (1987) Response of an endomycorrhizal fungus in *Allium porrum* L. to different concentrations of the systemic fungicides, metalaxyl (Ridomil®) and fosetyl-Al (Aliette®). Soil Biol Biochem 19:95–99

Jones MD, Hutchinson TC (1986) The effect of mycorrhizal infection on the response of *Betula papyerifera* to nickel and copper. New Phytol 102:429–442

Jones MD, Durall DM, Tinker PB (1991) Fluxes of carbon and phosphorus between symbionts in willow ectomycorrhizas and their changes with time. New Phytol 118:99–106

Juntunen ML, Kitunen V (2003) Leaching of propiconazole and chlorothalonil during protection of *Pinus sylvestris* seedlings in containers. Scand J For Res 18:45–53

Kelley WD, South DB (1980) Effects of herbicides on *in vitro* growth of mycorrhizae of Pine (*Pinus* spp.). Weed Sci 28:599–602

Kelley WD (1987) Effect of triadimefon on development of mycorrhizae from natural inoculum in loblolly pine nursery beds. Southern J Appl Forestry 11:49–52

Kreuger J (1998) Pesticides in stream water within an agricultural catchment in southern Sweden, 1990–1996. Sci Total Environ 216:227–251

Laatikainen T, Juntunen M-L, Heinonen-Tanski H (2000) The effect of nursery applied fungicides on the growth and mycorrhiza abundance of Scots pine seedlings. In: Lilja A, Sutherland JR (eds) Proceedings of the 4th Meeting of IUFRO Working Party 7.03.04 – Diseases and Insects in Forest Nurseries. The Finnish Forest Research Institute, Research Papers 781:177–187

Laatikainen T, Heinonen-Tanski H (2002) Mycorrhizal growth in pure cultures in the presence of pesticides. Microbiol Res 157:127–137

Laiho O (1965) Further studies on the ectendotrophic mycorrhizae. Acta For Fenn 79:1–34

Landis TD (1989) Mineral nutrients and fertilization. In: Landis TD, Tinus RW, McDonald SE, Barnett JP (eds) The container tree nursery manual, vol 4. Department of Agriculture, Forest Service, Washington DC, pp 1–67

Le Tacon F, Alvarez IF, Bouchard D, Henrion B, Jackson RM, Luff S, Parlade JI, Pera J, Stenström E, Villeneuve N, Walker C (1992) Variations in field response of forest trees to nursery ectomycorrhizal inoculation in Europe. In: Read DJ, Lewis DH, Fitter AH, Alexander IJ (eds) Mycorrhizas in ecosystems. CAB International, Wallingford, pp 119–132

Le Tacon F, Mousan D, Garbaye J, Bouchard D, Churin J-L, Argillier C, Amirault J-M, Généré R (1997) Mycorrhizes, pépinières et plantations forestières en France (Summary in English). Revue Forestière Française 49:131–154

Lekounougou S, Jacquot JP, Gérardin P, Gelhaye E (2008) Effects of propiconazole on extracellular enzymes involved in nutrient mobilization during *Trametes versicolor* wood colonization. Wood Sci Technol 42:169–177

Levanon D, Meisinger JJ, Codling EE, Starr JL (1994) Impact of tillage activity and the fate of pesticides in the upper soil. Water Air Soil Pollut 72:179–189

Manninen AM, Laatikainen T, Holopainen T (1998) Condition of Scots pine fine roots and mycorrhiza after fungicide application and low-level ozone exposure in a 2-year field experiment. Trees 12:347–355

Marin M (2009) Ectomycorrhizal fungi and its application in forest nursery. In: Rai M (ed) Advances in fungal biotechnology. IK International, New Delhi, pp 379–408

Marx DH, Rowan SJ (1981) Fungicides influence growth and development of specific ectomycorrhizae on Loblolly pine seedlings. Forest Sci 27:167–176

Marx DH, Cordell CH, France RC (1986) Effects of triadmefon on grown and ectomycorrhizal development on loblolly and slash pines in nurseries. Phytopathology 76:824–831

Massicote HB, Takaberry LE, Ingham ER, Thies WG (1998) Ectomycorrhizae establishment on Douglas-fir seedlings following chloropicrin treatment to control laminated-root rot disease: assessment of 4 and 5 years after outplanting. Appl Soil Ecol 10:117–126

McLaughlin A, Mineau P (1995) The impact of agricultural practices on biodiversity. Agric Ecosyst Environ 55:201–212

Meharg AA, Cairney JWG (2000) Ectomycorrhizas - extending the capabilities of rhizosphere remediation? Rev Soil Biol Biochem 32:1475–1484

Meharg AA, Cairney JWG, Maguire N (1997a) Mineralization of 2, 4-dichlorophenol by ectomycorrhizal fungi in axenic culture and in symbiosis with pine. Chemosphere 34:2495–2504

Meharg AA, Denis GR, Cairney JWG (1997b) Biotransformation of 2, 4, 6-trinitrotoluene (TNT) by ectomycorrhizal basidiomycetes. Chemosphere 35:513–521

Metcalf RL (1971) The chemistry and biology of pesticides. In: White-Stevens R (ed) Pesticides in the environment, Part 1, vol 1. Marcel Dekker, New York, pp 1–144

Montanini B, Marco Betti M, Márquez AJ, Balestrini R, Bonfante P, Ottonello S (2003) Distinctive properties and expression profiles of glutamine synthetase from a plant symbiotic fungus. Biochem J 373:357–368

Nerg A-M, Kasurinen A, Holopainen T, Julkunen-Tiitto R, Neuvonen S, Holopainen JK (2008) The significance of ectomycorrhizas in chemical quality of silver birch foliage and aboveground insect herbivore performance. J Chem Ecol 34:1322–1330

Niini SS, Raudaskoski M (1993) Response of ectomycorrhizal fungi to benomyl and nocodazole: growth inhibition and microtubule depolymeration. Mycorrhiza 3:83–91

O'Neill JJM, Mitchell DT (2000) Effects of benomyl and captan on growth and mycorrhizal colonization of Sitka-spruce (*Picea sitchensis*) and ash (*Fraxinus excelsior*) in Irish nursery soil. For Pathol 30:165–174

Page-Dumroese DS, Harvey AE, Jurgensen MF, Larsen MJ (1996) Ponderosa pine seedling response to planting-site soil fumigation and fungicide application. Northwest Sci 70:139–146

Pawuk WH, Ruehle IE, Marx DH (1980) Fungicide drenches affect ectomycorrhizal development of container-grown *Pinus palustris* seedlings. Can J For Res 10:61–64

Pedersen CT, Sylvia DM (1997) Limitations to use benomyl in evaluating mycorrhizal functioning. Biol Fertil Soils 25:163–168

Peltola A (2001) Finnish statistical yearbook of forestry 2000. Jyväskylä, Agriculture, Forestry and Fishery, vol 52. p 374 (in Finnish with English summary)

Perry DA, Molina R, Amaranthus MP (1987) Mycorrhizae, mycorrhizospheres, and reforestation: current knowledge and research needs. Can J For Res 17:929–940

Pimentel D, Tort M, D'Anna L, Krawic A, Berger J, Rossman J, Mugo F, Doon N, Shriberg M, Howard E, Lee S, Talbot J (1998) Ecology of increasing disease. Bioscience 48:817–826

Reddy MS, Natarajan K (1994) Effect of a synthetic pyrethroid on the growth of ectomycorrhizal fungi and mycorrhiza formation in *Pinus patula*. Mycorrhiza 5:115–117

Reddy MS, Natarajan K (1995) Effects of the fungicide dithane M-45 on the growth and mycorrhizal formation of *Pinus patula* seedlings. Soil Biol Biochem 27:1503–1504

Reitveld WJ (1989) Transplanting stress in bareroot conifer seedlings: its development and progression to establishment. North J Appl For 6:99–107

Rostad CE (1997) Concentration and transport of chlordaner and nonachlor associated with suspended sediments in the Missisippi river, May 1988 to June 1990. Arch Environ Contam Toxicol 33:369–377

Russell MH (1995) Recommend approaches to assess pesticide mobility in soil. In: Roberts TR, Kearney PC (eds) Environmental behaviour of agrochemicals. Wiley, Chichester, pp 57–129

Sánchez F, Honrubia M, Torres P, Díaz G, García G, Pérez P (1994) Biodiversity and ecological distribution of ectomycorrhizal fungi in mediterranean forests of the Sistema Ibérico Mountains. In: Azcon-Aguilar C, Barea JM (eds) Mycorrhizas in integrated systems from genes to plant development. Proceedings of the fourth European symposium on mycorrhizas, Granada, Spain

Schüepp H, Bodmer M (1991) Complex response of VA-mycorrhizae to xenobiotic substances. Toxicol Environ Chem 30:193–199

Schweiger PF, Jakobsen I (1998) Dose–response relationships between four pesticides and phosphorus uptake by hyphae of arbuscular mycorrhizas. Soil Biol Biochem 30:1425–1432

Sidhu SS, Chakravarty P (1990) Effect of select forestry herbicides on ectomycorrhizal development and seedling growth of lodgepole pine and white spruce under controlled and field enviroment. Eur J For Pathol 20:77–94

Singer R (1986) The Agaricales in modern taxonomy, 4th edn. Koeltz Scientific Books, Koenigstein, Germany, pp 829–839

Smith SE, Read DJ (1997) Mycorrhizal symbiosis, 2nd edn. Academic, London

Smith MD, Hartnett DC, Rice CW (2000) Effects of long-term fungicide application on microbial properties in tallgrass prairie soil. Soil Biol Biochem 32:935–946

Tarkka M, Nehls V, Hampp R (2005) Physiology of ectomycorrhiza (ECM). Prog Bot 66:247–276

Teste FP, Schmidt MG, Berch SM, Bulmer C, Egger KN (2004) Effects of ectomycorrhizal inoculants on survival and growth of interior Douglas-fir seedlings on reforestation sites and partially rehabilitated landings. Can J For Res 34:2074–2088

Teste FP, Karst J, Jones MD, Simard SW, Durall DM (2006) Methods to control ectomycorrhizal colonization: effectiveness of chemical and physical barriers. Mycorrhiza 17:51–65

Theodorou C, Skinner MF (1976) Effects of fungicides on seed inocula of basidiospores of mycorrhizal fungi. Aust For Res 7:53–58

Torstensson L, Wessén B (1984) Interactions between the fungicide benomyl and soil microorganisms. Soil Biol Biochem 16:445–452

Trappe JM, Molina R, Castellano M (1984) Reactions of mycorrhizal fungi and mycorrhiza formation to pesticides. Ann Rev Phytopathol 22:331–359

Tu CM (1978) Effect of some pesticides on acetylene reduction and microorganisms in sandy loam. Soil Biol Biochem 10:451–456

Tu CM (1991) Effect of some technical and formulated insecticides on microbial activities in soil. J Environ Sci Health B 26:557–573

Tu CM (1993) Effect of fungicides, captafol and chlorothalonil, on microbial and enzymatic activities in mineral soil. J Environ Sci Health B 28:67–80

Unestam T, Chakravarty P, Damm E (1989) Fungicides: *in vitro* tests not useful for evaluating effects on ectomycorrhizae. Agric Ecosyst Environ 28:535–538

Van der Heijden MGA, Sanders IR (2002) Mycorrhizal ecology. Ecological studies 157. Springer, Berlin, p 469

Whipps JM (2004) Prospects and limitations for mycorrhizas in biocontrol of root pathogens. Can J Bot 82:1198–1227

Wilde SA (1958) Forest soils-their properties and relation to silviculture. Ronald, New York, p 537

Zambonelli A, Govi G (1991) Competizione fra *Tuber albidum* ed altri funghi. 3° contributo. Micol Ital 10:5–12

Zambonelli A, Iotti M (2001) Effects on fungicide on *Tuber borchii* and *Hebeloma sinapizans* ectomycorrhizas. Mycol Res 105:611–614

Zambonelli A, Iotti M, Rossi I, Hall I (2000) Interaction between *Tuber borchii* and other ectomycorrhizal fungi in a field plantation. Mycol Res 104:698–702

Chapter 15
Metal-Chelating Agents from Ectomycorrhizal Fungi and Their Biotechnological Potential

Ángela Machuca

15.1 Introduction

Ectomycorrhizal (ECM) fungi have been investigated for many years concerning the inoculation of plants for the recovery of degraded terrestrial ecosystems. Inocula production and mycorrhization techniques have been enhanced over time due to the proven importance of symbiotic associations on the growth and development of plants. Furthermore, ECM fungi that exhibit an unusual capacity to regulate the bioavailability of metal ions in their natural environments by either increasing or decreasing them are of particular interest in the revegetation of disturbed ecosystems, poor in mineral nutrients or contaminated with metal ions (O'Dell et al. 1993; Haselwandter and Bowen 1996; Leyval et al. 1997; Prasad and Freitas 1999; Brunner 2001).

Both the ECM fungi and their host plants can interact with metal ions, essential (e.g., Ca, Fe, Cu, Zn) or nonessential (e.g., Cd, Pb, Hg) for growth, through physical or chemical mechanisms and/or transport systems of various specificities, either to acquire ions for their nutrition or to avoid the toxic effects that high concentrations of certain ions can cause, depending on the environmental conditions (Jentschke and Godbold 2000; Hall 2002; Meharg 2003). ECM fungi can benefit the growth of the host plant, promoting the uptake of mineral nutrients or increasing the plant's tolerance to certain metal ions in soils. The beneficial effects, however, depend on the fungal species, the plant species, the metal involved, and the soil conditions that surround the mycorrhizal association and that affect the bioavailability of the metal ions (Galli et al. 1994; Prasad and Freitas 1999, Meharg 2003). Despite the numerous investigations in the area of the amelioration of metal toxicity by ECM associations, the results are contradictory regarding the protective role of ECM

Á. Machuca
Department of Plant Science and Technology, Universidad de Concepción, Campus Los Ángeles, J.A.Coloma 0201 Los Ángeles, Chile
e-mail: angmachu@udec.cl

M. Rai and A. Varma (eds.), *Diversity and Biotechnology of Ectomycorrhizae*, Soil Biology 25, DOI 10.1007/978-3-642-15196-5_15,

fungi on their host plants. Some studies clearly describe a protective effect of ECM fungi in contaminated environments, while others describe an increase in metal ion uptake in the mycorrhizal roots, thereby increasing the toxic effect on the plants (Godbold et al. 1998; Jentschke and Godbold 2000). It is likely that the limited amount of knowledge that exists regarding the mechanisms involved in the modification of the bioavailability of metal ions by ECM fungi in association with their host plants is responsible for the contradictory results.

The mechanisms used by living organisms to mobilize essential metal ions from cultures or natural environments to the interior of the cells have been very well described for most bacterial, plant, and some fungal species, but the studies are limited to only a few species when it comes to ECM fungi. Something similar occurs when trying to determine the mechanisms that allow ECM fungi to deal with elevated, potentially toxic concentrations of certain essential or nonessential metal ions, metalloids, or radionuclides. In this way, a fungus is able to use a combination of various strategies to take up or detoxify metal ions, through mobilization or immobilization, extracellularly or intracellularly, depending on culture conditions or environment. Independently of the fungus using these mechanisms to take up or avoid the potentially toxic effect of metal ions, most of these mechanisms are based on chelation capacity. Chelation occurs when a bi- or multidentate ligand molecule, organic in nature, binds or sequesters a metal ion by forming multiple coordinated bonds. The ligands are also known as chelators or chelating agents and the complex formed as a chelate; the ions of most transition metals form stable chelates. A very important group of chelating agents with different affinities for metal ions includes the siderophores and low molecular weight organic acids (citric, oxalic, malic, etc.), widely distributed in a variety of organisms (Gadd 1999; Renshaw et al. 2002; Bellion et al. 2006). Another important group is formed by the peptides metallothioneins (MTs), phytochelatins (PCs), and reduced glutathione (GSH), some of which are restricted to certain organisms (Gadd 1993; Mejáre and Bülow 2001; Bellion et al. 2006). Although metal binding to the fungal cell wall (adsorption) is among the mechanisms that many ECM fungi use for metal detoxification (Gadd 1993; Bellion et al. 2006), this will not be treated here since this chapter focuses on those chelation mechanisms that depend on the metabolic activity of the fungi.

15.2 Specific Metal-Chelating Agents: Siderophores

Despite the abundance of iron in the Earth's crust (5% by weight), in aerobic environments it exists mainly as iron oxyhydroxides (e.g., goethite) of very low solubility at neutral pH (~10^{-38} M), which causes a bioavailability problem for most living organisms whose metabolism depends on this element. Therefore, most plants, fungi, and bacteria produce an efficient and highly specific system for iron acquisition, comprised of siderophores (Greek "iron carriers") that act as iron scavengers or chelators of high affinity (Hider 1984; Neilands 1995; Renshaw et al. 2002; Boukhalfa and Crumbliss 2002; Kraemer 2004). However, other mechanisms such as reductive assimilation and protonation are also used by some organisms with the aim of solubilizing iron from their environments (Guerinot 1994; Kosman 2003). Although plants (grasses) produce these compounds (phytosiderophores), the chemical structures and chelation mechanisms differ from those of fungal and bacterial origin (Sugiura and Nomoto 1984; Kraemer et al. 2006; Crowley 2006).

Siderophores are iron-chelating compounds of low molecular weight (500–1,500 Da), with the ability to form stable complexes of high affinity constants ($K_f > 10^{30}$) with the ferric form of iron (Fe^{3+}), but not with the ferrous form (Fe^{2+}). These compounds are excreted into the environment by microorganisms and after chelating the Fe^{3+} they return to the cell (Hider 1984; Guerinot 1994; Renshaw et al. 2002). The synthesis of siderophores and the proteins involved in their recognition and, in some cases, in their transport to the interior of the cells, is strictly regulated at the molecular level by the concentration of iron perceived by the cell, causing an induction of these systems only under iron stress. Depending on the organism, concentrations higher than 1 mM of Fe^{3+} can dramatically reduce or completely repress siderophore biosynthesis (van der Helm and Winkelmann 1994).

Siderophores are synthesized by cells as metal-free ligands (desferrisiderophores), which are excreted into the extracellular environment where they solubilize the iron by chelation and return to the cell as ferrisiderophores, releasing the iron through different mechanisms (Hider 1984; Guerinot 1994; Neilands 1995; Renshaw et al. 2002; Kraemer 2004). At least four different mechanisms have been proposed for the transport of ferrisiderophores to the cell, mostly based on the recognition of ferric complexes by specific transport systems present in the cell membrane. Once inside the cell, the Fe^{3+} is reduced to Fe^{2+}, which is released from the siderophore. A different mechanism performs the reduction of siderophore-transported Fe^{3+} by a reductase enzyme present in the membrane. The Fe^{2+} and not the siderophore is subsequently transported to the interior of the cell (van der Helm and Winkelmann 1994; Renshaw et al. 2002). Despite the existing knowledge on the ferrisiderophore transport, this has been much more studied in bacteria than in fungi. Although siderophores are produced by the majority of fungi and bacteria, there are some exceptions where the presence of these compounds has not been demonstrated. However, some microorganisms that do not produce siderophores are capable of using exogenous siderophores (xenosiderophores), produced by other species of bacteria or fungi (Guerinot 1994; Kosman 2003).

The chemical nature of siderophores varies a great deal, but it is possible to distinguish three large groups according to the functional groups involved in chelation: hydroxamates, catecholates (phenolates), and hydroxycarboxylates. In addition to these groups, characterized most frequently in microorganisms, siderophores with other structures or a combination of the structures previously mentioned have also been described. In bacteria, the production of siderophores mainly of the catecholate and hydroxamate types has been reported. In most fungi, by contrast, it has only been possible to detect and isolate hydroxamate-type siderophores of the ferrichrome, fusarinine, and coprogen families and, in a few cases, hydroxycarboxylates (Hider 1984; Renshaw et al. 2002; Haselwandter and Winkelmann 2007). Independent of the chemical nature, the kinetic and thermodynamically more stable complexes are obtained by the formation of hexadentate or six-coordinate siderophores with Fe^{3+}, which allows them to act as efficient iron scavengers from the insoluble oxyhydroxides (Hider 1984; Boukhalfa and Crumbliss 2002; Kraemer 2004). Although catechol-type siderophores can form complexes with a higher Fe^{3+} affinity than hydroxamates, they are very susceptible to oxidation, depending on the pH. By contrast, hydroxamates form complexes of lower affinity, but very stable over a broad pH range (Hider 1984; Boukhalfa and Crumbliss 2002).

The siderophore production by mycorrhizal fungi has been demonstrated in a limited number of investigations and for very few fungal species; however, less research has been done to illustrate the chemical nature of siderophores isolated from ECM fungi. Nevertheless, those studies demonstrate that siderophores produced by ectomycorrhizal, ericoid, orchidaceous, and ectendomycorrhizal fungi are hydroxamate ligands: mainly of the ferrichrome structural family (Haselwandter 1995; Haselwandter and Winkelmann 2007). The ferrichromes are cyclic peptides containing a tripeptide of *N*-acyl-*N*-hydroxy-ornithine and combinations of the amino acids glycine, serine, or alanine (Renshaw et al. 2002). The few studies of some species of ECM fungi (Table 15.1) have detected the presence of siderophores in axenic cultures using (a) the reagent chrome azurol S (CAS) in liquid or solid medium (Schwyn and Neilands 1987; Milagres et al. 1999), (b) the chemical assays to detect hydroxamate structures or the Csáky test (Csáky 1948), and catecholate structures or the Arnow test (Arnow 1937), or (c) through bioassays using the *Aureobacterium* (*Arthrobacter*) *flavescens* JG-9 strain, a hydroxamate siderophore auxotrophic soil organism (Neilands 1984). *Cenococcum geophilum* (only one ascomycete in Table 15.1) was the first ECM fungus described, from which siderophores were isolated and characterized through HPLC, mass spectrometry, and NMR spectra (Haselwandter and Winkelmann 2002). This fungus produced mainly ferricrocin-type hydroxamates; other lower concentration compounds were found and although their structures were not completely identified, seemed to correspond to ferrichrome, fusarinine, fusigen, and coprogen.

On the other hand, if the studies showing the production of siderophores by ECM fungi are few, studies describing siderophore production by ECM fungi in association with roots are even fewer. Nevertheless, van Hees et al. (2006), in a careful and precise study demonstrated the simultaneous release of siderophores and organic acids by hyphae of the extraradical mycelium of *Hebeloma crustuliniforme* in association with *Pinus sylvestris* seedlings. The siderophores characterized corresponded mainly to ferricrocin; ferrichrome also appeared in the exudates, but in a much lower concentration. Oxalate in concentrations 10,000 times higher than ferricrocin was also detected in exudates. The authors suggest that the combination of hyphal exudates, siderophores and oxalic acid, can significantly alter soil conditions through mineral dissolution (van Hees et al. 2006). Using the CAS assay, the Csáky and Arnow assays, and HPLC, the presence of iron-chelating compounds and organic acids was detected in cultures of *Suillus luteus*, *Rhizopogon luteolus* and *Scleroderma verrucosum*, collected from pine plantations (Machuca et al. 2007).

Table 15.1 Ectomycorrhizal fungi described by production and/or characterization of siderophores

Fungal species	Fungal growth (solid, liquid, symbiosis)	Demonstration of siderophore production	Purification/characterization of siderophores	References
Amanita muscaria *Boletus edulis* *Suillus brevipes* *S. lakei* *S. punctipes* *S. granulatus* *S. tomentosus* *Pisolithus tinctorius* *Cenococcum geophilum*	Pure culture Hagen medium (agar and broth), deferrated, without Fe^{3+} addition, or with 50 ng mL^{-1} Fe^{3+}	Bioassay with *Aureobacterium flavescens* JG-9, specific for hydroxamates	Only for *B. edulis* were the compounds partially purified and characterized. Czaky assay confirmed hydroxamate nature of compounds. Thin layer chromatography (TLC) revealed compounds mixture that contains ferricrocin	Szaniszlo et al. (1981)
P. tinctorius	Pure culture Iron-deficient liquid culture (iron concentration not specified)	No assay was carried out for siderophore detection	Siderophores isolated using ion-exchange resin and TLC. Siderophores partially purified were radiolabeled with ^{59}Fe. Structural characteristics not specified	Leyval and Reid (1991)
S. granulatus	Pure culture Pachlewsky broth, in presence of goethite, biotite and pyrite	Hydroxamate detection by Czaky assay in goethite, but not in biotite or pyrite presence. Iron solubilization from goethite was attributed to siderophore production more than to organic acids	No	Watteau and Berthelin (1994)
C. geophilum	Pure culture LIM-1 medium (Szaniszlo et al. 1981) and deferrated medium	Siderophore detection in culture filtrates by CAS assay	Siderophore isolation using ion-exchange resin and size exclusion chromatography. Ferricrocin was identified by HPLC, mass spectrometry, and ^{1}H–^{13}C-NMR spectra	Haselwandter and Winkelmann (2002)
Laccaria laccata *P. tinctorius* *Rhizopogon vulgaris* *R. luteolus*	Pure culture Modified Melin-Norkans (MMN) broth, with and without Fe^{3+} addition	Siderophore detection by CAS-agar assay and hydroxamate detection by chemical assay	No	Gupta and Satyanarayana (2002)

(*continued*)

Table 15.1 (continued)

Fungal species	Fungal growth (solid, liquid, symbiosis)	Demonstration of siderophore production	Purification/characterization of siderophores	References
Hebeloma crustuliniforme	Fungus in symbiosis with *Pinus sylvestris*, using aseptic multicompartment dishes	Release of siderophores and organic acids from extraradical mycelium	Siderophore isolation using ion-exchange resin. Hydroxamates were purified by HPLC and identified by mass spectrometry (ESI-MS/MS), mainly as ferricrocin, but ferrichrome was also detected. Organic acids were also detected, mainly oxalic acid	van Hees et al. (2006)
S. luteus *R. luteolus* *Scleroderma verrucosum*	Pure culture MMN broth, without Fe^{3+} addition and with 35 μmol L^{-1} Fe^{3+}.	Siderophore detection by CAS assay. Hydroxamate and catecholate detection by Czaky and Arnow assays, respectively	No	Machuca et al. (2007)

The Arnow assay revealed the presence of catecholate-type structures in the three species in higher concentrations than the hydroxamates; however, these compounds were not proven to be true siderophores (Machuca et al. 2007). Phenolic compounds with Fe^{3+}-chelating capacity have already been described in a variety of wood-rotting basidiomycetes (Goodell et al. 1997; Milagres et al. 1999; Machuca et al. 2001; Arantes and Milagres 2006); but their function as siderophores in the transport of Fe^{3+} has not been proven. Moreover, phenolic pigments produced by these species, responsible for the brown staining of the mycelia and culture medium, may be responsible for the positive reaction in the Arnow assay. Some pigments of this type (e.g., melanins) located in the cell wall, extracellular phenolic polymers, and other phenolic metabolites with metal-chelating properties have been involved in processes of metal uptake and detoxification by fungi (Gadd 1993; Fogarty and Tobin 1996; Bellion et al. 2006). Other ECM fungi showed iron-chelating compound production in liquid and/or solid medium through CAS assay: *S. bellinii*, *S. granulatus*, *Lactarius deliciosus*, *Paxillus filamentosus* and *Amanita* sp.. Unexpectedly, the ascomycete *Tuber borchii* was the only species we tested that has not shown any positive CAS reaction, either in liquid or solid medium.

The exudation of compounds with the capacity to chelate and mobilize Fe^{3+} has also been confirmed in the ECM root tips of *Xerocomus* sp. and *L. subdulcis,* collected in forest soils (Rineau et al. 2008). On the other hand, the presence of hydroxamate siderophores, mainly ferricrocin and ferrichrome, in forest soil in concentrations that vary between 2 and 12 nM has also been described (Holmström et al. 2004; Essén et al. 2006). The origin of these siderophores must be linked to the microorganisms that inhabit terrestrial ecosystems, among them ECM fungi. This has considerable ecological implications, since both microorganisms (fungi and bacteria) and plants could benefit from the production of siderophores by other species and use them as exogenous siderophores along with their own in order to increase the uptake of iron from the environment. The ability to utilize exogenous siderophores gives some plants and microorganisms a competitive advantage in environments deficient in this essential micronutrient (Crowley et al. 1991; Guerinot 1994; Winkelmann 2007; Johnson 2008).

Furthermore, the role of hydroxamate siderophores in iron uptake from soils has been shown using Fe^{3+}-minerals (e.g., goethite), from which siderophores release iron through dissolution of the mineral, acting alone or in combination with organic acids (Kraemer et al. 1999; Cervini-Silva and Sposito 2002; Borer et al. 2005; Reichard et al. 2007). The solubilization of goethite by *S. granulatus* was attributed more to the production of siderophores than to organic acids; and most of the Fe^{3+} mobilized was accumulated in the mycelium (Watteau and Berthelin 1994). *S. granulatus* also dissolved biotite and mobilized Fe^{3+} and Al^{3+}, but siderophores were not produced in the presence of this mineral, and the authors attribute this to Fe^{3+} being more available in biotite. Peña et al. (2007) proved the participation of siderophores in the dissolution of a Mn mineral (hausmannite), through strong chelation of Mn^{3+}. These studies confirm the importance of siderophores in the biogeochemical cycle of Fe and their connection to other cycles like the Mn cycle (Duckworth et al. 2009). Most of the research in this area has been undertaken using the commercial hydroxamate siderophore desferrioxamine B, while the studies with ECM fungi producing siderophores are limited.

Despite the high affinity of siderophores for Fe^{3+}, these compounds can also form stable complexes with other metals, such as Al, Cd, Cr, Cu, Pb, Mn, Zn, and some actinides like Pu, Th, and U (Hider 1984; Brainard et al. 1992; Renshaw et al. 2003; Zou and Boyer 2005). Ferricrocin, the main siderophore produced by most of the ECM fungi investigated (Table 15.1), which is also present in forest soils, can form stable mononuclear complexes in aqueous solution with trivalent ions (Al^{3+}, Cr^{3+}, Ga^{3+}, and Fe^{3+}); and with divalent ions (Cu^{2+}, Zn^{2+}), it can form multinuclear complexes (Zou and Boyer 2005). Although siderophores can form complexes with other metal ions besides iron, these complexes often do not complete their cycle of returning to the cell and delivering the metal ion because they are not recognized, or because metal ion reduction does not occur (Hider 1984; Clarke et al. 1987; Zou and Boyer 2005). These results are of great importance in view of the role that siderophores may play for ECM fungi and their host plants in metal-contaminated terrestrial ecosystems, where they might regulate the transport and mobility of these metal ions as part of detoxification mechanisms. Nevertheless, research is still necessary to determine the conditions under which siderophores can prevent the entrance of certain metal ions into the cell, thereby reducing the toxic effect on the fungus and its host plant, or the conditions under which the mobilization of metal ions could be increased, producing an adverse toxic effect as has been observed in other microorganisms (Arceneaux et al. 1984). Studies into the effects of ECM fungi on their host plants growing in metal-contaminated soils have often been contradictory; some illustrate the beneficial effect of ECM fungi on the protection of their host plants, while many others indicate an increase in metal ion uptake by mycorrhized plants (Godbold et al. 1998; Jentschke and Godbold 2000). Perhaps the contradictory effects described in the literature are related to the type of siderophore or other metal-chelating agents being produced by a certain fungal species and to the characteristics of the metal complexes formed.

From these results, it seems evident that there is a gap in our knowledge regarding the production and characterization of siderophores by ECM fungi, in the absence and/or presence of symbiotic association, since to date very few species have been evaluated (Table 15.1). The role that siderophores play as chelators of other metal ions besides Fe when EcM fungi are growing in contaminated environments must be investigated as part of possible detoxification mechanisms. Studies are also needed in relation to the effect of other metal ions besides Fe on siderophore synthesis regulation in EcM fungi. In addition, investigations that contribute to increasing our understanding of the role of fungal siderophores in the metabolism of iron in plants are necessary to determine, for example, whether ECM fungal siderophores are used by host plants, under what conditions and what mechanisms are involved.

15.3 Non-specific Metal-Chelating Agents: Low Molecular Weight Organic Acids

Low molecular weight (LMW) organic acids, along with siderophores, are among the main mediators of the biological weathering of soils caused by living organisms, leading to the dissolution of a variety of minerals through a double effect of metal cation chelation and anion displacement by protonation (Landeweert et al. 2001; Gadd 2007; van Schöll et al. 2008). In soils, organic acids come mainly from plants that release them into the rhizosphere by exudation from their roots, as well as from fungi and bacteria, and chemically they are carboxylic acids, with one or more functional groups per molecule (e.g., formic, malic, oxalic, citric, succinic, etc.). Those acids that present more than one carboxylic group (di- or tricarboxylic) are important in the formation of complexes with metal ions and can be considered as metal-chelating agents. Under the normally existing pH conditions in biological systems, these acids are found in the form of carboxylate anions (e.g., oxalate, citrate, etc.).

Citric (tricarboxylic) and oxalic (dicarboxylic) acids have been the most investigated in plants and fungi, mainly in saprophytes and pathogens, due not only to their nutritional and physiological importance for these organisms, but also to their industrial importance. Oxalate can form complexes with a variety of trivalent (e.g., Fe^{3+}, Al^{3+}, Cr^{3+}) and divalent (e.g., Ca^{2+}, Cu^{2+}, Mg^{2+}, Zn^{2+}, Pb^{2+}) metal ions and also with actinides and lanthanides; some oxalates are soluble, like that of Fe^{3+}, but others can be insoluble like those of Ca^{2+} and Pb^{2+}. Similarly, the citrate can form different complexes with a variety of metal ions (e.g., Ca^{2+}, Cd^{2+}, Cu^{2+}, Pb^{2+}, Fe^{3+}) (Dutton and Evans 1996; Jones 1998; Gadd 1999).

LMW organic acids perform important and varied metabolic roles for living cells and also participate in the mobilization and uptake of a series of mineral nutrients (e.g., Fe and P), essential to the metabolism of plants and most microorganisms in the soil. LMW organic acids are directly involved in the biogeochemical cycles of a variety of metal elements; they serve as pathogenic agents for certain fungi and as wood biodeterioration agents for others; and they have been involved in the detoxification of metal ions in plants, fungi, and bacteria (Dutton and Evans 1996; Jones 1998; Gadd 1999, 2007; Landeweert et al. 2001). Although LMW organic acids are important agents for the mobilization of mineral nutrients from inorganic sources, the strategy most utilized by ECM fungi for the mobilization of N and P from organic sources is enzyme production. Both strategies are used with more or less efficiency depending on the fungal species (Chalot and Brun 1998; Landeweert et al. 2001).

Malate and citrate are the main acids released into the rhizosphere from many plant roots under conditions of iron or phosphorus deficiency. In iron deficiency, it has been suggested that citrate may participate in the dissolution and uptake of Fe^{3+} by forming Fe^{3+}-citrate complexes in some dicotyledons (Jones 1998). In grasses, phytosiderophores excreted by the roots may be responsible for the solubilization of Fe^{3+} (Kraemer et al. 2006; Crowley 2006). On the other hand, oxalate, malate, and citrate have been involved in the detoxification of Al in plants through the formation of Al^{3+}-organic acid complexes, which are of lower toxicity for plants than free Al^{3+} (Jones 1998; Schöttelndreier et al. 2001).

LMW organic acids can regulate the speciation and bioavailability of certain metal ions in soils through mobilization by acidification (protonation), formation of soluble complexes (complexolysis), or immobilization by formation of insoluble complexes. These mechanisms are of considerable importance in the acquisition of essential metal ions or metal detoxification by LMW organic acids. The formation of complexes, however, depends not only on the concentration of carboxylate anions, but also on the type and concentration of metal ion, the pH of the medium, and the stability constants of the complexes formed (Jones 1998; Gadd 1999; Meharg 2003). Citrate, malate, and oxalate form stable complexes of high affinity constants with trivalent ions like Fe^{3+} and Al^{3+}. Oxalate also forms insoluble complexes with Ca^{2+} and this is one of the main forms in which it is found in soils, fungi, plants, and animals (Jones 1998; Gadd 1999). Despite the high affinities for Fe^{3+}, the stability constants of the complexes with organic acids are inferior to those obtained with hydroxamate or catecholate (Hider 1984; Kraemer 2004), and therefore the role of siderophores in the uptake and transport of Fe^{3+} must be of higher relevance than chelation by organic acids for living organisms dependent on these metal-chelating agents.

LMW organic acids have been studied in ECM fungi much more than siderophores, mainly in relation to the weathering of minerals. However, as in the case of siderophores, most detailed studies are limited to a few fungal species. Organic acid exudation by the hyphae of ECM fungi produces the solubilization and mobilization of a variety of nutrients (P, K, Ca, and Mg) directly from solid mineral substrates far from the rhizosphere, insoluble and inaccessible to the roots, thereby contributing to the nutritional status of the host plants (Landeweert et al. 2001; Gadd 2007; van Schöll et al. 2008). The formation of tubular pores in certain minerals has been attributed to organic acid exudation by the hyphae of ECM fungi inside mineral particles, and for this reason they are called rock-eating fungi. In this way, the dissolution of minerals by ECM fungi to obtain essential ions is not limited to an attack on particle surfaces, but inside the mineral particles as well (Jongmans et al. 1997; Landeweert et al. 2001; van Schöll et al. 2008).

The exudation of LMW organic acids, mainly oxalate, by different ECM fungi has been described in vitro in the absence of symbiotic association (Lapeyrie 1988; Lapeyrie et al. 1991; Arvieu et al. 2003; Machuca et al. 2007), in the presence of symbiosis (Ahonen-Jonnarth et al. 2000; Casarin et al. 2003; van Hees et al. 2005, 2006; van Schöll et al. 2006a), and also by ECM root tips harvested in forest ecosystems (Rineau et al. 2008). Some of these studies have shown a greater production of oxalate by mycorrhized compared to nonmycorrhized plants cultivated under axenic conditions (Ahonen-Jonnarth et al. 2000; Casarin et al. 2003). *P. involutus*, one of the most studied species in relation to oxalate production, can use bicarbonate ($NaHCO_3$) as a source of C and ammonium (NH_4^+), or nitrate (NO_3^-) as N sources to produce the acid and biosynthesis can happen directly from oxaloacetate or via citrate, isocitrate, and glyoxylate (Lapeyrie 1988;

Lapeyrie et al. 1991). In the presence of NO_3^-, calcium carbonate ($CaCO_3$) and different concentrations of orthophosphate (Pi), *P. involutus*, *H. cylindrosporum*, *R. roseolus* and, *S. collinitus* released oxalate and protons (H^+) to the culture medium (Arvieu et al. 2003). When carboxylates are exuded from the cells, the negative charges must be balanced by a simultaneous efflux of positive charges (like H^+), leading to a reduction of pH outside the cells (Jones 1998; Casarin et al. 2003). The increase in the excretion of oxalate and H^+ in the presence of $CaCO_3$ suggests the importance that these fungal species may have in the mobilization of nutrients in calcareous soils. In the rhizosphere of the associations between *R. roseolus* and *H. cylindrosporum* with *P. pinaster*, it was possible to detect the release of oxalate in presence of *R. roseolus* with simultaneous rhizosphere acidification. However, no excretion of oxalate was obtained with *H. cylindrosporum* and an alkalinization of the rhizosphere was observed (Casarin et al. 2003).

The excretion of citrate and succinate, in addition to oxalate, was detected in a culture medium of *R. luteolus*, *S. verrucosum* and *S. luteus*. *S. luteus* was the species that produced the highest concentrations of all the organic acids and the only one that also produced malonate. A strong pH reduction was also recorded in the cultures of all the species (Machuca et al. 2007). It must be emphasized that even though these species were not subjected to a nutrient deficit (P, N, or C) and used NH_4^+ as the source of N, they were able to excrete high organic acid concentrations and H^+. Strong acidification was also produced by these species in solid medium in the presence of high concentrations of Cu^{2+} and Zn^{2+}, but not in the presence of Cd. This suggests that the capacity to produce organic acids and acidify culture media is an intrinsic characteristic of these species. These species were harvested in forest plantations of *P. radiata*, where the fruiting bodies were found frequently and in large quantities in the different regions studied (Machuca et al. 2007). Van Schöll et al. (2008) indicate that species of the genera *Rhizopogon* and *Suillus* are phylogenetically related to species of the genera belonging to brown rot fungi like *Serpula*, *Coniophora*, and *Hygrophoropsis*, known to produce large amounts of oxalate.

LMW organic acids have also been investigated with respect to the solubilization of mineral nutrients from a great variety of insoluble substrates and their connection to plant nutrition via ECM fungi. The mobilization of P and Ca from the mineral apatite has been described (Wallander 2000a; Blum et al. 2002; Wallander et al. 2003), as well as K from biotite, microcline, and phlogopite (Paris et al. 1995, 1996; Wallander and Wickman 1999; Wallander 2000b). Using pot experiments, *P. sylvestris* plants mycorrhized with *P. involutus*, *Piloderma croceum*, and *H. longicaudum* were cultivated with mineral muscovite as the only source of K and hornblende as the source of Mg (van Schöll et al. 2006b). Under these conditions, *P. involutus* was the only species able to solubilize the muscovite, but not the hornblende; mobilizing K for its host plants. The authors emphasize that ECM fungi can indeed increase the dissolution of minerals in response to a deficiency in nutrients, but the efficiency of the process is clearly species-specific.

The exudation of LMW organic acids has been related to metal ion detoxification mechanisms in ECM fungi (Jentschke and Godbold 2000; Meharg 2003;

Bellion et al. 2006) and plants (Jones 1998). LMW organic acids may act as extracellular chelators, forming complexes with metal ions and preventing them from entering the cell (avoidance mechanisms) in the same way that may occur with siderophores (Jentschke and Godbold 2000; Bellion et al. 2006). In this respect, there is a great deal of controversy, given that when LMW organic acids produce the dissolution of certain minerals in soil, together with increasing the bioavailability of essential metal ions, they may also cause the mobilization of potentially toxic metal ions. Furthermore, the efflux of H^+ that can occur with the exudation of carboxylate anions may decrease the medium pH, increasing the bioavailability of certain metal ions, making them toxic to fungi and plants (Gadd 2007).

When *P. sylvestris* seedlings mycorrhized with *S. variegatus*, *R. roseolus*, and *P. involutus* were exposed to high concentrations of metals, a significant increase in oxalate production was observed in the presence of *S. variegatus* and *R. roseolus* exposed to Al. In addition, Cu exposure stimulated the production of oxalate in the presence of *S. variegatus* and *P. involutus,* and, by contrast, Ni and Cd had no effect on oxalate production (Ahonen-Jonnarth et al. 2000). Using several insoluble minerals, the solubilization capacity was attributed to the strong decrease in extracellular pH caused by different species of ECM fungi (Rosling et al. 2004), and this decrease was also related to metal tolerance (Fomina et al. 2005). Ray and Adholeya (2009) determined in vitro organic acid exudation by *P. tinctorius* and *S. verrucosum* using coal ash pond. Formic, malic, and succinic acids were produced by the fungi and a variety of metals (Al, As, Cd, Cr, Ni, and Pb) were detected in their mycelia. No oxalic acid production was described in this paper. Large differences were observed between the strains regarding the patterns of LMW organic acids exuded and metal accumulation. A correlation between organic acid production and a metal accumulation in fungal mycelia was also demonstrated (Ray and Adholeya 2009).

The presence of Ca oxalate crystals has been observed in the cultures of ECM fungi, mycorrhized roots, in the rhizosphere, and around the hyphae of the extraradical mycelium (Cromack et al. 1979; Lapeyrie et al. 1990; Allen et al. 1996; Mahmood et al. 2001; Tuason and Arocena 2009) and it has been suggested that these crystals constitute a reservoir of Ca in the ecosystems, where they can also affect phosphate availability (Dutton and Evans 1996; Gadd 1999, 2007). The formation of these crystals by reprecipitation of solubilized Ca may serve as a fungal detoxification mechanism when the plants grow in soils with high concentrations of Ca; this has also been suggested for other metals that form insoluble oxalates (Cromack et al. 1977; Dutton and Evans 1996; Gadd 2007).

Even though metal tolerance has been extensively researched in ECM fungi, the results with respect to the benefits of the fungi on their host plants can be contradictory at times (Godbold et al. 1998; Jentschke and Godbold 2000; Gadd 2007). This may be related to the lack of experimental evidence clearly showing which mechanisms are involved and how much they contribute to tolerance in fungi. The participation of chelating agents such as siderophores and LMW organic acids has often been suggested among these tolerance mechanisms. However, to date, there seem to be no conclusive studies that are able to correlate the degree of metal tolerance with the type and concentration of chelators produced by an ECM fungus when exposed to high concentrations of one or several metals simultaneously. The situation is of greater complexity when attempting to analyze the fungus–plant system in contaminated soils where the mycorrhized roots might be exposed to more than one metal ion and several other factors that affect the bioavailability of metals. This knowledge is essential if the possible application of certain ECM fungi is to be considered in bioremediation projects or degraded environment reforestation.

15.4 Other Metal-Chelating Agents: Thiol-Peptides

Another group of agents with metal-chelating properties has been described in some species of ECM fungi, but unlike siderophores and LMW organic acids, they are more related to detoxification mechanisms than to metal ion nutrition or mineral solubilization. This group includes MTs, PCs, and GSH, all the thiol peptides (Gadd 1993; Meharg 2003; Pocsi et al. 2004; Bellion et al. 2006). These peptides act intracellularly; once the metal ions have entered the cell, they regulate or prevent the activity of metal ions in the cytosol through complexation. Thus, unlike siderophores and LMW organic acids that can represent avoidance mechanisms in relation to metal detoxification, these peptides represent intracellular detoxification mechanisms.

MTs are peptides of low molecular weight (6–7 kDa), rich in cysteine (Cys), highly ubiquitous – given that they have been identified in animals, plants, fungi, and some bacteria – with the capacity to chelate metals through the thiol (–SH) groups. Some MTs can contain up to 60 amino acids, of which fewer than half correspond to Cys; they do not exhibit aromatic amino acids or histidine. MTs can form clusters with a large amount of divalent or monovalent metal ions through the domains rich in Cys (Robinson et al. 1993). The biosynthesis of MTs is regulated at transcriptional level by a variety of factors, among them some metals: Cd, Zn, Hg, Cu, Au, Ag, Co, Ni, and Bi (Kägi and Shäffer 1988); and of these Cu, Zn, Cd, Hg, Bi, Ag, Au, as well as Pb and Pt can be indeed chelated (Kägi 1991). However, the preference for one or the other metal, both in the regulation of biosynthesis and in chelation, depends exclusively on the organism (Robinson et al. 1993; Hall 2002). MTs were described in ECM fungi in 1986 and their production was induced by the presence of Cd^{2+} in a *P. tinctorius* culture medium (Morselt et al. 1986). Later, the production of MTs by *P. involutus* and *L. laccata* was shown, regulated by high Cd^{2+} concentrations (Howe et al. 1997). Using HPLC, several thiol-containing compounds were investigated in *P. involutus* (Courbot et al. 2004), with GSH, γ-glutamylcysteine, and a compound of low molecular weight being detected that the authors suggested to correspond to MTs; all of these increased by exposing the fungus to Cd^{2+}. The genes that codify for MTs were characterized in *P. involutus* (Pimt1 gene) (Bellion et al. 2007); the gene expression was regulated by Cu and Cd, but not by Zn, and the overexpression of Pimt1 in a strain transformed from

H. cylindrosporum gave it an increased tolerance to Cu. Recently, these genes were also characterized in *H. cylindrosporum* and their regulation by Cu and Cd was shown, but not by Zn, Pb, or Ni (Ramesh et al. 2009).

PCs are peptides rich in Cys, with a general (γ-Glu-Cys)$_n$-Gly ($n = 2$–11) structure, of low molecular weight, which in fungi does not seem to exceed 2 kDa. Unlike MTs, PCs are not codified in the genome and their presence has been shown mainly in plants, where they have been related to the detoxification of Cd (Cobbett 2000; Mejáre and Bülow 2001; Hall 2002). They are synthesized from GSH, through the action of an enzyme (PC synthase), whose activity is regulated by metal ions such as Hg, Ag, Cu, Ni, Au, Pb, Zn, but mainly by Cd, which is the strongest inducer of PC synthase (Gadd 1993; Cobbett 2000; Mejáre and Bülow 2001). Until recently, there were no reports of their presence in ECM fungi, and even the synthesis of these peptides has been shown in only a very few fungal species (Bellion et al. 2006). Nevertheless, Collin-Hansen et al. (2007) described for the first time the presence of PCs in the well known edible wild mushroom *Boletus edulis*. In this fungus, metal exposure induced the production of thiol compounds such as GSH, and an accumulation of Cd in the caps of this macromycete was observed due to the formation of complexes with the PCs. When the roots of *Picea abies* mycorrhized with *L. laccata* and the pure cultures of fungus were exposed to Cd, an increase was detected in the GSH, but not in the MTs (Galli et al. 1993). Recently, the role of tripeptide GSH (γ-Glu-Cys-Gly) was established as a key agent in the responses to various stress situations in fungi (Pocsi et al. 2004); and in the case of metal stress, GSH can act as a precursor to the synthesis of PCs, as a scavenger of reactive oxygen species or as a metal-chelating agent (Ott et al. 2002; Courbot et al. 2004; Hegedüs et al. 2007).

PCs, like MTs, have not only been linked to metal detoxification, but also to the regulation of the homeostasis of essential metal ions and other biological processes, acting as xenobiotic detoxifying agents, or antioxidants (Gadd 1993; Rauser 1995; Cobbett 2000; Bellion et al. 2006). Furthermore, these thiol peptides have also been related to the accumulation capacity of metal ions in the fruiting bodies of some basidiomycetes when they grow in contaminated environments, and this represents an enormous risk to human health due to the high consumption of edible wild mushrooms in many countries (Gadd 1993). Although there is a large number of studies in the literature on the accumulation of metals, metalloids, and radionuclides by several mushrooms (saprophyte and ECM fungi), most aim at determining the metal concentrations in the caps and stipes of the macrofungi and in their growth substrates. Nevertheless, the physiological and biochemical mechanisms involved in the accumulation are still not well understood, and the participation of metal-chelating agents like siderophores, MTs, PCs, and GSH have been suggested and in some cases proven, as with *B. edulis* (Collin-Hansen et al. 2007). On the other hand, a correlation between the GSH concentration and the Hg^{2+} and Cd^{2+} contents was described in the fruiting bodies of various ECM fungi (Kojo and Lodenius 1989). Recently, *Amanita stroboliformes* and *A. solitaria* were described as hyperaccumulators of Ag, accumulating 2,500 times more Ag in their fruiting bodies than in the growth soil (Borovicka et al. 2007). Despite Ag accumulation mechanisms not being investigated in this study, it may be similar to that described for saprophyte *Agaricus bisporus*, where the participation of MTs was suggested (Byrne and Tusek-Znidaric 1990). The accumulation of ^{137}Cs has been observed in several species of saprophytes and ECM fungi, edible wild mushrooms (Gaso et al. 2000). Among these, the ECM fungus *Clavariadelphus truncatus* was described as an accumulator of ^{137}Cs, Rb, and Pb; and this was related to metal-chelating compounds (siderophore-type) detected in its fruiting bodies (Gaso et al. 2007).

Given the metal-accumulation capacity of some macromycetes, their use as bioindicators of contaminated environments has been suggested by several authors. However, the variability in accumulation among fungal species is enormous and depends on the developmental stage of the fruiting bodies, the type of metal, and a series of environmental factors (Mejstrik and Lepsova 1993; Gadd 2007). Moreover, some ECM fungi may accumulate large concentrations of metals in their fruiting bodies with respect to growth soils, even when they are grown in non-contaminated environments. By contrast, other species exclude or do not accumulate metals in their fruiting bodies despite growing in close proximity to mine tailings. For this reason, the use of some saprophytic or ECM macromycetes as bioindicators of a metal-contaminated terrestrial ecosystem must be considered with caution.

Some of the metal-chelating agents or mechanisms described here for ECM fungi have also been described for arbuscular mycorrhizal fungi (AMF) (Khan et al. 2000; Göhre and Paszkowski 2006). However, given the symbiotic dependency required in AMF, the experiments are often of greater complexity. Regarding the metal-chelating mechanisms, there is a great difference between both groups of fungi, which is represented by glomalin, described to date exclusively in AMF. Glomalin is a glycoprotein produced by the hyphae, capable of chelating metals such as Cu, Pb, and Cd, and because of this some researchers have highlighted the significant role of the glomalin-producing AMF in the stabilization of the contaminated soils and in the protection of their host plants (González-Chávez et al. 2004; Khan 2006).

15.5 Biotechnological Potential of ECM Fungi Producing Metal-Chelating Agents

The biotechnological potential of ECM fungi producing metal-chelating agents is related to the isolated chelators with specific aims, or to the use of the fungi in the presence or absence of symbiotic association. Due to their extremely efficient metal-chelating properties, for many years the main application of siderophores has been in chelation therapy to treat Fe and Al overload in humans; its possible use in the decorporation of actinides is also being investigated. The main siderophore used for this purpose is desferrioxamine B, a hydroxamate obtained from *Streptomyces pilosus* (Messenger and Ratledge 1985; Renshaw et al. 2002; Ansoborlo et al. 2007). However, siderophores can have other important applications related to the environment; in the uptake of metals from industrial waste, low-grade ores, serpentine soils, contaminated terrestrial and aquatic environments, tailings of abandoned mines, etc. The uptake of metals can serve to remediate an environment and/or recover metals for recycling. In addition, the capacity to chelate actinides (Pu, U, Np, and Th) has been demonstrated in siderophores and for this reason their application has been proposed for the remediation of radioactive waste and the reprocessing of nuclear fuel (Hernlem et al. 1999; Renshaw et al. 2002, 2003). The majority of these studies have been conducted with commercial hydroxamate siderophores like desferrioxamine B and some with siderophore-producing soil microbes (John et al. 2001; Keith-Roach et al 2005; Mullen et al 2007). Therefore, this is an area where other fungal hydroxamate siderophores, like those produced by ECM fungi, could have a great potential for application. Another important possible application of siderophores is in the treatment of asbestos, a carcinogenic fibrous mineral with varied industrial applications that is prohibited in many countries. Although the toxicity mechanisms are not well understood, it is believed that the Fe present on the surface of asbestos fibers cause cell damage by generating free radicals. It was recently demonstrated that hydroxamates and soil fungi-producing siderophores, among them ECM fungi, inactivate fibers by removing the Fe from the material surfaces (Martino et al. 2003, 2004; Daghino et al. 2008). These fungi represent a potential tool for the bioremediation of asbestos waste, contaminated waters and soils, and asbestos abandoned mines. It must be pointed out that due to their chelating properties, LMW organic acids could have most of the environmental applications found for siderophores, as previously described.

The identification and selection of new strains of ECM fungi, efficiently producing siderophores and/or LMW organic acids, is fundamental to the mycorrhization of plants used in programs to recover degraded or destabilized forest ecosystems, poor in mineral nutrients either from natural causes or through anthropogenic action. The production of these chelators by ECM roots must contribute to the solubilization and uptake of mineral nutrients into the rhizosphere, facilitating the establishment and development of plants in these ecosystems. In addition, the modifications in the speciation and bioavailability of metals and nonmetals as a result of the chelators released by the hyphae beyond the rhizosphere promotes the restoration in the mineral nutrient balance, favoring not only vegetal species, but also all microorganisms inhabiting these ecosystems (Gadd 2007; Van Schöll et al. 2008).

The revegetation of metal-contaminated soils is another important application of plants mycorrhized with ECM fungi producing a wide range of chelators: siderophores, LMW organic acids, MTs, PC, and GSH. Given that these chelators contribute to the detoxification of metals through intracellular or extracellular mechanisms, the fungal species that produce them must protect their host plants in contaminated environments. This protection consists of excluding metals or accumulating them in their fruiting bodies, using both ways to prevent their entry into the mycorrhized roots. The use of plants in the bioremediation of soils contaminated by metals, aided by synthetic chelators like EDTA and called assisted phytoremediation, has been described in the literature (Khan et al. 2000; Lasat 2002; Wenzel 2009). Plants inoculated with ECM fungi producing metal-chelating agents could be used in the same way to aid in the increased uptake of metals by the roots, and at the same time improve the nutritional status of the plants, and with it the production of biomass for phytoremediation.

15.6 Conclusions

The results described illustrate the fundamental role that extracellular metal-chelating agents like siderophores and LMW organic acids play in the acquisition of essential metal ions for many ECM fungi and their host plants. At the same time, siderophores and LMW organic acids, together with the intracellular thiol-peptides metallothioneins, phytochelatins, and glutathione, have been related to detoxification mechanisms and metal ion accumulation, when the ECM fungi are growing in contaminated substrates or natural environments. Despite their biotechnological potential, very few species of ECM fungi have been investigated in relation to the production and characterization of metal-chelating agents. Moreover, the simultaneous production of more than one type of metal-chelating agent by the same fungal species has not often been considered, even though the efficiency of the combined action of these agents in the solubilization of minerals has been shown. Along with this, the demonstration that the transfer of the genes that codify for the metallothioneins provides increased copper tolerance between fungal species indicates the need to extend our knowledge regarding the genes that codify metal-chelating agents and the factors that regulate their expression in ECM fungi. Because of that, the urgent search for new species of ECM fungi producing more than one type of metal-chelating agent efficiently is necessary, since these species could be converted into valuable tools for applications in the bioremediation of substrates or contaminated environments or in the rehabilitation of disturbed forest ecosystems, alone or in association with their host plants. Nevertheless, success in the application of ECM fungi based on metal-chelating agent production depends on much-needed prior research that considers which type of metal-chelating agents are being produced by the ECM fungi, alone or in symbiotic association under axenic conditions. This is because, depending on the chelators produced, ECM fungi can promote the mobilization of metals towards the interior of the roots, or they can promote the immobilization of metals in the rhizosphere or the interior of their cells, preventing entry into the roots. The results of these studies will make it possible to select the most appropriate fungal partner for a particular vegetal species under a certain metal-stress condition (by excess or deficit), before applying the ECM fungi and/or ectomycorrhized plants in field assays.

Acknowledgments The author wishes to thank Dr. Adriane M.F. Milagres and Dr. Carolin Córdova for their helpful comments and revisions of the manuscript.

References

Ahonen-Jonnarth U, Van Hees PAW, Lundström US, Finlay RD (2000) Organic acids produced by mycorrhizal *Pinus sylvestris* exposed to elevated aluminium and heavy metal concentrations. New Phytol 146:557–567

Allen MF, Figueroa C, Weinbaum BS, Barlow SB, Allen EB (1996) Differential production of oxalate by mycorrhizal fungi in arid ecosystems. Biol Fertil Soils 22:287–292

Ansoborlo E, Amekraz B, Moulin C, Moulin V, Taran F, Bailly T, Burgada R, Hengé-Napoli M-H, Jeanson A, Den Auwer C, Bonin L, Moisy P (2007) Review of actinide decorporation with chelating agents. CR Chim 10:1010–1019

Arantes V, Milagres AMF (2006) Evaluation of different carbon sources for production of iron-reducing compounds by *Wolfiporia cocos* and *Perenniporia medulla-panis*. Process Biochem 41:887–891

Arceneaux JEL, Boutwell ME, Byers BR (1984) Enhancement of copper toxicity by siderophores in *Bacillus megaterium*. Antimicrob Agents Chemother 25:650–652

Arnow LE (1937) Colorimetric determination of the components of 3, 4-dihydroxy-phenylalanine tyrosine mixtures. J Biol Chem 118:531–537

Arvieu J-C, Leprince F, Plassard C (2003) Release of oxalate and protons by ectomycorrhizal fungi in response to P-deficiency and calcium carbonate in nutrient solution. Ann For Sci 60:815–821

Bellion M, Courbot M, Jacob C, Blaudez D, Chalot M (2006) Extracellular and cellular mechanisms sustaining metal tolerance in ectomycorrhizal fungi. FEMS Microbiol Lett 254:173–181

Bellion M, Courbot M, Jacob C, Guinet F, Blaudez D, Chalot M (2007) Metal induction of a *Paxillus involutus* metallothionein and its heterologous expression in *Hebeloma cylindrosporum*. New Phytol 174:151–158

Blum JD, Klaue A, Nezat CA, Driscoll CT, Johnson CE, Siccama TG, Eagar C, Fahey TJ, Linkens GE (2002) Mycorrhizal dissolution of apatite as an important calcium source in base-poor forest ecosystems. Nature 417:729–731

Borer PM, Sulzberger B, Reichard PU, Kraemer SM (2005) Effect of siderophores on the light-induced dissolution of colloidal iron (III) (hydr)oxides. Mar Chem 73:179–193

Borovicka J, Randa Z, Jelínek E, Kotrba P, Dunn CE (2007) Hyperaccumulation of silver by *Amanita strobiliformis* and related species of the section *Lepidella*. Mycol Res 111:1339–1344

Boukhalfa H, Crumbliss AL (2002) Chemical aspects of siderophore mediated iron transport. Biometals 15:325–339

Brainard JR, Strietelmeier BA, Smith PH, Langsten-Unkefer PF, Barr ME, Ryan RR (1992) Actinide binding and solubilization by microbial siderophores. Radiochim Acta 58 (59):357–363

Brunner I (2001) Ectomycorrhizas: their role in forest ecosystems under the impact of acidifying pollutants. Perspect Plant Ecol Evol Syst 4:13–27

Byrne AR, Tusek-Znidaric M (1990) Studies of the uptake and binding of trace metals in fungi, Part I: accumulation and characterisation of mercury and silver in the cultivated mushroom, *Agaricus bisporus*. Appl Organomet Chem 4:43–48

Casarin V, Plassard C, Souche G, Arvieu JC (2003) Quantification of oxalate ions and protons released by ectomycorrhizal fungi in rizosphere soil. Agronomie 23:461–469

Cervini-Silva J, Sposito G (2002) Steady-state dissolution kinetics of aluminum-goethite in the presence of desferrioxamine-B and oxalate ligands. Environ Sci Technol 36:337–342

Chalot M, Brun A (1998) Physiology of organic nitrogen acquisition by ectomycorrhizal fungi and ectomycorrhizas. FEMS Microbiol Rev 22:21–44

Clarke SE, Stuart J, Sanders-Loehr J (1987) Induction of siderophore activity in *Anabaena* spp. and its moderation of copper toxicity. Appl Environ Microbiol 53:917–922

Cobbett CS (2000) Phytochelatins and their roles in heavy metal detoxification. Plant Physiol 123:825–832

Collin-Hansen C, Pedersen SA, Andersen RA, Steinnes E (2007) First report of phytochelatins in a mushroom: induction of phytochelatins by metal exposure in *Boletus edulis*. Mycologia 99:161–174

Courbot M, Diez L, Ruotolo R, Chalot M, Leroy P (2004) Cadmium-responsive thiols in the ectomycorrhizal fungal *Paxillus involutus*. Appl Environ Microbiol 70:7413–7417

Cromack K, Sollins P, Todd RL, Fogel R, Todd AW, Fender WM, Crossley ME, Crossley DA (1977) The role of oxalic acid and bicarbonate in calcium cycling by fungi and bacteria: some possible implications for soil animals. Ecol Bull 25:246–252

Cromack K, Sollins P, Graustein WC, Speidel K, Todd AW, Spycher G, Li CY, Todd RL (1979) Calcium oxalate accumulation and soil weathering in mats of the hypogeous fungus *Hysterangium crassum*. Soil Biol Biochem 11:463–468

Crowley DE (2006) Microbial siderophores in the plant rhizosphere. In: Barton LL, Abadía J (eds) Iron nutrition in plants and rhizospheric microorganisms. Springer, Dordrecht, The Netherlands, pp 169–198

Crowley DE, Wang YC, Reid CPP, Szaniszlo PJ (1991) Mechanisms of iron acquisition from siderophores by microorganisms and plants. Plant Soil 130:179–198

Csáky TZ (1948) On the estimation of bound hydroxylamine in biological materials. Acta Chem Scand 2:450–454

Daghino S, Martino E, Vurro E, Tomatis M, Girlanda M, Fubini B, Perotto S (2008) Bioweathering of chrysolite by fungi isolated in ophiolitic sites. FEMS Microbiol Lett 285:242–249

Duckworth OW, Bargar JR, Sposito G (2009) Coupled biogeochemical cycling of iron and manganese as mediated by microbial siderophores. Biometals 22:605–613

Dutton MV, Evans CS (1996) Oxalate production by fungi: its role in pathogenicity and ecology in the soil environment. Can J Microbiol 42:881–895

Essén SA, Bylund D, Holmström SJM, Moberg M, Lundström US (2006) Quantification of hydroxamate siderophores in soil solutions of podzolic soil profiles in Sweden. Biometals 19:269–282

Fogarty RV, Tobin JM (1996) Fungal melanins and their interactions with metals. Enzyme Microb Technol 19:311–317

Fomina MA, Alexander IJ, Colpaert JV, Gadd GM (2005) Solubilization of toxic metal minerals and metal tolerance of mycorrhizal fungi. Soil Biol Biochem 37:851–866

Gadd GM (1993) Interactions of fungi with toxic metals. New Phytol 124:25–60

Gadd GM (1999) Fungal production of citric and oxalic acid: importance in metal speciation, physiology and biogeochemical process. Adv Microb Physiol 41:47–92

Gadd GM (2007) Geomycology: biogeochemical transformations of rocks, minerals, metals and radionuclides by fungi, bioweathering and bioremediation. Mycol Res 111:3–49

Galli U, Meier M, Brunold C (1993) Effects of cadmium on nonmycorrhizal and mycorrhizal Norway Spruce seedlings *Picea abies* (L.) Karst and its ectomycorrhizal fungus *Laccaria laccata* (Scop. ex Fr.) Bk. & Br. Sulfate reduction, thiols and distribution of the heavy metal. New Phytol 125:837–843

Galli U, Schuepp H, Brunold C (1994) Heavy metal binding by mycorrhizal fungi. Physiol Plant 92:364–368

Gaso MI, Segovia N, Morton O, Cervantes ML, Godinez L, Peña P, Acosta E (2000) ^{137}Cs and relationships with major and trace elements in edible mushrooms from Mexico. Sci Total Environ 262:73–89

Gaso MI, Segovia N, Morton O, Lopez JL, Machuca A, Hernandez E (2007) Radioactive and stable bioaccumulation, crystalline compound and siderophore detection in *Clavariadelphus truncatus*. J Environ Radioactiv 97:57–69

Godbold DL, Jentschke G, Winter S, Marschner P (1998) Ectomycorrhizas and amelioration of metal stress in forest trees. Chemosphere 36:757–762

Göhre V, Paszkowski U (2006) Contribution of the arbuscular mycorrhizal symbiosis to heavy metal phytoremediation. Planta 223:1115–1122

González-Chávez MC, Carrillo-González R, Wright SF, Nichols KA (2004) The role of glomalin, a protein produced by arbuscular mycorrhizal fungi, in sequestering potentially toxic elements. Environ Pollut 130:317–323

Goodell B, Jellison J, Liu J, Daniel G, Paszczynski A, Fekete F, Krishnamurthy S, Jun L, Xu G (1997) Low molecular weight chelators isolated from wood decay fungi and their role in the fungal biodegradation of wood. J Biotechnol 53:133–162

Guerinot ML (1994) Microbial iron transport. Annu Rev Microbiol 48:743–772

Gupta V, Satyanarayana T (2002) Production of extracellular siderophores by ectomycorrhizal fungi. Indian J Microbiol 42:107–110

Hall JL (2002) Cellular mechanisms for heavy metal detoxification and tolerance. J Exp Bot 53:1–11

Haselwandter K (1995) Mycorrhizal fungi: siderophore production. Crit Rev Biotechnol 15:287–291

Haselwandter K, Bowen GD (1996) Mycorrhizal relations in trees for agroforestry and land rehabilitation. For Ecol Manag 18:1–17

Haselwandter K, Winkelmann G (2002) Ferricrocin – an ectomycorrhizal siderophore of *Cenococcum geophilum*. Biometals 15:73–77

Haselwandter K, Winkelmann G (2007) Siderophores of symbiotic fungi. In: Varma A, Chincholkar SB (eds) Soil biology, vol 12. Spriger, Berlin, pp 91–103

Hegedüs N, Emri T, Szilágyi J, Karányi Z, Nagy I, Penninckx MJ, Pócsi I (2007) Effect of heavy metals on the glutathione status in different ectomycorrhizal *Paxillus involutus* strains. World J Microbiol Biotechnol 23:1339–1343

Hernlem BJ, Vane LM, Sayles GD (1999) The application of siderophores for metal recovery and waste remediation: examination of correlations for prediction of metal affinities. Water Res 33:951–960

Hider RC (1984) Siderophore mediated absorption of iron. Struct Bonding 58:25–87

Holmström SJM, Lundström US, Finlay RD, Van Hees AW (2004) Siderophores in forest soil solution. Biogeochemistry 71:247–258

Howe R, Levans R, Ketteridge SW (1997) Copper-binding proteins in ectomycorrhizal fungi. New Phytol 135:123–131

Jentschke G, Godbold DL (2000) Metal toxicity and ectomycorrhizas. Physiol Plant 109:107–116

John SG, Ruggiero CE, Hersman LE, Tung C-S, Neu MP (2001) Siderophore mediated plutonium accumulation by *Microbacterium flavescens* (JG-9). Environ Sci Technol 35:2942–2948

Johnson L (2008) Iron and siderophores in fungal–host interactions. Mycol Res 112:170–183

Jones DL (1998) Organic acids in the rhizosphere – a critical review. Plant Soil 205:25–44

Jongmans AG, van Breemen N, Lundström US, van Hees PAW, Finlay RD, Srinivasan M, Unestam T, Giesler R, Melkerud P-A, Olsson M (1997) Rock-eating fungi. Nature 389:682–683

Kägi JHR (1991) Overview of metallothioneins. Methods Enzymol 205:613–626

Kägi JHR, Shäffer A (1988) Biochemistry of metallothionein. Biochemistry 27:8509–8515

Keith-Roach MJ, Vetriburatti M, Worsfold PJ (2005) Thorium complexation by hydroxamate siderophores in perturbed multicomponent systems using flow injection electrospray ionization mass spectrometry. Anal Chem 77:7335–7341

Khan AG (2006) Mycorrhizoremediation – an enhanced form of phytoremediation. J Zhejiang Univ Sci B 7:503–514

Khan AG, Kuek C, Chaudhry TM, Khoo CS, Hayes WJ (2000) Role of plants, mycorrhizae and phytochelators in heavy metal contamined land remediation. Chemosphere 21:197–207

Kojo MR, Lodenius M (1989) Cadmium and mercury in macrofungi-mechanisms of transport and accumulation. Angew Bot 63:279–292

Kosman DJ (2003) Molecular mechanisms of iron uptake in fungi. Mol Microbiol 47:1185–1197

Kraemer SM (2004) Iron oxide dissolution and solubility in the presence of siderophores. Aquat Sci 66:3–18

Kraemer SM, Cheah S-F, Zapf R, Xu JD, Raymond KN, Sposito G (1999) Effect of hydroxamate siderophores on Fe release and Pb(II) adsorption by goethite. Geochim Cosmochim Acta 63:3003–3008

Kraemer SM, Crowley DE, Kretzschmar R (2006) Geochemical aspects of phytosiderophore-promoted iron acquisition by plants. Adv Agron 91:1–46

Landeweert R, Hoffland E, Finlay RD, Kuyper TW, van Breemen N (2001) Linking plants to rocks: ectomycorrhizal fungi mobilize nutrients from minerals. Trends Ecol Evol 16:248–254

Lapeyrie F (1988) Oxalate synthesis from soil bicarbonate by the mycorrhizal fungus *Paxillus involutus*. Plant Soil 110:3–8

Lapeyrie F, Picatto C, Gerard J, Dexheimer J (1990) TEM study of intracellular and extracellular calcium oxalate accumulation by ectomycorrhizal fungi in pure culture or in association with *Eucalyptus* seedlings. Symbiosis 9:163–166

Lapeyrie F, Ranger J, Vairelles D (1991) Phosphate solubilizing activity of ectomycorrhizal fungi *in vitro*. Can J Bot 69:342–346

Lasat MM (2002) Phytoextraction of toxic metals: a review of biological mechanisms. J Environ Qual 31:109–120

Leyval C, Reid CPP (1991) Utilization of microbial siderophores by mycorrhizal and non-mycorrhizal pine roots. New Phytol 119:93–98

Leyval C, Turnau K, Haselwandter K (1997) Effect of heavy metal pollution on mycorrhizal colonization and function: physiological, ecological and applied aspects. Mycorrhiza 7:139–153

Machuca A, Napoleao D, Milagres AMF (2001) Detection of metal-chelating compounds from wood-rotting *Trametes versicolor* and *Wolfiporia cocos*. World J Microbiol Biotechnol 17:687–690

Machuca A, Pereira G, Aguiar A, Milagres AMF (2007) Metal-chelating compounds produced by ectomycorrhizal fungi collected from pine plantations. Lett Appl Microbiol 44:7–12

Mahmood S, Finlay RD, Erland S, Wallander H (2001) Solubilisation and colonisation of wood ash by ectomycorrhizal fungi isolated from wood ash fertilized spruce forest. FEMS Microbiol Ecol 35:151–161

Martino E, Prandi L, Fenoglio I, Bonfante P, Perotto S, Fubini B (2003) Soil fungal hyphae bind and attack asbestos fibers. Angew Chem Int Ed 42:219–222

Martino E, Cerminara S, Prandi L, Fubini B, Perotto S (2004) Physical and biochemical interactions of soil fungi with asbestos fibers. Environ Toxicol Chem 23:938–944

Meharg AA (2003) The mechanistic basis of interactions between mycorrhizal associations and toxic metal cations. Mycol Res 107:1253–1265

Mejáre M, Bülow L (2001) Metal-binding proteins and peptides in bioremediation and phytoremediation of heavy metals. Trends Biotechnol 19:67–73

Mejstrik V, Lepsova A (1993) Applicability of fungi to the monitoring of environmental pollution by heavy metals. In: Markert B (ed) Plants as biomonitors. Indicators for heavy metals in the terrestrial environment. VCH, Weinheim, pp 365–378

Messenger AJM, Ratledge C (1985) Siderophores. In: Moo-Young M (ed) Comprehensive biotechnology: the principles, applications and regulations of biotechnology in industry, agriculture and medicine, vol 3. Pergamon, New York, pp 275–295

Milagres AMF, Machuca A, Napoleao D (1999) Detection of siderophores production from several fungi and bacteria by a modification of chrome azurol S (CAS) agar plate assay. J Microbiol Methods 37:1–6

Morselt AFW, Smits WTM, Limonard T (1986) Histochemical demonstration of heavy metal tolerance in ectomycorrhizal fungi. Plant Soil 96:417–420

Mullen L, Gong C, Czerwinski K (2007) Complexation of uranium(VI) with the siderophore desferrioxamine B. J Radioanal Nucl Chem 273:683–688

Neilands JB (1984) Methodology of siderophores. Struct Bonding 58:1–24

Neilands JB (1995) Siderophores: structure and function of microbial iron transport compounds. J Biol Chem 270:26723–26726

O'Dell TE, Castellano MA, Trappe JM (1993) Biology and application of ectomycorrhizal fungi. In: Metting BF (ed) Soil microbial ecology: applications in agriculture and environmental management. Dekker, New York, pp 379–415

Ott T, Fritz E, Polle A, Schutzendubel A (2002) Characterisation of antioxidative systems in the ectomycorrhiza-building basidiomycete *Paxillus involutus* (Barsch)Fr. and its reaction to cadmium. FEMS Micrbiol Ecol 42:359–366

Paris F, Bonnaud P, Ranger J, Lapeyrie F (1995) *In vitro* weathering of phlogopite by ectomycorrhizal fungi 1. Effect of K^+ and Mg^{2+} deficiency on phyllosilicate evolution. Plant Soil 177:191–201

Paris F, Botton B, Lapeyrie F (1996) In vitro weathering of phlogopite by ectomycorrhizal fungi 2. Effect of K^+ and Mg^{2+} deficiency and N sources on accumulation of oxalate and H^+. Plant Soil 179:141–150

Peña J, Duckworth OW, Bargar JR, Sposito G (2007) Dissolution of hausmannite (Mn_3O_4) in the presence of the trihydroxamate siderophore desferrioxamine B. Geochim Cosmochim Acta 71:5661–5671

Pocsi I, Prade RA, Penninckx MJ (2004) Glutathione, altruistic metabolite in fungi. Adv Microb Physiol 49:1–76

Prasad MNV, Freitas HMO (1999) Feasible biotechnological and bioremediation strategies for serpentine soils and mine spoils. Electron J Biotechnol 2:35–50

Ramesh G, Podila GK, Gay G, Marmeisse R, Reddy MS (2009) Copper and cadmium metallothioneins of the ectomycorrhizal fungus *Hebeloma cylindrosporum* have different patterns of regulation. Appl Environ Microbiol 75:2266–2274

Rauser WE (1995) Phytochelatins and related peptides. Plant Physiol 109:1141–1149

Ray P, Adholeya A (2009) Correlation between organic acid exudation and metal uptake by ectomycorrhizal fungi grown on pond ash *in vitro*. Biometals 22:275–281

Reichard PU, Kretzschmar R, Kraemer SM (2007) Dissolution mechanisms of goethite in the presence of siderophores and organic acids. Geochim Cosmochim Acta 71:5635–5650

Renshaw JC, Robson GD, Trinci APJ, Wiebe MG, Livens F, Collison D, Taylor RJ (2002) Fungal siderophores: structures, functions and applications. Mycol Res 106:1123–1142

Renshaw JC, Halliday V, Robson GD, Trinci APJ, Wiebe MG, Livens FR, Collison D, Taylor RJ (2003) Development and application of an assay for uranyl complexation by fungal metabolites, including siderophores. Appl Environ Microbiol 69:3600–3606

Rineau F, Courty P-E, Uroz S, Buée M, Garbaye J (2008) Simple microplate assays to measure iron mobilization and oxalate secretion by ectomycorrhizal tree roots. Soil Biol Biochem 40:2460–2463

Robinson NJ, Tommey AM, Kuske C, Jackson PJ (1993) Plant metallothioneins. Biochem J 295:1–10

Rosling A, Lindahl BJ, Taylor AFS, Finlay RD (2004) Mycelial growth and substrate acidification of ectomycorrhizal fungi in response to different minerals. FEMS Microbiol Ecol 47:31–37

Schöttelndreier M, Norddahl MM, Ström L, Falkengren-Grerup U (2001) Organic acid exudation by wild herbs in response to elevated Al concentrations. Ann Bot 87:769–775

Schwyn B, Neilands JB (1987) Universal chemical assay for the detection and determination of siderophores. Anal Biochem 160:46–56

Sugiura Y, Nomoto K (1984) Phytosiderophores. Structures and properties of mugineic acids and their metal complexes. Struct Bonding 58:107–135

Szaniszlo PJ, Powell PE, Reid CPP, Cline GR (1981) Production of hydroxamate siderophore iron chelators by ectomycorrhizal fungi. Mycologia 73:1158–1174

Tuason MMS, Arocena JM (2009) Calcium oxalate biomineralization by *Piloderma fallax* in response to various levels of calcium and phosphorus. Appl Environ Microbiol 75:7079–7085

Van der Helm D, Winkelmann G (1994) Hydroxamates and polycarboxylates as iron transport agents (siderophores) in fungi. In: Winkelmann G, Winge D (eds) Metal ions in fungi. Dekker, New York, pp 39–98

Van Hees PAW, Jones DL, Jentschke G, Godbold DL (2005) Organic acid concentrations in soil solution: effects of young coniferous trees and ectomycorrhizal fungi. Soil Biol Biochem 37:771–776

Van Hees PAW, Rosling A, Essén S, Godbold DL, Jones DL, Finlay RD (2006) Oxalate and ferricrocin exudation by the extramatrical mycelium of an ectomycorrhizal fungus in symbiosis with *Pinus sylvestris*. New Phytol 169:367–378

Van Schöll L, Hoffland E, van Breemen N (2006a) Organic anion exudation by ectomycorrhizal fungi and *Pinus sylvestris* in response to nutrient deficiencies. New Phytol 170:153–163

Van Schöll L, Smits MM, Hoffland E (2006b) Ectomycorrhizal weathering of the soil minerals muscovite and hornblende. New Phytol 171:805–814

Van Schöll L, Kuyper TW, Smits MM, Landeweert R, Hoffland E, van Breemen N (2008) Rock-eating mycorrhizas: their role in plant nutrition and biogeochemical cycles. Plant Soil 303:35–47

Wallander H (2000a) Uptake of P from apatite by *Pinus sylvestris* colonized by different ectomycorrhizal fungi. Plant Soil 218:249–256

Wallander H (2000b) Use of strontium isotopes and foliar K content to estimate weathering of biotite induced by pine seedlings colonized by ectomycorrhizal fungi from two different soils. Plant Soil 222:215–229

Wallander H, Wickman T (1999) Biotite and microcline as potassium sources in ectomycorrhizal and non-mycorrhizal *Pinus sylvestris* seedlings. Mycorrhiza 9:25–32

Wallander H, Mahmood S, Hagerberg D, Johansson L (2003) Elemental composition of ectomycorrhizal mycelia identified by PCR-RFLP analysis and grown in contact with apatite or wood ash in forest soil. FEMS Microbiol Ecol 44:57–65

Watteau F, Berthelin J (1994) Microbial dissolution of iron and aluminium from soil minerals: efficiency and specificity of hydroxamate siderophores compared to aliphatic acids. Eur J Soil Biol 30:1–9

Wenzel WW (2009) Rhizosphere processes and management in plant-assisted bioremediation (Phytoremediation) of soils. Plant Soil 321:385–408

Winkelmann G (2007) Ecology of siderophores with special reference to the fungi. Biometals 20:379–392

Zou G, Boyer GL (2005) Synthesis and properties of different metal complexes of the siderophore desferriferricrocin. Biometals 18:63–74

Chapter 16
Ectomycorrhiza and Secondary Metabolites

Hanna Dahm and Patrycja Golińska

16.1 Introduction

Approximately 6,000 species of ectomycorrhizal (EM) fungi have been described, considerably more than arbuscular (AM) fungi. This has led to assumption that EM fungi are most host specific than AM fungi. Host plant diversity, species composition, and age do have a role in regulation mycorrhizal communities. Some studies also suggest that plant secondary metabolites (SM) which are in part under plant genetic control can affect EM colonization.

The main function of SM is defense against herbivores and microbes; some SM are signal and attract compounds for seed dispersing animals and some play a role in the symbiotic relationships with plants and microorganisms.

Early in the twentieth century, it was considered that SM arise either spontaneously or with the aid of nonspecific enzymes. Now, there is good evidence that biosynthetic enzymes are highly specific. As a consequence of specific enzymatic synthesis, final products always have a distinct stereochemistry (Wink 2008). Only the enzymes that are involved in the degradation of SM (glucosidases, esterases, and other hydrolases) are less substrate specific.

SM are not functionless waste products, but are important substances for the symbiotic organisms. Precursor for SM synthesis usually derive from basic metabolic pathways such as glycolytic, Krebs cycle, and shikimate pathway. These bioprocesses may lead to synthesis of glucosinates, cyanogenic glucosides, alkaloids, nonprotein aminoacids, amines, flavonoids, terpenes, quinoline, indole, pyrrolidine, pyrrolizidine, alkaloids, cumarins, mono-, sesqui-, and triterpenes.

Some of the genes that encode biosynthetic enzymes have already been isolated and characterized. Wink (2008) consider question of when, where, and how the plant genes evolved that encode enzymes of SM biosynthesis, as well as those of transport, storage, and turnover.

H. Dahm (✉) and P. Golińska
Faculty of Biology and Earth Sciences, Department of Microbiology, Institute of General and Molecular Biology, University of Nicolaus Copernicus, 87-100 Toruń Gagarina 9, Poland
e-mail: dahm@biol.uni.torun.pl

M. Rai and A. Varma (eds.), *Diversity and Biotechnology of Ectomycorrhizae*, Soil Biology 25, DOI 10.1007/978-3-642-15196-5_16,

Theoretically some scenarios can be considered:

- Secondary metabolism could be a young phenomenon and modern plants have developed their pathways independently.
- Alternatively, secondary metabolism is an old innovation, which was developed early in the evolution of land plants and was inherited by modern plants.
- Plants could have developed the genes of SM from their own genes of primary metabolism. Starting with duplication of a gene, the new gene became mutated, exhibited new metabolic functions and was established by natural selection.
- Plant might have inherited some of the genes in early evolution by horizontal gene transfer from their bacterial symbionts, which later developed into modern mitochondria and plastids. Bacteria, especially *Actinomyces*, *Streptomyces*, cyanobacteria produce a wide diversity of SM, showing similar structures as plant SM (antraquinones, terpenoids, and alkaloids).

About 80% of modern plants live in symbiosis with fungi (endo-, ectomycorhiza). These fungi could directly have supplied its host with SM or might have transferred (horizontal) the genes to the host's genome.

Environmental factors (biotic and abiotic) control and regulate the biosynthesis of SM in plants (Laitinen et al. 2005; Zhi-lin et al. 2007).

As a defense reaction plants evolved bioactive compounds, which repel, deter, or poison herbivores and which can inhibit growth and development of bacteria, fungi, and viruses. Some of the defense compounds are constitutive, while others can be induced under stress conditions. Several SM (phytoalexins) and defense proteins are synthesized de novo when plant is invaded by microorganisms (Wink 2008).

Type of SM	Estimated numbers[a]
Nitrogen containing SM	
Alkaloids	21,000
Nonprotein amino acids (NPAAs)	700
Amines	100
Cyanogenic glycosides	60
Glucosinolates	100
Alkylamides	150
Lectins, peptides, polypeptides	2,000
SM without nitrogen	
Monoterpenes (C10)[b]	2,500
Sesquiterpenes (C15)[b]	5,000
Diterpenes (C20)[b]	2,500
Triterpenes, steroids, saponins (C30, C27)[b]	5,000
Tetraterpenes (C40)[b]	500
Flavonoids, tannins	5,000
Phenylpropanoids, lignin, coumarins, lignans	2,000
Polyacetylenes, fatty acids, waxes	1,500
Polyketides	750
Carbohydrates, simple acids	400

[a]Approximate number of known structures
[b]Total number of all terpenoides exceeds 22,000 at present

In the plant–microbe interaction, coevolution between plants and their microbial partners are mediated via plant chemical defense. Plant SM usually act as signal molecules or respond to pathogen and symbiont colonization. Mycorrhizal associations are the most important mutualist symbiosis which involve three-way interactions between plants, mycorrhizal fungi, and soil factors. Interactions in the mycorrhizal associations between macro- and microsymbiont in contrast to plant–pathogen interactions are for both profitable.

In presymbiotic phase, plant and their fungal partner secret signals into soil, mostly SM, recognized by roots and mycelium, inducing morphological and physiological modifications.

According to some investigators (Baron and Zambryski 1995; Garcia-Garrido and Ocampo 2002) signal perception and transductions proceed via similar pathways between symbiosis and pathogenesis of plants. However, the defense response in plant-mycorrhizal is probably weak.

The nature of signaling molecules, signal perception, and transduction in mycorrhiza are unknown or mistakenly denied (Martin et al. 2001). In the first stage, host plants release into the rhizosphere metabolites that are able to trigger basidiospore germination, growth of hyphae toward the roots and the early steps of mycorrhizal formation.

According to Kottke and Oberwinkler (1987), Horan et al. (1988), Lagrange et al. (2001), Martin et al. (2001), molecules that control the interactions between symbionts can be classified as follows:

- Tropism of hyphae for root tissues (rhizospheric signals)
- Attachment and penetration of host tissues by hyphae (adhesions, hydrolases)
- Induction of organogenetic programs in both fungal and root cells (hormones and secondary signals)
- Facilitating survival of the mycobiont despite plant defense responses
- Coordinating strategies for exchanging carbon and other metabolites for plant

Signals secreted into the rhizosphere can include flavonoids, terpenes, hormones, and various nutrients. These substances stimulate growth and modified hyphal morphology.

Some of these substances might be produced and released into rhizosphere by bacteria, namely mycorrhization helper bacteria (MHBs). Root exudates enhanced accumulation of fungal molecules such as hypaphorine, the betaine of tryptophan (Martin et al. 2001). This fungal alkaloid is the major indole compound produced in larger amounts by some EM fungi (e.g., *Pisolithus* sp.) during mycorrhiza formation and development (Béguiristain and Lapeyrie 1997; Martin et al. 2001).

Hypaphorine induces morphological changes in root hairs, which lead to a decreased rate of elongation and transitory swelling of the apex of the root hair (Ditengou et al. 2000).

Root hairs are a significant site for microbial interaction in the rhizosphere and it has been suggested that interaction between the EM fungus and root hairs may play a role in the symbiosis development. Growth in root hairs is associated with an apex-high cytosolic free Ca^{2+} gradient generated by a local Ca^{2+} influx at the tip.

Some investigators suggest that hypaphorine-induced cytoskeleton changes are related to interaction with calcium channels, cofilin/actin depolymerizing proteins and auxin signaling pathways (Ditengou et al. 2000).

The aim of this chapter was to point out a presence of various SM released both the micro- and macrosymbiont to the mutual interactions zone.

Any such chemical compounds (both the volatile and nonvolatile ones) affect either positively or negatively the mycorrhiza symbiosis formation and functioning as well as such chemical compounds interact between themselves. The role of some metabolites in these processes is better known (auxins), however any importance of the majority of them is still unknown and requires future, detailed studies.

16.2 Flavonoids

Flavonoids are derived from y-pyrone. They are either 2-phenylbenzopyrone or 3-phenylbenzopyrone. More than 1,300 different flavonoid compounds have been isolated from plants. Individual flavonoids in a group differ from each other by the number and position of the hydroxyl, methoxy, and sugar substituents.

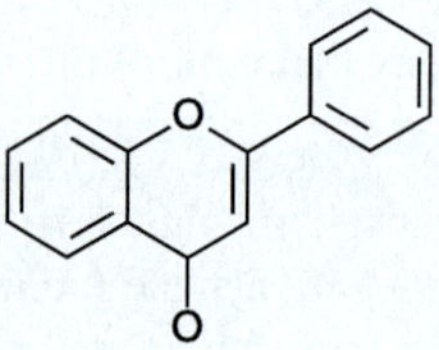

2-Phenyl-1,4-benzopyrone

Flavonoid compounds occur in plants as glycosides, with hexoses such as glucose, galactose and rhamnose, and pentoses such as arabinose and xylose as the most commonly found sugars. The sugars can be attached singly or in combination with each other.

Flavonoids are synthesized by the phenylpropanoid metabolic pathway in which the amino acid phenylalanine is used to produce 4-coumaroyl-CoA. This can be combined with malonyl-CoA to yield of compounds called chalcones, which contain two phenyl rings. Conjugate ring-closure of chalcones results in the familiar form of flavonoids, the three-ringed structure of a flavone.

Flavonoids form a large and heterogeneous group of SM having a bioactive role in the major processes of plants (e.g., attraction of seed disperses, defense reaction against predators, pathogens and abiotic stress condition [Taylor and Grotewold 2005; Niemi et al. 2007]).

Some authors suggest the role of flavonoids in modulating cell signaling pathways including polar auxin transport (Brown et al. 2001; Peer et al. 2004; Kakiuchi et al. 2006).

Despite increasing evidence the role of flavonoids and other phenolic substances in plant development, their role in EM symbioses is contradictory. Studies on the changes in the concentrations flavonoids in Scots pine seedlings during the establishment of the EM symbiosis with *Suillus variegatus* showed that in contrast to shoots, the concentrations of catechin and condensed tannins showed a tendency to decrease in the roots of both noninoculated and inoculated seedlings and regardless of high mycorrhiza frequencies the fungi caused hardly changes in the flavonoid concentrations of the roots.

Weiss et al. (1997, 1999) suggested that catechin and epicatechin accumulated in the inner part of the cortex to prevent the growth of the EM fungus into the inner cortex.

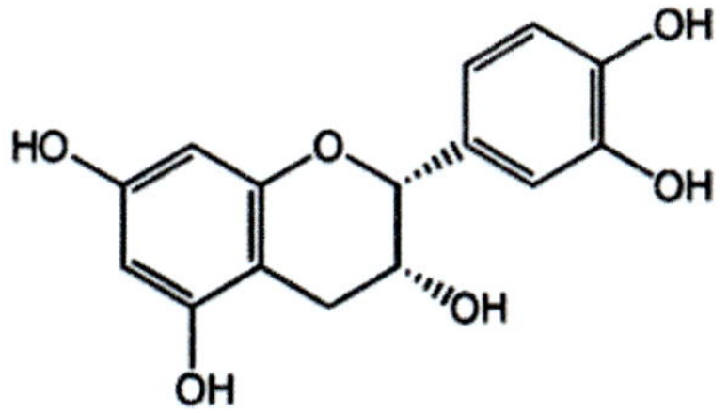

Epicatechin

In contrast, Beyler and Heyser (1997) reported that reduction of catechin and epicatechin in the root tips is a prerequisite for rapid mycorrhization. Similarly, Schützendübel and Polle (2004) showed that Scots pine short root tips covered by the mycelium of *Pisolithus tinctorius* contained less catechin than nonmycorrhizal ones.

16.3 Terpenes

Terpenes are widespread in nature, mainly in plants as constituents of essential oils. Many terpenes are hydrocarbons, but oxygen-containing compounds such as alcohols, aldehydes, or ketones (terpenoids) are also found. Their building block is the hydrocarbon isoprene, $CH_2 = C(CH_3)–CH = CH_2$.

Terpene hydrocarbons are classified according to the number of isoprene units: monoterpenes (2 isoprene units [i.u.]), sesquiterpenes (3 i.u.), diterpenes (4 i.u.), triterpenes (6 i.u.), tetraterpenes (8 i.u.). Examples of monoterpenes are: pinene, nerol, citral, camphor, menthol. Examples of sesquiterpenes are: nerolidol, farnesol. Examples of diterpenes are: phytol, vitamin A_1. Squalene is an example of a triterpene and carotene is a tetraterpene.

When terpenes are modified chemically, such as by oxidation or rearrangement of the carbon skeleton, the resulting compounds are generally referred to as terpenoids. Terpenoids are also known as isoprenoids.

Terpenes are produced by a wide variety of plants, particularly conifers. They are the major components of resin and of terpentine produced from resin. In conifers, biosynthesis and infiltration of tissues with resins are involved in defense system in response to wounding and subsequent inhabitation of the wounds by fungi and insects (Higuchi 1985; Werner 1993; Napierała-Filipiak et al. 2002). The response is nonspecific and similar after wounding and infection.

Members of Pinaceae produce two types of resins – oleoresin and parenchyma resins. Oleoresin is a super saturated solution of resin acids in liquid terpenes and is located in resin ducts and surrounding epithelial cells. Parenchyma or medullary resins are composed mainly of fatty acids (Prior 1976; Napierała-Filipiak et al. 2002).

Pine oleoresin contain 60–70% resin acids. The rest is comprised of volatile and other terpenoids. Among monoterpens, α-pinene, Δ^3-carene, and β-pinene dominate in the volatile fraction (Asiegbu et al. 1998).

The monoterpenes are toxic to wood-fungi, whereas the resin acids display low toxicity and function mainly as mechanical barrier (Prior 1976; Napierała-Filipiak et al. 2002). According to Mekin and Krupa (1971) and Krupa et al. (1973) volatile substances may play a role in the initiation and development of the mycorrhizal symbiosis, because colonization of pine roots by different mycorrhizal fungi results in quantitative changes in the concentration of the individual volatiles.

Phenolics and volatiles are significant factor root exudates and influence the activity of the rhizosphere microorganisms. Due to their volatility, the monoterpenes and some sesquiterpenes can display significant effect on fungi composition structure (Smith 1987). They can regulate competitive or antagonistic interactions and finally create environment stimulating the symbiotic associations.

Forest litter reach in volatile substances may be too a significant factor in the interactions between microorganisms in soil and rhizosphere. Koide et al. (1998) found that á-pinene and â-pinene showed differential effects on the growth of various EM fungi.

Varese et al. (1996) observed enhanced fungal growth due to volatiles; however, the stimulation was seldom significant as some of the substances, when present in sufficient concentration, may cause inhibition of the vegetative growth of mycorrhizal and pathogenic fungi outside the roots.

Differences in the sensitivity of EM fungi to several volatile substances might characterize of their ability to induce host reaction and consequently the ability to initiate symbiosis (Mekin and Krupa 1971; Napierała-Filipiak et al. 2002). Increased production of volatile and nonvolatile substances might be a mechanism of control over growth of mycorrhizal fungi in tissues of macrosymbiont (Molina and Trappe 1982).

Although the major volatiles identified in nonmycorrhizal and mycorrhizal roots of pine, the degree of accumulation of several compounds varied among fungal treatments. This suggests that each mycorrhizal fungus may elicit a different response in trees (Napierała-Filipiak et al. 2002). Also in the previous study by Krupa et al. (1973), the diverse levels of Δ^3-carene and β-phellandrene in roots of *Pinus echinata* inoculated with *P. tinctorius* and *Cenococcum graniforme* were explained in terms of the different ability to elicit a specific host response of the two fungi.

16.4 Plant Growth Regulating Substances (phytohormones)

It is assumed that hormones of plant and fungal origin may take part information and functioning of mycorrhizae (Gogala 1991). Phytohormones, SM synthesis by plant and EM fungi include auxins, cytokinins, GAs, abscisic acid (ABA), ethylene as well as alkaloids and phenylglycoside (Gogala 1991). The soil pool of phytohormones might have partially originated from plants released into the rhizosphere as root exudates and/or synthesized by soil microorganisms. These biomolecules respond to exchange of rhizospheric signals between microorganisms and plants. Signal perception may culminate in the induction of down-stream target gene products whose expressions are physiological and/or development responses (Martin et al. 2001).

16.4.1 Auxins

Several naturally occurring auxins include indole-3-acetic acid (IAA), its halogenated derivatives (4-Cl-IAA), and indole-3 butyric acid (IBA). On molecular level, auxins have an aromatic ring and a carboxylic acid group.

Auxins play an essential role in coordination of many growth and behavioral processes in the plant life cycle. They act in concert with (or opposition) other plant hormones. For example, the ratio of auxin to cytokinin in certain plant tissues determines initiation of root vs. shoot buds.

The plant hormones stimulate cell elongation. Auxin induces new root formation by breaking root apical dominance induced by cytokinins. However, high concentration of auxin inhibits root elongation and instead enhances adventitious root formation. In low concentration, auxin can inhibit ethylene formation and transport of precursors in plant; however, high concentration of auxin can induce the synthesis of ethylene.

Auxin production is widespread among many mycorrhizal fungi (Gay 1986; Gay and Debaud 1987; Frankenberger and Poth 1987; Kampert and Strzelczyk 1989).

Several studies have demonstrated increased auxin content (hyperauxiny) in response to mycorrhizal infection, which may indicate a role of auxin in EM symbiosis. Studies initiated by Slankis (1950) had shown that auxins as well as cytokinins are necessary for the formation of mycorrhizal structures.

Auxins added to the synthetic media inhibited elongation of pine seedlings roots. The roots became thicker and dichotomically branched devoid of roots hairs and caps, structures characteristic for the nonmycorrhizal roots.

IAA

Many studies indicate that changes in auxin balance are a prerequisite for mycorrhiza organogenesis (Gay et al. 1994; Martin et al. 2001). EM fungi enhance proliferation of short roots and the presence of plant-derived tryptophan in the root exudates could be sufficient for EM fungi to enhance the biosynthesis of fungal IAA (Rupp et al. 1989).

Although the structure of ectomycorrhizae in a natural habitat may show considerable variation, common features are a swollen appearance, lack of root hairs, and variable radial growth of cortical cells within the swollen region (Slankis 1973).

The fact that roots are very sensitive to auxins and that auxins take part in many physiological and metabolic processes can be expected that the presence of excess auxin in mycorrhizae would profoundly affect their physiology and metabolism. EM roots morphology reflect a specific physiological and metabolic state which is necessary for the functioning of the symbiosis.

According to Slankis (1973), the hyperauxiny in mycorrhizal roots is more likely to result from the host plant's endogenous auxins than from the fungus auxins. However, seedlings of pine inoculated with mutant strain of *Hebeloma crustuliniforme* that overproduced IAA generated an increased number of EM roots (Gay et al. 1994).

16.4.2 Cytokinins

Cytokinins are N6-substituted aminopurines, including ribosides, ribotides, and glucosides. These are adenine derivatives characterized by their ability to induce cell division in tissue culture in the presence of auxins. The most common cytokinin in plants is zeatin, which is converted to other cytokinins. Over 40 cytokinins have been characterized in plant tissues (McGaw and Burch 1995).

Cytokinins are responsible for the translocation of carbohydrates to the mycorrhizal roots. They also act indirectly on the activity of auxins (Gogala 1991).

Several mycorrhizal fungi have been shown to be capable of producing cytokinins in vitro. However, it is unclear whether these fungi that are capable of producing cytokinins also do so in association with the macrosymbiont (Arshad and Frankerberger 1998). No direct, unequivocal evidence indicates that cytokinins are a prerequisite for the formation of mycorrhizae. However, higher cytokinin levels in mycorrhizal plants have been reported, but the source of increased cytokinin levels in mycorrhizal plants is somewhat unresolved.

Allen et al. (1980) reported higher cytokinin activity in mycorrhizal plants compared with noninfected (control) plants. Similarly, Thiagarajan and Ahmad (1994) reported significantly greater cytokinin content (156%) in mycorrhizal roots compared to nonmycorrhizal roots.

Several other studies confirmed these findings and provided evidence that inoculation with mycorrhizal fungi results in increasing the endogenous cytokinin contents of host plants (Dixon 1989; Danneberg et al. 1992).

In the plant root zone (rhizosphere), there are also plant growth regulators elaborated by microorganisms accompanying mycorrhizae (Strzelczyk and Pokojska-Burdziej 1984) and those originating from the root exudates (Gogala 1991).

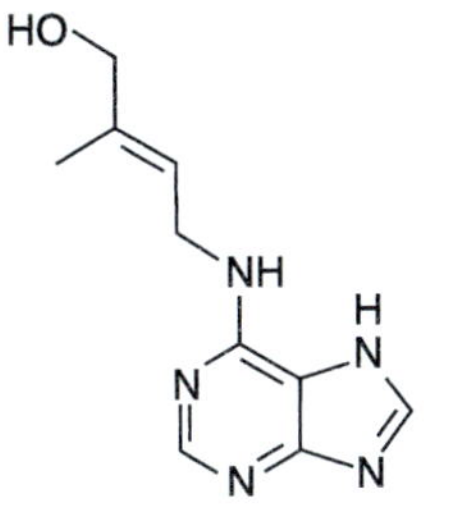

Zeatin

Little is known about the direct effects of these compounds on mycorrhizal fungi. However, Pokojska et al. (1993) showed differences in the effects of plant regulators on mycorrhizal fungi (*H. crustuliniforme*, *Laccaria laccata*, *Rhizopogon vinicolor*) depending upon the kind of hormones, its concentration, and the kind of fungus.

Kinetin inhibited biomass production by *L. laccata* in a liquid medium but it did not inhibit the linear growth of this fungus on agar medium. Reverse results were observed with *R. vinicolor*.

Auxins did not affect the growth of *L. laccata*, but some of them exhibited both inhibitory and stimulatory effects on the growth of *H. crustuliniforme* and *R. vinicolor* depending upon the concentration and type of the medium.

Gogala and Pohleven (1976) have shown that cytokinins promoted the mycelial growth of *S. variegatus* and affected the content of K, Ca, P, and Na in the mycelium of this fungus. In the presence of kinetin, the uptake of Cd, Zn, P by some EM fungi increased significantly (Stegnar et al. 1978).

The importance of auxins and cytokinins in plant growth and development is known. The role of these substances in microorganisms is not elucidated as yet. According to the data obtained from the literature, it can be assumed that auxins and cytokinins do not play the role of hormonal factor in microorganisms (Gogala and Pohleven 1976; Pohleven and Gogala 1986; Gogala 1989; Pokojska et al. 1993).

16.4.3 Gibberellins (GAs)

GAs are tetracyclic diterpenoid acids with an ent-gibberellane ring system. Mevalonic acid is the primary precursor of GAs biosynthesis in plants.

Although the most widely recognized gibberellin is GA3 (gibberellic acid) which is a fungal product, the most active GA in plants is GA1 which is primarily responsible for stem elongation (Arshad and Frankerberger 1998).

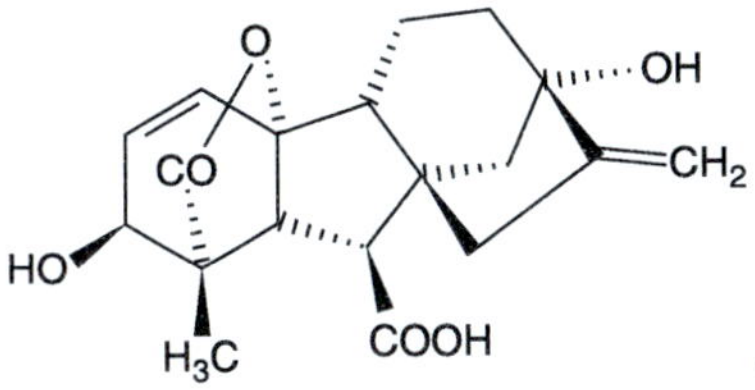

Gibberelic acid

Very little work has been conducted on the detection of GAs released by mycorrhizal fungi. Gogala (1971) detected gibberellin-like substances in culture medium of the EM fungus *Boletus edulis* and Ho (1987) in culture of *P. tinctorius*. Strzelczyk et al. (1975) found gibberellin-like substances produced by *Suillus bovinus*, *H. crustuliniforme*, and *C. graniforme*.

16.4.4 Ethylene

Ethylene is synthesized from methionine in many plant tissues, mostly in response to stress (Arshad and Frankerberger 1998). It is the only hydrocarbon (C_2H_4) with a pronounced effect on plants and is involved in developmental processes, from germination of seeds to senescence of various organs.

$$H_2C{=}CH_2$$

Graham and Linderman (1980) found that EM fungi produced ethylene when grown in medium containing methionine. DeVries et al. (1987) noted an apparent correlation between C_2H_4 production and morphological effects, such as stimulation of lateral root formation by mycorrhizal fungi.

16.4.5 Abscisic Acid

ABA is a sesquiterpene, derived from mevalonic acid. ABA appears to act as much as a promotor (e.g., storage protein), as an inhibitor, and a more open attitude toward its overall role in plant development is warranted (Davies 1995).

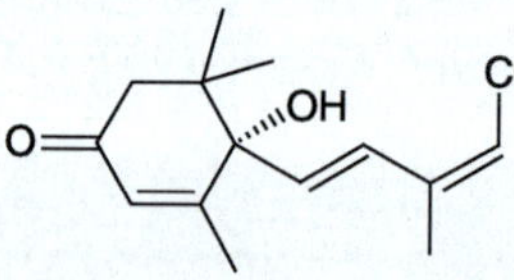

Abscisic acid (ABA)

Production of ABA by mycorrhizal fungi has not been demonstrated as yet; however, few studies have investigated the alteration in ABA levels in mycorrhizal-infected plants.

16.5 Sterols

Sterols play an essential role in the physiology of eukaryotic organisms. Sterols are also known as steroid alcohols. They are a subgroup of steroids with a hydroxyl group. They are amphipathic lipids synthesized from acetyl-coenzymes.

Ergosterol is a component of fungal cell membranes, serving the same function that cholesterol serves in animal cells. Ergosterol is used as an indicator of fungal biomass in soil.

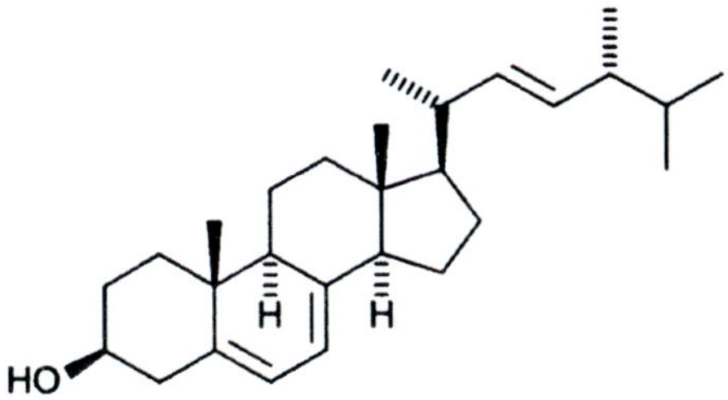

Ergosterol

The composition of fatty acids and sterols in soil lipid fraction is often used as an indicator for the changes of soil microorganisms.

Laczko et al. (2004) performed an experiment in which seedlings of *Pinus sylvestris* and EM fungus *P. tinctorius* were grown separately or combined to form ectomycorrhiza. Fatty acids of the neutral lipid fraction (NLFAs) and the phospholipids fraction (PLFAs) as well as sterol were identified. When grown separately, the two organisms differed strongly with respect to the sterol composition. Sterols had a much higher relative abundance in the fungus in comparison with the plant and the two main fungal sterols, ergosterol and 24-ethyllanosta-8,24(24)-diene-3 beta, 22 zeta-chiol (Et lano 8.24) as well as six minor fungal sterols were not found in the plant roots. When the fungus and plant were brought together, there was a drastic change in the lipid composition of the root.

It was detected that in symbiosis, the fungus transports plant lipids from the symbiotic interface to the extramatrical mycelium. Concerning sterols, the extramatrical mycelium acquired only a small amount of plant-specific sterols. However, its ergosterol content steadily decreased whereas the content of Et lano 8,24 remained high, causing the ratio of these two sterols to decrease from 1:70 to 1:20, whereas in the EM roots, the opposite phenomenon occurred, so that the ratio increased to a value of almost 1:1.

These results showed that an EM fungus may display markedly different lipid composition in its intraradical and extraradical part and highlight a potential role of plant lipid transfer from the root to the fungus in the functioning the of EM symbiosis.

16.6 Conclusions and Future Perspectives

Production of SM is widespread among plants and rhizosphere microorganisms including mycorrhizal fungi. However, conditions in the rhizosphere are often quite variable and many factors, mainly availability of nutrients composition and amount of root exudates as well as interaction between rhizosphere microorganisms, can affect the synthesis of SM by plants and microorganisms associated with plant root.

Research on the role of SM in the initiation development function of mycorrhiza is mainly concerned with two tasks: the determination of their role in the metabolism, growth, and development of the mycorrhizal fungi and the determination of their role in root morphology in the growth of the entire plant and in causing metabolic changes in plants.

Widespread ability of mycorrhizal fungi to produce plant hormones in culture media and induction of mycorrhizal-like changes in response to exogenous application of plant hormones favor the speculation that fungal hormones may have a role in establishment of the symbiotic relation and in the physiology of mycorrhizal plants.

However, physiology of hormones released by the microsymbiont and the role of these metabolites in the symbiotic association are still poorly understood. Very little is known about the molecules regulating the interaction between plants and EM fungi during root colonization.

The role of fungal auxin in ectomycorrhiza has repeatedly been suggested and questioned, suggesting that, if fungal auxin controls some steps of colonized root development, its activity might be tightly controlled in time and in space by plant and/or fungal regulatory mechanisms.

Increase in auxin synthesis or auxin accumulation was noted in most plant–microbe interactions in plant tissues. However, in some interactions (e.g., *P. tinctorius* and *Eucalyptus*) downregulation of the auxin activity in the host plant was observed.

It is assumed that hypaphorine (betaine of tryptophan) might be the specific IAA antagonist.

Despite increasing evidence, the role of flavonoids and other phenolic substances in EM symbiosis is yet contradictory. The change in the balance of plant hormones and other SM have yet to be examined for plant development and first signal for initiation of mutualistic symbiosis. Understanding of these problems could be of great ecological benefit to the agriculture and forestry industry.

A large number of studies have verified that multiplicity of signals and diversity of signaling pathways exist during the establishment of mycorrhizal associations with regulation of symbiosis-specific genes expression.

In presymbiotic phase, plant and their fungal partner secrete signals into soil, mostly SM inducing morphological and physiological changes. The nature of the signals released by the EM symbionts and processes triggering the expression of genes that participate and regulate symbiosis in partner recognition are only the beginning to be understood. Although many genes have been identified in various EM association, the product of which play role in recognition and attachment of the mycobiont on the root surface remains unknown. Many questions concerning the differentiation of plant and fungal symbiotic structure are also poorly recognized. It is interesting to be analyzed how is elicitor's signal achieved depending on the activation factors and which substances participate in this signaling network.

References

Allen MF, Moore TS, Christensen M (1980) Phytohormone changes in *Bouteloua gracilis* infected by vesicular-arbuscular mycorrhizae: I. Cytokinin increases in the host plant. Can J Bot 58: 371–374

Arshad M, Frankerberger WT (1998) Plant growth-regulating substances in the rhizosphere: microbial production and functions. Advan Agron 62:45–151

Asiegbu FO, Johansson M, Woodward S, Hütterman A (1998) Biochemistry of the host–parasite interaction. In: Woodward S, Stenlid J, Karialajnen R, Hütterman A (eds) *Heterobasidion annosum*: Biology, ecology, impact and control. University Press, Cambridge, pp 167–193

Baron C, Zambryski PC (1995) The plant response in pathogenesis, symbiosis, and wounding: variations on a common theme? Ann Rev Genet 29:107–129

Béguiristain T, Lapeyrie F (1997) Host plant stimulates hypaphorine accumulation in *Pisolithus tinctorius* hyphae during ectomycorrhizal infection while excreted fungal hypaphorine controls root hair development. New Phytol 136:525–532

Beyler M, Heyser W (1997) The influence of mycorrhizal colonization on growth in the greenhouse and on catechin, epicatechin and procyanidin in roots of *Fagus sylvatica* L. Mycorhhiza 7:171–177

Brown DE, Rashotte AM, Murphy AS, Normanly J, Tague BW, Peer WA, Taiz L, Muday GK (2001) Flavonoids act as negative regulators of auxin transport in vivo in *Arabidopsis*. Plant Physiol 126:524–535

Danneberg G, Latus C, Zimmer W, Hundeshagen B, Schneider-Poetsch H, Bothe H (1992) Influence of vesicular-arbuscular mycorrhiza on phytohormone balances in maize (*Zea mays* L.). J Plant Physiol 141:33–39

Davies PJ (1995) The plant hormones: their nature, occurence, and functions. In: Davies PJ (ed) Plant hormones: physiology, biochemistry and molecular biology. Kluwer Academic Publishers, Dordrecht, The Netherlands, pp 1–12

DeVries HE, Mudge KW, Lardner JP (1987) Ethylene production by several ectomycorrhizal fungi and effects on host root morphology. In: Sylvia EM, Hung LL, Graham JH (eds) Mycorrhizae in the next decade. Institute of Food and Agricultural Sciences, University of Florida, Gainesville, p 245

Ditengou FA, Béguiristain T, Lapeyrie F (2000) Root hair elongation is inhibited by hypaphorine, the indole alkaloid from the ectomycorrhizal fungus *Pisolithus tinctorius*, and restored by IAA. Planta 211:722–728

Dixon RK (1989) Cytokinin activity in *Citrus jambhiri* seedlings colonized by mycorrhizal fungi. Agric Ecosystems Environ 29:103–106

Frankenberger WT, Poth M (1987) Biosynthesis of indole-3-acetic acid by the pine ectomycorrhizal fungus *Pisolithus tinctorius*. Appl Environ Microbiol 53:2908–2913

Garcia-Garrido JM, Ocampo JA (2002) Regulation of the plant defence response in arbuscular mycorrhizal symbiosis. J Exp Bot 53:1377–1386

Gay G (1986) Effect of glucose on indole-3-acetic production by the ectomycorrhizal fungus *Hebeloma hiemale* in pure culture. Physiol Veg 24:185–192

Gay G, Debaud JC (1987) Genetic study on indole-3-acetic acid production by ectomycorrhizal *Hebeloma* species: inter- and intraspecific variability in homo- and dikaryotic mycelia. Appl Microbiol Biotechnol 26:141–146. doi:10.1007/BF00253898

Gay G, Normand L, Marmeisse R, Sotta B, Debaud JC (1994) Auxin overproducer mutants of *Hebeloma cylindrosporum* Romagnési have increased mycorrhizal activity. New Phytol 128: 645–657

Gogala N (1971) Growth substances in mycorrhiza of the fungus *Boletus pinicola* Vitt. and the pine-tree *Pinus sylvestris* L. Razprave 14:123–202

Gogala N (1989) Growth substances in root exudate *Pinus sylvestris* – their influence on mycorrhizal fungi (effect of jasmonic acid). Agric Ecosystems Environ 28:151–154

Gogala N (1991) Regulation of mycorrhizal infection by hormonal factors produced by hosts and fungi. Experimentia 47:331–339

Gogala N, Pohleven F (1976) The effect of cytokinins and auxins on the growth of mycorrhizal fungus *Suillus variegatus*. Acta Bot Croat 35:129–134

Graham JH, Linderman RG (1980) Ethylene production by ectomycorrhizal fungi, *Fusarium oxysporum* f. sp. *pini*, and by aseptically synthesized ectomycorrhizae and *Fusarium*-infected Douglas fir roots. Can J Microbiol 26:1340–1347

Higuchi T (1985) Biosynthesis and biodegradation of wood components. Academic Press, Orlando, p 679

Ho I (1987) Comparison of eight *Pisolithus tinctorius* isolates for growth rate, enzyme activity, and phytohormone production. Can J For Res 17:31–35

Horan DP, Chilvers GA, Lapeyrie F (1988) Time sequence of the infection process in eucalypt ectomycorrhizas. New Phytol 109:451–458

Kakiuchi Y, Gàlis I, Tamogami S, Wabiko H (2006) Reduction of polar auxin transport in tobacco by the tumorigenic *Agrobacterium tumefaciens* AK-*6b* gene. Planta 223:237–242

Kampert M, Strzelczyk E (1989) Effect of amino acids on cytokinin-like substances production by micorrhizal fungi of pine (*Pinus sylvestris* L.). Agric Ecosyst Environ 28:219–228

Koide RT, Suomi L, Stevens CM, McCormick L (1998) Interactions between needles of *Pinus resinosa* and ectomycorrhizal fungi. New Phytol 140:539–547

Kottke I, Oberwinkler F (1987) The cellular structure of the Hartig net: coenocytic and transfer cell-like organization. Nordic J Bot 7:85–95

Krupa S, Andersson J, Marx DH (1973) Studies on ectomicorrhizae of pine IV. Volatile organic constituents in micorrhizal and nonmycorrhizal root systems of Pinus echinata Mill. Eur J For Pathol 3:194–200

Laczko E, Boller T, Wiemken V (2004) Lipids in roots of *Pinus sylvestris* seedlings and in mycelia of *Pisolithus tinctorius* during ectomycorrhiza formation: changes in fatty acid and sterol composition. Plant Cell Environ 27:27–40

Lagrange H, Jay-Allemand C, Lapeyrie F (2001) Rutin, the phenolglycoside from *Eucalyptus* root exudates, stimulates *Pisolithus* hyphal growth at picomolar concentrations. New Phytol 150:349–355

Laitinen ML, Julkunen-Tiitto R, Tahvanainen J, Heinonen J, Rousi M (2005) Variation in birch (*Betula pendula*) shoot secondary chemistry due to genotype, environment, and ontogeny. J Chem Ecol 31:697–717

Martin F, Duplessis S, Ditengou F, Lagrange H, Voiblet C, Lapeyrie F (2001) Developmental cross talking in the ectomycorrhizal symbiosis: signals and communication genes. New Phytol 151:145–154

McGaw BA, Burch LR (1995) Cytokinin biosynthesis and metabolism. In: Davies PJ (ed) Plant hormones: physiology, biochemistry and molecular biology. Kluwer Academic Publishers, Dordrecht, The Netherlands, pp 98–117

Mekin E, Krupa S (1971) Studies on ectomicorrhizae of pine. II. Growth inhibition of mycorrhizal fungi by volatile organic constituents of *Pinus silvestis* L. (Scots pine) roots. Physiol Plantarum 25:337–340

Molina E, Trappe JM (1982) Patterns of ectomycorrhizal host specificity and potential among pacific northwest conifers and fungi. Forest Sci 28:423–458

Napierała-Filipiak A, Werner A, Mardarowicz M, Gawdzik J (2002) Concentrations of terpenes in micorrhizal roots of Scots pine (*Pinus sylvestris* L.) seedlings grown in vitro. Acta Physiol Plant 24:137–143

Niemi K, Julkunen-Tiitto R, Häggman H, Sarjala T (2007) *Suillus variegatus* causes significant changes in the content of individual polyamines and flavonoids in Scots pine seedlings during mycorrhiza formation in vitro. J Exp Bot 58:391–401. doi:10.1093/jxb/erl209

Peer WA, Bandyopadhyay A, Blakeslee AA, Makam SN, Chen RJ, Masson PH, Murphy AS (2004) Variation in expression and protein localization of the PIN family of auxin efflux facilitator proteins in flavonoid mutatnts with altered auxin transport in *Arabidopsis thaliana*. Plant Cell 16:1898–1911

Pohleven F, Gogala N (1986) The influence of natural cytokinins on the content of K, P, Ca and Na in the mycelium of the mycorrhizal fungus *Suillus variegatus*. Biol Vestn 34:79–88

Pokojska A, Strzelczyk E, Li CY, Rozycki H, Szablewska M (1993) Effect of plant growth regulators on growth of ectomycorrhizal fungi. Cryptogam Bot 4:8–13

Prior C (1976) Resistance by Corsican pine to *Fomes annosus*. Ann Bot 40:261–279

Rupp LA, Mudge KW, Negm FB (1989) Involvement of ethylene in ectomycorrhiza formation and dichotomous branching of roots of mugo pine seedlings. Can J Bot 67:477–482

Schützendübel A, Polle A (2004) Plant responses to abiotic stress: heavy metal-induced oxidative stress and protection by mycorrhization. J Exp Bot 53:1351–1365

Slankis V (1950) Effect of α napthaleneacetic acid on dichotomus branching of isolated roots of *Pinus silvestris*. Physiol Plantarum 3:40

Slankis V (1973) Hormonal relationships in mycorrhizal development. In: Marks GC, Kozlowski TT (eds) Ectomycorrhizae: their ecology and physiology. Academic Press, New York, pp 231–298

Smith WH (1987) The atmosphere and the rhizosphere: linkages with potential significance for forest tree health. In: Blasser RO (ed) Technical Bulletin of national council of the paper industry for air and stream improvements, vol. 527. New York, pp 30–94

Stegnar P, Gogala N, Pohleven F (1978) The uptake of cadmium, zinc, phosphorus, and plant hormone kinetin by ectomycorrhizal fungi. Acta Bot Croat 37:67–73

Strzelczyk E, Pokojska-Burdziej A (1984) Production of auxins and gibberellin-like substances production by mycorrhizal fungi, bacteria and actinomycetes isolated from soil and mycorrhizosphere of pine (*Pinus sylvestris* L.). Plant Soil 81:185–194

Strzelczyk E, Sitek JM, Kowalski S (1975) Production of gibberelin-like substances by fungi isolated from mycorrhizae of pine (*Pinus silvestris* L.). Acta Microbiol Pol 7:145–153

Taylor LP, Grotewold E (2005) Flavoniods as developmental regulators. Curr Opinion Plant Biol 8:317–323

Thiagarajan TR, Ahmad MH (1994) Phosphatase activity and citokinin content in cowpeas (*Vagna unguiculata*) inoculated with a vesicular-arbuscular mycorrhizal fungus. Biol Fertil Soils 17:51–56

Varese TC, Portinaro S, Trotta A, Scannerini S, Luppi-Mosca AM, Martinotti MG (1996) Bacteria associated with *Suillus grevillei* sporocarps and ectendomycorrhizae and their effects on in vitro growth of the mycobiont. Symbiosis-Rehovot 21:113–129

Weiss M, Mikolajewski S, Peipp H, Schmitt U, Schmidt J, Wray V, Strack D (1997) Tissue-specific and developmental-dependent accumulation of phenylpropanoids in larch mycorrhizas. Plant Physiol 114:15–27

Weiss M, Schmidt J, Neumann D, Wray V, Crist R, Strack D (1999) Phenylpropanoids in mycorrhizas of the *Pinaceae*. Planta 208:491–502

Werner A (1993) Pathological anatomy of thin woody roots of Scots pine invaded by *Heterobasidion annosum* (Fr.) Bref. Arboretum Kórnickie 38:113–129

Wink M (2008) Plant secondary metabolism: diversity, function and its evolution. Nat Prod Communicat 3:1205–1216

Zhi-lin Y, Chuan-chao D, Lian-qing C (2007) Regulation and accumulation of secondary metabolites in plant-fungus symbiotic system. Afr J Biotechnol 6:1266–1271

Chapter 17
C:N Interactions and the Cost:Benefit Balance in Ectomycorrhizae

Ana Corrêa and Maria-Amélia Martins-Loução

17.1 Introduction

Mycorrhizal (M) symbioses are known to improve plant nutrient uptake due to the fine exploration of the substrate by the fungal hyphae, while in return the plant supplies the fungus with carbohydrates that are essential for the completion of its life cycle.

The fungal hyphae constitute an extension of the nutrient absorption surface since they penetrate the soil more extensively than the root. The nutrients can be translocated through the depletion zones surrounding the roots, and inaccessible sources of nutrients can be mobilized. Due to their high surface-to-volume ratio and capacity for rapid exponential growth, fungal hyphae are superior to roots in the acquisition of immobile nutrient ions from the low and both spatially and temporally variable nutrient concentrations in natural soils. Fungi may also have access to other nutrient sources than plant roots, namely through production of degradative extracellular enzymes or organic acids (Leake and Read 1990; Zhu et al. 1994).

In particular, the role of ECM in improving the nitrogen (N) nutrition of the host plant is of central importance in this association, as ECM fungi evolved, and ECM interactions predominate, in N-limiting ecosystems (Martin and Botton 1993; Read and Pérez-Moreno 2003).

At the other end of the mycorrhizal partnership, the fungus is also dependent on the carbon (C) supply by the plant. A substantial part of the C fixed through

A. Corrêa (✉)
Departamento de Microbiología del Suelo y Sistemas Simbióticos, Estación Experimental del Zaidín, CSIC, Profesor Albareda 1, 18008 Granada, Spain
and
Universidade de Lisboa, Faculdade de Ciências, Centro de Biologia Ambiental, Campo Grande, 1749-016 Lisboa, Portugal
e-mail: ambcorrea@gmail.com

M.-A. Martins-Loução
Universidade de Lisboa, Faculdade de Ciências, Centro de Biologia Ambiental, Campo Grande, 1749-016 Lisboa, Portugal

M. Rai and A. Varma (eds.), *Diversity and Biotechnology of Ectomycorrhizae*, Soil Biology 25, DOI 10.1007/978-3-642-15196-5_17,

photosynthesis is allocated to the fungus, for formation, maintenance, and functioning of mycorrhizal structures (Jones et al. 1991; Wu et al. 2002; Heinemeyer et al. 2006). Host-derived carbohydrates are necessary for the mycorrhizal fungus as precursors for the synthesis of fungus-specific carbohydrates and sugar-alcohols like trehalose, glycogen or mannitol (Fajardo-López et al. 2007; Deveau et al. 2008), and fruit-body production has been shown to be dependent on the current photoassimilates provided by the host plant (Lamhamedi et al. 1994).

This exchange of nutrients for C is the very core of the association. Unraveling the C and nutrient fluxes between symbionts is therefore essential for its understanding.

17.2 Nitrogen

The ability of ECM fungi to take up inorganic nitrogen and transport nitrogen-containing solutes to their host plant is well established (Chalot and Brun 1998; Chalot et al. 2002). M roots and external hyphae have been found to have more transporters and higher N uptake rates than nonmycorrhizal (NM) roots (Javelle et al. 1999; Selle et al. 2005).

More specifically, ECM fungi can help increase ammonium (NH_4^+) uptake. Most of the tree species present in ECM-dominated ecosystems, as well as most ECM fungi, prefer NH_4^+ to nitrate (NO_3^-) as N source, presenting higher NH_4^+ uptake rates and growth than the ones measured for NO_3^- (Eltrop and Finlay et al. 1992; Marschner 1996; Anderson et al. 1999; Plassard et al. 2000; Fig. 17.1). The ECM fungus *Laccaria bicolor* was found to have a widely expanded NH_4^+ transporter family (AMT) when compared with other basidiomycetes, indicating a higher potential for its uptake (Lucic et al. 2008). Correspondingly, the uptake of NH_4^+ is generally improved in ECM plants, whereas the same has less frequently been observed for NO_3^- (Eltrop and Marschner 1996; Plassard et al. 2000). The importance of the external hyphae as NH_4^+-absorbing structures in ECM roots has been demonstrated. Hyphal NH_4^+ acquisition was observed to contribute with 45–73% of total plant N uptake under N deficiency (Brandes et al. 1998; Jentschke et al. 2001), and the ^{15}N ammonium uptake by birch seedlings was greatly decreased when the external mycelium was disrupted (Javelle et al. 1999).

Organic N may also play an important part in N nutrition of both ECM fungi and their host plants. In pure culture, some ECM fungi have been shown to readily assimilate amino acids (a.a.) and amides that appear to occur in significant amounts in soil solution such as glutamine, glutamate, and alanine (Anderson et al. 1999; Tibbett et al. 2000; Fig. 17.1). Recent evidence has been reported that this is also the case under natural conditions (Treseder et al. 2008). Genes encoding a.a. transporters have been isolated in *A. muscaria* (Nehls et al. 1999) and *Hebeloma cylindrosporum* (Wipf et al. 2002), and evidence has been obtained of the existence of a proton symport mechanism for a.a. uptake in *Paxillus involutus* (Chalot et al. 1996),

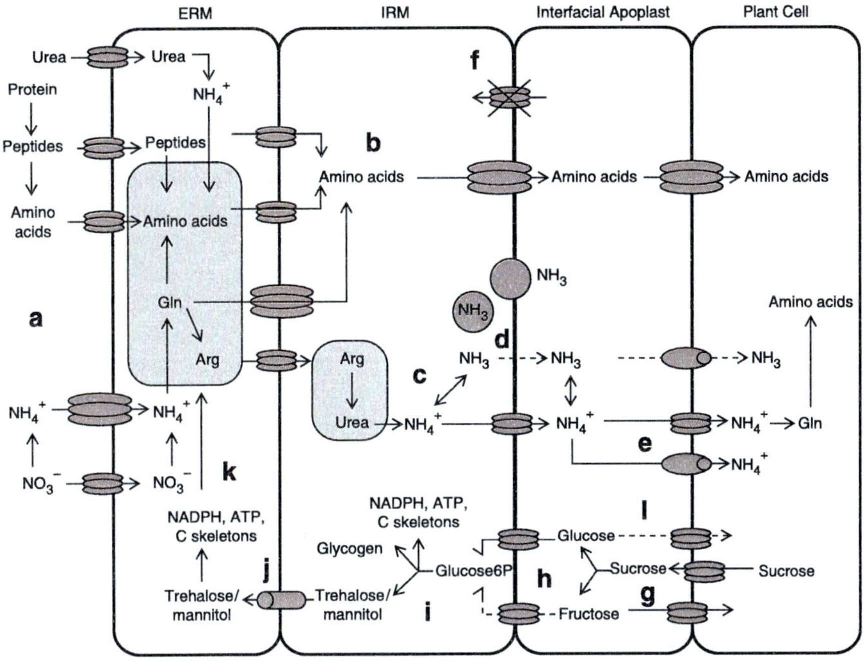

Fig. 17.1 Possible N and C transfer mechanisms, and their interactions. (**a**) Most ECM fungi prefer NH_4^+ as N source, which is taken up through plasma membrane fungal Amts. They can also take and use NO_3^- and several forms of organic N. Evidence has been obtained of a proton symport mechanism for a.a. uptake (Chalot et al. 1996), general a.a. permeases (Nehls et al. 1999; Chalot et al. 2002, 2006), and transporters of a.a. (Nehls et al. 1999; Wipf et al. 2002), peptides (Bendjia et al. 2006) and urea (Morel et al. 2008). (**b**) a.a. are thought to be the main sink for the assimilated N, and the N form that is transported inside the mycelium and to the mantle, where the absorption by the root system takes place. No transport mechanisms have been identified so far, but a.a. transfer from the fungus to the plant has been confirmed (e.g., Blaudez et al. 2001). (**c**) The direct transfer of arginine-derived NH_4^+ has also been suggested (Selle et al. 2005; Chalot et al. 2006). (**d**) Transfer of NH_4^+-derived NH_3 has also been hypothesized (Chalot et al. 2006). NH_3 transfer to the apoplast may happen through passive diffusion, driven by a favorable concentration gradient, or through ammonia-loaded vesicles which would fuse with the plasma membrane and release ammonia into the interfacial apoplast. (**e**) Transfer of NH_4^+ from the apoplast to the plant host cytoplasm might involve plant Amts or nonspecific channels such as aquaammoniaporins. (**f**) The number of fungal a.a. and NH_4^+ transporters may be decreased at the plasma membrane in symbiotic hyphae compared with those in extraradical hyphae. However, the fungus may also regulate the supply of N to the plant by increasing or decreasing the number of transporters, and the competition with the plant by N reuptake. (**g**) Sucrose is supposed to be carbohydrate released by the plant into the interfacial apoplast. The mechanism is still unknown. Sucrose is then hydrolysed into monosaccharides, either by plant-derived cell wall invertases, or nonenzymatic inversion due to the acidity of the apoplast. (**h**) From the resulting glucose and fructose, the fungus preferentially uses glucose. (**i**) Hexose phosphates are utilized for generation of energy/carbon skeletons or storage and transport compounds (trehalose/mannitol or glycogen) (Deveau et al. 2008). (**j**) Trehalose/mannitol are presumably used for long-distance carbohydrate transport towards ERM hyphae. In basidiomycetes, transport is supposed to be enabled by dolipore-guarded cell–cell connections. (**k**) The energy/carbon skeletons obtained from trehalose/mannitol in the ERM are, among other uses, used in N uptake and assimilation, which will partly be channeled towards the

and general a.a. permeases in *P. involutus* and *A. muscaria* (Chalot et al. 2002, 2006, Nehls et al. 1999).

Some fungal species have also been shown to be able to use peptides or protein as N source, namely through production of proteinases (Zhu et al. 1994; Anderson et al. 1999; Bendjia et al. 2006; Fig. 17.1). Recently, an urea transporter was isolated in *P. involutus* (Morel et al. 2008), and peptide transporters have been identified in *H. cylindrosporum* and shown to mediate dipeptide uptake (Bendjia et al. 2006).

Mycorrhization was observed to enhance the use of a.a., peptides or protein by the host plant (Blaudez et al. 2001; Finlay et al. 1992; Tibbett et al. 2000) or to enable its use in plants that are unable to do so when nonmycorrhizal (Abuzinadah and Read 1989). The uptake of ^{14}C glutamate by ECM plants has been observed to increase by up to eight times (Blaudez et al. 2001). In addition, expression and activity of the enzyme L-amino acid oxidase, with a potential role in N mineralization from a.a. at the ecosystem level, has been found in ECM fungi (Nuutinen and Timonen 2008).

17.2.1 N Assimilation, Transport, and Transference to the Plant

Following its uptake by the fungus, NH_4^+ is not accumulated as a free ion in the mycelial tissue but quickly assimilated at considerable distances of the fungal mantle. Several studies suggest a.a. as the main sink for the assimilated N, and the N form that is transported inside the mycelium and to the mantle, where the absorption by the root system takes place (Finlay et al. 1988; Ek et al. 1994; Fig. 17.1).

N assimilation in ECM fungi, similarly to plants, can be made through the same two main and alternative pathways: the NADP-glutamate-dehydrogenase (GDH, E.C. 1.4.1.4) and the glutamine synthetase (GS, E.C. 6.3.1.2)/glutamate synthase (GOGAT, E.C. 1.4.1.13) pathways. Both can be present in ECM fungi, the relative importance of each differing according to fungal species, the physiological state of the mycelium, and the mycorrhizal partner (Martin and Botton 1993; Rudawska et al. 1994). The uptake of both NH_4^+ and a.a. has been reported to involve GS/GOGAT, and this seems to be the main assimilation pathway similarly to what happens in plants (Wright et al. 2005; Morel et al. 2006).

Fig. 17.1 (continued) plant partner. (l) Rhizodermal cells can compete with fungal hyphae for glucose and/or can restrict fructose content in the common apoplast by hexose uptake. This competition for hexoses has been proposed to increase if the fungus fails to supply the plant with adequate amounts of nutrients, and therefore serve as a regulatory mechanism linking C transfer to the fungus with the fungal N supply (Nehls 2008). Modified from Chalot et al. (2006) and Nehls et al. (2010). *ERM* extraradical mycelium; *IRM* intraradical mycelium; *Gln* glutamine; *Arg* arginine

The main form of N transferred to the host plant is glutamine, followed by glutamate, alanine, and aspartate-asparagine (Finlay et al. 1988, 1992; Blaudez et al. 1998, 2001). The extension of the transference of other N compounds through the mantle is not clear, but the direct transfer of NH_4^+, arginine derived, as hypothesized for AM symbiosis has also been suggested for ECM (Chalot et al. 2006; Fig. 17.1). In support of this hypothesis, a NH_4^+ importer of poplar and the *L. bicolor* NH_4^+ transporter LbAMT2.2 were found to be highly upregulated in ECM roots and in symbiotic tissues, respectively (Couturier et al. 2007; Selle et al. 2005). The transfer of NH_3 has also been suggested (Chalot et al. 2006).

Both organic and inorganic N transfer may coexist. It has been suggested that the transfer of organic N may be larger in conditions of C sufficiency, when larger incorporation of NH_4^+ into C skeletons may be expected, and that transfer of N may be increased under C deficiency (Chalot et al. 2006).

The physiological processes involved in the transfer of N from fungus to plant within the symbiotic tissue are still poorly understood. N transporters from plant and fungal partners have been isolated (Javelle et al. 2003a, b; Lucic et al. 2008), but their role and regulation is still under investigation, as is their localization in mycorrhizal tissues (Müller et al. 2007; Fig. 17.1). On the other hand, little is known about the metabolic zonation of N assimilation in ECM. The GS-GOGAT cycle may take place entirely in the fungus, and the newly formed glutamine be exported into the plant where it is converted into the remaining amino acids. It may, however, also be separated, and GS could be active in the fungus and GOGAT in the root, in which case there could be a glutamine–glutamate shuttle across the interface (Martin and Botton 1993).

17.3 Carbon

ECM associations have a deep effect in plant C metabolism, generally increasing the C allocation to the roots (e.g., Jones et al. 1991; Rygiewicz and Andersen 1994; Wu et al. 2002), although this is not always the case (Jones et al. 1998). An important percentage of this C is allocated to mycorrhizae and fungal mycelia, indicating that the fungal partner is an important sink for the hosts' carbohydrates (Wu et al. 2002). Furthermore, mycorrhizae may impose a higher C cost on their hosts under field conditions than under laboratory conditions (Rygiewicz and Andersen 1994). Correspondingly, the expression of hexose importers from ECM fungi was found to be enhanced in symbiotic tissues, indicating enhanced fungal hexose uptake in symbiosis, and to be sugar-regulated, being activated by increased in vivo concentration of monosacharides at the plant–fungus interface (reviewed by Nehls 2008), while the expression of hexose transporters in spruce (Nehls et al. 2000) and birch (Wright et al. 2000) root cells was decreased in symbiosis.

As a result, mycorrhization can upregulate the host plant's rate of net photosynthesis and sucrose synthesis (Dosskey et al. 1990; Colpaert et al. 1996; Loewe et al. 2000). Mycorrhizal effects on photosynthesis are, however, very varied, and in some cases an explanation for them has not been found (Dosskey et al. 1990).

17.3.1 C Transfer and Use by the Fungus

Sucrose is the main carbohydrate transported by the host plant, but ECM fungi are unable to use it as C source. Its conversion into glucose and fructose seems to be a precondition for its use by the fungus. Hexoses are the main carbon compounds taken and used by ECM fungi, while the significance of other C sources is unclear (Buscot et al. 2000; Nehls 2008; Fig. 17.1). ECM fungi are able to take up and use both glucose and fructose, although they show a preference for glucose (Hampp et al. 1995; Wiese et al. 2000). The hexose transporters from ECM fungi investigated so far clearly favor glucose over fructose uptake (Nehls et al. 1998; Polidori et al. 2007), and in *A. muscaria* evidence was found that the same hexose transporters were responsible for carbohydrate uptake in symbiosis (Nehls et al. 1998).

Since so far ECM fungi have been found to have no sucrolytic capacity (Nehls 2008), it has been proposed that sucrolytic enzymes of the host plant are responsible for sucrose hydrolysis. Plants possess two major types of sucrolytic enzymes: invertases and sucrose synthase. Because in ECM the plasma membranes of the plant and fungus do not come into direct contact, and exchanges take place across the apoplast, the activity of cell wall-bound acid invertase has been hypothesized to have a potentially major role in this (Nehls 2008), and therefore expected to increase in M plants. In AM interactions, this has been observed repeatedly (e.g., Schaarschmidt et al. 2006), but not in ECM (Schaeffer et al. 1995; Wright et al. 2000; Corrêa et al. 2010). An alternative to a role of plant invertases could be that nonenzymatic inversion of sucrose takes place, since ECM formation leads to acidification of the already acidic apoplast (Martin and Selosse 2008).

The hexoses taken by the fungus are quickly converted to fungus-specific compounds, therefore maintaining an hexose concentration gradient between plant and fungal tissues, which is thought to maintain the sugar flow toward the fungus (Fajardo-López et al. 2007; Fig. 17.1). Increased carbohydrate export to the root and decreased levels of carbohydrates in ECM plants were observed, while simultaneously the synthesis and contents of fungus-specific compounds such trehalose and manitol, the two main carbohydrates accumulated in fungi, were increased (Hampp et al. 1995; Martin et al. 1998; Fajardo-López et al. 2007; Deveau et al. 2008; Corrêa et al. 2010). Trehalose biosynthesis was found to occur mainly in hyphae of the plant–fungus interface (Fajardo-López et al. 2007; Deveau et al. 2008; Fig. 17.1).

17.4 C–N Interactions

The supply of nutrients by the fungus to the plant is deeply connected to the supply of C to the fungus. The assimilation of inorganic N into a.a. is an important sink for carbohydrates in the fungal mycelium, and therefore dependent and closely regulated by the availability of C metabolites, which are ultimately supplied from host carbohydrates (Martin et al. 1998; Wallenda and Kottke 1998; Fig. 17.1). On the other hand, plant C metabolism is dependent on N uptake by the root, and the fungal partner, and its allocation to the shoot. Just as the integration of the plant C and N metabolisms involve extensive coregulation between the root and the shoot, and any change in activity of one of them implicates an adjustment in the other (Foyer and Noctor 2002), so must the integration of the exchanges of N and C between mycorrhizal partners.

N availability and the plant N needs seem to determinate the amount of C delivered to the fungus. Both the potential capacity of ECM fungi to colonize the roots and the growth of extramatrical mycelium increase with decreasing N availability, indicating that the supply of C to the fungus is higher at low N supply and plant productivity (Wallander and Nylund 1991, 1992; Högberg et al. 2003; Nilsson and Wallander 2003; Treseder 2004; Hobbie 2006; Corrêa et al. 2008, 2010). Because the degree of mycorrhizal colonization was observed to be strongly negatively correlated with shoot N concentration (Wallander and Nylund 1991; Corrêa et al. 2008), it has been proposed that the N status of the plant regulates the C supply to the fungus (Nilsson and Wallander 2003).

Increased N supply causes a switch from gluconeogenesis, that is, sucrose and starch formation, to glycolisis (Wallenda et al. 1996; Wingler et al. 1994), and the decreased sucrose production has been observed to negatively affect C delivery to the fungus, resulting in decreased fungal biomass (Wallenda et al. 1996). Furthermore, evidence has been found of a correlation between the transference of ^{14}C from the plant to the fungus and of ^{15}N from the fungus to the plant at various N availabilities (Kytöviita 2005).

A regulatory mechanism linking N supply to the plant with C transfer from the plant to the fungus has been proposed which assumes that the transport of N into the plant–fungus interface decreases the competition between the fungus and the plant for apoplastic hexoses, and that if the fungus fails to supply the plant with adequate amounts of nutrients, it will reduce the C supply to the fungus (Nehls 2008; Fig. 17.1). This control of the carbohydrate efflux by the host plant in symbiosis has been considered essential to avoid fungal parasitism (Nehls 2008). Several reports, however, indicate that ECM plants allocate more C to the fungus as the nutrient availability decreases, even if the amount of nutrients supplied by the fungus also decreases (Ingestad et al. 1986; Treseder and Allen 2002; Högberg et al. 2003; Hobbie 2006; Corrêa et al. 2008, 2010).

The effects of mycorrhization on C and N metabolisms are, however, not limited to changes in demand and supply of N and C. A readjustment of metabolic pathways has been observed to occur in mycorrhizal roots, starting very early following contact, as indicated by changes in free a.a. profile, and expression and activities of N metabolism enzymes. Namely, molecular and biochemical evidences have been obtained of a deactivation of the root metabolism, and an activation of the fungal metabolism, in response to mycorrhizae formation (Schaeffer et al. 1996; Blaudez et al. 1998; Johansson et al. 2004; Duplessis et al. 2005; Frettinger et al. 2007; Herrmann and Buscot 2007; Corrêa et al. 2010). In some cases, these changes have been considered a consequence of a metabolic shift, with some metabolic functions of the root being taken over by the fungus (Vézina et al. 1989; Johansson et al. 2004), namely N assimilation (Wingler et al. 1996). Those results have led to the idea that ECM symbioses allow plants to slow-down regulative pathways and that this may compensate the cost of photoassimilate transfer to the fungal partner. However, recent evidence has been found that decreased metabolic activities in mycorrhizal roots are not associated with decreased belowground C allocation (Corrêa et al. 2010). In addition, these metabolic readjustments may not be permanent but alternate between periods of plant metabolic deactivation/fungal activation and plant activation/fungal deactivation in a cyclic pattern (Corrêa et al. 2010).

17.5 Cost/Benefit and the Symbiotic Continuum

Mycorrhizae are traditionally accepted as being mutualistic symbioses, therefore resulting in a net beneficial outcome for both plant and fungal partners. However, in an increasing number of reported cases mycorrhization was found to decrease plant productivity (e.g., Dosskey et al. 1990; Colpaert et al. 1992, 1996; Eltrop and Marschner 1996; Plassard et al. 2000). This has led some authors to question the nature of the interaction, and to the growing belief that its effects on the host plant can vary from the traditionally accepted mutualistic through to antagonistic in a continuum of responses (Jonhson et al. 1997; Jones and Smith 2004). It has been proposed that in any particular host–plant/fungus combination the response can move along this continuum (Jonhson et al. 1997; Gange and Ayres 1999).

Central to this discussion is the evaluation of what are costs and benefits in mycorrhizae, and the degree to which the benefits exchanged are by-products or costly to the symbionts (Herre et al. 1999; Hoeksema and Schwartz 2002; Hoeksema and Kummel 2003; de Mazancourt et al. 2005). Mycorrhizae can have multiple nonnutritional effects on the host plant that may determine increased survival and fitness such as protection against root and shoot pathogens (e.g., Akema and Futai 2005), protection from toxic minerals (e.g., Hildebrandt et al. 2007; see Chaps. 14 and 18 in this volume), or resistance to drought (e.g., Bogeat-Triboulot et al. 2004). However, it is generally considered that the main benefit to the plant from mycorrhizae is an improved nutrition, while the cost is the C expended in growth and maintenance of the fungus. The outcome of the relationship is presumed to depend on the balance between the two, and negative effects of mycorrhizal colonization are expected to occur when the net C costs for fungal maintenance and growth exceed the net benefits obtained from improved nutrient uptake (Jonhson et al. 1997; Tuomi et al. 2001; Jones and Smith 2004). The C and N exchanges and balances in the association are, therefore, of central importance in the evaluation of costs and benefits in ECM.

The balance between C cost and nutrient benefit is expected to be reflected in plant productivity/growth, which can therefore be used as a benefit indicator. The meaningfulness of growth as benefit indicator has been questioned, and the advantages and disadvantages of this and other possible parameters have been previously reviewed (Jones and Smith 2004). Nonetheless, growth remains the parameter most often used to evaluate the plant response to mycorrhization, and in models that attempt to explain the variation in that response (Schwartz and Hoeksema 1998; Gange and Ayres 1999; Tuomi et al. 2001; Neuhauser and Fargione 2004; Morgan et al. 2005; Janos 2007).

The mycorrhizal response seems to vary according to the nutrient availability (Schwartz and Hoeksema 1998; Tuomi et al. 2001; Neuhauser and Fargione 2004; Janos 2007). Mycorrhizae have been found to improve N nutrition of the host plant when N was in limited supply, but not when it was abundant (Dickson et al. 1999; Bücking and Heyser 2000). In contrast, evidence was also found of negative effects of mycorrhization on N uptake and growth, which were more pronounced as N became more limited (Corrêa et al. 2008, 2010). Another factor that may determine the mycorrhizal response is the degree of mycorrhizal colonization (Gange and Ayres 1999; Tuomi et al. 2001). Negative correlations have been found between plant biomass and the extension of fungal development, the mass of the fungal mycelium produced or the relative growth rate of the mycelium (Dosskey et al. 1990; Colpaert et al. 1992). These two factors are, however, not independent, since decreased N availability leads to increased mycorrhization and extraradical fungal growth (Ingestad et al. 1986; Treseder and Allen 2002; Högberg et al. 2003; Nilsson and Wallander 2003; Corrêa et al. 2008). Some models have been proposed to explain the variation of mycorrhizal effects on plant productivity that considered it to depend on the nutrient availability (Morgan et al. 2005; Janos 2007), the balance between C cost and nutrient gain (Schwartz and Hoeksema 1998; Tuomi et al. 2001; Neuhauser and Fargione 2004), and the degree of mycorrhization, either alone or together with nutrient availability (Gange and Ayres 1999; Tuomi et al. 2001).

Two main possibilities have been advanced as the cause of negative effects of mycorrhizae on plant growth: (1) excessive fungal C drain and (2) nutrient retention by the fungus. ECM fungi have a high potential to take up, accumulate, and immobilize N (Colpaert et al. 1992, 1996; Wallenda and Kottke 1998; Plassard et al. 2000). However, negative effects on plant growth have generally been considered to be due to excessive C drain (Colpaert et al. 1992, 1996; Jonhson et al. 1997; Plassard et al. 2000). Namely, increased rates of photosynthesis that were not associated to increased growth relative to NM plants have been considered evidence of this (Colpaert et al. 1996; Conjeaud et al. 1996).

On the other hand, differences in plant performance upon mycorrhization have been proposed to be due to differences in cost efficiency, that is, nutrients acquired per C expended. Because mycorrhizae are considered to be most beneficial in conditions of nutrient deficiency, M plants have been proposed to be more cost efficient, and grow better, under nutrient poor conditions, and NM plants at high nutrient levels (Schwartz and Hoeksema 1998; Tuomi et al. 2001). However, whether M roots or fungal hyphae can be more cost efficient than NM roots is still a question. Cost efficiency has been calculated to be higher (Jones et al. 1991, 1998), equal (Tuomi et al. 2001) or lower (Jones et al. 1991) in M compared to NM roots. Some authors have found the construction and maintenance costs lower for fungal hyphae and M roots than NM roots, and this has been considered the reason for the greater cost efficiency of M plants (Jones and Smith 2004; Jones et al. 1998, 1991). However, others have found them higher (Colpaert et al. 1996). In addition, some results suggest that M roots may be less cost efficient than NM roots as P becomes more limited (Jones et al. 1991, 1998) and in conditions of N limitation (Colpaert et al. 1996).

Furthermore, although a lower cost efficiency at high levels of N could explain negative effects of mycorrhization in those conditions, it does not explain situations of decreased growth when N is limiting, which have also been observed (Ingestad et al. 1986; Colpaert et al. 1999; Hobbie 2006; Corrêa et al. 2008).

Recent results indicate that in N-limited conditions, the C cost of mycorrhizae, and by extension their cost efficiency, may have no influence on the growth of M plants, and that differences in growth are exclusively due to differences in N uptake (Corrêa et al. 2008; Hobbie et al. 2008). This is in accordance with the model of Tuomi et al. (2001), which predicts that under nutrient limitation, mycorrhizae may be beneficial and have a selective advantage, even when they are less cost efficient.

Under N-limiting conditions and testing a wide range of combinations between N availability, N concentration in plant tissues, and degree of mycorrhizal colonization, which resulted in a gradient of mycorrhizal effects from negative to positive plant growth responses, the mycorrhizal effects on plant growth and N uptake were very strongly and positively correlated, and no evidence was found of a C limitation to growth (Fig. 17.2a). This indicated that the negative effects of mycorrhization over host plant productivity were due to N retention by the fungal partner and not to excessive C drainage (Corrêa et al. 2008). This supports the idea that, under low nutrient supply, because growth is more limited by nutrients than by C, carbohydrates accumulate in plant organs, becoming more advantageous for plants to allocate more photosynthetic C belowground, and to the mutualistic partner, if by doing so they can acquire more of the nutrients they need (Kiers and Van der Heijden 2006). Furthermore, under nutrient-limited conditions, the accumulation of carbohydrates can lead to a downregulation of genes and enzymes responsible for photosynthesis, and the investment in the mycorrhizal symbiosis could even enhance the hosts' photosynthetic capacity (Kiers and Van der Heijden 2006). Under these conditions, the C supply to the fungus will not result in decreased growth, and will therefore not constitute a cost. This is in agreement with the suggestion that mycorrhizal associations are based on the exchange of excess resources or by-product benefits (Brundrett 2002; Kiers and Van der Heijden 2006).

Evidence was also found that the response of growth to mycorrhization varies according to the balance between the nutritional needs of the plant and the nutrient supply by the fungus (Corrêa et al. 2008). This balance was accessed by combining the mycorrhizal colonization (Myc: μg *P. tinctorius* DW/mg root DW) and the N availability (N relative adition rate; Ingestad et al. 1986) in a single parameter, the Mycorrhizal N Demand-Supply Balance (MDS_N; Fig. 17.2b). The variation in

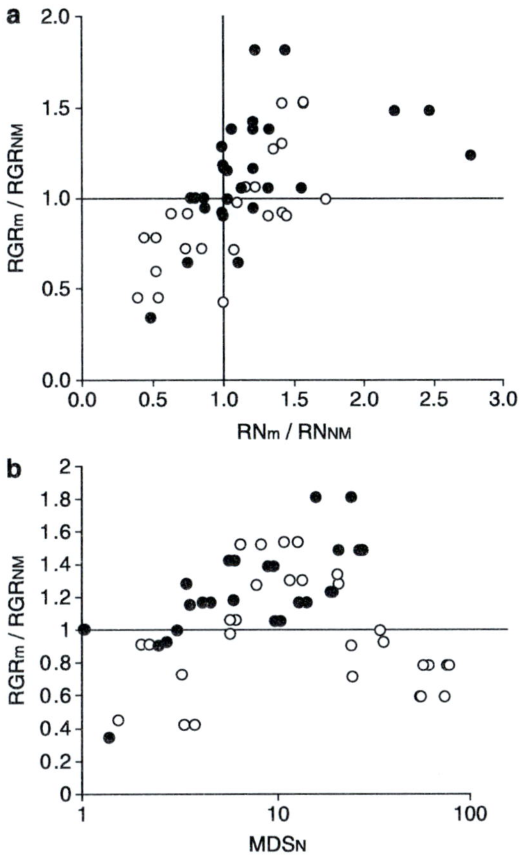

Fig. 17.2 (**a**) Correlations between the mycorrhizal profits on N uptake (RN_m/RN_{NM}) and on growth (RGR_m/RGR_{NM}). The mycorrhizal profit, or the net increase in total value generated by investing in mycorrhizae instead of in nonmycorrhizal roots, was calculated as the ratio between the parameter for an individual mycorrhizal plant and the mean value for non mycorrhizal plants grown in the same experimental conditions ($r = 0.58$, $n = 67$, $p < 0.001$). (**b**) Correlations between RGR_m/RGR_{NM} and the mycorrhizal demand supply balance (MDS_N). MDS_N was calculated as the ratio between the average level of mycorrhizal colonization (Myc; *P. tinctorius* μg DW root mg DW^{-1}) and the nitrogen relative addition rate (RAR_N; Ingestad and Lund, 1979) for the considered period. The plants were fed either 1.9 (*open circles*) or 3.8 mM NH_4^+ (*closed circles*) as N source. The *line dividing the graphic area* makes the division between negative (below 1) and positive (above 1) mycorrhizal growth profit. (Source: From Corrêa et al. (2008), with permission from the authors)

growth response had a curvilinear behavior, being negative at high and low values of MDS_N. Curvilinear correlations between degree of mycorrhizal colonization and host plant growth had been previously observed in AM mycorrhizal plants (Clapperton and Reid 1992) and predicted in several models, although the shape of the predicted curve has varied (Gange and Ayres 1999; Janos 2007). This is particularly important since it reveals that an evaluation of the mycorrhizal benefit or detriment cannot be based on only one, or a limited amount, of data points, but that M and NM plants should be compared over gradients of the conditions that are being tested.

17.6 Conclusions

Over the last years, new knowledge has been gained on C and N metabolisms and exchanges in ECM, mostly with the use of molecular tools. However, they remain mostly unknown to us. Pieces of the puzzle are slowly being discovered, but we are still far from fitting them into an integrative view of the system and its regulation. Away from the molecular approach, knowledge has also been gathering on how the C and N balances reflect on mycorrhizal cost:benefit, which will have important implications on our understanding of the regulation of exchanges. The time has come in which we need to start integrating the different approaches and results if we are to understand the function, nature, and evolution of the symbiosis.

References

Abuzinadah RA, Read DJ (1989) The role of proteins in the nitrogen nutrition of ectomycorrhizal plants IV. The utilization of peptides by birch (*Betula pendula* L.) infected with different mycorrhizal fungi. New Phytol 112:55–60

Akema T, Futai K (2005) Ectomycorrhizal development in a *Pinus thunbergii* stand in relation to location on a slope and effect on tree mortality from pine wilt disease. J For Res 10:93–99

Anderson IC, Chambers SM, Cairney JWG (1999) Intra- and interspecific variation in patterns of organic and inorganic nitrogen utilization by three Australian *Pisolithus* species. Mycol Res 103:1579–1587

Bendjia M, Rikirsch E, Müller T, Morel M, Corratgé C, Zimmermann S, Chalot M, Frommer WB, Wipf D (2006) Peptide uptake in the ectomycorrhizal fungus *Hebeloma cylindrosporum*: characterization of two di- and tri-peptide transporters (HcPTR2A and B). New Phytol 170: 401–410

Blaudez D, Chalot M, Dizengremel P, Botton B (1998) Structure and function of the ectomycorrhizal association between *Paxillus involutus* and *Betula pendula*. II. Metabolic changes during mycorrhizal formation. New Phytol 138:543–552

Blaudez D, Botton B, Dizengremel P, Chalot M (2001) The fate of [^{14}C] glutamate and [^{14}C] malate in birch roots is strongly modified under inoculation with *Paxillus involutus*. Plant Cell Environ 24:449–457

Bogeat-Triboulot M, Bartoli F, Garbaye J, Marmeisse R, Tagu D (2004) Fungal ectomycorrhizal community and drought affect root hydraulic properties and soil adherence to roots of *Pinus pinaster* seedlings. Plant Soil 267:213–223

Brandes B, Godbold DM, Kuhn AJ, Jentschke G (1998) Nitrogen and phosphorus acquisition by the mycelium of the ectomycorrhizal fungus *Paxillus involutus* and its effect on host nutrition. New Phytol 140:735–743

Brundrett MC (2002) Coevolution of roots and mycorrhizas of land plants. New Phytol 154:275–304

Bücking H, Heyser W (2000) Subcellular compartmentation of elements in non- mycorrhizal and mycorrhizal roots of *Pinus silvestris*: an X-ray microanalytical study. I. The distribution of phosphate. New Phytol 145:311–320

Buscot F, Munch JC, Charcosset JY, Gardes M, Nehls U, Hampp R (2000) Recent advances in exploring physiology and biodiversity of ectomycorrhizas highlight the functioning of these symbioses in ecosystems. FEMS Microbiol Rev 24:601–614

Chalot M, Brun A (1998) Physiology of organic nitrogen acquisition by ectomycorrhizal fungi and ectomycorrhizas. FEMS Microbiol Rev 22:21–44

Chalot M, Brun A, Botton B, Söderströmm B (1996) Kinetics, energetics and specificity of the general amino-acid transporter from the ectomycorrhizal fungus *Paxillus involutus*. Microbiology 142:1749–1756

Chalot M, Javelle A, Blaudez D, Lambilliote R, Cooke R, Sentenac H, Wipf D, Botton B (2002) An update on nutrient transport processes in ectomycorrhizas. Plant Soil 244:165–175

Chalot M, Blaudez D, Brun A (2006) Ammonia: a candidate for nitrogen transfer at the mycorrhizal interface. Trends Plant Sci 11:263–266

Clapperton MJ, Reid DM (1992) A relationship between plant growth and increasing VA mycorrhizal inoculum density. New Phytol 120:227–234

Colpaert JV, Assche JÁ, Luitjens K (1992) The growth of the extramatrical mycelium of ectomycorrhizal fungi and the growth response of *Pinus sylvestris* L. New Phytol 120:127–135

Colpaert JV, Van Laere A, Van Assche JA (1996) Carbon and nitrogen allocation in ectomycorrhizal and non-mycorrhizal *Pinus sylvestris* L. seedlings. Tree Physiol 16:787–793

Colpaert JV, Van Tichelen KK, Van Assche JÁ, Van Laere A (1999) Short-term phosphorus uptake rates in mycorrhizal and non-mycorrhizal roots of intact *Pinus sylvestris* seedlings. New Phytol 143:589–597

Conjeaud C, Scheromm P, Moussain D (1996) Effects of phosphorus and ectomycorrhiza on maritime pine seedlings (*Pinus pinaster*). New Phytol 133:345–351

Corrêa A, Strasser RJ, Martins-Loução MA (2008) Response of plants to ectomycorrhizae in N-limited conditions: which factors determine its variation? Mycorrhiza 18:413–427

Corrêa A, Magel E, Hampp R, Martins-Loução MA (2010) Carbon allocation in ectomycorrhizal plants at limited and optimal N supply: an attempt at unraveling conflicting theories. Mycorrhiza. doi:10.1007/s00572-010-0309-3

Couturier J, Montanini B, Martin F, Brun A, Blaudez D, Chalot M (2007) The expanded family of ammonium transporters in the perennial poplar plant. New Phytol 174:137–150

de Mazancourt C, Loreau M, Dieckman U (2005) Understanding mutualism when there is adaptation to the partner. J Ecol 93:305–314

Deveau A, Kohler A, Frey-Klett P, Martin F (2008) The major pathways of carbohydrate metabolism in the ectomycorrhizal basidiomycete *Laccaria bicolor* S238N. New Phytol 180: 379–390

Dickson S, Smith SE, Smith FA (1999) Characterization of two arbuscular mycorrhizal fungi in symbiosis with *Allium porrum*: colonization, plant growth and phosphate uptake. New Phytol 144:163–172

Dosskey MG, Linderman RG, Boersma L (1990) Carbon-sink stimulation of photosynthesis in Douglas fir seedlings by some ectomycorrhizas. New Phytol 115:269–274

Duplessis S, Courty P, Tagu D, Martin F (2005) Transcript patterns associated with ectomycorrhiza development in *Eucalyptus globulus* and *Pisolithus microcarpus*. New Phytol 165: 599–611

Ek H, Andersson S, Arnebrant K, Söderström B (1994) Growth and assimilation of NH_4^+ and NO_3^- by *Paxillus involutus* in association with *Betula pendula* and *Picea abies* as affected by substrate pH. New Phytol 128:629–637

Eltrop L, Marschner H (1996) Growth and mineral nutrition of non-mycorrhizal and mycorrhizal Norway spruce (*Picea abies*) seedlings grown in semi-hydroponic sand culture. I. Growth and mineral nutrient uptake in plants supplied with different forms of nitrogen. New Phytol 133: 469–478

Fajardo-López MF, Männer P, Willman A, Hampp R, Nehls U (2007) Increased trehalose biosynthesis in Hartig net hyphae of ectomycorrhizas. New Phytol 174:389–398

Finlay RD, Ek H, Odham G, Söderström B (1988) Mycelial uptake, translocation and assimilation of nitrogen from ^{15}N-labelled ammonium by *Pinus sylvestris* plants infected with four different ectomycorrhizal fungi. New Phytol 110:59–66

Finlay RD, Frostegård Å, Sonnerfeldt AM (1992) Utilization of organic and inorganic nitrogen sources by ectomycorrhizal fungi in pure culture and in symbiosis with *Pinus contorta* Dougl. ex Loud. New Phytol 120:105–115

Foyer CH, Noctor G (2002) Photosynthetic nitrogen assimilation: inter-pathway control and signalling. In: Foyer CH, Noctor G (eds) Photosynthetic nitrogen assimilation and associated carbon and respiratory metabolism. Advances in photosynthesis and respiration series. Kluwer, Dordrecht, pp 1–22

Frettinger P, Derory J, Herrmann S, Plomion C, Lapeyrie F, Oelmüller R, Martin F, Buscot F (2007) Transcriptional changes in two types of pre-mycorrhizal roots and in ectomycorrhizas of oak microcuttings inoculated with *Piloderma croceum*. Planta 225:331–340

Gange AC, Ayres RL (1999) On the relation between arbuscular mycorrhizal colonization and plant "benefit". Oikos 87:615–621

Hampp R, Schaeffer C, Wallenda T, Stülten C, Johann P, Einig W (1995) Changes in carbon partitioning or allocation due to ectomycorrhiza formation: biochemical evidence. Can J Bot 73(suppl 1):S548–S556

Heinemeyer A, Ineson P, Ostle N, Fitter AH (2006) Respiration of the external mycelium in the arbuscular mycorrhizal symbiosis shows strong dependence on recent photosynthates and acclimation to temperature. New Phytol 171:159–170

Herre EA, Knowlton N, Mueller UG, Rehner SA (1999) The evolution of mutualisms: exploring the paths between conflict and cooperation. Trends Ecol Evol 14:49–53

Herrmann S, Buscot F (2007) Cross talks at the morphogenetic, physiological and gene regulation levels between the mycobiont *Piloderma croceum* and oak microcuttings (*Quercus robur*) during formation of ectomycorrhizas. Phytochemistry 68:52–67

Hildebrandt U, Regvar M, Bothe H (2007) Arbuscular mycorrhiza and heavy metal tolerance. Phytochemistry 68:139–146

Hobbie EA (2006) Carbon allocation to ectomycorrhizal fungi correlates with belowground allocation in culture studies. Ecology 87:563–569

Hobbie EA, Colpaert JV, White MW, Ouimette AP, Macko SA (2008) Nitrogen form, availability, and mycorrhizal colonization affect biomass and nitrogen isotope patterns in *Pinus sylvestris*. Plant Soil 310:121–136

Hoeksema JD, Kummel M (2003) Ecological persistence of the plant-mycorrhizal mutualism: a hypothesis from species coexistence theory. Am Nat 162:S40–S50

Hoeksema JD, Schwartz MW (2002) Expanding comparative-advantage biological market models: contingency of mutualism on partners' resource requirements and acquisition trade-offs. Proc R Soc Lond B 270:913–919

Högberg MN, Bååth E, Nordgren A, Arnebrant K, Högberg P (2003) Contrasting effects of nitrogen availability on plant carbon supply to mycorrhizal fungi and saprotrophs- a hypothesis based on field observations in boreal forest. New Phytol 160:225–238

Ingestad T, Lund A (1979) Nitrogen stress in birch seedlings. I. Growth technique and growth. Physiol Plant 45:137–148

Ingestad T, Arveby AS, Kähr M (1986) The influence of ectomycorrhiza on nitrogen nutrition and growth of *Pinus sylvestris* seedlings. Physiol Plant 68:575–582

Janos DP (2007) Plant responsiveness to mycorrhizas differs from dependence upon mycorrhizas. Mycorrhiza 17:75–91

Javelle A, Chalot M, Söderström B, Botton B (1999) Ammonium and methylamine transport by the ectomycorrhizal fungus *Paxillus involutus* and ectomycorrhizas. FEMS Microbiol Ecol 30:355–366

Javelle A, André B, Marini AM, Chalot M (2003a) High-affinity ammonium transporters and nitrogen sensing in mycorrhiza. Trends Microbiol 11:53–55

Javelle A, Morel M, Rodríguez-Pastrana BR, Botton B, André B, Marini AM, Brun A, Chalot M (2003b) Molecular characterization, function and regulation of ammonium transporters (Amt) and ammonium metabolizing enzymes (GS, NADP-GDH) in the ectomycorrhizal fungus *Hebeloma cylindrosporum*. Mol Microbiol 47:411–430

Jentschke G, Godbold DL, Brandes B (2001) Nitrogen limitation in mycorrhizal Norway spruce (*Picea abies*) seedlings induced mycelial foraging for ammonium: implications for Ca and Mg uptake. Plant Soil 234:109–117

Johansson T, Le Quéré A, Ahren D, Söderström B, Erlandsson R, Lundeberg J, Uhlén M, Tunlid A (2004) Transcriptional responses of *Paxillus involutus* and *Betula pendula* during formation of ectomycorrhizal root tissue. MPMI 17:202–215
Jones MD, Smith SE (2004) Exploring functional definitions of mycorrhizas: are mycorrhizas always mutualisms? Can J Bot 82:1089–1109
Jones MD, Durall DM, Tinker PB (1991) Fluxes of carbon and phosphorus between symbionts in willow ectomycorrhizas and their changes with time. New Phytol 119:99–106
Jones MD, Durall DM, Tinker PB (1998) A comparison of arbuscular and ectomycorrhizal *Eucalyptus coccifera*: growth response, phosphorus uptake efficiency and external hyphal production. New Phytol 140:125–134
Jonhson NC, Graham JH, Smith FA (1997) Functioning of mycorrhizal associations along the mutualism-parasitism continuum. New Phytol 135:575–585
Kiers ET, Van der Heijden MGA (2006) Mutualistic stability in the arbuscular mycorrhizal symbiosis: exploring hypotheses of evolutionary cooperation. Ecology 87:1627–1636
Kytöviita M (2005) Role of nutrient level and defoliation on symbiotic function: experimental evidence by tracing 14C/15N exchange in mycorrhizal birch seedlings. Mycorrhiza 15:65–70
Lamhamedi MS, Godbout C, Fortin JA (1994) Dependence of *Laccaria bicolor* basidiome development on current photosynthesis of *Pinus strobus* seedlings. Can J For Res 24: 1797–1804
Leake JR, Read DJ (1990) Proteinase activity in mycorrhizal fungi. II. The effects of mineral and organic nitrogen sources on induction of extracellular proteinase in *Hymenoscyphus ericae* (Read) Korf, Kerman. New Phytol 116:123–128
Loewe A, Einig W, Shi L, Dizengremel P, Hampp R (2000) Mycorrhiza formation and elevated CO_2 both increase the capacity for sucrose synthesis in source leaves of spruce and aspen. New Phytol 145:565–574
Lucic E, Fourrey C, Kohler A, Martin F, Chalot M, Brun-Jacob A (2008) A gene repertoire for nitrogen transporters in *Laccaria bicolor*. New Phytol 180:343–364
Martin F, Botton B (1993) Nitrogen metabolism of ectomycorrhizal fungi and ectomycorrhiza. Adv Plant Pathol 9:85–102
Martin F, Selosse MA (2008) The *Laccaria* genome: a symbiont blueprint decoded. New Phytol 180:296–310
Martin F, Boiffin V, Pfeffer PE (1998) Carbohydrate and amino acid metabolism in the *Eucalyptus globulus-Pisolithus tinctorius* ectomycorrhiza during glucose utilization. Plant Physiol 118: 627–635
Morel M, Buée M, Chalot M, Brun A (2006) NADP-dependent glutamate dehydrogenase: a dispensable function in ectomycorrhizal fungi. New Phytol 169:179–190
Morel M, Jacob C, Fitz M, Wipf D, Chalot M, Brun A (2008) Characterization and regulation of PiDur3, a permease involved in the acquisition of urea by the ectomycorrhizal fungus *Paxillus involutus*. F Gen Biol 45:912–921
Morgan JAW, Bending GD, White PJ (2005) Biological costs and benefits to plant-microbe interactions in the rhizosphere. J Exp Bot 56:1729–1739
Müller T, Avolio M, Olivi M, Bendjia M, Rikirsch E, Kasaras A, Fitz M, Chalot M, Wipf D (2007) Nitrogen transport in the ectomycorrhiza association: the *Hebeloma cylindrosporum-Pinus pinaster* model. Phytochemistry 68:41–51
Nehls U (2008) Mastering ectomycorrhizal symbiosis: the impact of carbohydrates. J Exp Bot 59: 1097–1108
Nehls U, Wiese J, Guttenberg M, Hampp R (1998) Carbon allocation in ectomycorrhizas: identification and expression analysis of an *Amanita muscaria* monosaccharide transporter. MPMI 11:167–176
Nehls U, Ecke M, Hampp R (1999) Sugar- and nitrogen-dependent regulation of an *Amanita muscaria* phenylalanine ammonium lyase gene. J Bacteriol 181:1931–1933
Nehls U, Wiese J, Hampp R (2000) Cloning of a *Picea abies* monosaccharide transporter gene and expression- analysis in plant tissues and ectomycorrhizas. Trees 14:334–338

Nehls U, Göhringer F, Wittulsky S, Dietz S (2010) Fungal carbohydrate support in the ectomycorrhizal symbiosis: a review. Plant Biol 12:292–301

Neuhauser C, Fargione JE (2004) A mutualism-parasitism continuum model and its application to plant mycorrhizae interactions. Ecol Modell 177:337–352

Nilsson LO, Wallander H (2003) Production of external mycelium by ectomycorrhizal fungi in a norway spruce forest was reduced in response to nitrogen fertilization. New Phytol 158: 409–416

Nuutinen JT, Timonen S (2008) Identification of nitrogen mineralization enzymes, L-amino acid oxidases, from the ectomycorrhizal fungi *Hebeloma* spp. and *Laccaria bicolor*. Mycol Res 112:1453–1464

Plassard C, Bonafos B, Touraine B (2000) Differential effects of mineral and organic N sources, and of ectomycorrhizal infection by *Hebeloma cylindrosporum*, on growth and N utilization in *Pinus pinaster*. Plant Cell Environ 23:1195–1205

Polidori E, Ceccaroli P, Saltarelli R, Guescini M, Menotta M, Agostini D, Palma F, Stocchi V (2007) Hexose uptake in the plant symbiotic ascomycete *Tuber brochii* Vittadini: biochemical features and expression pattern of the transporter TBHXT1. Fungal Genet Biol 44:187–198

Read DJ, Pérez-Moreno J (2003) Mycorrhizas and nutrient cycling in ecosystems–a journey towards relevance? New Phytol 157:475–492

Rudawska M, Kieliszewska-Rokicka B, Debaud JC, Lewandowski A, Gay G (1994) Enzymes of ammonium metabolism in ectendomycorrhizal and ectomycorrhizal symbionts of pine. Physiol Plant 92:279–285

Rygiewicz PT, Andersen CP (1994) Mycorrhizae alter quality and quantity of carbon allocated below ground. Nature 369:58–60

Schaarschmidt S, Roitsch T, Hause B (2006) Arbuscular mycorrhiza induces gene expression of the apoplastic invertase LIN6 in tomato (*Lycopersicon esculentum*) roots. J Exp Bot 57: 4015–4023

Schaeffer C, Wallenda T, Guttenberger M, Hampp R (1995) Acid invertase in mycorrhizal and non-mycorrhizal roots of Norway spruce (*Picea abies* [L.] Karst.) seedlings. New Phytol 129:417–424

Schaeffer C, Johann P, Nehls U, Hampp R (1996) Evidence for an up-regulation of the host and a down-regulation of the fungal phosphofructokinase activity in ectomycorrhizas of Norway spruce and fly agaric. New Phytol 134:697–702

Schwartz MW, Hoeksema JD (1998) Specialization and resource trade: biological markets as a model of mutualisms. Ecology 79:1029–1038

Selle A, Willmann M, Grunze N, Geßler A, Weiß M, Nehls U (2005) The high-affinity poplar ammonium importer PttAMT1.2 and its role in ectomycorrhizal symbiosis. New Phytol 168: 697–706

Tibbett M, Hartley M, Hartley S (2000) Comparative growth of ectomycorrhizal basidiomicetes (*Hebeloma* spp.) on organic and inorganic nitrogen. J Basic Microbiol 40:393–395

Treseder KK (2004) A meta-analysis of mycorrhizal responses to nitrogen, phosphorus, and atmospheric CO_2 in field studies. New Phytol 164:347–355

Treseder KK, Allen MF (2002) Direct nitrogen and phosphorus limitation of arbuscular mycorrhizal fungi: a model and field test. New Phytol 155:507–515

Treseder KK, Czimczik CI, Trumbore SE, Allison SD (2008) Uptake of an amino acid by ectomycorrhizal fungi in a boreal forest. Soil Biol Biochem 40:1964–1966

Tuomi J, Kytöviita M, Härdling R (2001) Cost efficiency of nutrient acquisition and the advantage of mycorrhizal symbiosis for the host plant. Oikos 92:62–70

Vézina LP, Margolis HA, McAfee BJ, Delaney S (1989) Changes in the activity of enzymes involved with primary metabolism due to ectomycorrhizal symbiosis on jack pine seedlings. Physiol Plant 75:55–62

Wallander H, Nylund JE (1991) Effects of excess nitrogen on carbohydrate concentration and mycorrhizal development of *Pinus sylvestris* L. seedlings. New Phytol 119:405–411

Wallander H, Nylund JE (1992) Effects of excess nitrogen and phosphorus starvation on the extramatrical mycelium of ectomycorrhizas of *Pinus sylvestris* L. New Phytol 119:405–411
Wallenda T, Kottke I (1998) Nitrogen deposition and ectomycorrhizas. New Phytol 139:169–187
Wallenda T, Schaeffer C, Einig W, Wingler A, Hampp R, Seith B, George E, Marschner H (1996) Effects of varied soil nitrogen supply on Norway spruce (*Picea abies* (L.) Karst.). II. Carbon metabolism in needles and mycorrhizal roots. Plant Soil 186:361–369
Wiese J, Kleber R, Hampp R, Nehls U (2000) Functional characterization of the *Amanita muscaria* monosaccharide transporter, AmMst1. Plant Biol 2:278–282
Wingler A, Einig W, Schaeffer C, Wallenda T, Hampp R, Wallander H, Nylund J (1994) Influence of different nutrient regimes on the regulation of carbon metabolism in Norway spruce [*Picea abies* (L.) Karst.] seedlings. New Phytol 128:323–330
Wingler A, Wallenda T, Hampp R (1996) Mycorrhiza formation on Norway spruce (*Picea abies*) roots affects the pathway of anaplerotic CO_2 fixation. Physiol Plant 96:699–705
Wipf D, Ludewig U, Tegeder M, Rentsch D, Koch W, Frommer WB (2002) Conservation of amino acid transporters in fungi, plants and animals. Trends Biochem Sci 27:139–147
Wright DP, Scholes JD, Read DJ, Rolfe SA (2000) Changes in carbon allocation, expression of carbon transporter genes in *Betula pendula* Roth. colonized by the ectomycorrhizal fungus *Paxillus involutus* (Batsch) Fr. Plant Cell Environ 23:39–49
Wright DP, Johansson T, Le Quéré A, Söderström B, Tunlid A (2005) Spatial patterns of gene expression in the extramatrical mycelium and mycorrhizal root tips formed by the ectomycorrhizal fungus *Paxillus involutus* in association with birch (*Betula pendula*) seedlings in soil microcosms. New Phytol 167:579–596
Wu B, Kazuhide N, Hogetsu T (2002) Spatiotemporal transfer of carbon-14-labelled photosynthate from ectomycorrhizal *Pinus densiflora* seedlings to extraradical mycelia. Mycorrhiza 12:83–88
Zhu H, Dancik BP, Higginbotham KO (1994) Regulation of extracellular proteinase production in an ectomycorrhizal fungus *Hebeloma crustuliniforme*. Mycologia 86:227–234

Chapter 18
Ectomycorrhizal Interaction Between *Cantharellus* and *Dendrocalamus*

Rohit Sharma and Ram C. Rajak

18.1 Introduction

Studies on ectomycorrhizal mushroom diversity help in conservation of fungal and plant species, indigenous forests, and sustainable use of fungi as nontimber forest product. Considerable attention has been given to arbuscular mycorrhiza (AM) fungal research but knowledge on ectomycorrhizal mushroom is rare. Moreover, it has been studied scarcely in India. Sporadically, however researchers in India have conducted forays to study ectomycorrhizal fungi, accumulated knowledge and published.

Apart from two recognized hot spots viz., north eastern Himalayans and Western Ghats, India is bestowed with several regions rich in fungal biodiversity. From year 2004 to 2008, mycological studies undertaken to collect mushrooms from Central Indian dense forests of Madhya Pradesh and Chhattisgarh (under research project sponsored by Department of Biotechnology, New Delhi, Government of India) has been rewarding for authors, which yielded a large number of fleshy fungi found to be ectomycorrhizal with various tree species (Sharma 2008; Sharma et al. 2008a, 2009a, b, c, 2010a, b, c). Several investigators have also conducted studies on diversity of ectomycorrhizae (Lakhanpal 1996; Natarajan and Ravindran 2003a, b; Pande et al. 2004; Valentine et al. 2004; Kranabetter et al. 2005; Natarajan et al. 2005a, b; Riviere et al. 2007). About 61 ectomycorrhizal mushrooms were collected

R. Sharma (✉)
Microbial Culture Collection, Affiliated to National Centre for Cell Science, University of Pune, Ganeshkhind, Pune 411 007, Maharashtra, India
e-mail: rsmushrooms@yahoo.co.uk

R.C. Rajak
SGH Center for Biotechnology, B-96, Priyadarshini Housing Society, Dumna Road, Jabalpur 482001, Madhya Pradesh, India
e-mail: rcmykes@yahoo.com

M. Rai and A. Varma (eds.), *Diversity and Biotechnology of Ectomycorrhizae*, Soil Biology 25, DOI 10.1007/978-3-642-15196-5_18,

Fig. 18.1 *Cantharellus tropicalis* basidiomata in natural forest
Source: Adapted from Sharma et al. 2010b

(Sharma et al. 2009c) including *Russula*, *Lactarius*, *Scleroderma*, *Pisolithus*, *Geaster*, *Boletus*, *Cantharellus* and *Astreaus*. *Cantharellus tropicalis* (Fig. 18.1) is a popular edible and medicinal mushroom sold in the local market of Balaghat district, Madhya Pradesh, India. It is a high temperature species, which appears in late rainy season during August and September. It has been recommended as energy booster for pregnant women. *C. tropicalis* forms ectomycorrhiza (ECM) with *Dendrocalamus strictus* (bamboo) and consistently found associated with bamboo forest of Madhya Pradesh (Sharma 2008; Sharma et al. 2009c). The fruiting bodies are harvested in large amount from Balaghat, Lamta, Nainpur, and Baihar of Madhya Pradesh, India and nearby regions and often been taken to adjoining districts in large quantities. It is leathery in texture, with good taste and greater shelf-life. The seasonal collection may go up to 50 tons per year (based on personal observation). The number of described *Cantharellus* species worldwide exceeds 70 (Danell 1999). Other taxonomically related species viz. *Cantharellus minor*, *C. cibarius*, *C. cinereus*, *C. friesii*, *C. lutescens*, *C. melanoxeros*, *C. tubaeformis*, and *C. xanthopus* are commonly called chanterelle mushrooms known world over for edibility.

This chapter highlights various aspects of ectomycorrhizal association between *Cantharellus* (isolated from fresh mushroom body using stipe tissue which is aseptically transferred to ectomycorrhizal media) and bamboo species. It focusses on simple technique for synthesis of ECM, *in vitro* ectomycorrhization with *Dendrocalamus* seedlings, mass production of inoculum with low-cost substrate, enzymatic studies on the acid phosphatase which forms an important criteria for good ectomycorrhizal fungus and interaction studies with other microfungi isolated from rhizosphere of bamboo and its potential to increase growth of seedlings growing under green house conditions.

18.2 Ectomycorrhiza Formation and Growth Response on Host Plant

18.2.1 Growth in Culture

Cantharellus is relatively easy to isolate as compared to other mushrooms viz., *C. cibarius* wherein *Pseudomonas* lives in association (Danell 1994). To isolate, fruit body of a freshly collected mushroom is taken up for isolation. It is free from any insect or disease infestation. Usually young basidiomata is used for the same. Stipe of a clean basidiomata is broken vertically and a small tissue piece is aseptically transferred into Petri dish containing Melin–Norkrans Agar Medium (MNM) with the help of a sterile forcep (Sharma et al. 2009b). Wide mouth vials of 50ml capacity filled with moistened used tea-leaves (pre-sterilized) were also used. It prevents bacterial contamination as mushroom mycelia colonizes the cellulose, lignin rich substrate faster (Sharma 2008). We also observed that acidic pH is more suitable to *C. tropicalis*. When grown in liquid media *C. tropicalis* tends to bring final pH to acidic (Sharma unpublished). It grew faster when the medium (agar and liquid media) was supplemented with 0.2-0.5% malt extract. The mycelia of *C. tropicalis* is white and grow sticking to medium surface. However, at later stage some portion of it turns brown which may be due to aging (and should not be confused with contamination) (Fig. 18.2). The mycelia also forms clamp connections in artificial medium (Fig. 18.3). A detail study on physical and physiological factors affecting growth of *C. tropicalis* has already been conducted (Sharma et al. unpublished).

18.2.2 Simplified Technique of Ectomycorrhizal Synthesis

There is a need to enhance our knowledge on interactions within fungal community and explore potential of ectomycorrhizospheric environment for biotechnological

Fig. 18.2 Culture of *C. tropicalis* on modified FDA medium after 15 days

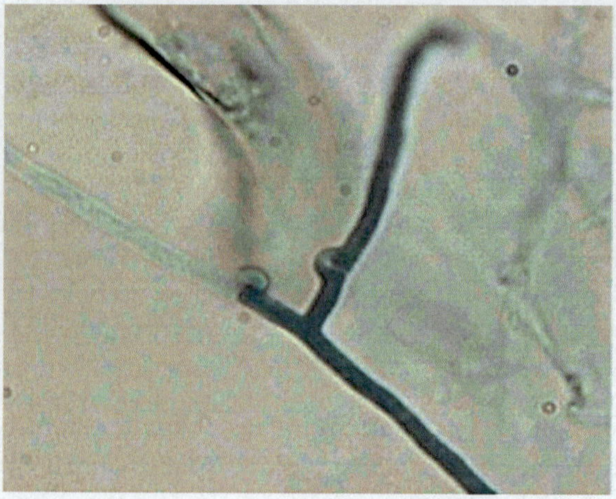

Fig. 18.3 Clamp connection of *C. tropicalis* mycelia in culture (100x)

purpose (Cairney and Meharg 2002; Whipps 2004; Cairney 2005). For this we need to study the process of mycorrhization in controlled environment under laboratory conditions. Before carrying out small- scale field experiments it is important to select a technique which can suit *Cantharellus* mycelia for symbiotic association with host plant seedlings. It is essential for development of an effective mycorrhizal symbiosis towards exploitation of this symbiosis. Several devices have been developed for *in vitro* ECM synthesis for performing physiological, biochemical, and structural experiments using various ectomycorrhizal mushrooms (Danell 1994; Cairney and Chambers 1999; Vaario et al. 1999, 2000; Yamada et al. 2001a; Danell 2002; Theodorou and Reddell 2006). But most of the equipments were found bulky and did not allow investigation of large number of experimental units at the same time.

Many aseptic methods involved large tubes and Petri dishes (Duddridge 1986) to maintain aseptic conditions for whole seedling (Chilvers et al. 1986) or for only roots (Duddridge 1986). Although there are numerous work on synthesis of pine ectomycorrhiza (Marx et al. 1982), less work has been done with angiosperm and none with monocots. Also techniques of synthesizing pine mycorrhiza are difficult to apply to angiosperms such as eucalypts, bamboo, sal, dipterocarps, etc. A filter paper-flask technique was developed by us, which was not bulky and requires limited space, forms rapid ECM, and yet permits observation of ectomycorrhiza formation (Sharma et al. 2009b). This technique enables bamboo shoots to grow well and roots to make contact with *Cantharellus* mycelium in a short time. Root and shoot parts extended within flask and formed typical ectomycorrhiza (Fig. 18.4). It keeps root and shoots system under aseptic conditions. Externally, ECM lacks root hairs and forms light-brown mantle with woven hyphae.

The filter paper-flask method permits observation of developing external mycelium before and after ectomycorrhizal formation. Fortin et al. (1980) suggested that exudation of substances stimulating the growth of *Pisolithus tinctorius* and

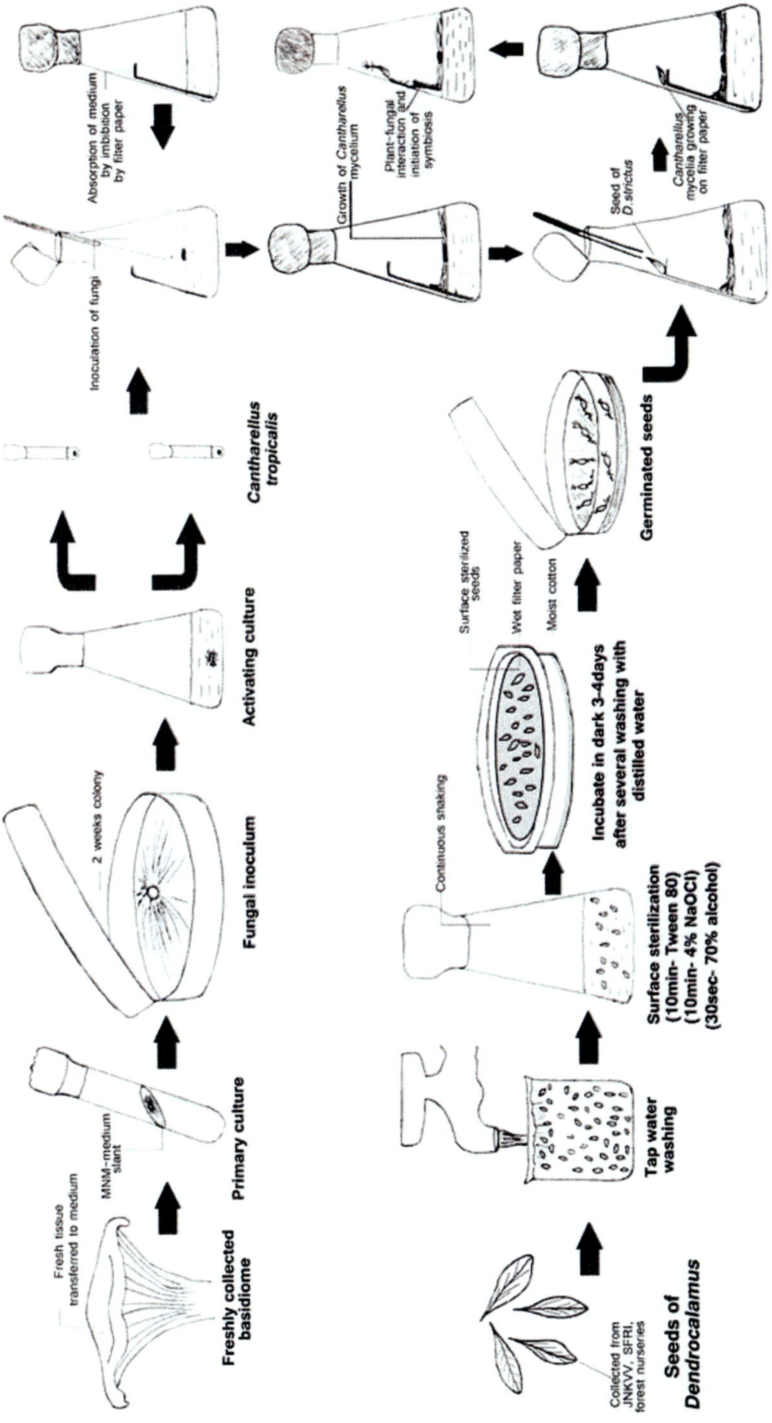

Fig. 18.4 Filter paper-flask technique used for *in vitro* ectomycorrhiza formation
Source: Adapted from Sharma et al. 2009b

Cenococcum graniforme which should also be investigated for this association in future studies. However, this technique has some features common with others including culture technique that used a Petri dish lined with paper (Chilvers et al. 1986); a plastic pouch technique (Fortin et al. 1980) uses paper to support and nourish roots; and a cellulose thimble technique (Littke et al. 1980) uses paper to help fungal growth and transmission. Moreover, Vaario et al. (2000) described a simple *in vitro* system for synthesis of *Abies firma–Cenococcum geophilum* ectomycorrhiza. This technique developed by us (Sharma et al. 2009b) is not only useful for the production of sufficient mycorrhizal seedlings for biochemical and physiological studies but can produce mycorrhizal material for practical forestry applications. Moreover, this technique permitted accurate evaluation of the colonization process of *C. tropicalis*. The host plant–fungus association will help in studies on development of its fruit bodies, physiological studies and other *in vitro* studies viz., effects of different organic sources of carbon and nitrogen on ectomycorrhizal formation capability of *Canthar-ellus*.

18.2.3 Ectomycorrhizal Interaction

In most instances, a fungus is considered ECM forming based on field observation. Consistent association of basidiomata with one or more tree species forms the basis of ECM forming fungi as an indirect method of ECM assessment. However, not all ectomycorrhizal fungi found associated with adult trees in field form mycorrhizae with young seedlings (Last et al. 1992). Pure culture synthesis technique has been modified by several workers for different mushroom species. Few edible ectomy-corrhizal mushrooms have been successfully cultivated under controlled condi-tions. Numerous studies on ECM formation between host plant seedlings and mushroom mycelia has been conducted and widely reviewed (Nezzar-Hocine et al. 1998; Vaario et al. 1999, 2000; Yamada et al. 1999, 2001b; Dahlstrom et al. 2000; Guerin-Laguette et al. 2000a, b). On the other hand, *Lyophillum shimeji* (Kawam) Hongo., *C. cibarius* Fr., and *Tuber melanosporum* Vitt. sporocarps have been produced under laboratory, green house, and field conditions, respectively (Danell and Camacho 1997). *Tricholoma* and *Lactarius* have been also successfully cultivated in laboratory conditions. Thus, establishment of artificial cultivation system for *C. tropicalis* and understanding its ectomycorrhizal interaction may form the basis for reforestation program mushroom cultivation.

The mushroom mycelia used for the study was isolated by excising tissue blocks from basidiomata stipe on MNM medium (Straatsma and van Griensven 1986). The test tube or flask system used in this experiment contained a sterilized mixture of sand + used tea leaves (50% v:v) moistened with sterile distilled water. It was cooled to room temperature and inoculated with *C. tropicalis* culture (Sharma et al. unpublished). The seedlings (produced by germinating seeds on Petri dishes) were inserted in growth system with *C. tropicalis* mycelia plug. Fungal colonization of bamboo roots was observed to confirm presence of intercellular Hartig Net confirming the symbiotic association (Harley and Smith 1983). After few weeks of incubation, seedlings of *D. strictus* showed yellow-white mycelium in substrate.

A mycorrhiza-like relation was observed from root morphology between roots of *D. strictus* and *C. tropicalis* and is a maiden example illustrating association between ectomycorrhizal fungus and a monocot plant. Mycorrhizal synthesis with other mushroom species and various host plants has also been achieved and reported (Molina and Trappe 1982; Yamada and Katsuya 1995; Dahlstrom et al. 2000). In a study, Theodorou and Reddell (2006) studied 11 species of mycorrhizal fungi from stands of *Eucalyptus* spp., *Allocasuarina* spp., or *Pinus radiata* D. Don and tested for their abilities to initiate ECM with *Allocasuarina littoralis* (Salisb.) L. Johnson, *Casuarina equisetifolia* L., and *C. cunninghamiana* Miq. in aseptic system. Moreover, Vaario et al. (1999) first reported *in vitro* mycorrhizal formation of *Abies firma* Sieb, et Zucc. with *P. tinctorius* (Pers.) Coker and Couch and improved the technique of ectomycorrhization of *A. firma* a slow-growing species and *P. tinctorius* using a novel culture medium with both sterilized and rerooted seedlings. Although *in vitro* mycorrhizal synthesis is difficult in the genus *Cantharellus*. In this experiment we tried to form *in vitro* ECM using low-cost substrate. *Pinus densiflora* formed ECM with 21 fungal species, including two species of *Russula*, which were also found difficult to manipulate *in vitro* (Yamada and Katsuya 1995). A mycorrhizal study indicated that *Morchella* formed mycorrhiza-like interaction with four tree species of Pinaceae (Dahlstrom et al. 2000).

Although different media are known to stimulate mycorrhization process, they have also detected to affect hyphal branching and several plant- microbe interactions. Extensive data are available concerning the effect of media or substrate on the process. In our study, the formation of ECM was also influenced by solid/liquid inoculum and environmental factors. Receptiveness was greatly influenced by the physical, chemical, and biological characteristics of substrate as confirmed by used tea leaves + sand as substrate. During the experiment, certain seedlings showed mortality (personal observation), which might be due to unsuitable conditions. Such results have been reported for *C. cibarius in vitro* synthesis, where CO_2 content, exogenous glucose content, water drainage, and host specificity have been discussed for such observations (Danell 1994). Most seedlings (including control) grew well throughout incubation and ectomycorrhizal seedlings showed similar or slightly better growth as compared to control plants (Sharma et al. unpublished).

Some fungi that form mycorrhizae in pure culture fail to form it in greenhouse or natural soil conditions (Molina and Trappe 1982; Duddridge 1986; Molina et al. 1997) Mycorrhizal succession (contamination) to another symbiont is problematic in cultivation of mycorrhizal mushrooms. Protection of inoculated fungus from competitive air borne mycorrhizal fungi in green house conditions is significant for their acclimatization. In present study, hairless or bifurcated root tips frequently exhibited mantle or hyphal penetration between the cortical cells. Soon after inoculation with pure cultures of Indian chanterelle, fungal, and/or bacterial contamination, or both appeared within some synthesis units. Insufficient surface sterilization period for seeds might be the cause. During our study, some contaminants appeared on seed cotyledon which affected mycorrhizae formation between chanterelle and seedlings and made system ineligible for further studies (personal observation).

18.2.4 Mass Multiplication of Inoculum

During plantation in forest areas/disturbed sites, it is observed that mycorrhizal fungi should accompany them for better survival in new microcosm in which they are introduced. Seedlings inoculated with effective mycorrhizal fungi in nursery can establish a healthy ECM system before outplanting into forests or mine sites. Inoculation of *P. tinctorius* significantly increased growth and survival of five southern pine species planted at different sites (Marx et al. 1982). Moreover, these can be stored at low temperature without damage (Lapeyrie and Bruchet 2006). During the study for large scale production the inoculum of *C. tropicalis* produced was kept at room temperature and when inoculated formed ECM with nursery seedlings of bamboo (Sharma et al. 2010a). The tea leaves based substrate conserved moisture which helped the fungus to grow actively (personal observation). Previously several types of natural or laboratory-produced inocula (seedlings with ectomycorrhiza or excised ectomycorrhiza, spores, crushed basidiomata) and various methods of application have proved successful through the years (Marx 1980). *T. melanosporum* has been established in nursery beds with ECM formed under laboratory conditions. Lamb and Richards (1974) demonstrated the use of basidiospores of *Rhizopogon*, *Scleroderma* and *Pisolithus* as inoculum. Basidiospore inoculum of *P. tinctorius* and *Rhizopogon* successfully forms ECM in pine seedlings (Molina and Trappe 1982; Bruns et al. 2009). However, Trappe (1977) had recommended pure mycelial or vegetative inoculum of ECM fungi for forest inoculation.

Pure culture inoculum of ectomycorrhizal fungi poses many difficulties for wide-scale application thus restricting it to laboratory or green house experiments. However, nursery beds of *Pinus cembra* have been successfully inoculated with pure cultures of *Suillus plorans*. Theodorou (1971) and Theodorou and Bowen (1973) inoculated *P. radiata* with isolates of *Rhizopogon luteolus*, *Suillus granulatus*, *S. luteolus*, and *Cenococcum geophilum*. Danell (1994) used peat/quartz sand mixture (25%/75% v/v) and 20 ml ingested solution for outplanting *in vitro*-formed *Cantharellus* mycorrhizae. Moreover, with four decades of mycorrhizal research, peat moss, and vermiculite remain major substrate for inoculum preparation (Garbaye et al. 1988; Nezzar-Hocine et al. 1998; Yamada et al. 2001a). Lapeyrie and Bruchet (2006) have used liquid fermentation in airlift bioreactor to produce inoculum of *Pisolithus microcarpus*.

Synthesis of *Cantharellus* ECM with *Dendrocalamus* has been achieved with *in vitro* germinated seedlings in laboratory (Sharma et al. 2009b; Sharma et al. unpublished) and green house (Sharma et al. 2008b). Moreover, mycelia of

Fig. 18.5 *Cantharellus* inoculum produced on sterilized waste tealeaves + sand
Source: Adapted from Sharma et al. 2010a

Cantharellus forms mycorrhiza with *laboratory grown seedlings* and exhibited positive effect on growth of *Dendrocalamus* seedlings (Sharma et al. 2008b; Sharma et al. 2009b; Sharma et al. unpublished). Mass production of *C. tropicalis* inoculum is essential for field application of *D. strictus*. A ratio of 1:1 used tea leaves and sand moistened with a volume of sterile distilled water equal to approximately half the volume of dry substrate proved best (Sharma et al. 2010a). The mycelia grew very fast through the substrate producing good fungal inoculums (Fig. 18.5). Used tea leaves have previously been successfully used by Sharma et al. (2003) for pure culturing of *Pleurotus* species in tubes, which reduced bacterial contamination and other noncellulolytic fungus.

According to Kumar and Satyanarayana (2002), selection of mycorrhizal fungi and performance of seedlings is dependent upon plantation site and plant species. Ectomycorrhizal fungi are suitable inoculants for various trees (Table 18.1). A good substrate allows penetration of mycelia, retains moisture, and low in production cost also. Moreover, it allows smooth ECM formation by the fungus in host seedlings. Moser (1958) produced ectomycorrhizal inoculum for inoculating *P. cembra* with *S. plorans* (Rolland) Sing and also used pure culture of *Suillus placidus* (Bon.) Sing., *S. grevillei* (Klotzsch) Sing., *S. aeruginascens* (Secr.) Snell., *Paxillus involutus*, *Phlegmacium glaucopus*, *Amanita muscaria*, and *Lactarius porninsis* Rolland. Takacs (1961) inoculated sterilized germinated grains of cereals, cereal chaff, or peat moss to produce inoculum of *A. verna* (Bull. ex Fr.) Lamarck, *S. granulatus*, *S. lutea*, *Hebeloma crustuliniforme*, *Russula* sp., *Scleroderma verrucosum* (Bull.) Pers., and *S. vulgare*. Park (1984) also grew mycelial cultures of *S. granulatus* and *C. geophilum* in cereal grains. Several attempts to produce effective inoculum of these fungi in wheat grains resulted in failure (Marx 1980). Vermiculite and peat moss moistened with modified MNM medium was found to be an excellent substrate by Marx and Bryan (1975). A successful commercial formulation of *P. tinctorius* mycelia has been developed by USDA,

Table 18.1 Commercial availability of vegetative mycelial inoculum of ectomycorrhizal fungi

Process	Inoculum form	Species (strain)	Commercial source
Solid substrate fermentation	Vermiculite (MycoRhiz[R])	*Pisolithus tinctorius*	Abbott Laboratories, USA
	Vermiculite	*Hebeloma crustuliniforme* *Laccaria laccata* *Paxillus involutus*	Somycel and INRA, France
	Vermiculite	*Laccaria laccata* *Paxillus involutus*	Sylvanspawn, Worthigton, PA
	Vermiculite	*Pisolithus tinctorius*	Mycorr Tech Inc., Pittsberg, PA
Submerged fermentation	Alginate	*Hebeloma crustuliniforme* *Laccaria laccata* *Paxillus involutus* *Thelephora terrestris*	Biotal Ltd., Cordiff, UK
	Alginate	*Hebeloma crustuliniforme*	Rhone-Poulec and INRA, France
	Alginate (Mycobead[R])	11 ECM strains	Biosynthetica Pvt. Ltd., Australia

Source: Kumar and Satyanarayana (2002)

Forest Service, and Abbott Laboratories (Marx et al. 1982). This inoculum trademarked as MycoRhiz®, is also grown in vermiculite–peat moss nutrient medium. Pure cultures of fungi viz., *S. granulatus*, *R. luteolus*, *Thelephora terrestris* (Ehrh.) Fr., and *P. tinctorius* improved survival and growth of seedlings (Marx 1980; Marx et al. 1982).

During course of our studies, it was observed that mycelia of *C. tropicalis* completely covered the substrate rich in cellulose and lignin within short incubation period. Sand helps in breaking the inoculum into pieces during application. Good ECM formation can be attributed to better mixing of inoculum to soil with sand allowing good aeration. However, Molina and Trappe (1982) have stressed on the difficulty of transporting, spreading, and mixing of vermiculite inoculum as it becomes heavy due to water saturation. Sand + used tea leaves inoculum was found promising as *Cantharellus* colonized entire root system and substrate. Moreover, high quality tree seedlings cannot be produced unless conventional nursery is substituted by modern forest nursery production techniques (Lamhamedi et al. 2009). This is possible when there is increase in technology transfer from researchers to nursery personnel.

18.2.5 Growth Response of Host Seedlings

Ectomycorrhizal mushrooms help in survival and growth of host seedlings. This association increases plant ability for nutrient and water uptake (Whipps 2001).

It has been successfully demonstrated through various experiments that *Pisolithus* and *Rhizopogon* increased growth of host seedlings (Cairney and Chambers 1999). Apart from these, several other ectomycorrhizal mushrooms viz., *Tuber*, *Lactarius*, *Laccaria*, *Scleroderma*, *Canharellus*, *Paxillus*., *Hebeloma*, and *Tricholoma* have been reported experimentally to enhance growth of host seedlings. During our studies in green house condition it was observed that although *C. tropicalis* formed ECM with *Dendrocalamus strictus*, *D. asper* and *Bambusa nutans* seedlings they were not stable, whereas fungus failed to form ECM with *B. vulgaris*. Further, Sharma et al. (2008b) reported that uninoculated seedlings conditioned to cyclic drought were smaller than mycelia inoculated seedlings. Natarajan et al. (1995) while studying ECM with *Acacia nilotica* found that *Laccaria fraternal* and *P. tinctorius* improved its growth. Chen et al. (2006) reported that in South China inoculation of *Scleroderma* spores increased the growth of eucalypt plants. Similar results have been reported by Khosla and Reddy (2008) with *Pisolithus albus* and *Eucalyptus tereticornis*. Pande et al. (2007) reported that oak and pine seedlings when inoculated with ectomycorrhizal fungi showed more growth. Since *Cantharellus* is comparatively easily isolated in pure culture and have moderate growth rate, they are relatively good for inoculation purpose. Current interest by researchers in production of ectomycorrhizal inoculum in forestry programs focusses on selection of potential strain for this purpose and also test inoculum effectiveness before field inoculations. The mushroom mycelia forms numerous rhizomorphs in soil which are essential for nutrient uptake and also help in stress tolerance during drought and hence increase seedling growth.

18.3 Ecological Studies

18.3.1 General Ecology

C. tropicalis has been reported from older natural forests and plantations but has not been reported from nurseries. However, during past several years their has been overharvesting of *Cantharellus* from the bamboo forests of Madhya Pradesh (Sharma unpublished). The observations are based on the sites visited by the author during 2004–2008. This may lead to decrease in productivity of *Cantharellus* in the region. The need for restricted harvesting and technique for harvesting has been emphasized as the tribals and villagers lack the knowledge of sustainable harvesting. It disturbs the below ground mycelia of mushroom and affects successive sporocarp formation. Danell (1999) has highlighted that whatever estimates we make is based on the above ground basidiomata and decrease in below ground mycelia and ECM should also be emphasized.

The basidiomata are observed in the late rainy season when their is high temperature and humidity. During laboratory experiments also it required high humidity i.e. above 80% for sporocarp formation in controlled environment chamber

(Sharma et al. unpublished). In another study optimum growth pH of *C. tropicalis* was acidic which can be correlated to low soil pH value which may be due to secretion of metabolites due to litter degradation. Although their is lack of information on ecology of *C. tropicalis* in Central India, increased interests among researchers will trigger more work on it.

18.3.2 Antagonistic Interactions with Rhizosphere Fungi

The mycobiota of forest soils consist of AM fungi, ECM and saprotrophic decomposer fungi which supply nutrients to trees and decomposite woody plant litter. Saprotrophic basidiomycetes are also abundant in bamboo forests. Ectomycorrhizal fungal mycelia (found in forest soils associated with host trees) with its special physiology can use either inorganic nutrients or utilize organic sources. ECM fungi provide an increased surface area for absorption of nutrients and interactions with other microorganisms (Smith and Read 1997). These interactions may be inhibitory, stimulatory, competitive, mutualistic and is important in biogeochemical cycling in ecosystems of forests. According to Leake and Johnson (2004), saprotrophs obtain most of their C from decaying organic matter while ectomycorrhizal fungi obtain it from their host plants. Antagonistic interactions between rhizosphere microorganisms and mycorrhizal fungi have significant function on mycorrhizal systems (Stark and Kytöviita 2005). Moreover, exudation and reabsorption of fluid droplets at ECM hyphal tips influence vicinity environment (Sun et al. 1999).

The interactions of plants with soil microorganisms, both pathogens (nematodes and fungi) and mutualists (nitrogen-fixing bacteria) are also influenced by mycorrhizal fungi. Pathogenic fungi invade roots and mycorrhizal fungi and can alter host response to these pathogens. Fitter and Garbaye (1994) observed that *Laccaria bicolor* prevented spread of *Fusarium oxysporum* in Douglas–fir roots as a result of flavonoids wall infusion. Wu et al. (2003) explored interactions between saprotrophic microbes and ECM fungi using a protein–tannin complex as N source by red pine (*Pinus resinosa*). In a similar study, Olsson (1999) showed the role of fatty acids for determination of distribution and interactions of mycorrhizal fungi in soil. Protection of root system from endemic pathogens (such as *Fusarium* spp.) causing root infection has been due to reduced phosphorus uptake. There is direct evidence that mycorrhizal fungi may reduce the incidence and severity of root diseases (Whipps 2004). In past role of ectomycorrhizal fungi in controlling plant diseases has been realized. Moreover, tree seedlings has been infected by several pathogens and hence these interactions can be exploited to restrict them (Finlay 2004).

Soil mycoflora was recovered from rhizosphere soil samples of *D. strictus* collected from three sites of bamboo forests in Balaghat, Mandla, and Shahdol, Madhya Pradesh, India. Pairwise combinations were made by plating mycelial plugs of ECM and soil microfungi on opposite corners of Petri dish. Radial growth toward other mycelium was determined by measuring colony radius. The cross inoculation method (Figs. 18.6 and 18.7) showed that *C. tropicalis* was active

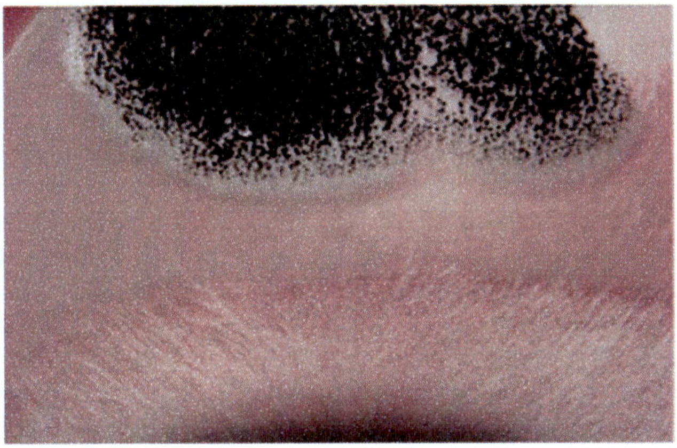

Fig. 18.6 Dual culture interaction between *C. tropicalis* and *Aspergillus niger*
Source: Sharma et al. 2010c

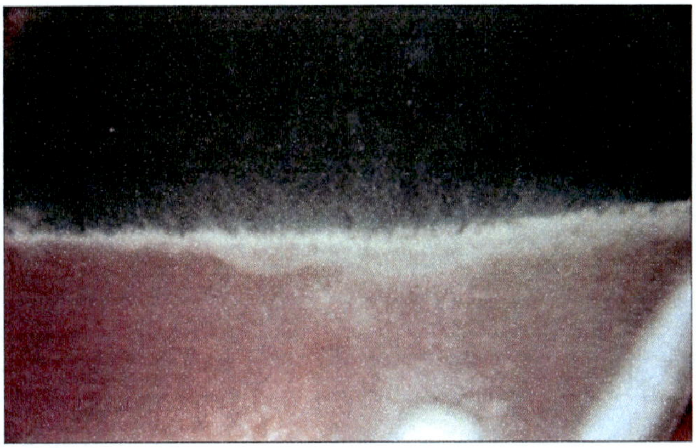

Fig. 18.7 Dual culture interaction between *C. tropicalis* and *Alternaria* sp.
Source: Sharma et al. 2010c

against rhizosphere soil microorganism (Sharma et al. 2010c). This resulted in different types of interactions between fungi, but also in differences in the degree of interactions. Overgrowth was the most common (45%) interaction, followed by inhibition at a distance (29%), intermingling (17%), and contact inhibition (13%) (Sharma et al. 2010c).

The inhibition of soil microfungi, mostly at a distance by *C. tropicalis* suggested that it prevented invasion by potential competitors (Sharma et al. 2010c). The inhibition of soil microfungi by *C. tropicalis* might be caused by production of secondary metabolites. In some instances, antibiotics produced by ectomycorrhizal fungi *Amanita*, *Boletus*, and *Cenococcum* spp. have been reported (Santoro and Casida 1962). There are reports of *in vitro* inhibition of pathogenic fungi by several ectomycorrhizal mushroom mycelia. Shaw et al. (1995) reported growth suppression of *R. roseolus* by several saprotrophic basidiomycetes. Furthermore growth of *S. granulatus* (L.:Fr.) Rouss, has been shown to be inhibited by rhizoplane fungi of *Pinus halepensis* (Girlanda et al. 1995). Baar and Stanton (2000) attributed low investment of N in mycelial biomass for reduced competition of some ECM fungi. Hardly any sporocarps of saprotrophic basidiomycetes found to occur in bamboo forest but species of *Ramaria*, *Clavaria*, and *Clitocybe* have been collected from bamboo forests and can be studied for their competitiveness with chanterelle. In a study, *Clitocybe marginella* Harmaja inhibited the growth of *C. geophilum* and *L. bicolor* (Baar and Stanton 2000). Leake et al. (2001) observed limited effect of mycorrhiza on growth of saprotroph. In our studies, ectomycorrhizal fungi suppressed soil microfungi indicating that *Cantharellus* mycelia has higher competitiveness than soil microfungi, which may be due to alkaloids, terpenes, polysaccharides produced by *Cantharellus* mycelia. Low competitiveness of some microfungi viz., *Curvularia*, *Alternaria*, *Mucor*, *Fusarium*, and *Penicillium*, etc. suggest that they were unable to compete with *Cantharellus* in the acidified soil.

Ectomycorrhizal fungi show inhibitory effects on root pathogenic fungi but as Johansson et al. (2004) puts in, their interactions with saprophytic fungi have received little attention. In another study, *Tricholoma* sp., *P. involutus*, *Hebeloma cylindrosporum*, and *L. bicolor* demonstrated inhibition of a range of pathogens *in vitro* (Morin et al. 1999). When grown in coculture Werner and Zadworny (2003) observed suppression of *Mucor hiemalis* by *L. laccata* and inhibition of growth of *Trichoderma virens* in coculture (Werner et al. 2002). Antifungal and antibacterial action of ECM fungi *Pisolithus* and *Scleroderma* was tested *in vitro* against eight fungi and six bacteria. Both showed higher activity against all fungi except some *Aspergillus* spp. (Vaidya et al. 2005). Significant progress has been made toward understanding of interaction (Zak 1971; Stark and Kytöviita 2005; Sampangiramaiah and Perrin 1990; Natarajan and Govindasamy 1990) but more extensive research is required to enhance our knowledge on interactions within fungal community and exploring potential for manipulating ectomycorrhizosphere environment for biotechnological purposes (Bruns and Bidartondo 2002; Cairney and Meharg 2002). The intensity of interactions between different soil fungi and ECM fungus *C. tropicalis* highlights the potential importance of interactions on functioning of these organisms in forest ecosystems.

18.3.3 Acid Phosphatase Production

Growth enhancements of host plant associated with ectomycorrhizal fungi has been correlated with increased nutrient uptake by ectomycorrhiza. They solubilize insoluble forms of nutrients and have a significant role in carbon, nitrogen, and phosphorus cycling in forested ecosystems (Cullings et al. 2008). Activities of phosphatase, laccase, glucuronidase, cellobiohydrolase, *N*-acetyl-glucosamine, leucine aminopeptidase, xylosidase, and β- glucosidase were found responsible for increased nutrient uptake (Courty et al. 2007; Mosca et al. 2007), an advantage for enhancing nutrient acquisition (Cameron et al. 2006). There are several reports of acid phosphatase activity of *Amanita, Hebeloma, Tricholoma*, etc. (Alvarez et al. 2005; Buée et al. 2005, 2007; Courty et al. 2005) which help in selecting effective mycorrhizal symbiont for inoculation in reforestation. According to Antibus et al. (1986) external factors affect production and activity of acid and alkaline phosphatase and hence efficiency of ectomycorrhizal fungi but inadequate data is known on enzyme activities (Courty et al. 2005).

C. tropicalis was found to use a broad range of phosphorus sources during *in vitro* studies conducted by Sharma (2008). When it was grown in defined media, it releases phosphatase (probably acid phosphatase) as media has acidic pH. Activity of acid phosphatase in ectomycorrhizal fungus *C. tropicalis* under controlled conditions has been studied by Baghel et al. (2009). The pH had strong effect on production of wall-bound acid phosphatase and maximum was observed at pH 5 followed by 7 (Sharma et al. 2010d). Culture pH strongly influences extracellular acid phosphatase production. Generally, ectomycorrhizal phosphatase has a pH optimum approaching that of native soil (Antibus et al. 1986). Acid phosphatase production was not affected by various temperatures tested except at 40 ± 2°C which showed marked reduction in enzyme production. Of nine carbon sources used, citric acid supported highest acid phosphatase production (Sharma et al. 2010d). All tested nitrogen sources supported enzyme production and highest with yeast extract. Among heavy metals and trace elements studied, ferric chloride did not inhibit acid phosphatase production. Potassium di-hydrogen phosphate and di-ammonium hydrogen phosphate produced significantly high enzyme when supplemented as sole phosphorus source.

Different temperature environments have considerable effect on the physiological and ecological consequences of ectomycorrhizal associations (Tibbett and Cairney 2007). Like other secondary metabolites, acid phosphatase production is directly related with mycelial growth of fungus and regulated by several abiotic and biotic factors. Moreover, trace elements reduce enzyme activity by interacting with the enzyme–substrate complex, by denaturing the enzyme protein, or interacting with the protein active group (Nannipieri 1995). Trace elements (metal ions) are assumed to inactivate enzymes by reacting with sulfhydryl groups, a reaction analogous to formation of a metal sulfide. It has been generally recognized that copper and cadmium are more toxic than other metals (Hattori 1992). However, Gibson and Mitchell (2005) showed that copper has no effect on the wall-bound phosphatase activity up to a level. Baxter and Dighton (2005) found that phosphatase enzymes are differentially expressed under contrasting phosphorus condition by different ectomycorrhizal fungi. *Piloderma* has also shown species-specific substrate preferences in response to organic and inorganic sources of phosphorus (Rosling and Rosenstock 2008). However, ECM fungi differ greatly in their capacity to produce acid phosphatase due to differential potentiality to utilize phosphorus, which can also be affected by season and succession (Meyselle et al. 1991). Moreover, Courty et al. (2006) found less seasonal differences in ectomycorrhizal acid phosphatase activity while working with *Lactarius quietus*, *Cortinarius anomalus*, and *Xerocomus chrysenteron*.

18.3.4 Activity of Acid Phosphatase

Ectomycorrhizal fungi use organic forms of soil nutrients through production of extracellular enzymes (Aučina et al. 2007) as an adaptation for plants to colonize soils (Read and Perez-Moreno 2003). In forest ecosystems, P is one of the most important growth-limiting nutrients for plants. In soil, there may be many sources of P viz., in solution as orthophosphate, ionically bound or bound in organic compounds. The major part of soil P (sometimes as much as 90%) is sequestered in the organic compounds phosphomonoesters and phosphodiesters (Nygren 2008). The P uptake by forest trees has been shown to be greatly enhanced in plants colonized by ectomycorrhizal fungi (Conn and Dighton 2000; Courty et al. 2005). The phosphatase enzyme capabilities of ectomycorrhizal fungi are continuously distributed between species rather than restricted to a particular taxonomic group (Nygren 2008). Sheathing mycorrhizal fungi have been shown to possess phosphatase enzymes which can hydrolyze inositol hexaphosphate. Phosphatase production by basidiomycete fungi in liquid culture is independent of P in the medium. Saprophytic basidiomycetes tend to incorporate hydrolyzed phosphate into their biomass. In contrast, mycorrhizal fungi release more hydrolyzed phosphate into solution than they absorb (Dighton 1983).

Ectomycorrhizal fungi are able to secrete hydrolytic enzymes involved in the degradation of organic matter (Burns and Dick 2002; Lindahl et al. 2005). Acid phosphatases solubilize insoluble forms of P not readily available to uninfected plant roots (Tibbett et al. 1998). These enzymes are generally bound to the outer cell walls (Rast et al. 2003). Phosphatase activities of ectomycorrhizal fungi can vary between species, resulting in different efficiency of P utilization of host plant (Dighton 1983). These enzymes are in direct contact with soil environment but are able to adapt to various soil conditions and maintain activity. However, it is known that soil components, pH, and trace elements modify the conformation of enzymes and affect their activities. Activities of acid phosphatase are found to differ significantly amongst ECM synthesized with different fungi and among different species of the same fungi (Buée et al. 2005, 2007; Courty et al. 2006).

C. tropicalis showed maximum growth in acidic culture medium (Sharma 2008; Baghel et al. 2009). The p-nitrophenol phosphatase (p-NPPase) activity of *C. tropicalis* isolate showed a stable activity at a pH range of 3–4 with optimum being 4 even though the mycelial biomass production was less as compared with pH 3 (Table 18.2). The activity of p-NPPase for *C. tropicalis* isolate dropped significantly above pH 4.0. Experiments conducted at higher pH up to 12 to detect any alkaline phosphatase activity yielded negative results (Baghel et al. 2009). Baghel et al. (2008) have also studied the acid phosphatase activity of soil of these bamboo forests. Surface-bound phosphoesterases activities of *P involutus*,

Table 18.2 Effects of pH and temperature on the acid phosphatase activity of *C. tropicalis*

pH study		Temperature study	
Enzyme incubation pH	Enzyme activity (mg g^{-1} mdw)	Incubation temperature (±2 C)	Enzyme activity (mg g^{-1} mdw)
1	0.30 ± 0.01 a	5	3.73 ± 0.05 a
2	0.97 ± 0.03 b	10	3.57 ± 0.05 b
3	3.90 ± 0.28 cd	15	4.50 ± 0.14 c
4	4.16 ± 0.54 d	20	5.67 ± 0.33 d
5	3.71 ± 0.15 cf	25	6.66 ± 0.08 e
6	3.81 ± 0.26 cf	30	11.13 ± 0.81 f
7	3.55 ± 0.18 f	35	8.80 ± 0.33 g
8–12	–	40	18.48 ± 0.55 h

Initial pH was 5.5; culture was 15 days old; results are average of mycelial dry weights (mdw) with standard deviations; values within a column followed by the same alphabet (viz., a, b, c, etc.) do not differ significantly at $p < 0.05$
Source: Adapted from Baghel et al. (2009)

Austropaxillus boletinoides, *Descolea antartica*, *C. geophilum*, and *P. tinctorius* have been reported at pH 3–7 (Alvarez et al. 2005). Investigations by Antibus et al. (1986) indicated that *Hebeloma*, *Paxillus*, *Entoloma*, and *C. geophilum* were typified by sharp decrease in phosphatase activity above pH 5.

18.4 Conclusion

Several devices have been developed for *in vitro* ectomycorrhizal synthesis but most of them have been found to be bulky and does not allow handling of several experimental units simultaneously. Although considerable attention has been given to synthesis of pine ECM, the techniques used were found difficult to apply to angiosperms. A filter paper-flask technique developed by the authors was quite handy, requires limited space, forms rapid ECM, and permits observation of ECM formation. The technique was found highly successful in a maiden attempt of ectomycorrhization in a monocot plant, *D. strictus* by *C. tropicalis* (Sharma et al. 2009b). The technique not only appears to be useful for the production of sufficient mycorrhizal seedlings for biochemical and physiological studies but can also produce desired quantity of mycorrhizal material for practical forestry applications. In depth, understanding of the structure of ECM and its functioning in the performance of tailored seedlings in nursery and field conditions are the ultimate objectives of research on *in vitro* ectomycorrhization. However, further development of the techniques involved in artificial synthesis of ECM is desirable.

Ectomycorrhization was influenced by solid or liquid inoculum, physical, chemical and biological characteristics of the substrate as confirmed by used tea leaves and sand mixture. The substrate stored sufficient amount of moisture which helped the fungus to grow actively. Several types of natural or laboratory produced inocula and various methods of application have been reported to be successful through the years. Ectomycorrhization helps in survival and growth of host seedlings and also increases host ability for nutrient and water uptake. The mycobiota of forest soils consists of AM fungi, ECM fungi, and saprotrophic decomposers and pathogens. ECM fungi interact with other microorganisms, the outcome of which may be inhibitory or stimulatory, some competitive, while others mutualistic, which are important in biogeochemical cycling in forest ecosystems. Antagonistic interactions with mycorrhizal fungi have significant effect on mycorrhizal systems. In the present studies, *C. tropicalis* was found highly antagonistic to some rhizosphere soil microorganisms. Growth enhancement of host plant associated with ECM fungi has been often correlated with increased nutrient uptake by ectomycorrhizal which is due to secretion of surface-bound extracellular enzymes including phosphatases, laccases, xylosidase, cellulohydrolase, etc. There are several reports of acid phosphatase activity of ECM fungi, which help in selecting effective mycorrhizal symbionts for field inoculation of seedlings in reforestation program. However, there are inadequate data on diversity and distribution of enzymes activities in native ectomycorrhizal communities.

Work in future will focus on several aspects which are essential for complete study of the mushroom. Bacteria affect functioning of ectomycorrhizal symbioses including establishment, mobilization of minerals, nitrogen fixation, and antagonism of pathogens (Frey-Klett et al. 2007). Few attempts have been made to characterize bacterial communities associated with ECM (Mogge et al. 2000; Bertaux et al. 2005; Burke et al. 2008). In future studies, the bacterial population associated with this mushroom has to be studied along with its role which will highlight their beneficial role in the life cycle of *Cantherellus*. Moreover, Danell et al. (1993) has already demonstrated the role of *Pseudomonas* on culturing of *C. cibarius*. ECM fungi are able to alleviate the stress for plants caused by heavy metal contamination of soil. Analyzes of molecular response of ECM fungi to these pollutants will lead to new insights in the study of *C. tropicalis*. There is a need to investigate the kinetics, energetics, and specificity of amino acid transporter from the ECM fungus *C. tropicalis*.

The amplification of internal transcribed spacer (ITS) of ribosomal RNA genes using PCR and subsequent RFLP or sequence analysis offers promise for improved species or strain level identification of ectomycorrhizal fungi. Intrageneric homogeneity and intraspecific heterogeneity of these characters are however, frequent. Therefore, there is a need to study intraspecific variation among various strains of *Cantharellus* which holds the future of our research. The size and spatial distribution of mycelial individuals of *C. tropicalis* could be studied using RAPD and microsatellite data and genetically distinct individuals can be recognized which will help in knowing its genetic makeup in the region. In fact, the present trend is to use the advance molecular tools and built-up capability to manipulate the symbionts to our advantage (Krishna 2005). There is a need to study field plantation and study its interaction with other fungi and organisms. All these studies will help to gather information which will be suitable to study the biology of *C. tropicalis*.

Finally, sustainable forest management and the establishment of plantation of fast growing tree species are key issues in present day tropical forestry. Bamboos are the most important feature of tropical forests along with dipterocarps and other broad leaf trees. Thus, management of the existing natural forests for sustained timber supply, research in to the role of ectomycorrhizas in seedling establishment, and growth of plant is need of hour.

References

Alvarez M, Godoy R, Heyser W, Hartel S (2005) Anatomically physiological determination of surface bound phosphatase activity of ectomycorrhizae of *Nothofagus oblique*. Soil Biol Biochem 37:125–132

Antibus RK, Kroehler CJ, Linkins AE (1986) The effects of external pH temperature and substrate concentration on acid phosphates activity of ectomycorrhizal fungi. Can J Bot 64:2383–2387

Aučina A, Rudawska M, Leski T, Skridaila A, Riepsas E, Iwanski M (2007) Growth and mycorrhizal community structure of *Pinus sylvestris* seedlings following the addition of forest litter. Appl Environ Microbiol 73:4876–4883

Baar J, Stanton NL (2000) Ectomycorrhizal fungi challenged by saprotrophic basidiomycetes and soil micro fungi under different ammonium regimes *in vitro*. Mycol Res 104:691–697

Baghel RK, Sharma R, Pandey AK (2008) Acid phosphatase activity in rhizospheric soils of tropical bamboo forests. J Mycol Plant Pathol 38(3):596–598

Baghel RK, Sharma R, Pandey AK (2009) Activity of acid phosphatase in the ectomycorrhizal fungus *Cantharellus tropicalis* under controlled conditions. J Trop For Sci 21(3):222–226

Baxter JW, Dighton J (2005) Phosphorus source alters host plant response to ectomycorrhizal diversity. Mycorrhiza 15:513–523

Bertaux J, Schmid M, Hutzler P, Hartmann A, Garbaye J, Frey-Klett P (2005) Occurrence and distribution of endobacteria in the plant associated mycelium of the ectomycorrhizal fungus *Laccaria bicolor* S238N. Environ Microbiol 7:1786–1795

Bruns TD, Bidartondo MI (2002) Molecular windows into the below-ground interactions of ectomycorrhizal fungi. Mycologist 16:47–50

Bruns TD, Peay KG, Boynton PJ, Grubisha LC, Hynson NA, Nguyen NH, Rosenstock NP (2009) Inoculum potential of *Rhizopogon* spores increases with time over the first 4 yr of a 99-yr spore burial experiment. New Phytol 181(2):463–470

Buée M, Vairelles D, Garbaye J (2005) Year-round monitoring of diversity and potential metabolic activity of the ectomycorrhizal community in a beech forest subjected to two thinning regimes. Mycorrhiza 15:235–245

Buée M, Courty PE, Mignot D, Garbaye J (2007) Soil niche effect on species diversity and catabolic activities in an ectomycorrhizal fungal community. Soil Biol Biochem 39:1947–1955

Burke DJ, Dunham SM, Kretzer AM (2008) Molecular analysis of bacterial communities associated with the roots of Douglas fir (*Pseudotsuga menziesii*) colonized by different ectomycorrhizal fungi. FEMS Microbiol Ecol 65:299–309

Burns GB, Dick RP (2002) Enzymes in the environment; activity ecology and applications. Marcel Dekker, New York

Cairney JWG (2005) Basidiomycete mycelia in forest soils: dimensions, dynamics and roles in nutrient distribution. Mycol Res 109(1):7–20

Cairney JWG, Chambers SM (1999) Ectomycorrhizal fungi-key genera in profile. Springer-Verlag, Berlin

Cairney JWG, Meharg AA (2002) Interactions between ectomycorrhizal fungi and soil saprotrophs: implications for decomposition of organic matter in soils and degradation of organic pollutants in the rhizosphere. Can J Bot 80(8):803–809

Cameron W, Margaret SP, Johnston E, Ramage CM, Edwards DG, Cawthray GR, Lambers H (2006) Functional significance of dauciform roots: exudation of carboylates and acid phosphatase under phosphorus deficiency in *Caustis blakei* (Cyperaceae). New Phytol 170:491–500

Chen YL, Kang LH, Dell B (2006) Inoculation of *Eucalyptus urophylla* with spores of *Scleroderma* in a nursery in South China: comparison of field soil and potting mix. For Ecol Manage 222:439–449

Chilvers GA, Douglas PA, Lapeyrie FF (1986) A paper sandwich technique for rapid synthesis of ectomycorrhizas. New Phytol 103:397–402

Conn C, Dighton J (2000) Litter quality influences on decomposition, ectomycorrhizal community structure and mycorrhizal root surface acid phosphatase activity. Soil Biol Biochem 32:489–496

Courty PE, Pritsch K, Schloter M, Hartmann A, Garbaye J (2005) Activity profiling of ectomycorrhiza communities in two forest soils using multiple enzymatic tests. New Phytol 167 (1):309–319

Courty PE, Pouysegur R, Bùee M, Garbaye J (2006) Laccase and phosphatase activities of the dominant ectomycorrhizal types in a lowland oak forest. Soil Biol Biochem 38(6):1219–1222

Courty P-E, Bréda N, Garbaye J (2007) Relation between oak tree phenology and the secretion of organic matter degrading enzymes by *Lactarius quietus* ectomycorrhizas before and during bud break. Soil Biol Biochem 39(7):1655–1663

Cullings K, Ishkhanova G, Hensen J (2008) Defoliation effects on enzyme activites of the ectomycorrhizal fungus *Suillus granulatus* in a *Pinus contorta* (lodgepole pine) stand in Yellostone National Park. Oecologia 158(1):77–83

Dahlstrom JL, Smith JE, Weber NS (2000) Mycorrhiza like interaction by *Morchella* with species of the Pinaceae in pure culture synthesis. Mycorrhiza 9:279–285

Danell E (1994) Formation and growth of the ectomycorrhiza of *Cantharellus cibarius*. Mycorrhiza 5:89–97

Danell E (1999) *Cantharellus*. In: Cairney JWG, Chambers SM (eds) Ectomycorrhizal fungi-key genera in profile. Springer-Verlag, Berlin, pp 253–267

Danell E (2002) Current research on chantharelle cultivation in Sweden. In: Hall I, Wang Y, Danell E, Zambonelii A (eds) Edible mycorrhizal mushrooms and their cultivation. Crop & Food Research, Christchurch, pp 1–4

Danell E, Camacho FJ (1997) Successful cultivation of the golden chanterelle. Nature 385:303

Danell E, Sadhna A, Ternstrom A (1993) *Pseudomonas fluorescens* in association with fruit bodies of the ectomycorrhizal mushroom *Cantharellus cibarius*. Mycol Res 97(9):1148–1152

Dighton J (1983) Phosphatase production by mycorrhizal fungi. Plant Soil 71:455–462

Duddridge JA (1986) The development and ultrastructure of ectomycorrhizae III. Compatible and incompatible interactions between *Suillus grievllei* (Klotzsch) Sing, and species of ectomycorrhizal hosts *in vitro* in the absence of exogenous carbohydrate. New Phytol 103:457–464

Finlay RD (2004) Mycorrhizal fungi and their multifunctional role. Mycologist 18(2):91–96

Fitter AH, Garbaye J (1994) Interactions between mycorrhizal fungi and other soil organisms. Plants Soil 159:123–132

Fortin JA, Piche Y, Lalonde M (1980) Technique for the observation of early morphological changes during ectomycorrhiza formation. Can J Bot 58:361–365

Frey-Klett P, Garbaye J, Tarkka M (2007) The mycorrhiza helper bacteria revisited. New Phytol 176:22–36

Garbaye J, Delwaulle JC, Diangana D (1988) Growth response of eucalypts in the Congo to ectomycorrhizal inoculation. For Ecol Manage 24(2):151–157

Gibson BR, Mitchell DT (2005) Phosphatses of ericoid mycorrhizal fungi: kinetic properties and the effect of copper on activity. Mycol Res 109(4):478–486

Girlanda M, Varese GC, Luppi-Mosca AM (1995) *In vitro* interactions between saprotrophic microfungi and ectomycorrhizal symbionts. Allionia 33:81–86

Guerin-Laguette A, Plassard C, Mousain D (2000a) Effects of experimental conditions on mycorrhizal relationships between *Pinus sylvestris* and *Lactarius deliciosus* and unprecedented fruit-body formation of the Saffron milk cap under controlled soil less conditions. Can J Microbiol 46:790–799

Guerin-Laguette A, Vaario L, Gill WM, Lapeyrie F, Matsushita N, Suzuki K (2000b) Rapid *in vitro* ectomycorrhizal infection on *Pinus densiflora* roots by *Tricholoma matsutake*. Mycoscience 41:389–393

Harley JL, Smith SE (1983) Mycorrhizal symbiosis. Academic Press Inc., London, p 483

Hattori H (1992) Influence of heavy metals on soil microbial activities. Soil Sci Plant Nutr 38:93–100

Johansson J, Paul L, Finlay RD (2004) Microbial interactions in the mycorrhizosphere and their significance for sustainable agriculture. FEMS Microbiol Ecol 18:1–13

Khosla B, Reddy S (2008) Response of ectomycorrhizal fungi on the growth and mineral nutrition of *Eucalyptus* seedlings in bauxite mined soil. Am Eurasian J Agric Environ Sci 3(1):123–126

Kranabetter JM, Friesen J, Gamiet S, Kroeger P (2005) Ectomycorrhizal mushroom distribution by stand age in western hemlock–lodgepole pine forests of North-Western British Columbia. Can J For Res 35:1527–1539

Krishna KR (2005) Mycorrhizas: a molecular analysis. Oxford and IBH Publishing Co. Pvt. Ltd., New Delhi

Kumar S, Satyanarayana T (2002) Production of inoculum of ectomycorrhizal fungi. In: Mukerji KG, Manoharachary C, Chamola BP (eds) Techniques in mycorrhizal studies. Kluwer Academic Publishers, Netherlands, pp 143–166

Lakhanpal TN (1996) Mushrooms of India Boletaceae. In: Mukerji KG (ed) Studies in cryptogamic botany, vol I. APH Publishing Corporation, Delhi

Lamb RJ, Richards BN (1974) Survival of potential of sexual and asexual spores of ectomycorrhizal fungi. Trans Br Mycol Soc 62:181–191

Lamhamedi MS, Abourouh M, Fortin JA (2009) Technological transfer: the use of ectomycorrhizal fungi in conventional and modern forest tree nurseries in northern Africa. In: Khasa D, Piche Y, Coughlan AP (eds) Advances in mycorrhizal science and technology. NRC Research Press, Ottawa, p 197

Lapeyrie FF, Bruchet G (2006) Some factors influencing viability of ectomycorrhizal fungal inoculum. New Phytol 100(4):585–593

Last FT, Natarajan K, Mohan V, Mason PA (1992) Sequences of sheathing (ecto)mycorrhizal fungi associated with man-made forests, temperate and tropical. In: Read DJ, Lewis DH, Fitter AH, Alexander IJ (eds) Mycorrhizas in ecosystems. CAB International, Wallingford, pp 214–219

Leake JR, Johnson D (2004) Networks of power and influence: the role of mycorrhizal mycelium in controlling plant communities and agro ecosystem functioning. Can J Bot 82(8):1016–1045

Leake JR, Donnelly DP, Saunders EM, Boddy L, Read DJ (2001) Rates and quantities of carbon flux to ectomycorrhizal mycelium following ^{14}C pulse labeling of *Pinus sylvestris* L. seedlings: effects of litter patches and interaction with a wood-decomposer fungus. Tree Physiol 21:71–82

Lindahl BD, Finlay RD, Cairney JWG (2005) Enzymatic activities of mycelia in mycorrhizal fungal communities. In: Dighton J (ed) The fungal community, its organization and role in the ecosystem. Marcel Dekker, New York, pp 331–348

Littke WR, Bledsoe CS, Nadkarni NM, Edmonds RL (1980) Technique for rapid mycorrhizal colonization of container-grown Douglas-fir by *Hebeloma crustuliniforme*. Soil Biol Biochem 12:575–578

Marx DH (1980) Ectomycorrhiza fungus inoculations, a tool for improving forestation practices. In: Mikola P (ed) Tropical mycorrhiza research. Oxford University Press, Oxford, pp 13–71

Marx DH, Bryan WC (1975) Growth and ectomycorrhizal development of loblolly pine seedlings in fumigated soil infested with the fungal symbiont *Pisolithus tinctorius*. For Sci 21:245–254

Marx DH, Ruehle JL, Kennedy DS, Cordell CE, Riffle JW, Ruehle JW, Molina RJ, Pawac WH, Nauratil S, Tinus RW, Goodwin OC (1982) Commercial vegetative inoculum of *Pisolithus tinctorius* and inoculation techniques for development of ectomycorrhizae on container grown seedlings. For Sci 28:373–400

Meyselle JP, Gay G, Debaud JC (1991) Intraspecific genetic variation of acid phosphatase activity in monokaryotic and dikaryotic populations of ectomycorrhizal fungus *Hebeloma cylindrosporum*. Can J Bot 69:8089–8113

Mogge B, Loferer C, Agerer R, Hutzler P, Hartmann A (2000) Bacterial community structure and colonization patterns of *Fagus sylvatica* L. ectomycorrhizospheres as determined by fluorescent *in situ* hybridization and confocal laser scanning microscopy. Mycorrhiza 9:271–278

Molina R, Trappe JM (1982) Applied aspects of ectomycorrhiza. In: Subbarao NS (ed) Advances in agricultural microbiology. Oxford and IBH Publishing Co., New Delhi, pp 305–324

Molina R, Smith JE, McKay D, Melville LH (1997) Biology of the ectomycorrhizal genus *Rhizopogon*. New Phytol 137:519–528

Morin C, Samson J, Dessureault M (1999) Protection of black spruce seedlings against *Cylindrocladium* root rot with ectomycorrhizal fungi. Can J Bot 77(1):169–174

Mosca E, Montecchio L, Scattolin L, Garbaye J (2007) Enzymatic activities of three ectomycorrhizal types of *Quercus robur* L. in relation to tree decline and thinning. Soil Biol Biochem 39 (11):2897–2904

Moser M (1958) Die Kiinstliche Mycorrhizaimp fung an Forstpflanzen. I. Erfahrungen bei der Reinkulture von Mycorrhizapilzen. Forest wiss Centralbl 77:32–40

Nannipieri P (1995) The potential use of soil enzymes as indicators of productivity, sustainability and pollution. In: Pankhurst CE, Doube BM, Gupta VVSR, Gracel PR (eds) Soil biota: management in sustainable farming systems. CSIRO Publishing, East Melbourne, pp 238–244

Natarajan K, Govindasamy V (1990) Antagonism of ectomycorrhizal fungi to some common root pathogens. In: Jalali BL, Chand H (eds) Current trends in mycorrhizal research. Sankat Mochan Art Press, Hisar, India, pp 98–99

Natarajan K, Ravindran C (2003a) Two new species of the genus *Entoloma* from South India. Mycotaxon 85:143–146

Natarajan K, Ravindran C (2003b) Two new species of the genus *Pholiota* from South India. Mycotaxon 85:271–275

Natarajan K, Nagarajan G, Sudhakara Reddy M (1995) *In vitro* mycorrhization and growth response of *Acacia nilotica* seedlings by inoculation with ectomycorrhizal fungi. Ind J Microbiol 35(1):35–38

Natarajan K, Senthilarasu G, Kumaresan V, Riviere T (2005a) Diversity in ectomycorrhizal fungi of a dipterocarp forest in Western Ghats. Curr Sci 88(12):1893–1895

Natarajan K, Narayan K, Ravindran C, Kumaresan V (2005b) Biodiversity of agarics from Nilgiri Biosphere Reserve, Western Ghats, India. Curr Sci 88(12):1890–1893

Nezzar-Hocine H, Pen-In R, Halli-Hargas R, Chevalier G (1998) Ectomycorrhizal associations with *Cedrus atlantica* (Endl) Manetti ex Carriere-I, mycorrhizal synthesis with *Tricholoma tridentinum* Singer var. *cedretorum* Bon. Mycorrhiza 8:47–51

Nygren C (2008) Functional diversity in nutrient acquisition by ectomycorrhizal fungi. PhD thesis, Swedish University of Agricultural Sciences, Uppsala

Olsson PA (1999) Signature fatty acids provide tools for determination of the distribution and interactions of mycorrhizal fungi in soil. FEMS Microbiol Ecol 29(4):303–310

Pande V, Palni UT, Singh SP (2004) Species diversity of ectomycorrhizal fungi associated with temperate forest of Western Himalaya: a preliminary assessment. Curr Sci 86(12):1619–1623

Pande V, Palni UT, Singh SP (2007) Effect of ectomycorrhizal fungal species on the competitive outcome of two major forest species. Curr Sci 92(1):80–84

Park JE (1984) Inoculum potential of ectomycorrhizal fungi in forest soil from South-West Oregon and Northern California. For Sci 30:300–304

Rast DM, Baumgartner D, Mayer C, Hollenstein GO (2003) Cell wall-associated enzymes in fungi. Phytochemistry 64:339–366

Read DJ, Perez-Moreno J (2003) Mycorrhizas and nutrient cycling in ecosystems – a journey towards relevance? New Phytol 157:475–492

Riviere T, Diedhiou AG, Diabate M, Senthilarasu G, Natarajan K, Verbeken A, Buyck B, Dreyfus B, Bena G, Ba Amadou M (2007) Genetic diversity of ectomycorrhizal basidiomycetes from African and Indian tropical rain forests. Mycorrhiza 17(5):415–428

Rosling A, Rosenstock N (2008) Ectomycorrhizal fungi in mineral soil. Minerol Mag 72 (1):127–130

Sampangiramaiah K, Perrin R (1990) Interactions between isolates of ectomycorrhizal *Laccaria* spp. and root rot fungi of conifers. In: The proceedings of national conference on mycorrhiza, Harayana Agricultural University, Hissar, India, pp 124–125

Santoro T Jr, Casida LE (1962) Elaboration of antibiotics by *Boletus luteus* and certain other mycorrhizal fungi. Can J Microbiol 8:43–48

Sharma R (2008) Studies on ectomycorrhizal mushrooms of M.P. and Chhattisgarh. PhD thesis, R.D. University, Jabalpur, India

Sharma ND, Nema S, Rai BK (2003) A simple technique for obtaining pure culture of *Pleurotus* spp. J Mycopathol Res 41(1):119–120

Sharma R, Rajak RC, Pandey AK (2008a) Some ectomycorrhizal mushrooms of Central India – I. *Russula*. J Mycopathol Res 46(2):201–212

Sharma R, Rajak RC, Pandey AK (2008b) Growth response of *Dendrocalamus* seedlings by inoculation with ectomycorrhizal fungi. Middle East J Sci Res 3(4):200–206

Sharma R, Rajak RC, Pandey AK (2009a) Some ectomycorrhizal mushrooms of Central India – II. *Lactarius*. J Mycopathol Res 47(1):43–47

Sharma R, Rajak RC, Pandey AK (2009b) Simple technique for ectomycorrhizal formation between *Cantharellus* and *Dendrocalamus strictus*. Taiwan J For Sci 24(2):141–148

Sharma R, Rajak RC, Pandey AK (2009c) Ectomycorrhizal mushrooms in Indian tropical forests. Biodiversity 10(1):25–30

Sharma R, Rajak RC, Pandey AK (2010a) Mass multiplication of ectomycorrhizal *Cantharellus* inoculum for large scale tailoring nursery inoculations of bamboo seedlings. Asian J Sci Res, published online

Sharma R, Rajak RC, Pandey AK (2010b) Some ectomycorrhizal mushrooms of Central India-V. *Pisolithus, Scleroderma, Geastrum, Cantharellus*. J Mycopathol Res, in press

Sharma R, Rajak RC, Pandey AK (2010c) Evidence of antagonistic interactions between rhizosphere and mycorrhizal fungi associated with *Dendrocalamus strictus* (Bamboo). J Yeast Fungal Res 1(7):112–117

Sharma R, Baghel RK, Pandey AK (2010d) Dynamics of acid phosphatase production of the ectomycorrhizal mushroom *Cantharellus tropicalis*. Afr J Microbiol Res 4(18), in press

Shaw TM, Dighton J, Sanders FE (1995) Interactions between ectomycorrhizal and saprotrophic fungi on agar and in association with seedlings of lodgepole pine (*Pinus contorta*). Mycol Res 99:159–165

Smith SE, Read DJ (1997) Mycorrhizal symbiosis, 2nd edn. Academic Press, London

Stark S, Kytöviita M-M (2005) Evidence of antagonistic interactions between rhizosphere microorganisms and mycorrhizal fungi associated with birch (*Betula pubescens*). Acta Oecologica 28(2):149–155

Straatsma G, van Griensven LJLD (1986) Growth requirements of mycelial cultures of the mycorrhizal mushroom, *Cantharellus cibarius* Fr. Trans Br Mycol Soc 87:135–141

Sun YP, Unestam T, Lucas SD, Johanson KJ, Kenne L, Finlay R (1999) Exudation-reabsorption in a mycorrhizal fungus, the dynamic interface for interaction with soil and soil microorganisms. Mycorrhiza 9:137–144

Takacs EA (1961) Inoculation de especies de pinos con hongos formadores de micorizas. Silvicultura (Uruguay) 15:5–17

Theodorou C (1971) Introduction of mycorrhizal fungi into soil by spore inoculation of seed. Aust For 35:17–22

Theodorou C, Bowen GD (1973) Introduction of seeds and soil with basidiospores of mycorrhizal fungi. Soil Biol Biochem 5:765–771

Theodorou C, Reddell P (2006) *In vitro* synthesis of ectomycorrhizas on Casuarinaceae with a range of mycorrhizal fungi. New Phytol 118(2):279–288

Tibbett M, Chambers SM, Cairney JWG (1998) Method for determining extracellular and surface-bound phosphatase activities in ectomycorrhizal fungi. In: Varma A (ed) Mycorhhiza manual. Springer, New York, pp 217–226

Tibbett M, Cairney WG (2007) The cooler side of mycorrhizas: their occurrence and functioning at low temperatures. Can J Bot 85(1):51–62

Trappe JM (1977) Selection of fungi for ectomycorrhizal inoculation in nurseries. Annu Rev Phytopathol 15:203–222

Vaario L, Tanaka M, Ide Y, Gill WM, Suzuki K (1999) *In vitro* ectomycorrhiza formation between *Abies firma* and *Pisolithus tinctorius*. Mycorrhiza 9:177–183

Vaario L, Gill WM, Tanaka M, Ide Y, Suzuki K (2000) Aseptic ectomycorrhizal synthesis between *Abies firma* and *Cenococcum geophilum* in artificial culture. Mycoscience 41:395–399

Vaidya GS, Shrestha K, Wallander H (2005) Antagonistic study of ectomycorrhizal fungi isolated from Baluwa forest (Central Nepal) against with pathogenic fungi and bacteria. Sci World 3(3):49–52

Valentine LL, Fiedler TL, Hart AN, Petersen CA, Berninghausen HK, Southworth D (2004) Diversity of ectomycorrhizas associated with *Quercus garryana* in southern Oregon. Can J Bot 82:123–135

Werner A, Zadworny M (2003) *In vitro* evidence of mycoparasitism of the ectomycorrhizal fungus *Laccaria laccata* against *Mucor hiemalis* in the rhizosphere of *Pinus sylvestris*. Mycorrhiza 13(1):41–47

Werner A, Zadworny M, Idzikowska K (2002) Interaction between *Laccaria laccata* and *Trichoderma virens* in co-culture and in the rhizosphere of *Pinus sylvestris* grown in vitro. Mycorrhiza 12(3):139–145

Whipps JM (2001) Microbial interactions and biocontrol in the rhizosphere. J Exp Bot 52:487–511

Whipps JM (2004) Prospects and limitations for mycorrhizas in biocontrol of root pathogens. Can J Bot 82:1198–1227

Wu T, Sharda JN, Koide RT (2003) Exploring interactions between saprotrophic microbes and ectomycorrhizal fungi using a protein-tannin complex as an N source by red pine (*Pinus resinosa*). Special issue: functional genomics of plant–pathogen interactions. New Phytol 159(1):131–139

Yamada A, Katsuya K (1995) Mycorrhizal association of isolates from sporocarps and ectomycorrhizas with *Pinus densiflora* seedlings. Mycoscience 40:455–463

Yamada A, Maeda K, Ohmasa M (1999) Ectomycorrhiza formation of *Tricholoma matsutake* isolates on seedlings of *Pinus densiflora in vitro*. Mycoscience 40:455–463

Yamada A, Ogura T, Ohmasa M (2001a) Cultivation of mushrooms of edible ectomycorrhizal fungi associated with *Pinus densiflora* by *in vitro* mycorrhizal synthesis. I. Primordium and basidiocarp formation in open-pot culture. Mycorrhiza 11:59–66

Yamada A, Ogura T, Ohmasa M (2001b) Cultivation of mushrooms of edible ectomycorrhizal fungi associated with *Pinus densiflora* by *in vitro* mycorrhizal synthesis. II. Morphology of mycorrhizas in open-pot soil. Mycorrhiza 11:67–81

Zak B (1971) Characterization and classification of mycorrhizae of Douglas-fir-II, *Pseudostuga menziesii* + *Rhizopogon vinicolor*. Can J Bot 49:1079–1084

Chapter 19
Edible Ectomycorrhizal Fungi: Cultivation, Conservation and Challenges

Alka Karwa, Ajit Varma and Mahendra Rai

19.1 Introduction

A mycorrhiza (Gk: *fungus roots*) is a symbiotic association between a fungus and the roots of a vascular plant (Frank 1885; Brundrett 2004). In a mycorrhizal association, the fungus colonizes roots of the host plants, either intracellularly as in endomycorrhizal fungi, also known as arbuscular mycorrhiza (AM), or extracellularly as in ectomycorrhizal fungi (ECM). Most of the plants are equally dependent on mycorrhizal fungi, and without them, the plants become stunted and yellow, often due to a lack of phosphorus (Tarafdar and Marschner 1994; Schweiger et al. 1995; Kahiluoto and Vestberg 1998; Redecker et al. 2000). "Plants without mycorrhizal fungi are competitively inferior. Graham stated that mycorrhizal fungi function as an auxiliary root system to provide additional nutrients" (Harvey et al. 1987; Graham et al. 1994, 1999). The mutualistic association provides the fungus with relatively constant and direct access to carbohydrates, such as glucose and sucrose supplied by the plant (Cook 1977; Harrison 2005). The carbohydrates are translocated from their source, usually leaves, to root tissue and on to the fungal partners. In return, the plant gains the benefits of the higher absorptive capacity of mycelium for water and mineral nutrients due to comparatively large surface area of mycelium:root ratio (Tarafdar and Marschner 1994; Schweiger et al. 1995; Kahiluoto and Vestberg 1998), thus improving the plant's mineral absorption capabilities (Selosse et al. 2006). Plant roots alone may be incapable of taking up phosphate ions that are demineralized, for example, in soils with a basic pH (Abbott and Robson 1991). The mycelium of the mycorrhizal fungus can, however, access these phosphorus sources and make them available to the plants they colonize (Li et al. 2006). Thus, the mechanisms of increased absorption are both physical

A. Karwa (✉) and M. Rai
Department of Biotechnology, SGB Amravati University, Amravati 444 602, Maharashtra, India
e-mail: alka0404@gmail.com

A. Varma
Amity Institute of Microbial Technology, Amity University, Noida, Uttar Pradesh, India
e-mail: ajitvarma@aihmr.amity.edu

M. Rai and A. Varma (eds.), *Diversity and Biotechnology of Ectomycorrhizae*,
Soil Biology 25, DOI 10.1007/978-3-642-15196-5_19,

and biochemical. Mycorrhizal mycelia are much smaller in diameter than the smallest root hairs, and thus can explore a greater volume of soil, providing a larger surface area for absorption (Tuomi et al. 2001). Also, the cell membrane chemistry of fungi is different from that of plants (Ogawa 1985; Agerer 1995; Unestam and Sun 1995) that include organic acid excretion which aids in ion displacement (Griffiths and Caldwell 1992; http://cropsoil.psu.edu/sylvia/mycorrhiza.htm). Mycorrhizae are thus especially beneficial for the plant partner in nutrient-poor soils (Malajczuk et al. 1982).

19.2 Occurrence

The concept that plants have varying degrees of dependence on mycorrhizal associations has gained acceptance long back (Janos 1980; St John 1980; Brundrett 1991; Marschner 1995). Mycorrhizae are present in 95% of plant families (Trappe 1987; Wang and Qiu 2006), with endomycorrhizae being the ancestral and predominant form (Wang and Qiu 2006) and indeed the most prevalent symbiotic association found in the entire plant kingdom (Harrison 2005). The structure of endomycorrhizae has been highly conserved since their first appearance in the fossil record (Pirozynski and Dalpé 1989; Remy et al. 1994; Redecker et al. 2000; Dotzler et al. 2006; Strullu-Derrien and Strullu 2007) with both the development of ectomycorrhizae and the loss of mycorrhizae, evolving convergently on multiple occasions (Wang and Qiu 2006). Both endomycorrhiza and ectomycorrhiza have the ability to be mycorrhizal fungi. In endomycorrhiza, the hyphae of the fungus penetrate the outer cells of the plant root and extend into the surrounding soil (Schultze et al. 1997; Addy et al. 2005; Schultz and Boyle 2005). Whereas in ectomycorrhiza, the hyphae surround, but do not penetrate the roots. Endomycorrhiza are much more common than ectomycorrhiza. The fungal component of endomycorrhiza is a zygomycete. While only about 30 species of zygomycetes are known to be involved in endomycorrhizal relationships, the zygomycetes are associated with more than 200,000 species of plants (85%) (Wang and Qiu 2006).

Basidiomycetes are the most common fungal components of ectomycorrhiza, although some ascomycetes also form ectomycorrhizal relationships (Hosaka et al. 2007; Wilson et al. 2007). More species of fungi are involved in ectomycorrhiza (at least 5,000), but most are only associated with a single species of plant. Furthermore, the total number of plants involved in ectomycorrhiza is limited to a few thousand (only 10%). The most common plants associated with ectomycorrhiza are trees and shrubs growing in temperate regions. These trees include teak, shorea, bamboo, acacia, pines, firs, oaks, beeches, and willows (Rose 1980; Malloch and Malloch 1981; Harley and Harley 1987; Brundett et al. 1990; Redhead 1997; Tedersoo et al. 2007). These plants tend to be more resistant to extreme temperatures, drought, and other harsh environmental conditions. Some ectomycorrhizal fungi may provide protection from acidic precipitation. Mycorrhizal plants are often more resistant to diseases, such as those caused by microbial soil-borne pathogens, and to the effects of drought (Brundrett and Kendrick 1988). These effects are perhaps due to the improved water and mineral uptake in mycorrhizal plants.

19.3 Role of Ectomycorrhizas

Ectomycorrhizas consist of a hyphal sheath, or mantle, covering the root tip and a hartig net of hyphae surrounding the plant cells within the root cortex. In some cases, the hyphae may also penetrate the plant cells, in which case the mycorrhiza is called an ectendomycorrhiza. Outside the root, the fungal mycelium forms an extensive network within the soil and leaf litter (Meyer 1973; Harvey et al. 1976; Garbaye 1994). Nutrients can be shown to move between different plants through the fungal network (sometimes called the wood wide web). Carbon has been shown to move from birch trees into fir trees, thereby promoting succession in ecosystems (Simard et al. 1997). The ectomycorrhizal symbiosis represents one of the most prominent and ecologically crucial mutualistic associations in terrestrial habitats (Marks et al. 1968; Vogt et al. 1981; Hunt and Fogel 1983). ECMs occur in most of the temperate and boreal ecosystems and in large forested areas of tropical and subtropical regions (Smith and Read 1997; Cairney and Chambers 1999; Verbeken and Buyck 2001; Comandini et al. 2006; Wang and Qiu 2006; Rinaldi et al. 2008; Sharma 2008; Baghel et al. 2009). The hyphae of most ECM fungal species proliferate in the duff layer of the forest floor, but some also inhabit mineral soil, and still others prefer decaying wood as a substrate (Goodman and Trofymow 1998). Some ECM fungi require large amounts of carbohydrate, which they acquire from their plant hosts, and so are dependent on mature trees that can meet their carbohydrate demands (Deacon and Fleming 1992).

19.4 Mycorrhizal Mushrooms

There is a wide taxonomic and structural diversity of higher fungi. The reproductive structures of larger fungi include epigeous mushrooms, puffballs, coral fungi, crust fungi, etc., and subterranean (hypogeal) fungi called truffles, or truffle-like fungi. Most of these categories contain ectomycorrhizal mushrooms. The fact to be noted is that each species of host plant is not restricted to just one mycorrhizal fungus. For example, Douglas fir can form mycorrhizas with hundreds of different mycorrhizal fungi. Similarly, some oaks can form ectomycorrhizas with a wide range of fungi, such as *Amanita caesarea*, *Amanita phalloides*, *Boletus edulis*, as well as the Périgord black truffle *Tuber melanosporum*, and it is not unusual to find half a dozen different ECM fungi competing for space on the roots of a suitable host (Hall and Wang 1998). A list of putative 955 mycorrhizal edible and/or medicinal mushrooms has been provided by Hall et al. (2003). This suggests that most mushrooms are ectomycorrhizal. Examples of these mushrooms are *A. caesarea*, *Astraeus hygrometricus*, *B. edulis*, *Cantharellus cibarius*, *Lactarius deliciosus*, *Russula virescens*, and *Tricholoma matsutake*. The mushrooms mentioned in Table 19.1 belong to the clades as classified by Moncalvo et al. (2002) and families

Table 19.1 Epigeous and hypogeous mushrooms with confirmed ectomycorrhizal relations

Sr no	Order/family	Epigeous mushrooms (gilled mushrooms, puffballs, etc.)	Hypogeous and semihypogeous mushrooms (truffles, sclerodermas, etc.)
1	*Amanitaceae*	*Amanita, Limacella*	*Ammarrendia, Torrendia*
2	*Boletales*	*Boletellus, Boletochaete, Boletus, Austroboletus, Aureoboletus, Psiloboletinus, Rubinoboletinus, Fuscoboletinus, Paxillus, Astropaxillus, Gomphidus, Gyroporus, Chroogomphus, Calostoma, Leccinum, Heimielia, Paragyrodon, Phylloporus, Phlebopus, Pisolithus, Poryphyrellus, Scleroderma, Strobilomyces, Tylopilus, Suillus, Xanthoconium, Xerocomus*	*Austrogaster, Austrogautieria, Astaeus, Gastroboletus, Alpova, Gymnopaxillus, Gymnogaster, Horakiella, Melanogaster, Rhizopogon, Sclerogaster, Scleroderma, Chamonixia, Hysterogaster, Mycoamaranthus, Octaviania, Velligaster, Truncocolumella*
3	*Cantharellaes*	*Cantharellus, Cantharellula, Cratrellus*	
4	*Clavariaceae*	*Clavaria, Clavicorna, Clavulina, Clavulinopsis, Clavaridelphus, Ramariopsis, Aphelaria*	
5	*Cortinariaceae*	*Cortinarius, Naucoria, Rozites, dermocybe*	*Cortinarius, Hymenogaster, Quadrispora, Stephanopus, Protoglossum, Destuntzia, Thaxeterogaster, Setchelliogaster*
6	*Elaphomycetaceae*		*Elaphomyces, Pseudotulostoma*
7	*Endogonaceae*		*Endogone, Densospora, Yougimyces, Peridiospora*
8	*Entolomataceae*	Some species of *Entoloma*	*Rhodogaster, Richoniella*
9	*Gomphales, Hysterangiales*	*Gomphus, Boletopsis, Bankera, Clavaridelphus, Hydnum, Hydnellum, Sarcodon, Ramaria, Phellodon*	*Aroramyces, Hysterangium, Chondrogaster, Gautieria, Gummiglobus, Mesophellia, Malajczukia, Austrogautiera, Gallacea, Protrubera, Trappea, Phallogater, Catoreum*
10	*Hydnangiaceae*	*Laccaria*	*Hydnangium, Podohydnangium, Gigaspera*
11	*Hygrophoraceae*	*Hygrophorus, Gliophorus, Humidicutis*	
12	*Hymenogastraceae*	*Hebeloma, Phaeocollybia*	*Hymenogaster*
13	*Inocybaceae*	*Inocybe, Auritella*	*Auritella*
14	*Lyophyllum*	*Lyophyllum*	
15	*Pezizales*	*Geopora, Helvella, Hydnotyra, Pulvinulla, Geopora, Sarcosphaera, Tirmania, Tricharina, Sphaerozone, Genea, Wilcoxina*	*Tuber, Kalaharituber , Terfezia, Turmania, Dingleya, Labyrinthomyces, Pachyphloeus, Redellomyces, Eremiomyces*
16	*Russulales*	*Russula, Lactarius*	*Stephanospora, Cystangium, Arcangeliella, Leucogaster, Gymnomyces, Zelleromyces*
17	*Thelephorales*	*Thelephora, Sarcodon, Phellodon, Bankera*	
18	*Tricholomataceae*	*Tricholoma, Leucopaxillus*	

as described by Matheny et al. (2006) and Hibbett et al. (2007). A survey of literature provides evidence that many researchers have contributed in the field of identity and taxonomy of ectomycorrhizal fungi (Arora 1986, 1991; LoBuglio et al. 1996; Erland and Taylor 1999; Kõljalg et al. 2000; Moncalvo et al. 2000; Bougher and Lebel 2001; Humpert et al. 2001; Peinter et al. 2001; Binder and Bresinsky 2002; Selosse et al. 2002; Tehler et al. 2003; Urban et al. 2003; Binder et al. 2005; Douhan and Rizzo 2005; Ferdman et al. 2005; Henkel et al. 2006; Matheny and Bougher 2006; Matheny et al. 2006; Tedersoo et al. 2006; Trocha et al. 2006; Watling 2006; Barroetaveña et al. 2007; Hibbett et al. 2007; Hosaka et al. 2007; Wilson et al. 2007). Many of them are the confirmed ECM associations by fungal isolation and resynthesis under controlled conditions (Warcup 1980, 1985; Kropp and Trappe 1982; Malajczuk et al. 1982; Molina and Trappe 1982; Bougher and Malajczuk, 1985; Godbout and Fortin 1985; Reddell and Milnes 1992; Massicotte et al. 1994; Thomson et al. 1994; Cripps and Miller 1995; McGee 1996; Kawai 1997; Lu et al. 1998; Reddell et al. 1999; Baxter and Dighton 2001; Yamada et al. 2001a, b; Yun and Hall 2004; Brundret et al. 2005; Henkel et al. 2006). However, the medicinal properties of mycorrhizal mushrooms have not been well-investigated.

Mycorrhizal mushrooms, some of which are among the world's most expensive foods, are of considerable interest to the chef and the gourmet (Table 19.2). Most of these ectomycorrhizal mushrooms are found only in the North of the equator and fruit for short periods during the year. There has been a marked decline in the harvests of some edible mycorrhizal mushrooms over the past century (Hall et al. 2003). Possible reasons for this decline include deforestation, the loss of host plants within forests due to pests or disease, changed forest management practices such as planting more densely than occurs in natural forests, the replacement of natural forests with plantations of species that are poor hosts for edible mycorrhizal mushrooms, global warming since the last ice age, soil compaction by hordes of pickers, acid rain, and, for truffles, the loss of expertise during two World Wars as to

Table 19.2 Market value of commercially important ectomycorrhizal edible mushrooms

Mushroom species	Common name	Estimated world production (t)	Approx. price US$/kg
B. edulis	Porcini	20,000–100,000	20–200 (Hall et al. 1998a; http://en.wikipedia.org/wiki/Boletus_edulis, 2010)
C. cibarius	Chanterelle	220,000	8–19 (Danell 2000; Warner 2010, http://basiceating.blogspot.com/2010/04/Chantharelle-cantharellus.cibarius.html)
T. matsutake	Matsutake	1,000	90–2,000 (Ashkenazi and Jacob 2003, http://en.wikipedia.org/wiki.matsutake, 2010)
Tuber melanogaster	Black Truffle	150	1,000–3,490 (Maison de la 2010)
T. magnatum	White Truffle	10,200	10,200 (Maison de la 2010)

where and how to find them. There has been a spectacular increase of interest and commercial activity concerned with dietary supplements, functional foods, and other products that are "more than just food." On the contrary, there has been a marked decline in the harvests of some edible mycorrhizal mushrooms over the past century, which is illustrated by the official figures of Eurasian countries. This makes the cultivation of edible mycorrhizal mushrooms all the more attractive.

19.5 Production and Forest Management

Wild edible mushroom harvest generates millions of dollars each year and consists largely of ectomycorrhizal fungi, such as pine mushrooms (*Tricholoma magnivelare*), chanterelles (*Cantharellus formosus* and *Cantharellus subalbidus*), and boletes (*B. edulis*) (Danell and Camacho 1997). Pine mushrooms are the most commercially important wild forest mushroom and are exported exclusively to Japan (de Geus 1995), while chanterelles, boletes, and others are primarily exported to parts of North America and Europe (de Geus 1995). Known commercial mycorrhizal mushroom (particularly tubers) sites are located across all the regions of France (de Geus 1995; Freeman 1997; Trowbridge and Macadam 1999; Ehlers and Frederickson 2000; Berch and Wiensczyk 2001; Kranabetter et al. 2002), in forests from 20 to more than 200 years old (Hosford and Ohara 1995; Norvell 1995; Redhead 1997; Pilz et al. 1998). Forest practices, such as logging, site preparation, tree selection, fire, fertilization, pesticide use, brushing and spacing, and grazing, influences mushroom presence, reproduction, and productivity. Ectomycorrhizal fungi require living roots, and therefore living trees, to survive. As a result, timber harvesting, particularly clearcutting, profoundly reduces mushroom production (Smith et al. 2002) until the mature forest becomes reestablished. In some areas, gap area size significantly affects the production of fruiting bodies in forests (Durall et al. 1999, 2006). Sporocarp diversity declines significantly in forests as soil compaction from machinery and trampling can damage the mycelium and reduce mushroom productivity (Colgan et al. 1999).

Forest management techniques that promote mushroom production have been studied in many countries. To encourage matsutake mushroom production in Japanese forests, for example, various silviculture treatments have been applied. Overstorey trees are thinned, tree species composition is altered, nonhost understorey shrubs and herbs are cut, and organic litter is removed from the forest floor (Hosford et al. 1997). In North America, such intense management of forests for pine mushroom production does not occur. Studies in Europe show that nitrogen deposits from air pollution (Arnolds 1991) and applications of nitrogen fertilizers (Termorshuizen 1993) reduce the productivity of edible ectomycorrhizal fungi. Information on the effects of pesticide application or grazing on edible mushrooms is currently not available for any country. More research is required to determine how silviculture techniques could be used to promote the fruiting of economically important fungi in forests across the globe.

19.5.1 Ectomycorrhizal Mushrooms and Wildlife Association

A highly evolved beneficial relationship exists between ectomycorrhizal fruiting bodies, host trees, and wildlife. Fruiting bodies of ectomycorrhizal fungi are an important food source for many temperate forest mammals and invertebrates (Fogel 1975; Fogel and Trappe 1978; Bruns 1984; Lawrence 1989; Cazares and Trappe 1994; Johnson 1994; Janos and Sahley 1995; North and Trappe 1996). In the process of consuming the fruiting body, fungal inoculum is dispersed throughout the animal's range, thereby exposing the ectomycorrhizal host trees to a higher diversity of inocula. Aboveground sporocarps generally disperse their spores through the forest by means of air and water currents. Below-ground fruiting bodies, by contrast, depend on animals for spore dispersal. Animals are attracted to the aromatic compounds produced by truffles and false-truffles (i.e., true truffles belong to the class Ascomycete; false-truffles belong to the class Basidiomycete), which lead them to excavate and consume these mushrooms. The spores in the fruiting bodies ingested by the animal pass through the gut and are deposited with fecal pellets (Cazares and Trappe 1994; Johnson 1994; Currah et al. 2000). Some small mammals, such as the northern flying squirrel and the California red-backed vole, use truffles as their primary food source (Amaranthus et al. 1994). These mammals, in turn, are important prey for other species of animals (Forsman et al. 1984). Cavities in downed and standing dead wood are commonly used by small mammals to store food when foraging (Bunnell et al. 1999) and are therefore important not only for wildlife survival but also as sources of ectomycorrhizal fungal inocula for the surrounding forest. Retaining standing and downed coarse woody debris is thus important for the dispersal of spores from both above- and belowground ectomycorrhizal fruiting bodies (Klironomos and Hart 2001).

19.6 Cultivation of Edible Mycorrhizal Mushrooms

As is evident, about half of the world's species of edible mushrooms are mycorrhizal, and among them are some of the world's most expensive foods. A few of these mushrooms have well-established worldwide markets measured in billions of dollars. All of the mycorrhizal mushrooms are seasonal, best-eaten fresh, and do not preserve well (Hall et al. 2007). Of more than 300 ectomycorrhizal mushrooms eaten around the world, those held in the highest regard are the Périgord black truffle (Hall et al. 1994), Italian white truffle (Hall et al. 1998a), porcini (Hall et al. 1998b), chanterelle (*C. cibarius*; Danell 2001), and matsutake (*T. matsutake*; Wang et al. 1997). However, many of the other edible mycorrhizal mushrooms like Caesar's mushroom (*Amanita caesaria*), honshimeji (*Lyophilum shimeji*), Burgundy truffle (*Tuber uncinatum*), Oregon white truffle (*Tuber gibbosum*), and saffron milk cap (*L. deliciosus*) also have significant local markets. Figure 19.1 showed the different edible ectomycorrhizal mushrooms of high culinary demand.

Fig. 19.1 Edible ectomycorrhizal mushrooms of high culinary demand: (**a**) Perigord truffle *Tuber melanosporum*, (**b**) *Russula brevipes*m, (**c**) *Cantharellus cibarius*, (**d**) *Boletus edulis*, (**e**) *Tuber magnatum*, (**f**) *Russula emetica*, (**g**) *Cantharellus cibarius*, (**h**) *Boletus edulis*

19.6.1 Truffles

A truffle is a subterranean fungal fruiting body that develops underground and relies on mycophagy for spore dispersal (FEMAT 1993). There are 70 known varieties of truffles, 32 of which are found in Europe, but the fruiting bodies of some are highly prized as a food. They have a pungent, intense, earthy fragrance and lend a unique flavor to food, sometimes referred to as "Black gold" or "black diamonds" because of their scarcity and worth. Edible truffles are held in high esteem in international *haute cuisine*. They prefer argillaceous or calcareous soils, which are well drained and neutral or alkaline (Martin et al. 2010). Truffles fruit throughout the year, depending on the species, and can be found buried between the leaf litter and the soil. One of the best-known is the French truffles or the Périgord black truffle (*T. melanosporum*). This mushroom is found in the forests of southern France, northern Italy, and northeastern Spain on the roots of, for example, oaks and hazels. It is used very widely in gourmet cooking where it either imparts a flavor of its own to a dish or enhances the flavor of other foods it is cooked with. These truffles are considered as one of the great foods of the world, and when fresh in season, prices can be as high as $2,500 per kg.

At the beginning of the twentieth century, production of the Périgord black truffle has been estimated to have been between 1,000 and 2,000 tons, but in the last decade, it is reported to be less than 150 tons (Olivier 2000, Lefevre and Hall 2001). The earliest cultivation of truffles was achieved by Joseph Talon (father of truffle raising) in the early 1808s (Hall et al. 1994; Hall and Wang 1998). He found that transplanted oak seedlings from the rooting zones of trees produced Périgord black truffles. Despite its lack of sophistication, Talon's technique was widely used for 150 years. In some last decades, new attempts for mass production of truffles have been started. There are truffle-growing areas in the United States, Spain, Sweden, New Zealand, Australia, Chile, and the UK. Truffles have long eluded the modern techniques of domestication known as *trufficulture*. Although the field of trufficulture has greatly expanded since its inception in 1808, several species still remain uncultivated. Cultivation of Terfezia (Gutierrez, unpublished) (Morte et al. 2000) and various species of Tuber, including *T. borchii* (bianchetto) (Zambonelli et al. 2002), *T. melanosporum* (Lefevre and Hall 2001; Hall et al. 2003), and *T. uncinatum* (Chevalier and Frochot 1997) through sporal suspensions has been reported at different times.

Lately, the genome sequence of the Périgord black truffle has been published in March 2010 (Martin et al. 2010).

19.6.2 Russula

The genus *Russula* (Pers. Ex. Fr.) S.F. Gray accounts in a large measure for monsoon mushroom flora in forests. It is cosmopolitan and largely ectomycorrhizal genus with a wide range of Gymnosperms and Angiosperms (Richardson 1970; Agerer 2002). A number of *Russula* species form ectomycorrhizas with different tree species (Arora 1986). *Russula brevipes* and its morphotypes/ectomycorrhiza have been described and illustrated (Niazi et al. 2006) from moist temperate forests of Pakistan, associated with *Pinus wallichiana. Russula* species is widely distributed and mainly associated with species of *Abies, Picea, Tsuga, and Pseudosuga* (Stanis 1979; Pillukat and Agerer 1992; Kraigher et al. 1995; Kernaghan et al. 1997). It is more commonly found in Himalayan forests under conifers in late fall and can easily be identified by its large fruit size. Yamada, since 1997, has studied the cultivation of edible EMF. Among many species, *Russula nigiricans* and *Russula mariae* were successfully induced to synthesize mycorrhiza in vitro (Yamada 2003).

The mushroom contains very useful phytochemicals such as phenolics, flavonoids, ergosterol, and β-carotene. The sesquiterpene lactone named russulactarorufin along with lactarorufin-A and 24-ethyl-cholesta-7,22*E*-diene-3β,5α,6β-triol have been isolated and characterized (Suri et al. 1997). The wild species proved to have antioxidant potential and free radical scavenging activity. The combination of bioactive substances and rich nutritional composition (high contents in protein and carbohydrates, low content in fat) in the mushroom should be useful to consumers in encouraging them to utilize the nutritive potential of this edible wild mushroom (Chen et al. 2010).

19.6.3 Cantharellus

C. cibarius, commonly known as the golden chanterelle, is probably the best known species of the genus *Cantharellus*. It is orange or yellow, meaty and funnel-shaped with smooth cap, and gill-like ridges that run almost all the way down its stipe, which tapers down seamlessly from the cap. It has a fruity smell, reminiscent of apricots, and a mildly peppery taste and is considered an excellent food mushroom. Chanterelles are common in northern parts of Europe and North America, Asia, and in Africa (Boa 2004). Chanterelles are associated with conifers and oaks (Arora 1979; Metzler 1992). In Scotland, chanterelles grow in mixed forest (silver birch and scots pine), especially when the forest has plenty of moist, mossy undergrowth. Chanterelles are versatile and can be added as an ingredient to most dishes. They can also be pickled in brine and can last from 6 to 12 months. In most places, they are dry stored. Fresh chanterelles can generally be stored up to 10 days in a refrigerator.

Nils (1979) first proved that spores of *C. cibarius* can be germinated in vitro. Danell attempted to cultivate the species (Danell and Camacho 1997). Up to 6.7% of its tissues contain vitamin C; as for carotene, they can be as much as 23.1%. Research has suggested that the golden chanterelle may have potent insecticidal properties that are harmless against humans and yet protect the mushroom body against insects and other potentially harmful organisms (Philpot 1965).

19.6.4 Boletus

Boletus has a cosmopolitan distribution, concentrated in cool-temperate to subtropical regions (Marais and Kotzé 1977; Eicker 1990; Masuka 1996; Hall et al. 1998b; Hall et al. 2003; Orlovich et al. 2004; Oria-de-Rueda et al. 2008). It has been reported from Europe – from northern Scandinavia, south to the extremities of Greece and Italy – and North America, where its southern range extends as far south as Mexico (Tylukti 1987), China (Tylukti 1987), Nepal (Giri and Rana 2007), and India (Adhikary et al. 1999). The mushroom's habitat consists of areas dominated by *Pinus* spp.(Vozzo and Hackskaylo 1961; Froidevaux and Amiet 1975), spruce (*Picea* spp.) (Ceruti et al. 1988), hemlock (*Tsuga* spp.) fir (*Abies* spp.) trees (Gobl 1977), chestnut, chinquapin, beech, *Keteleeria* spp., *Lithocarpus* spp., and oak (Wang et al. 1995; Gross et al. 1998;

Quan and Lei 2000; Agueda et al. 2006, 2008; Fu et al. 2009). The mushroom has been noted to commonly cooccur with *Amanita muscaria* or *Amanita rubescens*, although it is unclear whether this is due to a biological association between the species or because of similarities in growing season, habitat, and ecological requirements (Hall et al. 1998b; Hall et al. 2003; Peinter et al. 2007).

Estimates suggest the total annual worldwide consumption of *B. edulis* and closely related species (*B. aereus*, *B. pinophilus*, and *B. reticulatus*) has increased from 20,000 to 100,000 tons in the current decade (Hall et al. 1998b, 2003; Agueda et al. 2008). They are widely exported and sold in dried form, reaching countries where they do not occur naturally, such as Australia and New Zealand. In autumn, the price of porcini typically ranges between \$20 and \$80 per kilogram, although the scarcity of fruit bodies sometimes elevates the wholesale price to over \$200 per kilogram (Sitta 2000; Hall et al. 2003; Boa 2004; Sitta and Floriani 2008; Drumeva and Gyosheva 2009).

As with other strictly mycorrhizal fungi, *B. edulis* has eluded attempts to cultivate it (Arora 1986; Chang and Miles 2004). The results of some studies suggest that unknown components of the soil microflora might be required for *B. edulis* to successfully establish a mycorrhizal relationship with the host plant (Veselkov 1975; Ceruti et al. 1988; Fitter and Garbaye 1994).

B. edulis fruit bodies contain ergosterol and ergosterol peroxide (Mattila et al. 2002; Ey et al. 2007; Ribeiroa et al. 2008), with a wide spectrum antimicrobial and anti-inflammatory activity, and cytotoxicity to various tumor cell lines grown in laboratory culture (Lucas et al. 1957; Krzyczkowskia et al. 2008). However, some investigations in the United States do not support this (Lamaison and Polese 2005). The mushroom also contains mitogenic lectin with antiviral properties against the human immunodeficiency virus enzyme reverse transcriptase (Zheng et al. 2007), *Vaccinia* virus (Kandefer-Szersen et al. 1980) and tobacco mosaic virus grown in culture (Piraino 2006; Li et al. 2009).

19.6.5 Cultivation of Other Species

While some plants can live in the absence of the fungal partner, at least under amended (fertilized) environments, mycorrhizal fungi cannot live in the absence of the host (Harley and Smith 1983). In poor, dry tropical soils, however, mycorrhizal associations are vital for plant growth and survival (Munyanziza 1994). The normal reproductive cycle of mushroom-producing mycorrhizal fungi involves the following stages: (1) spore germination, (2) mycelium growth, (3) infection of the specific host, and (4) fruiting-body (mushroom) production and sporulation. Without the right host, the fungus does not reach stages 3 and 4.

Plants infected following inoculation with spore suspensions (Hall and Wang 1998; Hall et al. 2002) or pure cultures prepared either from fruiting bodies (Sisti et al 1998) or mycorrhizal root tips (Kagan-Zur 2002) have led to the formation of edible mycorrhizal mushroom fruiting bodies in the field, including *L. deliciosus*

(Poitou et al. 1989; Wang et al. 2002a, b), *Lyophyllum shimeji* ("Kyoto scientists grow hon-shimeji mushrooms," http://www.kippo.or.jp/KansaiWindowhtml/News/1996e/19961119_NEWS.HTML), *Rhizopogon rubescens* (Wang et al. 2002a, b), *Suillus granulatus* (Poitou et al. 1989), Terfezia (Morte et al. 2000), and various species of Tuber (Chevalier and Frochot 1997; Lefevre and Hall 2001; Zambonelli et al. 2002; Hall et al. 2003); Danell and Camacho (1997) also produced *C. cibarius* in pots in the greenhouse. Despite these successes, fewer than a dozen of the many hundreds of edible mycorrhizal mushrooms have ever been cultivated with any degree of success (Wang et al. 2002a, b), and this includes important commercial species such as *B. edulis* (Hall et al. 1998a, b), *T. matsutake* (Yun and Hall 2004), and the Italian white truffle (*Tuber magnatum*) (Hall et al. 1998a; Gregori 2002). *T. matsutake* was first found to form a typical Hartig net showing ectomycorrhiza, which subsequently synthesized mycorrhiza. Other tested EMF also produced mycorrhizal seedlings. Among them, *Lactarius akahatsu*, *R. rubescens*, *Tricholoma portentosum*, and *T. saponaceum* were successfully induced to fruit with juvenile pine seedlings of around of 1-year old (Yamada 2003).

Consequently, supplies of most commercially important edible mycorrhizal mushrooms are still restricted to those that can be harvested from the wild during autumn or winter. For these, we are left to speculate on the reasons why we have not been able to establish viable infections on a suitable host plant or why fruiting bodies simply fail to form.

19.7 Conservation

In spite of their importance, discussions of ectomycorrhizal mushroom conservation are not widespread. Many factors may be influencing the more prominent decline in these mushrooms compared to saprophytic and lignicolous fungi (Leake et al. 2002). Mycorrhiza formation and especially fruit body formation is sensitive to soil disturbance. The amount of organic matter available in forest stands influences the types of mycorrhizae present on tree roots (Harvey et al. 1976). Similarly, chemical weeding, which is often done in forests or agroforestry (Amakiri 1977), may have consequences on mycorrhiza fungi and hence on mushroom production (Iloba 1980). A broader appreciation for the conservation of ectomycorrhizal mushrooms is needed as these fungi are crucial to many ecosystem functions and have great ecological and economic value. For the in vitro culture, usually isolation of the mycelia is difficult. Mycelial growth on synthetic medium is poor, and fruit bodies do not form easily in pure cultures. Many attempts to cultivate mycorrhizal mushrooms have failed. Fungal propagules including spores and hyphae do not remain viable for long periods. In nature, these fungi are more or less obligate root symbionts.

Much concern about the conservation of EMF diversity has come from Europe where populations have declined over the last three decades (Arnolds 1988;

Arnolds 1991). Ectomycorrhizal fungal spores are lost to predation, germination under unsuitable conditions, and weathering (Miller et al. 1994). The most dramatic fungal declines have been associated with increases in various types of pollution (Arnolds 1988). Though mushrooms like *Tricholoma* and *Lactarius* are not cultivable, *Cantharellus* and Truffles can be cultivated using symbiotic relationship as a technique forming mycorrhiza in the field or by applying the same methods used for cultivating saprophytic mushrooms to use substrates and nutrients in bottles or bags as culture medium. Environmental conditioning of the wooded area by thinning the undergrowth and by plowing the litter layer can also increase the number of mycelia. Controlled burning of trees to induce formation of new fungal colonies is yet another way practiced to grow these mushrooms.

In fact, in the tropics, about 95% tree species from endomycorrhizas do not form mushrooms (Lapeyrie and Högberg 1994; Mason and Wilson 1994). Clearly, in the domestication of mycorrhizal fungi for mushroom production on trees in agroforestry systems, the tree component has to be made of the species forming ectomycorrhizas. For increased mushroom diversity and sustained production, the domestication of mycorrhizal mushrooms will mean the creation of an agroforestry system having trees of different species and age groups supplied with the right spectrum of fungal partners (Mason et al. 1987; Termoshuizen 1991).

19.7.1 Challenges Ahead

One possible reason behind the failure in cultivating ectomycorrhizal mushrooms is simply the assumption that they are mycorrhizal when their relationship with their hosts is more complicated. Out of about 2,500 recorded species of edible mushrooms, the most expensive and sought after mushrooms belong to the mycorrhizal group and include *T. melanosporum*, *T. magnatum*, *T. matsutake*, *B. edulis*, *C. cibarius*, *Russula emetica*, and *A. caesarea*. Over the past century, harvests of many mycorrhizal mushrooms have declined dramatically, which has prompted interest in the development of methods for their cultivation. So far, only a few species of truffles have been produced in commercial quantities, although methods have been developed that may see the cultivation of species such as *C. cibarius*, *L. shimeji*, and *L. deliciosus*. Despite this, many of the most expensive mycorrhizal mushrooms, including *T. magnatum* Pico & Vitt. and *T. matsutake*, have defied cultivation (Yun and Hall 2004).

The links between wild edible fungi and tree hosts are well known for economically important species such as *B. edulis* and *Tuber* spp. *Cantharellus* spp. form mycorrhizae with many tree species in tropical countries. There is an expanding body of information about many other edible fungus–tree associations, but this has not been assembled in the form of a database that would, for example, allow for predictive searches. The search for *matsutake* in Asia was assisted by a knowledge of its tree hosts, and this approach would assist in prospecting for other wild edible fungi. Knowledge about the mycorrhizal partners of edible species of *Amanita*, *Lactarius*, and *Russula* is steadily increasing. However, our understanding of the ecological conditions favored by edible ectomycorrhizal mushroom is restricted to a handful of fungi and a paucity of publications (Chevalier and Frochot 1997; Wang et al. 1997; Hall et al. 1998a; Morte et al. 2000; Sbrana et al. 2000; Lefevre and Hall 2001; Hall et al. 2003). Clearly, there is a need for considerable study in this area to determine whether an edible mycorrhizal mushroom is found on young or old trees – the so-called early or late-stage mushrooms (Visser 1995; Redecker et al. 2001); define the optimum edaphic and climatic conditions tolerated by the fungus; monitor the effect of changing environmental conditions on mycorrhizal community structures (Lilleskov et al. 2002); their relationship with other belowground organisms (Duplessis et al. 2002); and identify the factors required to trigger fruiting (Hall et al. 2002). There is a considerable body of literature on truffles (Federation-Française-des-Trufficulteurs 2001). Researchers have focused much attention to the complex relationships between biological, social, and economic issues, a welcome move towards establishing a sound basis for sustainable production of wild edible fungi.

19.8 Conclusions

Ectomycorrhizal mushrooms constitute an important resource being exploited on a commercial scale. In future, ectomycorrhizal mushrooms may play important role as medicines and as bioindicators of sustainable ecosystems practices. Traditionally gathered for household consumption, they have won international attention. There has been a marked decline in the harvests of some edible mycorrhizal mushrooms over the past century. Possible reasons for this decline include deforestation, the loss of host plants within forests due to pests or disease, deforestation as well as changed forest management practices such as planting more densely than occurs in natural forests, the replacement of natural forests with plantations of species that are poor hosts for edible mycorrhizal mushrooms, global warming since the last ice age, soil compaction by hordes of pickers, acid rain, etc. There has been a spectacular increase of interest and commercial activity concerned with dietary supplements and functional foods. This makes the cultivation of edible mycorrhizal mushrooms all the more attractive. Biochemical and molecular studies also might prove crucial to our understanding of these issues. The domestication of edible ectomycorrhizal mushrooms is therefore gathering some momentum in a few parts of the world. This needs to build on the understanding of the relation between the partners and the environment under which the cultivation can be optimized. Cultivating ectomycorrhizal mushrooms requires special management of forestry systems. This will depend on incorporating appropriate tree species, practicing limited soil disturbance, understanding the mycorrhizal partnership, and forest dynamics.

References

Abbott LK, Robson AD (1991) Factors influencing the occurrence of vesicular-arbuscular mycorrhizas. Agric Ecosyst Environ 35:121–150

Addy HD, Piercey MM, Currah RS (2005) Microfungal endophytes in roots. Can J Bot 83:1–13

Adhikary RK, Baruah P, Kalita P, Bordoloi D (1999) Edible mushrooms growing in forests of Arunachal Pradesh. Adv Hortic For 6:119–123

Agerer R (1995) Anatomical characteristics of identified ectomycorrhizas: an attempt towards a natural classification. In: Varma A, Hock B (eds) Mycorrhiza. Springer Verlag, Berlin, pp 685–734

Agerer R (2002) Color atlas of ectomycorrhiza, 12th edn. Einhorn Verlag Edward, Dientenurger, Germany

Agueda B, Parlade J, Fernandez-Toiran LM, Cisneros O, de Miguel AM, Modrego MP, Martinez-Pena F, Pera J (2008) Mycorrhizal synthesis between *Boletus edulis* species complex and rockroses (*Cistus* sp.). Mycorrhiza 18(8):443–449

Agueda B, Parladé J, de Miguel AM, Martínez-Peña F (2006) Characterization and identification of field ectomycorrhizae of *Boletus edulis* and *Cistus ladanifer*. Mycologia 98(1):23–30

Amakiri MA (1977) Effect of herbicides on microbial populations and activities in soils under teak (*Tectona grandis* L.) plantation. East Afr Agric For J 42(4):420–426

Amaranthus MP, Trappe JM, Bednar L, Arthur D (1994) Hypogeous fungal production in mature Douglas-fir forest fragments and surrounding plantations and its relation to coarse woody debris and animal mycophagy. Can J For Res 24:2157–2165

Arnolds E (1988) The changing mycromycete flora in the Netherlands. Trans Br Mycol Soc 90 (3):391–406

Arnolds E (1991) Decline of ectomycorrhizal fungi in Europe. Agric Ecosyst Environ 35:209–244

Arora D (1979) Mushroom demystified. Ten Speed Press, Berkeley. ISBN 0-89815-009-4

Arora D (1986) Mushrooms demystified: a comprehensive guide to the fleshy fungi, 2nd edn. Ten Speed Press, Berkeley

Arora D (1991) All That the Rain Promises, and More: A Hip Pocket Guide to Western Mushrooms. Berkeley: Ten Speed Press, 1991. ISBN 0898153883

Ashkenazi M, Jacob J (2003) Food culture in Japan. Greenwood Publishing Group, p 49. ISBN 0313324387

Baghel RK, Sharma R, Pandey AK (2009) Activity of acid phosphatase in the ectomycorrhizal fungus *Cantharellus tropicalis* under controlled conditions. J Trop For Sci 21(3):218–222, ISSN 0128-1283

Barroetaveña C, Cázares E, Rajchenberg M (2007) Ectomycorrhizal fungi associated with ponderosa pine and Douglas-fir: a comparison of species richness in native western North American forests and Patagonian plantations from Argentina. Mycorrhiza 17:355–373

Baxter JW, Dighton J (2001) Ectomycorrhizal diversity alters growth and nutrient acquisition of grey birch (*Betula populifolia*) seedlings in host-symbiont culture conditions. New Phytol 152:139–149

Berch SM, Wiensczyk AM (2001) Ecological description and classification of some pine mushroom (*Tricholoma magnivelare*) habitat in British Columbia. B.C. Ministry of Forests and Southern Interior Forest Extension and Research Partnership. Victoria, B.C. Research Report No. 19

Binder M, Bresinsky A (2002) Derivation of a polymorphic lineage of gasteromycetes from boletoid ancestors. Mycologia 94:85–98

Binder M, Hibbett DS, Larsson KH, Larsson E, Langer E, Langer G (2005) The phylogenetic distribution of resupinate forms across the major clades of mushroom-forming fungi (Homobasidiomycetes). Syst Biodivers 3:113–157

Boa E (2004) Wild edible fungi: a global overview of their use and importance to people (non-wood forest products). Food & Agriculture Organization of the UN, Rome, Italy, pp 96–97

Bougher NL, Lebel T (2001) Sequestrate (Truffle-like) fungi of Australia and New Zealand. Aust Syst Bot 14:439–484

Bougher NL, Malajczuk N (1985) A new species of *Descolea* (Agaricales) from Western Australia and aspects of its ectomycorrhizal status. Aust J Bot 33:619–627

Brundret M, Malajczuk N, Mingqin G, Daping X, Snelling S, Dell B (2005) Nursery inoculation of *Eucalyptus* seedlings in Western Australia and Southern China using spores and mycelial inoculum of diverse ectomycorrhizal fungi from different climatic regions. For Ecol Manage 209:193–205

Brundrett MC (1991) Mycorrhizas in natural ecosystems. In: Macfayden A, Begon M, Fitter AH (eds) Advances in ecological research, vol 21. Academic Press, London, pp 171–313

Brundrett MC (2004) Diversity and classification of mycorrhizal associations. Biol Rev 79:473–495

Brundrett MC, Kendrick WB (1988) The mycorrhizal status, root anatomy, and phenology of plants in a sugar maple forest. Can J Bot 66:1153–1173

Brundrett MC, Murase G, Kendrick B (1990) Comparative anatomy of roots and mycorrhizae of common Ontario trees. Can J Bot 68:551–578

Bruns TD (1984) Insect mycophagy in the boletales: fungivore diversity and the mushroom habitat. In: Wheeler Q, Blackwell M (eds) Fungus–insect relationships: perspectives in ecology and evolution. Columbia University, New York, pp 91–129

Bunnell FL, Kremsater LL, Wind E (1999) Managing to sustain vertebrate richness in forests of the Pacific Northwest: relationships within stands. Environ Rev 7:97–146

Cairney JWJ, Chambers SM (1999) Ectomycorhizal fungi: key genera in profile. Springer, Germany

Cazares E, Trappe JM (1994) Spore dispersal of ectomycorrhizal fungi on a glacier forefront by mammal mycophagy. Mycologia 86(4):507–510

Ceruti A, Tozzi M, Reitano G (1988) Micorize di sintesi tra *Boletus edulis*, *Pinus sylvestris* e *Picea excelsa* (in Italian). Allionia (Turin) 28:117–124

Chang ST, Miles PG (2004) Mushrooms: cultivation, nutritional value, medicinal effect, and environmental impact, 2nd edn. CRC Press, Boca Raton, FL, p 131

Xin-Hua Chen, Le-Xian Xia, Hong-Bo Zhou, Guan-Zhou Qiu (2010) Chemical composition and antioxidant activities of Russula griseocarnosa sp. nov. J Agric Food Chem 58(11):6966–6971

Chevalier G, Frochot H (1997) La Truffe de Bourgogne. Pétrarque, Levallois-Perret

Colgan W III, Carey AB, Trappe JM, Molina R, Thysell D (1999) Diversity and productivity of hypogeous fungal sporocarps in a variably thinned Douglas-fir forest. Can J For Res 29:1259–1268

Comandini O, Contu M, Rinaldi AC (2006) An overview of *Cistus* ectomycorrhizal fungi. Mycorrhiza 16:381–395

Cook R (1977) The biology of symbiotic fungi. Wiley, London

Cripps CL, Miller OK (1995) Ectomycorrhizae formed in vitro by quaking aspen: including *Inocybe lacera* and *Amanita pantherina*. Mycorrhiza 5:1432–1890

Currah RS, Smreciu EA, Lehesvirta T, Niemi M, Larsen KW (2000) Fungi in the winter diets of northern flying squirrels and red squirrels in the boreal mixedwood forest of northeastern Alberta. Can J Bot 78:1514–1520

Danell E (2000) Cantharellus. In: Cairney JWG, Chambers SM (eds) Ectomycorrhizal fungi: key genera in profile. Springer, Berlin, pp 253–267

Danell E (2001) Current research on chanterelle cultivation in Sweden. Second international workshop on edible mycorrhizal mushrooms. http:/www.crop.cri.nz/whats_on/mushroom_conf/Index.htm

Danell E, Camacho FJ (1997) Successful cultivation of the golden chanterelle. Nature 385:303

de Geus N (1995) Botanical forest products in British Columbia: an overview. B.C. Ministry of Forests, Victoria

Deacon JW, Fleming LV (1992) Interactions of ectomycorrhizal fungi. In: Allen MF (ed) Mycorrhizal functioning: an integrative plant-fungal process. Chapman and Hall, New York, pp 249–300

Dotzler D, Krings M, Taylor TN, Agerer R (2006) Germination shields in *Scutellospora* (Glomeromycota: Diversisporales, Gigasporaceae) in the 400 million-year-old Rhynie chert. Mycol Prog 5:178–184

Douhan GW, Rizzo DM (2005) Phylogenetic divergence in a local population of the ectomycorrhizal fungus *Cenococcum geophilum*. New Phytol 166:263–271

Drumeva M, Gyosheva M (2009) The macromycetes fungi of Bulgaria. World Wildlife Fund. http://www.worldwildlife.org/bsp/publications/europe/bulgaria/bulgaria1.html

Duplessis S et al (2002) Living together underground, a molecular glimpse of the ectomycorrhizal symbiosis. In: Osiewacz HD (ed) Molecular biology of fungal development. Marcel Dekker, New York, pp 297–323

Durall DM, Gamiet S, Simard SW, Kudrna L, Sakakibara SM (2006) Effects of clearcut logging and tree species composition on the diversity and community composition of epigeous fruit bodies formed by ectomycorrhizal fungi Canadian Journal of Botany, 84:966–980

Durall DM, Jones MD, Wright EF, Kroeger P, Coates KD (1999) Species richness of ectomycorrhizal fungi in cutblocks of different sizes in the interior Cedar-Hemlock forests of northwestern British Columbia: sporocarps and ectomycorrhizae. Can J For Res 29:1322–1332

Ehlers T, Frederickson S (2000) An inventory and characterization of pine mushroom habitats in Galena and Fosthall small business enterprises. B.C. Ministry of Forests, Arrow Forest District, Small Business Forest Enterprise Program. Unpublished report

Eicker A (1990) Commercial mushroom production in South Africa. Bulletin 418. Department of Agricultural Development, Pretoria

Erland S, Taylor AFS (1999) Resupinate ectomycorrhizal fungl genera. In: Cairney JWG, Chambers SM (eds) Ectomycorrhizal fungi key genera in profile. Springer Verlag, Berlin, pp 347–363

Ey J, Schömig E, Taubert D (2007) Dietary sources and antioxidant effects of ergothioneine. J Agric Food Chem 55(16):6466–6474

Ferdman Y, Aviram S, Roth-Bejerano N, Trappe JM, Kagan-Zur V (2005) Phylogenetic Studies of *Terfezia pfeilii* and *Choiromyces echinulatus* (Pezizales) support new genera for southern African truffles: *Kalaharituber* and *Eremiomyces*. Mycol Res 109:237–245

Fitter AH, Garbaye J (1994) Interaction between mycorrhizal fungi and other organisms. In: Robson AD, Abbott LK, Malajczuk N (eds) Management of mycorrhizas in agriculture, horticulture and forestry. Kluwer Academic Publishers, Dordrecht, Netherlands

Fogel R (1975) Insect mycophagy: a preliminary bibliography. U.S. Department of Agriculture Forest Service, p 16

Fogel R, Trappe JM (1978) Fungus consumption (mycophagy) by small animals. Northwest Sci 52 (1):1–31

Forest Ecosystem Management Assessment Team [FEMAT] (1993) Forest ecosystem management: an ecological, economic, and social assessment. U.S. Department of Agriculture Forest Service, Pacific Northwest Region, Portland, Oregon

Forsman ED, Meslow EC, Wight HM (1984) Distribution and biology of the potted owl in Oregon. Wildlife Monogr 87:1–64

Frank AB (1885) Über die auf Würzelsymbiose beruhende Ehrnährung gewisser Bäum durch unterirdische Pilze. Berichte der Deutschen Botanischen Gesellschaft 3:128–145

Freeman S (1997) An estimate of pine mushroom production in the Nahatlatch watershed. Forest Renewal BC. Unpublished Report

Froidevaux L, Amiet R (1975) Ecto mycorrhizae endo mycorrhizae of *Pinus mugo* plus *Boletus edulis* ssp *edulis* and *Pinus cembra* plus *Suillus variegatus* formed in pure culture. Eur J For Pathol 5(1):57–61

Fu SC, Tan Q, Chen MJ, Shang XD, Cai LY, Zhang MY (2009) Mycorrhizal synthesis involving *Boletus edulis* and *Pinus massoniana*. Acta Edulis Fungi 16(1):31–41

Garbaye J (1994) Helper bacteria: a new dimension to the mycorrhizal symbiosis. Tansley review no. 76. New Phytol 128:197–210

Giri A, Rana P (2007) Some higher fungi from Sagarmatha National Park (SNP) and its adjoining areas, Nepal. Sci World 5(5):67–74

Gobl F (1977) Mycorrhiza in Austrian Douglas fir stands (in German). Centralblatt fur das Gesamte Forstwesen 94(4):185–194

Godbout C, Fortin JA (1985) Synthesized ectomycorrhizae of aspen: fungal genus level of structural characterization. Can J Bot 63:252–262

Goodman DM, Trofymow JA (1998) Comparison of communities of ectomycorrhizal fungi in old-growth and mature stands of Douglas-fir at two sites on southern Vancouver Island. Can J For Res 28:574–581

Graham RT, Harvey AE, Jurgensen MF, Jain TB, Tonn JR, Page-Dumroese DS (1994) Managing coarse woody debris in forests of the rocky mountains. U.S. Department of Agriculture Forest Service, Intermountain Research Station, Ogden, Utah. Research Paper INT-RP-477

Graham RT, Jain TB, Harvey AE (1999) Fuel: logs, sticks, needles, duff and much more. In: Proceedings, joint fire science conference and workshop. Crossing the millennium: integrating spatial technologies and ecological principles for a new age in fire management. L.F. Neunsch

Gregori G (2002) Problems and expectations with the cultivation of Tuber magnatum. In: Hall IR et al (eds) Edible mycorrhizal mushrooms and their cultivation, Proceedings of the second international conference on edible mycorrhizal mushrooms, cd-rom, New Zealand Institute for Crop & Food Research Limited

Griffiths RP, Caldwell BA (1992) Mycorrhizal mat communities in forest soils. In: Read DJ, Lewis DH, Fitter AH, Alexander IJ (eds) Mycorrhizas in ecosystems. CAB International, Wallingford, CT, pp 98–105

Gross E, Thomazini-Casagrande LI, Caetano FH (1998) A scanning electron microscopy study of ectomycorrhizae on *Pinus patula* Schiede and Deppe. Naturalia (Rio Claro) 23(1):93–101

Gutiérrez A, Morte A, Honrubia M (2003) Morphological characterization of the mycorrhiza formed by Helianthemum almeriense Pau with Terfezia claveryi Chatin and Picoa lefebvrei (Pat.) Maire. Mycorrhiza. 2003, 13(6):299–307

Hall IR, Wang Y (1998) Methods for cultivating edible ectomycorrhizal mushrooms. In: Varma A (ed) Mycorrhiza manual, Springer laboratory manual. Springer Verlag, Heidelberg, pp 99–114

Hall IR, Brown G, Byars J (1994) The black truffle: its history, uses and cultivation, 2nd edn. New Zealand Institute for Crop & Food Research Limited, Christchurch, New Zealand

Hall IR et al (1998a) Ectomycorrhizal fungi with edible fruiting bodies 3 Tuber magnatum. Econ Bot 52:192–200

Hall IR et al (1998b) Ectomycorrhizal fungi with edible fruiting bodies. 2. Boletus edulis. Econ Bot 52:44–56

Hall IR et al (2002) Edible Mycorrhizal mushrooms and their cultivation. Proceedings of the second international conference on edible Mycorrhizal mushrooms, CD-ROM, New Zealand Institute for Crop & Food Research Limited

Hall IR, Wang Y, Amicucci A (2003) Cultivation of edible ectomycorrhizal mushrooms. Trends Biotechnol 21(10):433–438

Hall IR, Brown G, Zambonelli A (2007) Taming the Truffle: the history, lore, and science of the ultimate mushroom. Timber Press, Portland, p 304

Harley JL, Harley EL (1987) A check-list of mycorrhiza in the British flora. New Phytol (Suppl) 105:1–102

Harley JL, Smith SE (1983) Mycorrhizal symbiosis. Academic Press, London, p 483

Harrison MJ (2005) Signaling in the arbuscular mycorrhizal symbiosis. Annu Rev Microbiol 59:19–42

Harvey AE, Larsen MJ, Jurgensen MF (1976) Distribution of ectomycorrhizae in a mature douglas-fir/larch soil in western Montana. For Sci 22:393–633

Harvey AE, Jurgensen MF, Larsen MJ, Graham RT (1987) Relationships among soil microsite, ectomycorrhizae, and natural conifer regeneration of old-growth forests in western Montana. Can J For Res 17:58–62

Henkel TW, James TY, Miller SL, Aime ML, Miller OK Jr (2006) The mycorrhizal status of *Pseudotulostoma volvata* (Elaphomycetaceae, Eurotiales, Ascomycota). Mycorrhiza 16:241–244

Hibbett DS, Binder M, Bischoff JF, Blackwell M, Cannon PF, Eriksson OE, Huhndorf S, James T, Kirk PM, Lücking R, Lumbsch HT, Lutzoni F, Matheny PB, McLaughlin DJ, Powell MJ, Redhead S, Schoch CL, Spatafora JW, Stalpers JA, Vilgalys R, Aime MC, Aptroot A, Bauer R, Begerow D, Benny GL, Castlebury LA, Crous PW, Dai YC, Gams W, Geiser DM, Griffith GW, Gueidan C, Hawksworth DL, Hestmark G, Hosaka K, Humber RA, Hyde KD, Ironside JE, Kõljalg U, Jyrtznabm CP, Larsson KH, Lichtwardt R, Longcore J, Miadlikowska J, Miller A, Moncalvo JM, Mozley-Standridge S, Oberwinkler F, Parmasto E, Reeb V, Rogers JD, Roux C, Ryvarden L, Sampaio JP, Schüßler A, Sugiyama J, Thorn RG, Tibell L, Untereiner WA, Walker C, Wang Z, Weir A, Weiß M, White MM, Winka K, Yao YJ, Zhang N (2007) A higher-level phylogenetic classification of the fungi. Mycol Res 111:509–547

Hosaka K, Castellano MA, Spatafora JW (2008) Biogeography of hysterangiales (phallomycetidae, basidomycota). Mycol Res 112:488–462

Hosford D, Ohara H (1995) Ecological study of *Tricholoma magnivelare* shiros in central Washington. In: Schnepf C (ed) Dancing with an elephant. Proceedings: the business and science of special forest products, Western Forestry and Conservation Association, Portland, Oregon, pp 111–116

Hosford D, Pilz D, Molina R, Amaranthus M (1997) Ecology and management of the commercially harvested American matsutake mushroom. U.S. Department of Agriculture Forest Service, Pacific Northwest Research Station, Portland, Oregon. General Technical Report PNW GTR-412

Humpert AJ, Muench EL, Giachini AJ, Castellano MA, Spatafora JW (2001) Molecular phylogenetics of *Ramaria* and related genera: evidence from nuclear large subunit and mitochondrial small subunit rDNA sequences. Mycologia 93:465–477

Hunt GA, Fogel R (1983) Fungal hyphal dynamics in a western Oregon douglas-fir stand. Soil Biol Biochem 15:614–649

Iloba C (1980) The use of chemicals for weed control in forest tree nurseries. In: Mikola P (ed) Tropical mycorrhiza research. Clarendon Press, Oxford, pp 121–123

Janos DP (1980) Vesicular arbuscular mycorrhizae affect lowland tropical rain forest plant growth. Ecology 61:151–162

Janos DP, Sahley CT (1995) Rodent dispersal of vesicular-arbuscular mycorrhizal fungi in Amazonian Peru. Ecology 76(6):1852–1858

Johnson CN (1994) Mycophagy and spore dispersal by a rat-kangaroo: consumption of ectomycorrhizal taxa in relation to their abundance. Funct Ecol 8:464–468

Kagan-Zur V (2002) Tuber melanosporum research in Israel. In: Hall IR et al (eds) Edible mycorrhizal mushrooms and their cultivation. Proceedings of the second international conference on edible mycorrhizal mushrooms, CD-ROM, New Zealand Institute for Crop & Food Research Limited

Kahiluoto H, Vestberg M (1998) The effect of arbuscular mycorrhiza on biomass production and phosphorus uptake from sparingly soluble sources by leek (*Allium porrum* L.) in Finnish field soils. Biol Agric Hortic 16:65–85

Kandefer-Szersen M, Kawecki Z, Salata B, Witek M (1980) Mushrooms as a source of substances with antiviral activity (in Polish). Acta Mycologica 16(2):215–220

Kawai M (1997) Artificial ectomycorrhiza formation on roots of air-layered *Pinus densiflora* saplings by inoculation with *Lyophyllum shimeji*. Mycologia 89:228–232

Kernaghan G, Currah RS, Bayer RJ (1997) Russulaceous ectomycorrhizas of *Abies lasicarpa* and *Picea engelmanii*. Can J Bot 75:1843–1850

Klironomos JN, Hart MM (2001) Animal nitrogen swap for plant carbon. Nature 410:651–652

Kõljalg U, Dahlberg A, Taylor AFS, Larsson E, Hallenberg N, Stenlid J, Larsson KH, Fransson PM, Kårén O, Jonsson L (2000) Diversity and abundance of resupinate thelephoroid fungi as ectomycorrhizal symbionts in Swedish boreal forests. Mol Ecol 9:1985–1996

Kraigher H, Agerer A, Javornik B (1995) Ectomycorrhizae of *Lactarius lygniotus* on Norway spruce, characterized by anatomical and molecular tools. Mycorrhiza 5:175–180

Kranabetter JM, Trowbridge R, Macadam A, McLennan D, Friesen J (2002) Ecological descriptions of pine mushroom (*Tricholoma magnivelare*) habitat and estimates of its extent in northwestern British Columbia. For Ecol Manage 158:249–261

Kropp BR, Trappe JM (1982) Ectomycorrhizal fungi of *Tsuga heterophylla*. Mycologia 74:479–488

Krzyczkowskia W, Malinowskaa E, Suchockib P, Klepsa J, Olejnikd M, Herold F (2008) Isolation and quantitative determination of ergosterol peroxide in various edible mushroom species. Food Chem 113(1):351–355

Lamaison JL, Polese JM (2005) The great encyclopedia of mushrooms. Könemann, Koln, Germany, p 28

Lapeyrie F, Högberg P (1994) Harnessing symbiotic associations: ectomycorrhizas. In: Leakey RRB, Newton AC (eds) Tropical trees: the potential for domestication and the rebuilding of forest resources. HMSO, London

Lawrence JF (1989) Mycophagy in the Coleoptera: feeding strategies and morphological adaptations. In Wilding, N., Collins, M., Hammond, P.M. and Webber, J.F., eds, Insect - Fungus Interactions. Royal Entomological Society Symposium Series. 14:1–23

Leake JR et al (2002) Interactions between ecto-mycorrhizal and saprotrophic fungi. In: Van der Heijden MGA, Sanders IR (eds) Mycorrhizal ecology, ecological studies analysis and synthesis. Springer, Berlin, pp 345–372

Lefevre C, Hall IR (2001) In: Mehlenbacher SA (ed) The global status of truffle cultivation. Proceedings of the fifth international congress on hazelnut (556), Acta Hortic, pp 513–520

Li H, Smith SE, Holloway RE, Zhu Y, Smith FA (2006) Arbuscular mycorrhizal fungi contribute to phosphorus uptake by wheat grown in a phosphorus-fixing soil even in the absence of positive growth responses. New Phytol 172(3):536–543

Li D, Zhao WH, Kong BH, Ye M, Chen HR (2009) Inhibition effects of the extract and polysaccharide in macrofungus on TMV. J Yunnan Agri Univ 24(2):175–180

Lilleskov EA et al (2002) Belowground ectomycorrhizal fungal community change over a nitrogen deposition gradient in Alaska. Ecology 83:104–115

LoBuglio KF, Berbee ML, Taylor JW (1996) Phylogenetic origins of the asexual mycorrhizal symbiont *Cenococcum geophilum* Fr. and other mycorrhizal fungi among Ascomycetes. Mol Phylogenet Evol 6:287–294

Lu X-H, Malajczuk N, Dell B, Dell B (1998) Mycorrhiza formation and growth of *Eucalyptus globulus* seedlings inoculated with spores of various ectomycorrhizal fungi. Mycorrhiza 8:81–86

Lucas EH, Montesono R, Pepper MS, Hafner M, Sablon E (1957) Tumor inhibitors in *Boletus edulis* and other holobasidiomycetes. Antibiot Chemother 7:1–4

Maison de la (2010) Truffle-Truffe fraiche Noire (in French). http://en.wikipedia.org/wiki/Truffle-(fungus)

Malajczuk N, Molina R, Trappe J (1982) Ectomycorrhiza formation in *Eucalyptus* I. Pure culture synthesis, host specificity and mycorrhizal compatibility with *Pinus radiata*. New Phytol 91:467–482

Malloch D, Malloch B (1981) The mycorrhizal status of boreal plants: species from northeastern Ontario. Can J Bot 59:2167–2172

Marais LJ, Kotzé JM (1977) Notes on ectotrophic mycorrhizae of *Pinus patula* in South Africa. South Afr For J 100:61–71

Marks GC, Ditchborne N, Foster RC (1968) Quantitative estimates of mycorrhizal populations in radiata pine forests. Aust For 32:26–38

Marschner H (1995) Mineral nutrition of higher plants. Academic Press, London

Martin F, Kohler A, Murat C, Balestrini R, Coutinho P M, Jaillon O, Montanini B, Morin E, Noel B, Percudani R, Porcel B, Rubini A, Amicucci A, Amselem J, Anthouard V, Arcioni S, Artiguenave F, Aury J-M, Ballario P, Bolchi A, Brenna A, Brun A, Buee M, et al. (2010) The Black Truffle Genome Uncovers Evolutionary Origins and Mechanisms of Symbiosis. Nature 464:1033–1038

Mason P, Last FT, Wilson J, Deacon JW, Fleming LV, Fox FM (1987) Fruiting and succession of ecto-mycorrhizal fungi. In: Fungal infection of plants. (Ed GF Pegg and PG Ayres) 253–268. Cambridge University Press

Mason PA, Wilson J (1994) Harnessing symbiotic associations: vesicular arbuscular mycorrhizas. In: Leakey RRB, Newton AC (eds) Tropical trees: the potential for domestication and the rebuilding of forest resources. HMSO, London, pp 165–175

Massicotte HB, Molina R, Luoma DL, Smith JE (1994) Biology of the ectomycorrhizal genus. Rhizopogon II. Patterns of host-fungus specificity following spore inoculation of diverse hosts grown in monoculture and dual culture. New Phytol 126:677–690

Masuka AJ (1996) Dynamics of mushroom (*Boletus edulis*) production in pine plantations in Zimbabwe. J Appl Sci Southern Afr 2(2):69–76

Matheny PB, Bougher NL (2006) The new genus *Auritella* from Africa and Australia (Inocybaceae, Agaricales): molecular systematics, taxonomy and historical biogeography. Mycol Progr 5:2–17

Matheny PB, Curtis JM, Hofstetter V, Aime MC, Moncalvo JM, ZaiWei G, ZhuLiang Y, Slot JC, Ammirati JF, Baroni TJ, Bougher NL, Hughes KW, Lodge DJ, Kerrigan RW, Seidl MT, Aanen DK, Denitis M, Daniele GM, Desjardin DE, Kropp BR, Norvell LL, Parker A, Vellinga EC, Vilgalys R, Hibbett DS (2006) Major clades of Agaricales: a multilocus phylogenetic overview. Mycologia 98:982–995

Mattila P, Lampi AM, Ronkainen R, Toivo J, Piironen V (2002) Sterol and vitamin D_2 contents in some wild and cultivated mushrooms. Food Chem 76(3):293–298

McGee PA (1996) The Australian zygomycetous mycorrhizal fungi: the genus *Densospora* gen. nov. Aust Syst Bot 9:329–336

Meyer FH (1973) Distribution of ectomycorrhizae in native and man-made forests. In: Marks GC, Kozlowski TT (eds) Ectomycorrhizae, their ecology and physiology. Academic Press, New York, pp 79–105

Miller SL, Torres P, McClean TM (1994) Persistence of basidiospores and sclerotia of ectomycorrhizal fungi and *Morchella* in soil. Mycologia 86:89–95

Molina R, Trappe JM (1982) Patterns of ectomycorrhizal host specificity and potential amoung pacific northwest conifers and fungi. Forest Sci 28:423–458

Moncalvo JM, Lutzoni FM, Rehner SA, Johnson J, Vilgalys R (2000) Phylogenetic relationships of agaric fungi based on nuclear large subunit ribosomal DNA sequences. Syst Biol 49:278–305

Moncalvo JM, Vilgalys R et al (2002) One hundred and seventeen clades of euagarics. Mol Phylogenet Evol 23:357–400

Morte A et al (2000) Effect of drought stress on growth and water relations of the mycorrhizal association Helianthemum almeriense – Terfezia claveryi. Mycorrhiza 10:115–119

Munyanziza E (1994) Miombo trees and mycorrhizae: ecological strategies, a basis for afforestation. Ph.D. thesis, Wageningen Agricultural University, The Netherlands, p 193

Niazi R, Iqbal SH, Khalid AN (2006) Biodiversity of Mushrooms and ectomycorrhizas. 1. Russula brevipes Peck, and its ectomycorrhiza – a new record from Himalayan moist temperate forest of Pakistan. Pakistan J Bot 38(4):1271–1277

Nils F (1979) Germination of spores of *Cantharellus cibarius*. Mycologia 71(1):216–219

North M, Trappe JM (1996) Relative abundance and animal consumption of fungal sporocarps in Pacific Northwest young- mature- and old-growth forests. In: Szaro TM, Bruns TD (eds) Proceedings, 1st international conference on mycorrhizae, Berkeley, California, p 93

Norvell LL (1995) Loving the chanterelle to death? The ten-year Oregon chanterelles project. McIlvainea 12(1):6–25

Ogawa M (1985) Ecological characters of ectomycorrhizal fungi and their mycorrhizae – an introduction to the ecology of higher fungi. JARQ 18:305–314

Oria-de-Rueda J, Martin-Pinto P, Olaizola J (2008) Bolete productivity of cistaceous scrublands in northwestern Spain. Econ Bot 62(3):323–330

Orlovich D, Stringer A, Yun W, Hall I, Prime G, Danell E, Weden C, Bulman S (2004) *Boletus edulis* Bull. Ex Fries in New Zealand. Aust Mycol Soc Newslett 1(1):6

Peinter U, Bougher NL, Castellano MA, Moncalvo J-M, Moser MM, Trappe JM, Vilgalys R (2001) Multiple origins of sequestrate fungi related to *Cortinarius* (Cortinariaceae). Am J Bot 88:2168–2179

Peinter U, Iotti M, Klotz P, Bonuso E, Zambonelli A (2007) Soil fungal communities in a *Castanea sativa* (chestnut) forest producing large quantities of *Boletus edulis sensu lato* (porcini): where is the mycelium of porcini? Environ Microbiol 9(4):880–889

Philpot R (1965) Viennese cookery. Hodder & Staughton, London, pp 139–140

Pillukat A, Agerer R (1992) Studies an Ektomycorrhizen XL. Vergleichende Untersuchungen zur bambezogenen variabilitat der Ektomycorrhizen von Russula ochroleuca. Z Mykol 58:211–242

Pilz D, Molina R, Liegel L (1998) Biological productivity of chanterelle mushrooms in and near the Olympic Peninsula Biosphere Reserve. Ambio 9:8–13

Piraino FF (2006) Emerging antiviral drugs from medicinal mushrooms. Int J Med Mushrooms 8 (2):101–114

Pirozynski KA, Dalpé Y (1989) Geological history of the Glomaceae with particular reference to mycorrhizal symbioses. Symbiosis 7:1–36

Poitou N et al (1989) Mycorrhization controlee et culture experimentale du champ de Boletus (¼ Suillus) granulatus et Lactarius deliciosus. In: Grabbe K, Hilber O (eds) Proceedings of the twelfth international congress on the science and cultivation of edible fungi, Braunschweig, pp 551–563

Quan L, Lei ZP (2000) A study on effect of ectomycorrhizae on promoting *Castanea mollissima* resistance to drought and its mechanism (in Chinese). For Res 13(3):249–256

Reddell P, Milnes AR (1992) Mycorrhizas and other specialised nutrient-acquisition strategies: their occurrence in woodland plants from Kakadu and their role in rehabilitation of waste rock dumps at a local uranium mine. Aust J Bot 40:223–242

Reddell P, Gordon V, Hopkins MS (1999) Ectomycorrhizas in *Eucalyptus tetrodonta* and *E. miniata* forest communities in tropical northern Australia and their role in the rehabilitation of these forests following mining. Aust J Bot 34:435–444

Redecker D, Kodner R, Graham LE (2000) Glomalean fungi from the Ordovician. Science 289:1920–1921

Redecker D et al (2001) Small genets of Lactarius xanthogalactus, Russula cremoricolor and Amanita francheti in late-stage ectomycorrhizal successions. Mol Ecol 10:1025–1034

Redhead SA (1997) The pine mushroom industry in Canada and the United States: why it exists and where it is going. In: Palm ME, Chapela IH (eds) Mycology in sustainable development: expanding concepts, vanishing borders. Parkway Publishers, Boone, NC, pp 15–54

Remy W, Taylor TN, Hass H, Kerp H (1994) 4 hundred million year old vesicular-arbuscular mycorrhizae. Proc Natl Acad Sci 91(25):11841–11843

Ribeiro Bárbara, Andrade PB, Silva BM, Baptista P, Seabra RM, Valentão P (2008) Comparative study on free amino acid composition of wild edible mushroom species. Journal of agricultural and food chemistry. 56(22):10973–9.

Richardson MJ (1970) Studies on *Russula emetica* and other agarics in a Scot pine plantation. Trans Br Mycol Soc 55:217–229

Rinaldi AC, Comadini O, Kuyper TW (2008) Ectomycorrhizal fungi diversity: separating the wheat from the chaff. Fungal Divers 33:1–45

Rose SL (1980) Mycorrhizal associations of some actinomycete nodulated nitrogen fixing plants. Can J Bot 58:1449–1454

Sbrana C et al (2000) Adhesion to hyphal matrix and antifungal activity of Pseudomonas strains isolated from Tuber borchii ascocarps. Can J Microbiol 46:259–268

Schulz B, Boyle C (2005) The endophytic continuum. Mycol Res 109:661–686

Schulze W, Schulze ED, Pate JS, Gillison AN (1997) The nitrogen supply from soil and insects during growth of the pitcher plants *Nepenthes mirabilis*, *Cephalotus follicularis* and *Darlingtonia californica*. Oecologia 112:464–471

Schweiger PF, Robson AD, Barrow NJ (1995) Root hair length determines beneficial effect of a *Glomus* species on shoot growth of some pasture species. New Phytol 131:247–254

Selosse MA, Bauer R, Moyersoen B (2002) Basal hymenomycetes belonging to the Sebacinaceae are ectomycorrhizal on temperate deciduous trees. New Phytol 155:183–195

Selosse MA, Richard F, He X, Simard SW (2006) Mycorrhizal networks: des liaisons dangereuses? Trends Ecol Evol 21(11):621–628

Sharma R (2008) Studies on ectomycorrhizal mushrooms of M.P. and Chhattisgarh. PhD thesis, R.D. University, Jabalpur

Simard SW, Perry DA, Jones MD, Myrold DD, Durall DM, Molina R (1997) Net transfer of carbon between ectomycorrhizal tree species in the field. Nature 388:579–582

Sisti D et al (1998) In vitro mycorrhizal synthesis of micropropagated Tilia platyphyllos Scop. plantlets with Tuber borchii Vittad. Mycelium in pure culture. Acta Hortic 457:379–387

Sitta N (2000) Presence of *Tylopilus* into dried "Porcini" mushrooms from China (in Italian). Micol Ital 29(1):96–99

Sitta N, Floriani M (2008) Nationalization and globalization trends in the wild mushroom commerce of Italy with emphasis on porcini (*Boletus edulis* and allied species). Econ Bot 62 (3):307–322

Smith SE, Read DJ (1997) Mycorrhizal symbiosis. Academic Press, London

Smith JE, Molina R, Huso MMP, Luoma DL, McKay D, Castellano MA, Lebel T, Valachovic V (2002) Species richness, abundance, and composition of hypogeous and epigeous ectomycorrhizal fungal sporocarps in young, rotation-age, and old-growth stands of Douglas-fir (*Pseudotsuga menziesii*) in the Cascade Range of Oregon, USA. Can J Bot 80(2):186–204

St John TV (1980) A survey of mycorrhizal infection in an Amazonian rain forest. Acta Amazon 10:527–533

Stanis VF (1979) Russula species occurring n the boreal region of Ontario and Quebec Canada. M.Sc. thesis, Department of Botany, University of Toronto, Toronto, Ontario

Strullu-Derrien C, Strullu DG (2007) Mycorrhization of fossil and living plants. CR Palevol 6:483–494

Suri OP, Shah R, Satti NK, Suri KA (1997) Russulactarorufin, a lactarane skeleton sesquiterpene from *Russula brevipes*. Phytochemistry 45(7):1453–1455

Susan M (1992) Texas mushrooms: a field guide, 1st edn. University of Texas Press, Austin

Olivier JM (2000) Progress in the cultivation of truffles. In: Van Griensven LJLD (ed) Mushroom science XV: science and cultivation of edible fungi, vol 2. Balkema, Rotterdam, pp 937–942

Tarafdar JC, Marschner H (1994) Efficiency of VAM hyphae in utilisation of organic phosphorus by wheat plants. Soil Sci Plant Nutr 40:593–600

Tedersoo L, Hansen K, Perry BA, Kjller R (2006) Molecular and morphological diversity of Pezizalean ectomycorrhiza. New Phytol 170:581–596

Tedersoo L, Suvi T, Beaver K, Kõljag U (2007) Ectomycorrhizal fungi of the Seychelles: diversity patterns and host shifts from the native *Vateriopsis seychellarum* (Dipterocarpaceae) and *Intsia bijuga* (Caesalpiniaceae) to the introduced *Eucalyptus robusta* (Myrtaceae), but not *Pinus caribea* (Pinaceae). New Phytol 175:321–333

Tehler A, Little DP, Farris JS (2003) The full-length phylogenetic tree from 1551 ribosomal sequences of chitinous fungi, Fungi. Mycol Res 107:901–916

Termorshuizen AJ (1993) The influence of nitrogen fertilisers on ectomycorrhizas and their fungal carpophores in young stands of *Pinus sylvestris*. For Ecol Manage 57:179–189

Termoshuizen AJ (1991) Succession of mycorrhizal fungi in stands of *Pinus sylvestris* in the Netherlands. J Veg Sci 2:255–264

Thomson BD, Grpe TS, Malajczuk N, Hardy GES (1994) The effectiveness of ectomycorrhizal fungi in increasing the growth of *Eucalyptus globulus* Labill in relation to root colonization and hyphal development in soil. New Phytol 126:517–524

Trappe JM (1987) Phylogenetic and ecologic aspects of mycotrophy in the angiosperms from an evolutionary standpoint. In: Safir GR (ed) Ecophysiology of VA mycorrhizal plants. CRC Press, Boca Raton, FL, pp 5–25

Trocha LK, Rudawska M, Leski T, Dabert M (2006) Genetic diversity of naturally established ectomycorrhizal fungi of norway spruce seedlings under nursery conditions. Microb Ecol 52:418–425

Trowbridge R, Macadam A (1999) Ecological description and classification of highly productive pine mushroom sites in Northwestern British Columbia. In: Moss P (ed) Studies of the pine mushroom (*Tricholoma magnivelare*) in the Skeena-Bulkley Region. Northwest Institute of Bioregional Studies, Smithers, BC

Tuomi J, Kytoviita M, Hardling R (2001) Cost efficiency of nutrient acquisition of mycorrhizal symbiosis for the host plant. Oikos 92:62–70

Tylukti EE (1987) Mushrooms of Idaho and the Pacific Northwest. Non-gilled Hymenomycetes, vol 2. The University of Idaho Press, Moscow, Idaho, pp 9–10

Unestam T, Sun YP (1995) Extramatrical structures of hydrophobic and hydrophilic ectomycorrhizal fungi. Mycorrhiza 5:301–311

Urban A, Weiss M, Bauer R (2003) Ectomycorrhizas involving sebacinoid mycobionts. Mycol Res 107:3–14

Verbeken A, Buyck B (2002) Diversity and ecology of tropical ectomycorrhizal fungi in Africa. In: Watling R, Frankland JC, Ainsworth AM, Isaac S, Robinson CH, eds. Tropical mycology, Vol. 1. Macromycetes. Wallingford, UK: CABI Publishing, 11–24

Veselkov IM (1975) Artificial propagation of *Boletus edulis* in forests. Растительные Ресурсы России 11:574–578

Visser S (1995) Ectomycorrhizal fungal succession in Jack pine stands following wildfire. New Phytol 129:389–401

Vogt KA, Edmonds RL, Grier CC (1981) Biomass and nutrient concentrations of sporocarps produced by mycorrhizal and decomposer fungi in *Abies amabilis* stands. Oecologia 50:170–175

Vozzo JA, Hackskaylo E (1961) Mycorrhizal fungi on *Pinus virginiana*. Mycologia 53 (5):538–539

Wang B, Qiu YL (2006) Phylogenetic distribution and evolution of mycorrhizas in land plants. Mycorrhiza 16:299–363

Wang Y, Sinclair L, Hall IR, Cole ALJ (1995) *Boletus edulis* sensu lato: a new record for New Zealand. New Zeal J Crop Hort Sci 23:227–231

Wang Y et al (1997) Ectomycorrhizal fungi with edible fruiting bodies 1. Tricholoma matsutake and related fungi. Econ Bot 51:311–327

Wang Y et al (2002a) Some ectomycorrhizal mushrooms. In: Hall IR et al (eds) Edible mycorrhizal mushrooms and their cultivation. Proceedings of the second international conference on edible mycorrhizal mushrooms, CD-ROM, New Zealand Institute for Crop & Food Research Limited

Wang Y et al (2002b) The cultivation of Lactarius deliciosus (saffron milk cap) and Rhizopogon rubescens (shoro) in New Zealand. In: Hall IR et al (eds) Edible mycorrhizal mushrooms and their cultivation. Proceedings of the second international conference on edible mycorrhizal mushrooms, CD-ROM, New Zealand Institute for Crop & Food Research Limited

Warcup JH (1980) Ectomycorrhizal associations of Australian indigenous plants. New Phytol 85:531–535

Warcup JH (1985) Ectomycorrhizal formation by *Glomus tubiforme*. New Phytol 99:267–272

Warner Jeremy (2010) http://basiceating.blogspot.com/2010/04/Chantharelle-cantharellus.cibarius.html retrieved on 02/06/10

Watling R (2006) The sclerodermoid fungi. Mycoscience 47:18–24

Wilson AW, Hobbie EA, Hibbet DS (2007) The ectomycorrhizal status of *Calostoma cinnabarinum* determined using isotopic, molecular and morphological methods. Can J Bot 85:385–393

Yamada A (2003) Studies in ecology and mycorrhizal taxonomy of ectomycorrhizal fungi in Pinus densiflora forests and cultivation of ectomycorrhizal mushrooms. Nippon Kin Gakkal Kaiho 44 (1):9–18

Yamada A, Ogura T, Ohmasa M (2001a) Cultivation of mushrooms of edible ectomycorrhizal fungi associated with *Pinus densiflora* by in vitro mycorrhizal synthesis I. Primordium and basidiocarp formation in open pot culture. Mycorrhiza 11:59–66

Yamada A, Ogura T, Ohmasa M (2001b) Cultivation of mushrooms of edible ectomycorrhizal fungi associated with *Pinus densiflora* by in vitro mycorrhizal synthesis II. Morphology of mycorrhizas in open pot soil. Mycorrhiza 11:67–81

Yun W, Hall IR (2004) Edible ectomycorrhizal mushrooms: challenges and achievements. Can J Bot 82(8):1063–1073

Zambonelli A et al (2002) Tuber borchii cultivation: an interesting perspective. In: Hall IR et al (eds) Edible mycorrhizal mushrooms and their cultivation. Proceedings of the second international conference on edible mycorrhizal mushrooms, cd-rom, New Zealand Institute for Crop & Food Research Limited

Zheng S, Li C, Ng TB, Wang HX (2007) A lectin with mitogenic activity from the edible wild mushroom *Boletus edulis*. Process Biochem 42(12):1620–1624

Index

M. Rai and A. Varma (eds.), *Diversity and Biotechnology of Ectomycorrhizae*,
Soil Biology 25, DOI 10.1007/978-3-642-15196-5,